Advances in
ORGANOMETALLIC CHEMISTRY

VOLUME 44

Advances in Organometallic Chemistry

EDITED BY

ROBERT WEST

DEPARTMENT OF CHEMISTRY
UNIVERSITY OF WISCONSIN
MADISON, WISCONSIN

ANTHONY F. HILL

DEPARTMENT OF CHEMISTRY
IMPERIAL COLLEGE OF SCIENCE,
TECHNOLOGY, AND MEDICINE
LONDON, ENGLAND

FOUNDING EDITOR

F. GORDON A. STONE

VOLUME 44

ACADEMIC PRESS

San Diego   London   Boston   New York
Sydney   Tokyo   Toronto

Academic Press
A Harcourt Science and Technology Company
525 B Street, Suite 1900, San Diego, California 92101-4495, USA
http://www.apnet.com

Academic Press
24-28 Oval Road, London NW1 7DX, UK
http://www.hbuk.co.uk/ap/

International Standard Book Number: 0-12-031144-5

PRINTED IN THE UNITED STATES OF AMERICA
99 00 01 02 03 04 EB 9 8 7 6 5 4 3 2 1

Contents

Element Derivatives of Sterically Encumbering Terphenyl Ligands

BRENDAN TWAMLEY, SCOTT T. HAUBRICH,
and PHILIP P. POWER

Hydrogen Donor Abilities of the Group 14 Hydrides

CHRYSSOSTOMOS CHATGILIALOGLU and
MARTIN NEWCOMB

Doubly Bonded Derivatives of Germanium

JEAN ESCUDIE and HENRI RANAIVONJATOVO

The Chemistry of Phosphinocarbenes

DIDIER BOURISSOU and GUY BERTRAND

Zwitterionic Pentacoordinate Silicon Compounds

REINHOLD TACKE, MELANIE PÜLM, and BRIGITTE WAGNER

Transition Metal Complexes of Vinylketenes

SUSAN E. GIBSON (NÉE THOMAS) and MARK A. PEPLOW

Contributors

Numbers in parentheses indicate the pages on which the authors' contributions begin.

GUY BERTRAND (175), Laboratoire de Chimie de Coordination du CNRS, 31077 Toulouse Cédex, France

DIDIER BOURISSOU (175), Laboratoire de Chimie de Coordination du CNRS, 31077 Toulouse Cédex, France

CHRYSSOSTOMOS CHATGILIALOGLU (67), I.Co.C.E.A., Consiglio Nazionalle delle Richerche, 40129 Bologna, Italy

JEAN ESCUDIE (113), Hétérochimie Fondamentale et Appliquée, UPRES A 5069, Université P. Sabatier, 31062 Toulouse Cédex 4, France

SUSAN E. GIBSON (NÉE THOMAS) (275), Centre for Chemical Synthesis, Department of Chemistry, Imperial College of Science, Technology, and Medicine, South Kensington, London SW7 2AY; and Department of Chemistry, Kings College, The Strand, London, United Kingdom WC2R 2LS

SCOTT T. HAUBRICH (1), Department of Chemistry, University of California, Davis, California 95616

MARTIN NEWCOMB (67), Department of Chemistry, Wayne State University, Detroit, Michigan 48202

MARK A. PEPLOW (275), Centre for Chemical Synthesis, Department of Chemistry, Imperial College of Science, Technology, and Medicine, South Kensington, London, United Kingdom SW7 2AY

PHILIP P. POWER (1), Department of Chemistry, University of California, Davis, California 95616

MELANIE PÜLM (221), Institut für Anorganische Chemie, Universität Würzburg am Hubland, D-97074 Würzburg, Germany

HENRI RANAIVONJATOVO (113), Hétérochimie Fondamentale et Appliquée, UPRES A 5069, Université P. Sabatier, 31062 Toulouse Cédex 4, France

REINHOLD TACKE (221), Institut für Anorganische Chemie, Universität Würzburg am Hubland, D-97074 Würzburg, Germany

BRENDAN TWAMLEY (1), Department of Chemistry, University of California, Davis, California 95616

BRIGITTE WAGNER (221), Institut für Anorganische Chemie, Universität Würzburg am Hubland, D-97074 Würzburg, Germany

ADVANCES IN ORGANOMETALLIC CHEMISTRY, VOL. 44

Element Derivatives of Sterically Encumbering Terphenyl Ligands

BRENDAN TWAMLEY, SCOTT T. HAUBRICH,
and PHILIP P. POWER

*Department of Chemistry
University of California
Davis, California 95616*

I

INTRODUCTION

A. *Sterically Crowding Ligands*

The use of sterically crowding ligands to manipulate reactivity and to effect the stabilization of previously unknown bonding types, geometries, or electron configurations is one of the major themes of inorganic and organometallic chemistry. Quite a large range of ligands has been employed for these purposes, and the introduction of new ligand types is a constant feature of the current literature. Most uninegative, bulky ligands use large organic (or closely related) substituents at a first-row element (e.g., carbon, nitrogen, or oxygen), through which they are bound to a desired site. Crowding ligands based on organosubstituted second-row elements silicon,[1a] phosphorus,[1b] or sulfur (or their heavier congeners)[1c] have also been employed extensively, not least on account of their different electronic properties. Nonetheless, the much smaller relative size of the first-row

1

elements (which keeps interatomic distances short) ensures that the pendant organic group is generally more sterically effective at shielding the reactive site than if it were attached through a larger second-row element. This consideration, together with factors such as reduced commercial availability, and sometimes more difficult synthetic accessibility, have maintained the widespread use of the first-row ligands.

The most common uninegative ligands are of the type -R, -NR$_2$, or -OR (R = alkyl or aryl group). Of these, the alkyl groups are the most sterically flexible, since the bound carbon may carry up to three organic substituents, as in the *tert*-butyl (-CMe$_3$) or trityl (-CPh$_3$) ligands. Next in steric order are the aryl and amide (-NR$_2$) ligands in which a carbon and nitrogen bear two substituents. In the -OR ligand, the oxygen atom has only one substituent. It may be noted that although the steric options decrease in the sequence -CR$_3$ > -NR$_2$ > -OR, this is partially compensated by the fact that the metal–ligand distances generally decrease in the same order.

Of essentially equal importance to the organo- (alkyl- or aryl-) substituted carbon, nitrogen, or oxygen ligands are the related silyl-substituted moieties. These may be exemplified by the widely employed -CH(SiMe$_3$)$_2$,[2] -C(SiMe$_3$)$_3$,[3] or -N(SiMe$_3$)$_2$,[4] ligands, the use of which has been responsible for numerous breakthroughs in the stabilization of hitherto inaccessible low coordination numbers and/or multiple bonds.

This brief survey of bulky ligands may be concluded by considering the diorganoboryl group -BR$_2$. These groups are usually employed as substituents on more electronegative ligand atoms such as oxygen, nitrogen, or phosphorus to afford ligands of the type -OBR$_2$,[5] -NRBR$_2$,[6a,b] or -PRBR$_2$.[6c] The replacement of an organic substituent by the π-acceptor diorgano boryl group alters the electronic properties by reducing its bridging tendency through interaction with lone pairs on oxygen, nitrogen, or phosphorus. The boron center is itself protected from attack by the use of bulky substituents such as Mes (mesityl) or -CH(SiMe$_3$)$_2$. Employment of bulky -BR$_2$ ligands attached directly to other elements such as transition metals is at an early stage of development but has already shown much promise.[7]

B. *Aryl Ligands*

As a ligand class, the aryls have usually been treated in the same context as the alkyls. In 1991, however, a review,[8] which dealt exclusively with aryl transition metal complexes, drew attention to their often unique structures and reactivity. It was also apparent that the influence of steric properties of aryl ligands had not been extensively investigated for transition-metal complexes. The bulkiest aryl ligand to be widely employed with these metals was the Mes group (Mes = $-C_6H_2$-2,4,6-Me$_3$). The bulkier ligands Trip ($-C_6H_2$-2,4,6-i-Pr$_3$) or Triph ($-C_6H_2$-2,4,6-Ph$_3$) had only been applied to a few metals, e.g., iron[9] and copper.[10] The reported use of the Triph ligand to synthesize monomeric MC_6H_2-2,4,6-Ph$_3$ (M = Cu or Ag)[11] has been questioned (vide infra).[12] In addition, some bulkier aryl derivatives such as MMes$_2^*$ (M = Mn or Fe)[13] or Me$_2$SCuMes*[14] (Mes* = $-C_6H_2$-2,4,6-t-Bu$_3$) have been reported. These species are notable for their two-coordinate geometry.

In sharp contrast, bulky aryl ligands have been employed extensively in main group chemistry. In addition to the abovementioned $-C_6H_2$-2,4,6-i-Pr$_3$ (Trip) ligand, perhaps the best known and most widely used is the so-called supermesityl ligand $-C_6H_2$-2,4,6-t-Bu$_3$ (Mes*). Although the Mes* group has been known for some time,[15] its use to stabilize the first diphosphene Mes*P = PMes* has had, perhaps, the greatest impact on subsequent work.[16] The investigation of the main group chemistry of its phenol derivative Mes*OH (or related bulky species such as HOC_6H_3-2,6-t-Bu$_2$ or HOC_6H_2-2,6-t-Bu$_2$-4-Me) as its aryloxide salt was initiated at about the same time.[17] These studies were then extended to include amide (-NHMes*),[18] thiolate (-SMes*),[19] or selenolate (-SeMes*)[20] derivatives. In addition, silyl-substituted ligands such as $-C_6H_3$-2,6-(SiMe$_3$)$_2$[21] or $-C_6H_2$-2,4,6-{CH(SiMe$_3$)$_2$}$_3$[22] have proven to be of considerable significance. For instance, the use of $-C_6H_2$-2,4,6-{CH(SiMe$_3$)$_2$}$_3$ has allowed the synthesis of the first compounds with Sb–Sb or Bi–Bi double bonds.[23]

C. Terphenyl and Related Ligands

In spite of the attention focused on these bulky aryl substituents, especially in main-group chemistry, aryl ligands in which crowding is caused by the presence of *ortho*-aryl substituents formerly received little attention. In contrast, related terphenyl-substituted phenolate derivatives (e.g., $-OC_6H_3$-2,6-Ph$_2$) of the early transition metals have been studied for many years, and zinc terphenolate derivatives have attracted attention (vide infra). A small number of similar aryl thiolate derivatives of transition metals (e.g., Mo, Fe, and Rh) have been characterized and published (vide infra). In addition, the related ligand $-C_6H_2$-2,4,6-Ph$_3$ (Triph) had been employed as a ring substituent in bulky porphyrins.[24]

The precursors to the bulky terphenyl ligands and their derivatives that are the subject of this review are also of relatively recent origin. They were investigated as part of a general approach to the formation of aryl–aryl bonds and, in addition, for their utility as a building block for concave reagents that combine rigidity and a concave shape with endofunctionality.

a

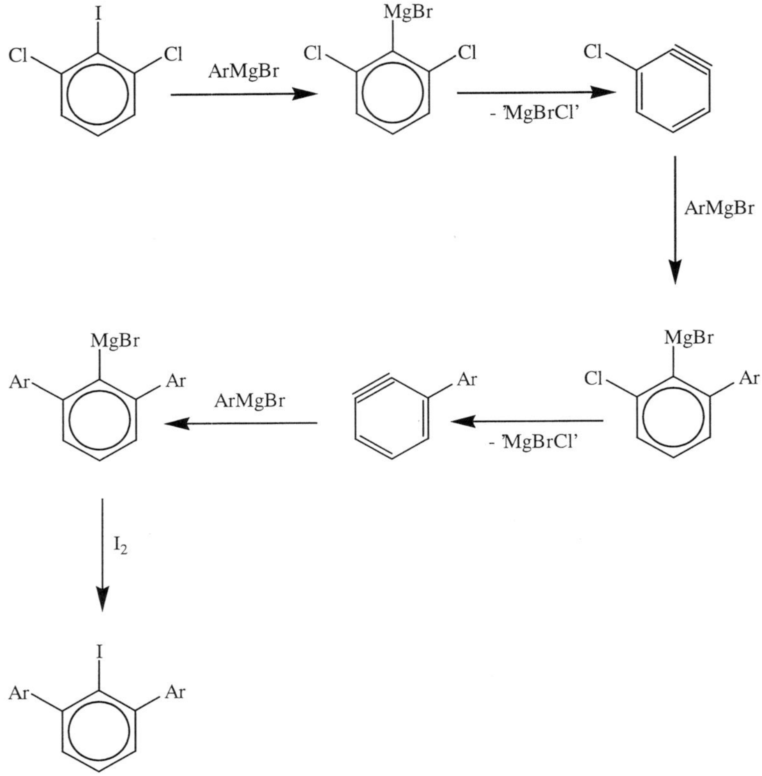

SCHEME 1.   Original (a) and shorter (b) synthetic routes to iodoterphenyl derivatives.[25]

The original synthesis of the key iodoterphenyl reagents involves the reaction sequence depicted in Scheme 1a, which was developed by Hart and co-workers.[25] The interesting steps in this reaction scheme involve the sequential creation and capture of two aryne intermediates with Grignard reagents. This reaction sequence has since been superseded[26] by a much shorter one-pot synthetic route (Scheme 1b) that involves the low-temperature generation of the lithium aryl LiC_6H_3-2,6-Cl_2 and its subsequent reaction with 2 equiv of aryl-Grignard reagent and quenching with iodine to give IC_6H_3-2,6-Ar_2. It was shown that a variety of *ortho*-aryl substituents ranging from a simple phenyl group to naphthyl or mesityl could be introduced in this manner. It has also been suggested that there appears to be a limit to the size of the group that can be added.[27] Thus, it proved possible to obtain the triisopropylphenyl (Trip) species IC_6H_3-2,6-$Trip_2$ or the closely

b

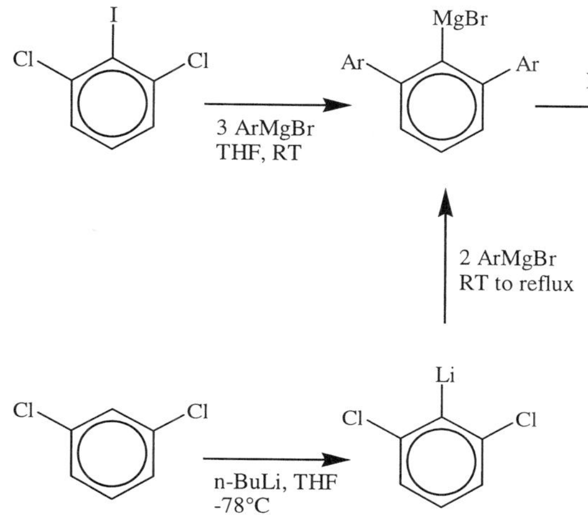

SCHEME 1. (*continued*)

related diisopropylphenyl (Dipp) compound $IC_6H_3-2,6-(2,6-i-Pr_2C_6H_3)_2$, i.e., $IC_6H_3-2,6-Dipp_2$. Attempts to obtain the bulkier $IC_6H_3-2,6-Mes_2^*$ resulted only in the synthesis of $1,3-Mes_2^*C_6H_4$.[27] This species was consistently obtained in spite of numerous attempts to effect halogenation and give the product $IC_6H_3-2,6-Mes_2^*$. It is not possible, at present, to say why the halide is not produced. One plausible explanation is that, in the last step, the Grignard reagent $XMgC_6H_3-2,6-Mes_2^*$ may be unstable and the Mg–C bond may be cleaved, in which case the hydrogen derivative may be formed by proton abstraction from the solvent. The crowding in the hydrogen derivative is supported by the crystal structure, which shows some deviation from the expected 120° angles at the substituted carbons of the central aryl ring. In addition, there are significant deviations (up to ca. 10°) of the Mes* ring planes from the line of the C–C bonds through which they are connected to the C_6H_4 aryl ring.

The terphenyl and related ligands of most relevance to this review are depicted in Fig. 1. The list is not exhaustive, but most of the work published to date has used the six ligands illustrated. The $-C_6H_2-2,4,6-Ph_3$ (Triph) ligand is included, although it is not, strictly speaking, a terphenyl species. However, the presence of two *ortho*-phenyl substituents render its steric properties very similar to those of related terphenyl ligands such as $-C_6H_3-2,6-Ph_2$ (Diph). Furthermore, among the ligands illustrated, it is the only one that is commercially available, i.e., in the form of $1,3,5-Ph_3C_6H_3$. This

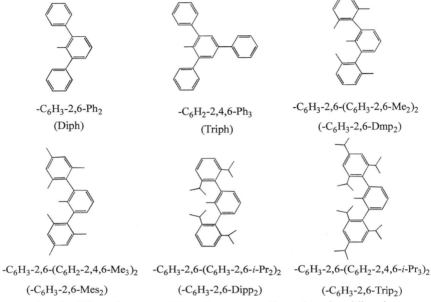

$-C_6H_3-2,6-Ph_2$
(Diph)

$-C_6H_2-2,4,6-Ph_3$
(Triph)

$-C_6H_3-2,6-(C_6H_3-2,6-Me_2)_2$
$(-C_6H_3-2,6-Dmp_2)$

$-C_6H_3-2,6-(C_6H_2-2,4,6-Me_3)_2$
$(-C_6H_3-2,6-Mes_2)$

$-C_6H_3-2,6-(C_6H_3-2,6-i-Pr_2)_2$
$(-C_6H_3-2,6-Dipp_2)$

$-C_6H_3-2,6-(C_6H_2-2,4,6-i-Pr_3)_2$
$(-C_6H_3-2,6-Trip_2)$

FIG. 1.   Schematic representation of some currently used terphenyl ligands.

arene can be readily converted to the brominated derivative[28] from which the lithium salt $(Et_2O)_2LiC_6H_2-2,4,6-Ph_3$[29] can be obtained in essentially quantitative yield (vide infra).

In this review of terphenyl derivatives of the elements, the literature has been scanned to late Spring 1998.

II

TERPHENYL DERIVATIVES OF *s*-BLOCK ELEMENTS

A. *Group 1 Terphenyl Derivatives*

After the halide starting materials themselves, the alkali metal salts are the most important terphenyl compounds. This is a result of their widespread and almost exclusive use as transfer agents for terphenyl ligands. They can be synthesized by the simple reaction of *n*-BuLi with the halide (usually bromide or iodide). This general reaction,

$$n\text{-BuLi} + XC_6H_3\text{-}2,6\text{-}Ar_2 \longrightarrow n\text{-BuX} + LiC_6H_3\text{-}2,6\text{-}Ar_2 \qquad (1)$$

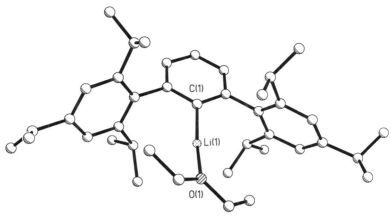

FIG. 2. Computer-generated drawing of the structure of $(Et_2O)LiC_6H_3-2,6-Trip_2$.[27]

which usually proceeds in high yield, is given in Eq. (1). The first such lithium salt to be structurally characterized was the compound $(Et_2O)_2Li$-Triph.[29] The monomeric structure is observed because of the crowding nature of the aryl group. It may be contrasted with the tetrameric structure seen for $\{(Et_2O)LiPh\}_4$[30] or the dimeric structures observed for

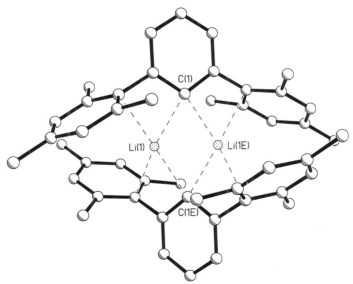

FIG. 3. Computer-generated drawing of the dimer $(LiC_6H_3-2,6-Mes_2)_2$.[33]

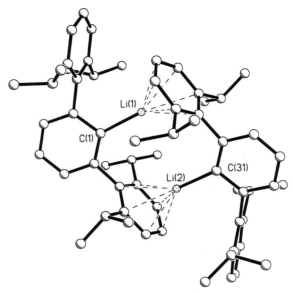

FIG. 4.   Computer-generated drawing of the dimer {LiC$_6$H$_3$-2,6-(C$_6$H$_3$-2,6-*i*-Pr$_2$)$_2$}$_2$.[34]

{(THF)$_2$LiMes}$_2$[31] and {(Et$_2$O)LiTrip}$_2$.[32] The use of the more crowding group -C$_6$H$_3$-2,6-Trip$_2$ also results in a monomeric structure for its lithium salt.[27] In this case, however, the coordination of only one Et$_2$O molecule is observed (Fig. 2), which results in an essentially linear coordination for the Li$^+$ ion and (nominally) the shortest Li–C bond observed in the solid state.

Lithium salts of terphenyl ligands also may be crystallized in the absence of donor solvents. The first such example was the salt (LiC$_6$H$_3$-2,6-Mes$_2$)$_2$,[33] which has a dimeric structure (Fig. 3). This compound also represented the first structural characterization of a Lewis base–free molecular lithium aryl species. The central aryl rings are bridged to each other by two Li$^+$ ions with Li–C(*ipso*-C$_6$H$_3$) distances in the range 2.16(2)–2.19(2) Å. Furthermore, there are considerably longer Li–C contacts (2.51(1)–2.56(1) Å) to the *ipso*-carbons of the *ortho*-Mes rings.

If the size of the *ortho*-aryl substituents is increased to -C$_6$H$_3$-2,6-*i*-Pr$_2$(Dipp), a different structural arrangement (Fig. 4) can be obtained for the Li$^+$ salt. In this compound each Li$^+$ ion is η^1-bound to the *ipso*-carbon of the central aryl ring of the -C$_6$H$_3$-2,6-Dipp$_2$ and η^6-bound to an *ortho*-Dipp substitutent of the other -C$_6$H$_3$-2,6-Dipp$_2$ aryl ligand. The average Li–C(η^1) distance is 2.068(6) Å and the LiC(η^6) distances range from 2.366(7) to 2.534(6) Å. The coordination sphere of Li$^+$ ions is distorted

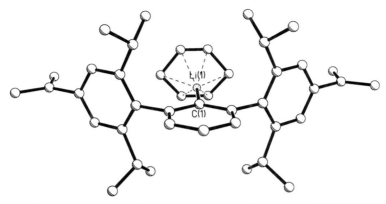

FIG. 5.　Computer-generated drawing of the monomeric structure of $(C_6H_6)LiC_6H_3$-2,6-Trip$_2$.[34]

from regularity; for example, the angles between the Li–C(*ipso*) bonds and the C(*ipso*)---C(*para*) vectors are 16.7° for Li(1) and 21.0° for Li(2). Furthermore, the Li–C(*ipso*) bonds subtend angles of 5.0° and 4.1° to the normals of the C(7) and C(49) aromatic rings. These distortions are most probably caused by the crowded nature of the dimeric structure. Further examination of the structural details reveal that the presence of *para-i-*Pr groups (e.g., at C(10) or C(52)) would have caused undue steric interactions of the C(19) or C(37) *ortho* aryl rings. This feature, together with the apparent weakness (Li–C = ca. 2.46 Å) of the Li–C(η^6) bonds, led to the conclusion that the use of *ortho*-Trip instead of *ortho*-Dipp substituents would effect dissociation to monomers. The structure of the compound $(C_6H_6)LiC_6H_3$-2,6-Trip$_2$ (Fig. 5)[34] showed that this hypothesis was essentially correct, but only if benzene is available to satisfy the coordination requirements of what would have been a one-coordinate Li$^+$ ion. Attempts to crystallize this compound in the absence of benzene led to decomposition and isolation of 1,3-Trip$_2$C$_6$H$_4$. In $(C_6H_6)LiC_6H_3$-2,6-Trip$_2$ the Li$^+$ ion is bound η^1 to the *ipso*-C of the central ring (Li–C = 2.03(2) Å) and η^6 to benzene (Li–C = 2.35(2) Å).

The sole heavier alkali metal derivative of a terphenyl ligand is the compound $(NaC_6H_3$-2,6-Mes$_2)_2$.[35] It was synthesized by the reaction of its lithium congener with

$$(LiC_6H_3\text{-}2,6\text{-}Mes_2)_2 + 2NaO\text{-}t\text{-}Bu \longrightarrow 2LiO\text{-}t\text{-}Bu + (NaC_6H_3\text{-}2,6\text{-}Mes_2)_2 \quad (2)$$

NaO-*t*-Bu as shown in Eq. (2). Its structure (Fig. 6) resembles that seen for the lithium dimer $(LiC_6H_3$-2,6-Mes$_2)_2$. However, in this case the metal ion (Na$^+$) interacts almost equally (Na(1)–C(1) = 2.572(2) Å,

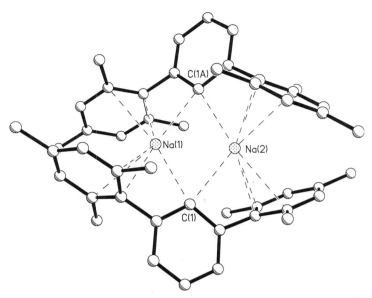

FIG. 6. Computer-generated drawing of the dimer $(NaC_6H_3\text{-}2,6\text{-}Mes_2)_2$.[35]

$Na(2)-C(1) = 2.609(2)$ Å) with the *ipso*-carbon of the central ring $C(1)$ and the *ipso*-carbons of the *ortho*-Mes substituents $C(7)$ ($Na(1)-C(7) = 2.595(2)$ Å) and $C(16)$ ($Na(2)-C(16) = 2.663(2)$ Å). Like its lithium counterpart, the sodium terphenyl salt was the first structure of a non-Lewis base complexed sodium aryl. Attempts to synthesize heavier potassium or rubidium congeners were unsuccessful owing to their very high reactivity.[35] For instance, when $(LiC_6H_3\text{-}2,6\text{-}Mes_2)_2$ is allowed to react with KO-*t*-Bu at room temperature, an orange solution is observed and a microcrystalline off-white solid is precipitated. Only $1,3\text{-}Mes_2C_6H_4$ was detected by ^1H NMR spectroscopy of the supernatant liquid. Even $(NaC_6H_3\text{-}2,6\text{-}Mes_2)_2$ slowly metallates benzene at room temperature, and 50% conversion is observed at 20°C in ca. 30 h.

B. *Group 2 Terphenyl Derivatives*

Main group 2 terphenyl compounds are of potential importance for a number of reasons. Like their lithium analogues, magnesium derivatives such as the Grignard reagents have obvious applications as ligand transfer agents. In addition, it has been suggested[36] that the use of terphenyl ligands may effect the stabilization of multiply bonded or catenated derivatives of the lightest element, beryllium, about which very little is currently known.

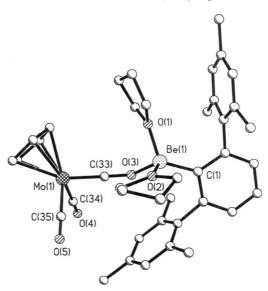

FIG. 7. Computer-generated drawing of the isocarbonyl species 2,6-Mes$_2$H$_3$C$_6$(THF)$_2$Be-(CO)$_3$Mo(η^5-C$_5$H$_5$).[36]

1. Beryllium Compounds

Several -C$_6$H$_3$-2,6-Mes$_2$ derivatives of beryllium have been reported.[36] These include the halides 2,6-Mes$_2$H$_3$C$_6$Be(X)(OEt$_2$) (X = Cl or Br), which were synthesized by the reaction of LiC$_6$H$_3$-2,6-Mes$_2$ with the appropriate beryllium dihalide in ether. Their crystal structures show that there is distorted trigonal planar coordination at beryllium with interligand angles that range from 111.8(3) to 129.7(4)°. Treatment of these compounds with 1 equiv of LiSMes*, LiN(H)Ph, LiN(H)SiPh$_3$, LiN(SiMe$_3$)$_2$, or Na{Mo(η^5-C$_5$H$_5$)(CO)$_3$} affords the products 2,6-Mes$_2$H$_3$C$_6$Be(SMes*)(OEt$_2$), (2,6-Mes$_2$H$_3$C$_6$BeNHPh)$_2$, 2,6-Mes$_2$H$_3$C$_6$Be{N(H)SiPh$_3$} (OEt$_2$), 2,6-Mes$_2$H$_3$C$_6$-BeN(SiMe$_3$)$_2$, or 2,6-Mes$_2$H$_3$C$_6$(THF)$_2$Be(OC)$_3$Mo(η^5-C$_5$H$_5$).[36] Both the thiolate 2,6-Mes$_2$H$_3$C$_6$Be(SMes*)(OEt$_2$) and 2,6-Mes$_2$H$_3$C$_6$Be{N(H)SiPh$_3$}-(OEt$_2$) have similar trigonal planar structures with Be–S and Be–N distances of 1.984(3) and 1.577(4) Å. The less sterically crowded amide compound {2,6-Mes$_2$H$_3$C$_6$BeN(H)Ph}$_2$ has a dimeric structure. Three-coordination for beryllium is retained, however, since there is no complexation to ether. The Be–N distances in the planar Be$_2$N$_2$ core are in the range 1.690(6)–1.734(5) Å and there are wider (ca. 95°) ring angles at beryllium than at nitrogen (ca. 85°).

The unusual isocarbonyl species 2,6-Mes$_2$H$_3$C$_6$(THF)$_2$Be(OC)$_3$Mo(η^5-

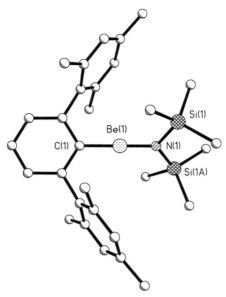

FIG. 8.   Computer-generated drawing of the monomer 2,6-Mes$_2$H$_3$C$_6$BeN(SiMe$_3$)$_2$.[38]

C$_5$H$_5$) (Fig. 7) was isolated from the reaction of the arylberyllium chloride with Na{Mo(η^5-C$_5$H$_5$)(CO)$_3$}. The beryllium is four-coordinate and is bound to the aryl group, two THFs, and an oxygen from one of the carbonyls. The Be–O(OC) distance, 1.657(5) Å, is slightly shorter than the Be–O(THF) distances, 1.682(5) and 1.718(5) Å, and the CO bond length (1.212(4) Å) in the beryllium-bound isocarbonyl is ca. 0.05 Å longer than in the other carbonyl ligands. By the same token, the Mo-C(33) length 1.875(3) Å is shorter than the Mo–C(34) and –C(35) distances of 1.946(4) and 1.949(5) Å. These changes in the Mo–C and C–O bond lengths are consistent with the normal synergistic bonding model for transition metal–carbonyls. Each carbonyl has almost linear carbon geometry, although the geometry at the beryllium-bound oxygen is bent with a Be–O(3)–C(33) angle of 147.1(3)°.

The structure of the compound 2,6-Mes$_2$H$_3$C$_6$BeN(SiMe$_3$)$_2$ (Fig. 8) is notable in that it is the first for a two-coordinate beryllium species in the solid state. The small number of known two-coordinate beryllium structures have been obtained in vapor phase.[37] Among these is Be{N(SiMe$_3$)$_2$}$_2$ (Be–N = 1.56(2) Å), which has a linear geometry and mutually perpendicular nitrogen coordination planes, suggesting an isolobal analogy with allene and the possibility of Be–N π-bonding.[37c] The Be–C bond in 2,6-Mes$_2$H$_3$C$_6$BeN-(SiMe$_3$)$_2$, 1.700(4) Å, is essentially identical to the 1.698(5) and 1.699(3) Å seen in BeMe$_2$ and Be(t-Bu)$_2$. These Be–C distances enable the covalent

(sp) radius of beryllium in two-coordinate beryllium species to be estimated by the Schomaker–Stevenson formula.[38] Using 1.70 Å for the Be–C bond length, 0.74 Å as the radius of sp^2-hybridized C, and a calculated ionic correction of 0.093 Å, a radius of 1.05 Å may be calculated for sp-hybridized beryllium. Thus, if the 0.70 Å value for an sp^2 radius of nitrogen is added to this number, a predicted Be–N bond length of 1.75 Å is obtained. This must be corrected for ionic character, which affords an adjusted distance of 1.62 Å. This predicted value is about 0.1 Å longer than the 1.519(4) Å measured in 2,6-$Mes_2H_3C_6BeN(SiMe_3)_2$, which suggests the presence of π-bonding in the Be–N bond. It may be noted that this Be–N bond is the shortest seen to date and may be compared to the 1.56(2) Å observed in $Be\{N(SiMe_3)_2\}_2$.[37] Although the relatively high standard deviation in the latter renders comparisons of limited value, the longer Be–N bond in this compound may be a result of competitive π-donation by the nitrogen lone pairs of the two amide groups.

2. Magnesium Compounds

At present, only two organomagnesium derivatives of terphenyl ligands have been well characterized. However, the first terphenyl magnesium compound was reported more than 60 years ago with the generation of BrMgTriph in solution and its subsequent reactions to give the carboxylic acid, the acetyl derivative, and triphenylbenzophenone.[28]

The reaction between IC_6H_3-2,6-Mes_2 and activated magnesium affords the Grignard compound $\{Mg(\mu\text{-Br})C_6H_3\text{-}2,6\text{-}Mes_2(THF)\}_2$.[39] The appearance of bromide instead of iodide in the product arises from the use of the activated magnesium, which is a mixture of magnesium metal and $MgBr_2$. The addition of 1,4-dioxane to this compound in order to obtain the as yet uncharacterized compound $Mg(C_6H_3\text{-}2,6\text{-}Mes_2)_2$ and the coproduct $MgBr_2$ $(1,4\text{-dioxane})_2$ was only partly successful. Nonetheless, evidence of reaction was apparent and an impure $Mg(C_6H_3\text{-}2,6\text{-}Mes_2)_2$ product was isolated. This behavior may be contrasted with that of $Mg(Br)Mes^*$ in THF, which, when treated with 1,4-dioxane, affords $MgMes_2^*$ cleanly and in good yield.[13]

The structure of $\{Mg(\mu\text{-Br})C_6H_3\text{-}2,6\text{-}Mes_2(THF)\}_2$ consists of noninteracting, centrosymmetric, dimeric molecules. The two metals are bridged by bromides to form an almost perfectly square Mg_2Br_2 core. Each magnesium is also bound to a THF and a $-C_6H_3$-2,6-Mes_2 group and is thus four-coordinated. The coordination at magnesium is very distorted from idealized tetrahedral. The sum of the interligand angles at magnesium involving the two bromides and the organo group is 341.7(2)°, suggesting that the magnesium geometry has a tendency to become trigonal pyramidal.

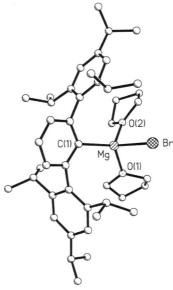

FIG. 9.   Computer-generated drawing of the structure of Mg(Br)(C$_6$H$_3$-2,6-Trip$_2$)(THF)$_2$.[42]

The dimeric structure bears some resemblance to those of the bridged dimers [Mg(μ-Br)Et{O(i-Pr)$_2$}]$_2$[40] or {Mg(μ-Br)Et(NEt$_3$)}$_2$.[41] The structure of the latter species has led to the suggestion that dimers of this kind are found only when the compounds are crystallized from a weakly coordinating solvent, i.e., O(*i*-Pr)$_2$ or NEt$_3$. It is thought that such solvents are incapable of effecting dissociation of the dimeric structure into monomers that are normally observed when Grignard reagents are crystallized from diethyl ether or tetrahydrofuran.

Dissociation of the dimeric terphenyl-substituted Grignard reagent can be effected if the size of the substituents is increased. Thus, the compound Mg(Br)(C$_6$H$_3$-2,6-Trip$_2$)(THF)$_2$ (Fig. 9)[42] has a monomeric structure in which the magnesium displays a distorted tetrahedral geometry. The Mg–C distance is 2.145(7) Å, which is very close to the 2.136(6) Å observed in the dimer. The Mg–Br bond length, 2.479(2) Å, is shorter than the average distance of 2.569(5) Å in the dimer, which is no doubt due to the terminal nature of the Mg–Br bonds in the monomeric structure. The coordination number 4 observed for the Mg^{2+} ion is somewhat surprising, especially in view of the coordination number 2 observed for the similarly sized Li$^+$ ion in (Et$_2$O)LiC$_6$H$_3$-2,6-Trip$_2$.

Currently, no terphenyl derivatives of the heavier group 2 elements, calcium, strontium, or barium have been well characterized.

III

TERPHENYL DERIVATIVES OF *d*-BLOCK AND GROUP 12 ELEMENTS

A. *Transition Metal Derivatives*

At present, there are very few complexes in which the metals in these 10 groups are directly σ-bound to terphenyl ligands. There are, however, several derivatives in which terphenyl groups are used to impart steric congestion to atoms or groups that ligate the transition elements. Among the best-known examples are the terphenolate ligands (e.g., -ODiph or -OTriph)[43–49] and their sulfur[50–54] or selenium[55] analogues. An interesting aspect of terphenolate and terphenolthiolate chemistry is that cyclometalla- tion of *ortho*-phenyl groups has been observed to occur readily at electro- philic main group centers[43] as well as a variety of high- and low-valent *d*- block metal systems.[44] In addition, π interactions with the *ortho*-phenyl groups have been observed in many complexes where the metal center is coordinatively unsaturated.[45] Although such complexes are beyond the scope of this review, we note that investigations of the terphenolate deriva- tives have focused primarily on the early transition metals[46] as well as lanthanide metals[47] with relatively few reports of terphenolate-supported main group complexes.[48] More recently, zinc compounds such as Zn-

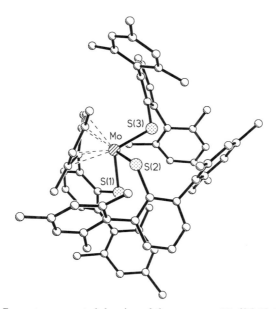

FIG. 10.   Computer-generated drawing of the monomer $Mo(SC_6H_3\text{-}2,6\text{-}Mes_2)_3$.[52]

$(ODiph)_2(OEt_2)_2$ have attracted attention for their catalytic activity in the copolymerization of CO_2 and epoxides.[49] Such compounds illustrate the ability of the bulky terphenolate ligands to stabilize monomeric Zn(II) complexes that contain accessible coordination sites.

Sterically hindered terphenylthiolate ligands have been investigated because they can afford complexes with unusual geometries and oxidation states and low coordination numbers.[50] Some of these complexes are capable of binding small substrate molecules because of their formal coordinative unsaturation. An example is the Mo(II) species $Mo(SDiph)_2CO$, in which Mo also interacts with a *ortho*-phenyl group from one of the -SDiph ligands.[51] The molybdenum complex $Mo(SC_6H_3-2,6-Mes_2)_3$ is monomeric owing to a similar interaction involving one of the *ortho*-Trip substituents[52] (Fig. 10). The bulky thiolate ligand $-SC_6H_3-2,6-Mes_2$ has been shown to be effective at stabilizing nominal two-coordination in the monomeric compounds $M(SC_6H_3-2,6-Mes_2)_2$ (M = Fe[53] or Zn[54]), whereas the related complexes of the -STriph and -SMes* ligands are dimers with thiolate bridging. Similarly, the monomeric selenolate derivatives $M(SeC_6H_3-Mes_2)_2$[55] (M = Mn or Zn) have been characterized.

Terphenyl ligands have also been utilized in the context of metalloporphyrin chemistry. For example, the use of four Triph substituents at the periphery of manganese and iron porphyrin derivatives dramatically in-

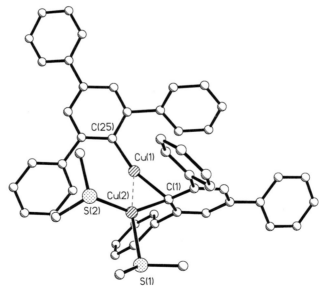

FIG. 11.   Computer-generated drawing of $TriphCu(\mu-Triph)Cu(SMe_2)_2$.[57]

creases the resistance of the metal center to oxidation.[24] A dramatic example of the effectiveness of terphenyl-substituted ligands in transition metal chemistry is in the complex $(\eta^5\text{-}C_5H_5)(CO)_2MoGeC_6H_3\text{-}2,6\text{-}Mes_2$, which has a Mo–Ge triple bond (vide infra).[56]

Terphenyl groups as ligands bound directly to transition metals were first applied to the metals copper and silver. The products of the reaction between MCl (M = Cu or Ag) and TriphMgBr in THF were structurally characterized and were described as the first examples of one-coordinate metals in the solid state—i.e., monomeric MTriph (M = Cu or Ag) compounds.[11] Subsequent work, however, has shown that the compounds were erroneously characterized.[12] Nonetheless, "CuTriph" can be stabilized as a dimethyl sulfide–stabilized adduct $TriphCu(\mu\text{-}Triph)Cu(SMe_2)_2$[57] (Fig. 11). This compound has a unique structure in which two copper atoms are symmetrically bridged by an *ipso*-C atom of one of the Triph groups, one of the coppers is terminally bound to the remaining Triph group, and the other copper is solvated by two Me_2S donors. In addition, there is a relatively short Cu---Cu distance of 2.443(1) Å. It has been suggested[57] that the unsymmetric structure of this compound is due to the fact that a minimum Cu---Cu separation of ca. 2.4 Å would impose a strongly bent C–Cu–C angle (ca. 106°) in a symmetric structure in which both Triph groups bridge the coppers. Apparently, such an angle is not readily accommodated by the σ-bonding requirements of copper.

If the size of the aryl group is increased from Triph to $\text{-}C_6H_3\text{-}2,6\text{-}Trip_2$, the monomeric complex $(Me_2S)CuC_6H_3\text{-}2,6\text{-}Trip_2$ is isolated.[27] The copper is two-coordinate with an interligand angle of 176.3(2)°. The Cu–C distance (1.894(6) Å) is quite short and may be compared to 1.916(3) Å observed in $(Me_2S)CuMes^*$.[57] The structures of both $TriphCu(\mu\text{-}Triph)Cu(SMe_2)_2$[57] and $(Me_2S)CuC_6H_3\text{-}2,6\text{-}Trip_2$[27] provide circumstantial evidence that the monomeric formulation of "CuTriph" is incorrect. For example, the aggregation of "CuTriph" in the presence of Me_2S is surprising since the addition of Lewis bases to metal complexes usually results in a lower, not higher, degree of aggregation. In addition, the almost identical Cu–C distances observed in "CuTriph" and $(Me_2S)CuC_6H_3\text{-}2,6\text{-}Trip_2$ might also be unexpected in view of the different coordination numbers of the metals. Unfortunately, efforts to crystallize "$CuC_6H_3\text{-}2,6\text{-}Trip_2$" in the absence of donor stabilization have not been successful to date. Nonetheless, it is probable that a one-coordinate copper species will be isolated eventually.

The only other known transition metal terphenyl derivative is the blue, crystalline cobalt species $\{Co(\mu\text{-}Br)C_6H_3\text{-}2,6\text{-}Mes_2(THF)\}_2$[39] (Fig. 12). It is isostructural to the corresponding magnesium derivative discussed earlier. It was synthesized via the reaction of $CoCl_2$ with the Grignard reagent,

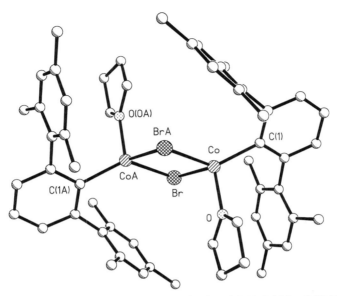

FIG. 12.   Computer-generated drawing of $\{Co(\mu\text{-}Br)C_6H_3\text{-}2,6\text{-}Mes_2(THF)\}_2$.[39]

and the presence of Br^- instead of Cl^- in the product is attributable to halide exchange. It represents a very rare instance of a structurally characterized Co(II) aryl derivative, the only other example being $MesCo(\mu\text{-}Mes)_2CoMes$,[58] which has a considerably shorter Co–C(terminal) distance (1.984(5) Å vs 2.053(8) Å) owing to the lower coordination number of the metals. The Co---Co distance, 3.520(2) Å, precludes Co–Co bonding and is ca. 1.0 Å longer than the Co–Co distance in $MesCo(\mu\text{-}Mes)_2CoMes$. A notable feature of the compound is its thermal stability (decomposes at 140°C), which may be contrasted with that of $MesCo(\mu\text{-}Mes)_2CoMes$, which decomposes slowly at room temperature. It is to be hoped that the apparent stability of at least some of the transition metal terphenyl derivatives will lead to synthesis of further examples.

B. *Group 12 Terphenyl Derivatives*

The sole representative of group 12 terphenyl derivatives is the mercury(II) compound $Hg(C_6H_3\text{-}2,6\text{-}Mes_2)_2$.[35] This species was synthesized from $HgBr_2$ and $LiC_6H_2\text{-}2,6\text{-}Mes_2$. It was intended for use as a reagent for transmetallation reactions with the heavier alkali metals sodium, potassium, or cesium. Experiments showed that it did not undergo any significant reaction with sodium or potassium in benzene at 75°C over a period of 14 h. A reaction was observed with cesium, but the cesium aryl product reacted rapidly with solvent benzene. The structure reveals no unusual features. The C–Hg–C angle (178.6(3)°) is essentially linear. The Hg–C bond length 2.073(4) Å (av.) is close to that seen in HgMes$_2^*$(2.077(6) Å).[59]

IV

TERPHENYL DERIVATIVES OF p-BLOCK ELEMENTS

A. Group 13 Terphenyl Derivatives

Currently, more terphenyl derivatives of group 13 elements are known than of any other group in the periodic table. Furthermore, derivatives of each element of the group have been reported. These include novel classes of compounds such as the cyclic dianionic salts $M_2\{GaC_6H_3\text{-}2,6\text{-}Mes_2\}_3$-(M = Na[60] or K[61]); the dimer $Na_2\{GaC_6H_3\text{-}2,6\text{-}Trip\}_2$[62] and 2,6-Trip$_2H_3C_6$-GaFe(CO)$_4$,[63] which have engendered considerable literature discussion of their bonding; and the compounds $MC_6H_3\text{-}2,6\text{-}Trip_2$ (M = In[64] or Tl[65]), which possess unique examples of one-coordinate metals in the solid state. An early use of terphenyl ligands in group 13 chemistry concerned the synthesis of a formal Ga–P analogue of borazine {TriphGaP(c-C$_6$H$_{11}$)}$_3$.[66] In the 1990s approximately 60 terphenyl derivatives of the group 13 elements have been synthesized and characterized.

1. Boron Compounds

Terphenyl derivatives of boron were originally synthesized with the object of preparing compounds with B–B triple bonds. It was thought that these aryl ligands would be effective in the stabilization of triple bonds in a similar manner to their success in the stabilization of double bonds. For example, a B–B triple-bonded boron analogue of a substituted acetylene might be accessible by reduction of an ArBX$_2$ or 1,2-dimethoxydiborane(4) species as shown by Eq. (3).

$$2ArBX_2 \xrightarrow[\text{ether solvent}]{6M}$$

$$Ar(MeO)B\text{-}B(OMe)Ar \xrightarrow{4M} \longrightarrow M_2^+ [R-B\equiv B-R]^{2-} + 4MX \text{ or } 2 \text{ MOMe} \quad (3)$$

(Ar = terphenyl group; M = alkali metal)

The B–B bonded 1,2-dimethoxydiborane(4) precursor[67] was selected on the basis that it could be obtained readily from $(MeO)_2B\text{-}B(OMe)_2$[68] by reaction with the lithium aryl. In contrast to diborane(4) tetrahalides,

$(MeO)_2B-B(OMe)_2$ is a fairly stable compound that can be obtained by a conventional (if tedious) synthetic route in useful amounts.[68] Sterically crowding terphenyl groups were selected as the aryl boron substituents, since it was felt that one such group would provide an amount of steric protection similar to that provided by two aryl groups in the doubly bonded $[R_2B{=}BR_2]^{2-}$ ions.[69] Accordingly, the monoaryl–boron dihalide derivatives were readily synthesized in good yield by the reaction of the appropriate boron halide with the lithium aryl in accordance with Eq. (4).[67] The large size of the substituent

$$BBr_3 + LiAr \longrightarrow ArBBr_2 + LiX \tag{4}$$

$$(Ar= -C_6H_3\text{-}2,6\text{-}Mes_2; \quad Ar = -C_6H_3\text{-}2,6\text{-}Trip_2)$$

precludes the further substitution of halides and the mixture of products normally expected in such reactions. The X-ray crystal structure data for the products indicate that the plane of the central phenyl ring of the terphenyl ligand is essentially perpendicular (Fig. 13) to the coordination plane at boron, in keeping with the large steric requirements of these ligands.

Reduction of the arylboron dihalide $2,6\text{-}Mes_2H_3C_6BX_2$ (X = Cl or Br) or $2,6\text{-}Trip_2H_3C_6BBr_2$ with various alkali metals is described in Scheme 2, where most of the products were characterized structurally.[67] Thus, reduction of $2,6\text{-}Mes_2H_3C_6BBr_2$ with excess lithium powder in Et_2O gave a deep red solution, from which the bis-lithium–9-borafluorenyl complex could be crystallized in a 74% overall yield (reduction of the dichloride $2,6\text{-}Mes_2H_3C_6BCl_2$ gave a lower yield). A similar reduction in benzene instead of Et_2O affords, upon workup in Et_2O/hexane, the related dimeric lithium–9-borafluorenyl complex, which is closely related to the former species

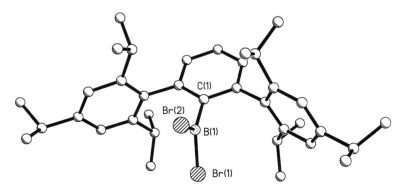

FIG. 13.   Computer-generated drawing of $2,6\text{-}Trip_2H_3C_6BBr_2$.[67]

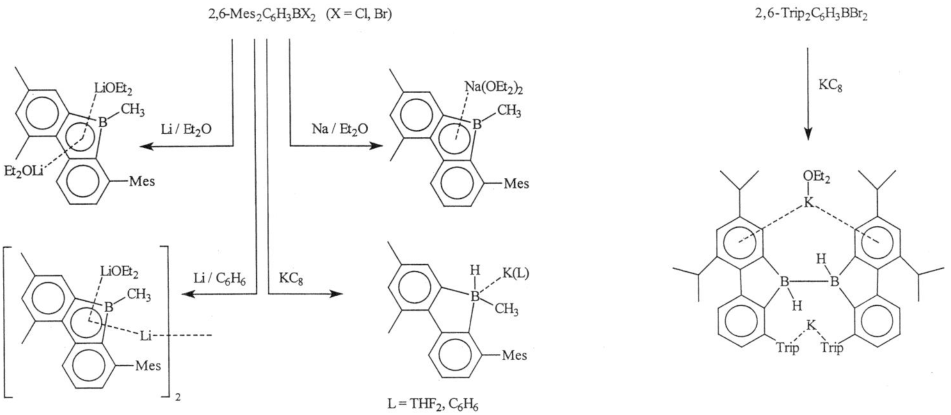

SCHEME 2. The products of the reduction of 2,6-Mes$_2$H$_3$C$_6$BX$_2$ (X = Cl, Br) and 2,6-Trip$_2$H$_3$C$_6$BBr$_2$.[67]

except that the reduced amount of Et$_2$O-Li$^+$ solvation has caused aggregation to dimers in which two Li$^+$ ions are sandwiched between two 9-borafluorenyl rings. Reduction of 2,6-Mes$_2$H$_3$C$_6$BBr$_2$ by KC$_8$ in THF or benzene also gave a deep red solution from which only colorless crystals of the borate products could be isolated.[67] Reduction of 2,6-Trip$_2$H$_3$C$_6$BBr$_2$ by

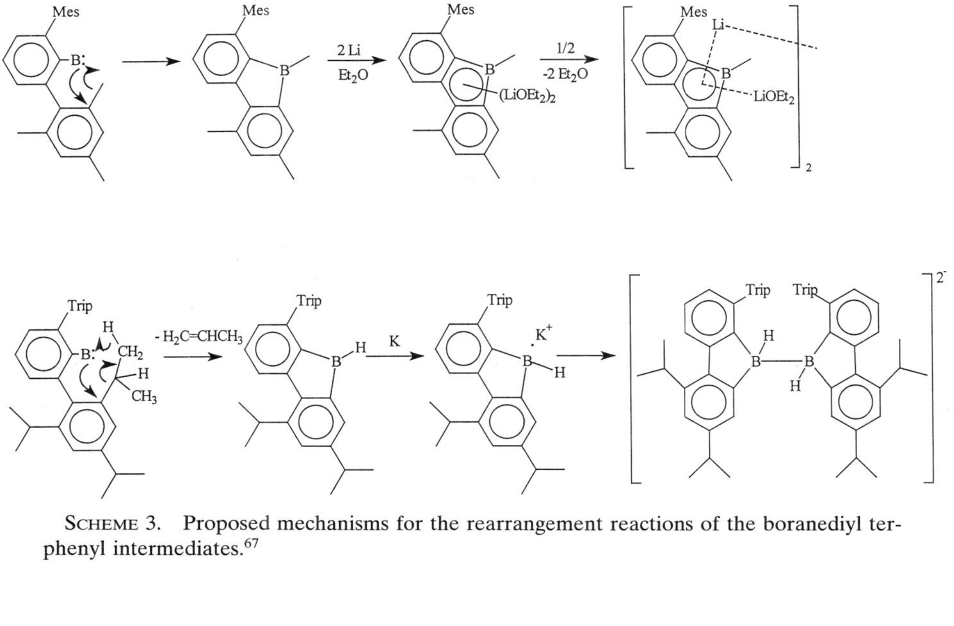

SCHEME 3. Proposed mechanisms for the rearrangement reactions of the boranediyl terphenyl intermediates.[67]

KC$_8$ also affords a borate salt, but in this case it is uniquely dimerized with a relatively long (1.83(2) Å) B–B bond, which is caused, no doubt, by the presence of adjacent negative charges and the four-coordination of the boron atoms. Reduction of 2,6-Trip$_2$H$_3$C$_6$BBr$_2$ with sodium gave the 9-borafluorenyl radical anion salt, which was characterized by EPR and elemental analysis.[67]

The range of products isolated from these reductions may be rationalized by assuming that the initial reduction step leads to a boranediyl (ArB:) intermediate that undergoes an unprecedented insertion into an *ortho*-Mearyl C–C bond with subsequent further reduction from the 9-borafluorenyl product. Details of the proposed mechanism are given in Ref. 67 (Scheme 3). The reductions in Schemes 2 and 3 represented the first instance of C–C bond activation by boranediyls. The formation of diborate salt can also be rationalized by assuming the initial formation of the boranediyl. However, in this case propene is eliminated with concomitant formation of a B–H bonded intermediate, which is subsequently reduced and dimerized.

The failure to obtain B–B multiply bonded products from these reactions led to the investigation of singly bonded B–B precursors derived from B$_2$(OMe)$_4$.[70] However, reaction of B$_2$(OMe)$_4$ with 2 equiv of LiAr (Ar = -C$_6$H$_3$-2,6-Mes$_2$ or -C$_6$H$_3$-2,6-Tips$_2$) affords only Ar(MeO)BB(OMe)$_2$.

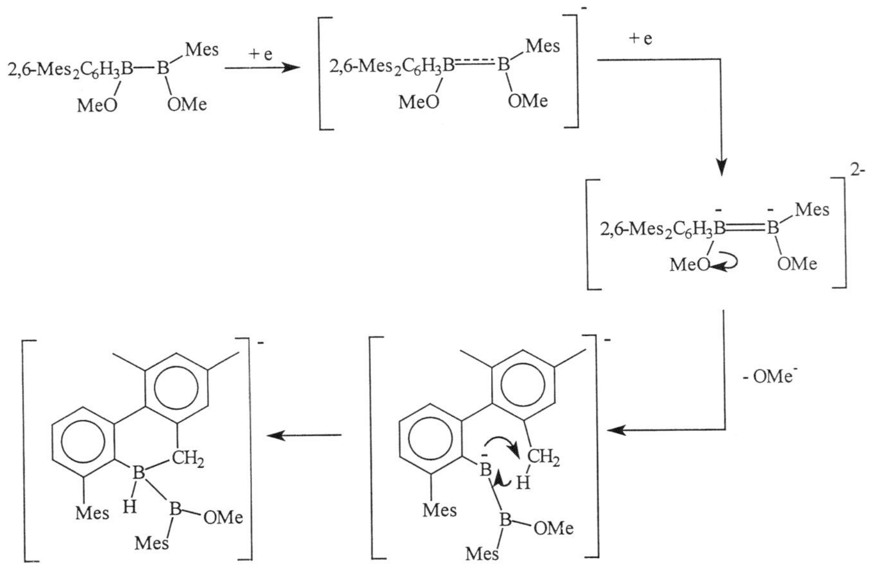

SCHEME 4. Proposed mechanism for the reduction and subsequent rearrangement of 2,6-Mes$_2$H$_3$C$_6$(MeO)BB(OMe)Mes.[70]

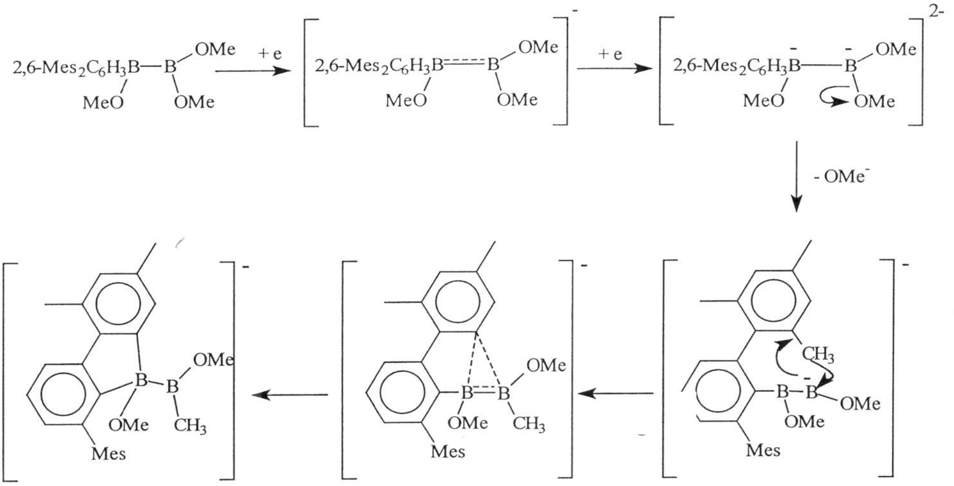

SCHEME 5. Proposed mechanism for the reduction and subsequent rearrangement of 2,6-Mes$_2$H$_3$C$_6$(MeO)BB(OMe)$_2$.[70]

Even at elevated temperature no 1,2-diaryl-1,2-dimethoxydiborane(4) product could be isolated. Seemingly, further substitution by Ar does not take place for steric reasons. Nonetheless, further substitution of MeO-groups is possible with less crowding groups such as mesityl to give Ar-(MeO)BB(OMe)Mes. The structure of 2,6-Mes$_2$H$_3$C$_6$(MeO)BB(OMe)$_2$ features a normal B–B distance of 1.718(5) Å, almost equal B–O bond lengths of ca. 1.36 Å, and a torsion angle of 64° between the boron planes. The structure of 2,6-Mes$_2$H$_3$C$_6$(MeO)BB(OMe)Mes is very similar except that one methoxy group is replaced by mesityl.[70] The reduction of this compound with excess lithium in diethyl ether gives a borate product as illustrated in Scheme 4.

The reduction of 2,6-Mes$_2$H$_3$C$_6$(MeO)BB(OMe)$_2$ results in the isolation of a product in which one of the methoxy groups at boron is replaced by a methyl group. A possible mechanism for this process is shown in Scheme 5. The two initial steps are sequential one-electron reductions, the second of which produces an unstable dianion (cf. reduction of Mes(MeO)BB-(OMe)Mes) from which methoxide is eliminated. The unstable intermediate contains a two-coordinate boron, possibly stabilized by π-overlap with the oxygen lone pair, which then inserts into the *ortho*-Me-Aryl C–C bond and rearranges through a nonclassical intermediate. The first three steps in the reduction of 2,6-Mes$_2$H$_3$C$_6$(MeO)BB(OMe)$_2$ are similar to one-electron reductions and subsequent elimination of methoxide to give a boranide intermediate, which then inserts into a methyl C–H group to give a boron heterocycle dihydrophenanthrene. Further details are in Ref. 70.

In summary, the attempted further reduction of 1,2-dimethoxydiborane compounds results, initially, in the elimination of a methoxide group. When this elimination occurs, however, there is an immediate insertion of the reactive boranide intermediate into a C–C or C–H bond before elimination of a second methoxide occurs.

2. Aluminum Compounds

Several terphenyl compounds of aluminum have been synthesized and structurally characterized. The simplest compounds are the monoarylhalides of the type $ArAlCl_2$ (Ar = $-C_6H_3\text{-}Mes_2$ or $-C_6H_3\text{-}Trip_2$).[71] However, the conventional synthetic route, which involves direct reaction of the LiAr salt with the aluminum halide, does not generally afford high yields of the desired product, although $TriphAlBr_2(OEt_2)$ could be synthesized in 36% yield by this method. Instead, the recently developed aluminum hydride route was used.[72] This involves the reaction sequence described in Eq. (5). The arylaluminum hydride may also be obtained by using 1, rather than 3, equiv of Mes_3SiCl as in Eq. (6).

$$LiC_6H_3\text{-}2,6\text{-}Mes_2 + AlH_3(NMe_3) \xrightarrow[-78 \longrightarrow -20\ °C]{Et_2O} 2,6\text{-}Mes_2H_3C_6AlH_3Li(OEt_2)_{1.5} + NMe_3$$

$$+3Me_3SiCl \downarrow \qquad\qquad (5)$$

$$LiCl + 3Me_3SiH + 2,6\text{-}Mes_2H_3C_6AlCl_2(OEt_2) \longleftarrow$$

$$2,6\text{-}Mes_2H_3C_6AlH_3Li(OEt_2)_{1.5} + Me_3SiCl \xrightarrow{Et_2O} 2,6\text{-}Mes_2H_3C_6AlH_2(OEt_2) + LiCl \qquad (6)$$
$$+ Me_3SiH$$

Both the halide and hydride etherate products of Eqs. (5) and (6) can be easily desolvated by heating to ca 90–120°C under reduced pressure [Eq. (7)]. The halide and hydride

$$2,6\text{-}Mes_2H_3C_6AlX_2(OEt_2) \xrightarrow{90\text{-}120\ °C} (2,6\text{-}Mes_2H_3C_6AlX_2)_2 + 2Et_2O \qquad (7)$$

$$(X = H\ or\ Cl)$$

etherate complexes are monomeric in the crystalline state. When they are desolvated, however, they become dimeric with either hydride or halide bridges. Their tendency to lose ether from the monomeric solvates is confirmed by the fact that both the hydride and to a lesser extent the halide are dissociated in solution. The behavior of these $-C_6H_3\text{-}2,6\text{-}Mes_2$ derivatives may be contrasted with that of the corresponding $-C_6H_3\text{-}2,4,6\text{-}t\text{-}Bu_3$ (Mes*) derivatives.[71,73] For example, Mes^*AlCl_2 does not form stable complexes with Et_2O^{73}; also, Mes^*AlH_2 exists exclusively as a hydride-bridged dimer $(Mes^*AlH_2)_2$, which does not form complexes with either Et_2O or THF.[72,73]

The corresponding $-C_6H_3$-2,6-$Trip_2$ derivatives behave-similarly to their $-C_6H_3$-2,6-Mes_2 counterparts.[71] The reaction of $(Et_2O)LiC_6H_3$-2,6-$Trip_2$ with $AlH_3(NMe_3)$ proceeded smoothly to give $\{2,6\text{-}Trip_2H_3C_6AlH_3Li\text{-}(OEt_2)\}_n$ (n is probably 2) in nearly quantitative yields. Removal of LiH by reaction with 1 equiv of Me_3SiCl gave $2,6\text{-}Trip_2H_3C_6AlH_2(OEt_2)$ in good yield. This can be readily desolvated to form the dimeric hydride $(H_2AlC_6H_3$-2,6-$Trip_2)_2$ in high yield. The dimeric structure is retained in solution even up to 85°C, and the ^1H NMR spectrum displays both bridging and terminal Al–H resonances. The reaction between $(H_2AlC_6H_3$-2,6-$Trip_2)_2$ and further equivalents of Me_3SiCl did not lead to $Cl_2AlC_6H_3$-2,6-$Trip_2$, but to a mixture of products involving bridging hydroxides. However, the halide $2,6\text{-}Trip_2C_6H_3AlCl_2(OEt_2)$ may be prepared by the direct reaction of $(Et_2O)LiC_6H_3$-2,6-$Trip_2$ with 2 equiv of $AlCl_3$ in toluene followed by crystallization from hexane/Et_2O to afford the product in 18% yield.[71]

These results for the monoarylaluminum hydrides and halides underline the different steric properties of the Mes^*, $-C_6H_3$-2,6-Mes_2, and $-C_6H_3$-2,6-$Trip_2$ ligands.[72] The Mes^* aluminum hydrides and halides do not form complexes with Et_2O, whereas the corresponding terphenyl derivatives do so readily. One of the major reasons for this difference is the lack of any significant interaction between the *ortho* aryl groups of the terphenyl group and the aluminum center. In contrast, there are close approaches (ca. 2.2 Å) between the hydrogens of the *ortho-t*-Bu groups of the Mes^* ligand in the case of its derivatives. A further reason appears to lie in the nature of the *ortho*-aryl groups themselves, which, being effectively two-dimensional and confined to a perpendicular orientation with respect to the central aryl ring, can arrange themselves so that effective steric and electronic masking of the vacant metal coordination site by a single ligand is not observed.

Some reactions of the hydride $(H_2AlC_6H_3$-2,6-$Mes_2)_2$ have been investigated.[74] Thus, its treatment with H_2EPh (E = N, P, or As) affords the products $[2,6\text{-}Mes_2H_3C_6Al\{\mu\text{-}E(H)Ph\}]_n$ (E = N, n = 2; E = P or As, n = 3) or the monomeric compound $2,6\text{-}Mes_2H_3C_6Al\{As(H)Ph\}(OEt_2)$. Reaction of 3 equiv of H_2NPh with $(H_2AlC_6H_3$-2,6-$Mes_2)_2$ gives the unsymmetrically substituted species $2,6\text{-}Mes_2H_3C_6Al\{N(H)Ph\}\{\mu\text{-}N(H)Ph\}_2Al(H)C_6H_3$-2,6-$Mes_2$. Heating the arsenic species $\{2,6\text{-}Mes_2H_3C_6Al\{As(H)Ph\}H\}_2$ affords the unusual As–As bonded product $2,6\text{-}Mes_2H_3C_6Al\{\mu\text{-}As(H)Ph\}_2\{\mu\text{-}As(Ph)As(Ph)\}AlC_6H_3$-2,6-$Mes_2$.

Reduction of the halides $2,6\text{-}Ar_2H_3C_6AlX_2(OEt_2)$ (Ar = Mes or Trip; X = Cl or Br) with alkali metals with the objective of synthesizing Al–Al bonded species leads to decomposition with deposition of aluminum met-

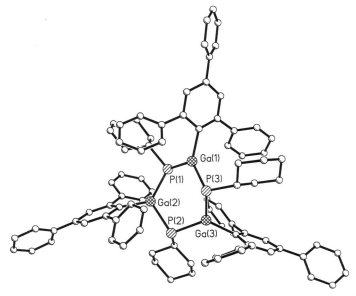

FIG. 14. Computer-generated drawing of the trimer {TriphGaP(*c*-C$_6$H$_{11}$)}$_3$.[66]

al.[74b] The arenes 1,3-Ar$_2$C$_6$H$_4$ (Ar = Mes or Trip) were isolated from the supernatant solution.

3. *Gallium Compounds*

As previously mentioned, terphenyl-type ligands were originally used to stabilize Ga–P six-membered ring analogues of benzenes or borazines. The compound {TriphGaP (*c*-C$_6$H$_{11}$)}$_3$[66] was synthesized in 45% yield via the reaction of TriphGaCl$_2$ (generated *in situ*) and the dilithium salt Li$_2$P (*c*-C$_6$H$_{11}$). Its structure (Fig. 14) involves a nonplanar Ga$_3$P$_3$ array. Both the gallium and phosphorus atoms are three-coordinate and the Ga–P bond lengths average ca. 2.3 Å. The geometry at the gallium centers is distorted trigonal planar, whereas the phosphorus atoms are all pyramidally coordinated with Σ°P varying from 315.7 to 331.1°. The puckering of the ring structure indicates that delocalization of the phosphorus lone pairs into the nonbonding Ga *p*-orbitals is not extensive. This view is also supported by the upfield chemical shift (δ = −61) observed in the ^{31}P NMR spectrum. A downfield shift (deshielding) might have been expected had the delocalization of the P lone pairs been large.

Several gallium halide derivatives of the -C$_6$H$_3$-2,6-Mes$_2$ ligand have been synthesized. The first of these was the compound ClGa(C$_6$H$_3$-2,6-Mes$_2$)$_2$[75]

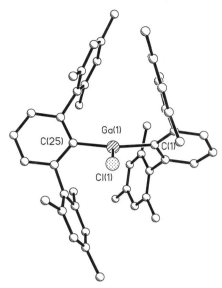

FIG. 15.   Computer-generated drawing of $Ga(Cl)C_6H_3$-2,6-$Mes_2)_2$, which has an extremely wide C–Ga–C angle of 153.3(5)°.[75]

(Fig. 15), which was shown to have a very wide C–Ga–C angle of 153.3(5)° (cf. 153.5(2) in $BrGa(C_6H_3$-2,6-$Mes_2)_2)$,[76] whereas the corresponding angle in $ClGaMes_2^*$ is 135.6(2)°.[77] The indium analogue $BrIn(C_6H_3$-2,6-$Mes_2)_2$[78] has a somewhat wider C–In–C angle of 157.3(8)°, which may be compared to the values of 144.6(3)° and 144.2(5)° seen in $BrInMes_2^*$[79] and $ClInMes_2^*$.[80] The monoaryls $\{Cl_2GaC_6H_3$-2,6-$Mes_2\}_2$[76] and $\{Cl_2InC_6H_3$-2,6-$Mes_2\}_2$[81] have also been reported. These compounds, however, are dimerized through two μ-Cl bridges. These dimeric structures are in sharp contrast to the gallium and indium Mes* derivatives X_2MMes^* (M = Ga or In; X = Cl or Br),[82] which are monomeric. The structures of these two sets of diaryl and monoaryl derivatives again underline the difference in steric properties of the Mes* and the -C_6H_3-2,6-Mes_2 ligands and are in harmony with the results for the aluminum compounds already discussed.[72]

Reduction of $Cl_2GaC_6H_3$-2,6-Mes_2 by either sodium or potassium gives the novel and unique doubly reduced cyclotrigallyl ring compounds $M_2\{GaC_6H_3$-2,6-$Mes_2\}_3$ (M = Na[60] or K[61]) (Fig. 16). The dianion core structure consists of an almost equilateral Ga_3 triangle with Ga–Ga distances in the range 2.419(5)–2.441(5) Å. The Na$^+$ or K$^+$ cations are located above and below the Ga_3 ring planes and are apparently also complexed by $ortho$-Mes rings. The doubly reduced cyclotrigallyl ring can be regarded

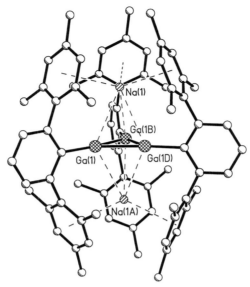

FIG. 16.   Computer-generated drawing of $Na_2(GaC_6H_3\text{-}2,6\text{-}Mes_2)_3$.[60]

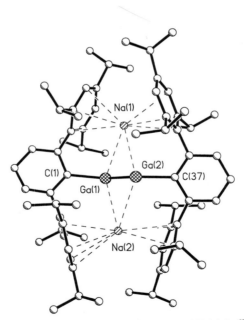

FIG. 17.   Computer-generated drawing of $Na_2(GaC_6H_3\text{-}2,6\text{-}Triph_2)_2$,[62] which was reported to have a Ga–Ga triple bond.

as having two electrons in a delocalized bonding π-orbital and thus conforms to the Hückel rule. Calculations on hypothetical $M_2\{GaH\}_3$ (M = Li, Na or K) compounds afford Ga_3 rings with significantly longer Ga–Ga distances near 2.53 Å and upfield ^1H NMR shifts consistent with the presence of a ring current. The calculated Ga–Ga bonds are very similar to Ga–Ga single-bond distances in sterically crowded digallium compounds R_2GaGaR_2-(R = -CH(SiMe$_3$)$_2$,[83] Trip,[84] and 2,2,6,6-tetramethylpiperidine[85]). Moreover, it can be argued that the upfield NMR shifts are consistent with the presence of a double negative charge on the $\{GaH\}_3{}^{2-}$ ring. The complexed Na$^+$ ions may also play a-role in shortening the Ga–Ga bond and/or stabilizing the compound as they appear to do in the dimeric compound Na$_2\{GaC_6H_3$-2,6-Trip$_2\}_2$ (vide infra).

An increase in the size of the aryl group to -C$_6$H$_3$-2,6-Trip$_2$ results in the isolation of the very interesting dimeric species Na$_2\{GaC_6H_3$-2,6-Trip$_2\}_2$ (Fig. 17).[62] In this case the dianion has a *trans*-bent structure with Ga–Ga–C angles of 128.5(4)° and 133.5(4)° and a shortened Ga–Ga bond length of 2.319(3) Å. It was described as having a triple bond—a view that has received some theoretical support.[86,87] Nonetheless, DFT and extended Hückel ab initio calculations on hypothetical species such as Na$_2\{Ga$-Diph$\}_2$[88] and Li$_2\{GaMe\}_2$[89] have indicated that one of the three M.O.s associated with the Ga–Ga bonding is primarily nonbonding or even slightly antibonding (for a contrary view, see Ref. 86) and that the nonbonding orbital has lone-pair character. This view is consistent with the bending of the geometry at the galliums. The DFT calculations have also indicated that the Na$^+$ ions that complex to the *ortho*-aryl substituents play a significant role in the shortening of the Ga–Ga bond and that, furthermore, the Ga–Ga length variation is associated with a shallow potential well.[88] These data suggest that the Ga–Ga bond is quite weak overall. The view that the Ga–Ga bond has double character is also consistent with theoretical DFT calculations on the ethylene-like dianion $\{Ga(H)Diph\}_2{}^{2-}$, which has a Ga–Ga distance very similar to that of the Na$_2\{GaDiph\}_2$ species.[88] The ab initio calculations indicate that the bending of the CGaGaC framework in Li$_2$[GaMe]$_2$ not only converts one of the original π-orbitals in the linear form to a lone-pair orbital, but also weakens the remaining σ- and π bonds considerably.[89]

Another interesting low-coordinate gallium species was obtained from the reaction between Cl$_2$GaC$_6$H$_3$-2,6-Trip$_2$ and Na$_2$Fe(CO)$_4$.[63] The product 2,6-Trip$_2$H$_3$C$_6$GaFe(CO)$_4$ was described as having a gallium–iron triple bond, i.e., as a "ferrogallyne." Consistent with this description, the gallium coordination is linear and there is a short Ga–Fe bond length of 2.2248(7) Å. However, DFT calculations[90] on the closely related analogue DiphGaFe(CO)$_4$ accurately reproduce the major structural features of 2,6-Trip$_2$H$_3$C$_6$GaFe(CO)$_4$, but do not support the existence of a triply bonded interaction. In fact, they indicate that π-orbital interactions between the gallium and iron are weak and that the bonding interaction between gallium and iron is dominated by a strong σ-donation of the gallium lone pair to iron. In this respect, the theoretical results are in close agreement with those obtained earlier for the related aluminum derivative (η^5-C$_5$Me$_5$)AlFe(CO)$_4$.[91] The iron carbonyl C–O stretching frequencies in the aluminum and gallium compounds are very close and are lower than those observed in the phosphine complex Ph$_3$PFe(CO)$_4$.[92] This indicates that the σ-donor/π-acceptor ratio of (η^5-C$_5$Me$_5$)Al and 2,6-Trip$_2$H$_3$C$_6$Ga ligands is less than that of the Ph$_3$P ligand, suggesting that a triaryl phosphine may be a superior π-acceptor to these ligands. The short Fe–Ga distance may be a consequence of strong σ-bonding of the *sp* hybridization at the gallium and its low coordination number.

4. *Indium and Thallium Compounds*

The indium compounds BrIn(C$_6$H$_3$-2,6-Mes$_2$)$_2$[78] and (Cl$_2$InC$_6$H$_3$-2,6-Mes$_2$)$_2$[81] have already been discussed in the context of their gallium counterparts. More recent studies have focused on the lower oxidation state In(I).[64] Thus, the reaction of (Et$_2$O)LiC$_6$H$_3$-2,6-Trip$_2$ with InCl in THF at ca. −78°C affords the compound InC$_6$H$_3$-2,6-Trip$_2$ (Fig. 18) in moderate yield as very air-sensitive orange crystals. It features a one-coordinate indium with an In–C distance of 2.256(4) Å.[64] This compound (along with its thallium congener) appears to be the first authenticated example of a one-coordinate metal in the solid state (cf. copper and silver derivatives dis-

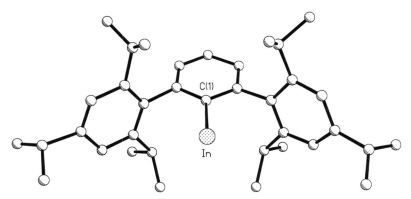

FIG. 18.   Computer-generated drawing of InC$_6$H$_3$-2,6-Trip$_2$ featuring a one-coordinate indium.[64]

cussed earlier[11,12]). The Tl(I) analogue TlC_6H_3-2,6-$Trip_2$ was synthesized by reaction of TlCl with $(Et_2O)LiC_6H_3$-2,6-$Trip_2$ in Et_2O near 0°C.[65] It is also monomeric and features a one-coordinate thallium with a Tl–C distance of 2.34(1) Å. The InC_6H_3-2,6-$Trip_2$ species can function as a Lewis base, and the complex $(\eta_5$-$C_5H_5)(CO)_2MnInC_6H_3$-2,6-$Trip_2$, which features an Mn–In distance of 2.4102(9)Å and an Mn–In–C angle of 175.39(9)°, has been synthesized and structurally characterized.[64] Infrared studies of this complex indicate that the indium ligand is primarily a good σ-donor with weak π-acceptor character.

B. *Group 14 Terphenyl Derivatives*

This section focuses on the heavier element group 14 terphenyl compounds. However, some carbon derivatives that have obvious synthetic importance are also described briefly.

1. *Carbon Compounds*

The synthetic approaches to *ortho*-substituted iodoterphenyl reagents, which are used as starting materials for many of the compounds in this review, are due largely to the work of Hart, Lüning, and co-workers.[25,93] These routes allow the incorporation of new functionalities into the 1-position with relative ease. Lithiation and subsequent reaction with elec-

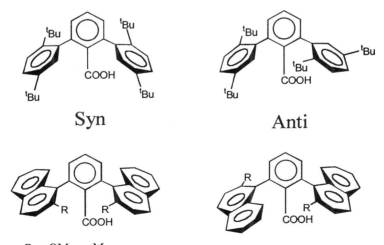

Syn   Anti

R = OMe or Me

FIG. 19. Schematic drawings of various configurations of terphenylcarboxylic acids.[94]

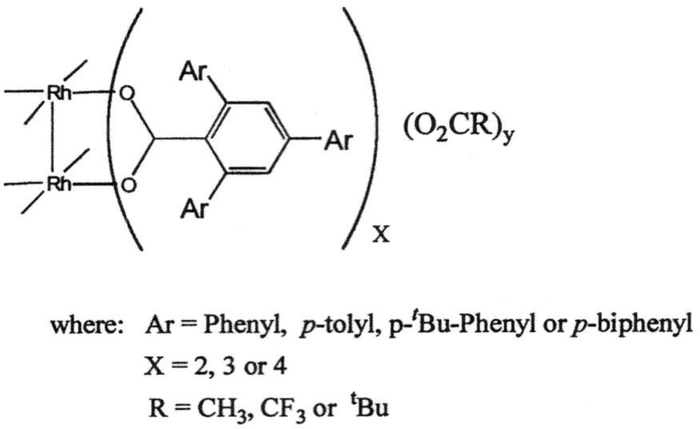

where: Ar = Phenyl, *p*-tolyl, p-*t*Bu-Phenyl or *p*-biphenyl
        X = 2, 3 or 4
        R = CH$_3$, CF$_3$ or *t*Bu

FIG. 20.   Schematic drawing of dirhodium complexes of terphenylcarboxylate ligands.[95]

trophiles produce aldehyde, carbinol, carboxylic acid, ester, oxime, alkene, cyano, and isocyano derivatives of $-C_6H_3$-2,6-Dmp$_2$, in moderate to good yields.[93] There is particular interest in terphenyl carboxylic acids, as they may be used as "building blocks" for the assembly of host molecules and for concave reagents.[93] Concave acids, such as $HOOCC_6H_3$-2,6-Dmp$_2$, have been used to govern diastereoselectivity and regioselectivity for the proton-ation of allyl and cyclohexyl anions. Terphenyl carboxylic acids are prepared from the direct reaction of the lithium salt with CO_2 followed by hydrolysis with dilute HCl. A series of chiral terphenyl carboxylic acids have been synthesized[94] using Hart's[25] method (Fig. 19). Kinetic epimerization experi-ments on these compounds show first-order behavior with $\Delta G^{\neq}$ values between 25 and 35 kcal/mol^{-1}.

The terphenyl carboxylic acid derivatives have also been employed as supporting ligands in transition metal chemistry. For example, several binu-clear rhodium(II) terphenyl carboxylates (Fig. 20) have been reported.[95] These were synthesized using the alternative route described in Eq. (8).[96]

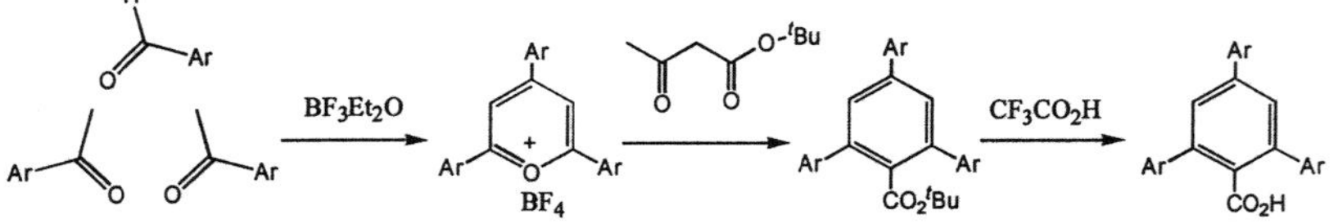

Finally, attention is drawn to the extensive work of Okazaki and co-workers, wherein two terphenyl units have been attached to a central functionality in order to create novel all-carbon molecular bowl-like molecules.[97] For example, the novel molecular bowl compound Ar′Br (Ar′Br = 2,6-bis($CH_2C_6H_3$-2,6-Dmp$_2$)-4-*tert*-butyl-1-bromophenyl), was synthesized by a cross-coupling reaction between the Grignard reagent of the iodoterphenyl

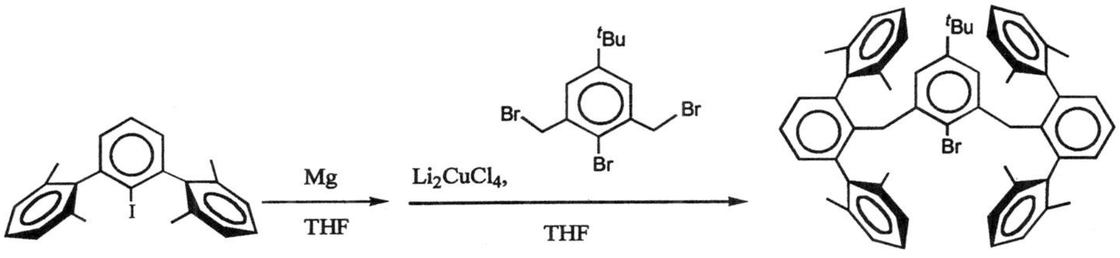

SCHEME 6. Synthetic route to a molecular bowl molecule with terphenyl substituents.[97]

derivative and the aryl tribromide reagent in the presence of a copper(I) catalyst (Scheme 6).[97]

2. *Silicon Compounds*

The synthesis of the primary terphenyl silyl halides, $Cl_3SiC_6H_2$-2,4,6-Ph_3, $Cl_3SiC_6H_3$-2,6-Mes_2, $Cl_3SiC_6H_3$-2,6-$Trip_2$, $HCl_2SiC_6H_3$-2,6-Mes_2, and $HCl_2SiC_6H_3$-2,6-$Trip_2$, can be accomplished by the reaction between $SiCl_4$ or $SiHCl_3$ and the appropriate lithium reagent, (yields = 60–80%).[98,99] Similarly, the disilane $Cl_3SiSiCl_2C_6H_3$-2,6-Mes_2 is prepared in an analogous manner from Si_2Cl_6.[98] Reduction of the halides, $Cl_3SiC_6H_3$-2,6-Mes_2 and $Cl_3SiC_6H_3$-2,6-$Trip_2$, with $LiAlH_4$ yields the silanes $H_3SiC_6H_3$-2,6-Mes_2 and $H_3SiC_6H_3$-2,6-$Trip_2$, respectively.[98]

Reaction of $Cl_3SiC_6H_2$-2,4,6-Ph_3 with ammonia affords the tris(amino)-silane compound $(H_2N)_3SiC_6H_2$-2,4,6-Ph_3 in 71% yield [Eq. 9].[99] The X-ray crystal structure was the first for a species in which three NH_2 groups were bound to one silicon atom. There is pyramidal geometry at the nitrogen atoms. The structural characterization of a series of such Si–NH_2 compounds led to the conclusion that N → Si p–d π-overlap was not significant in the Si–N bonding.[99]

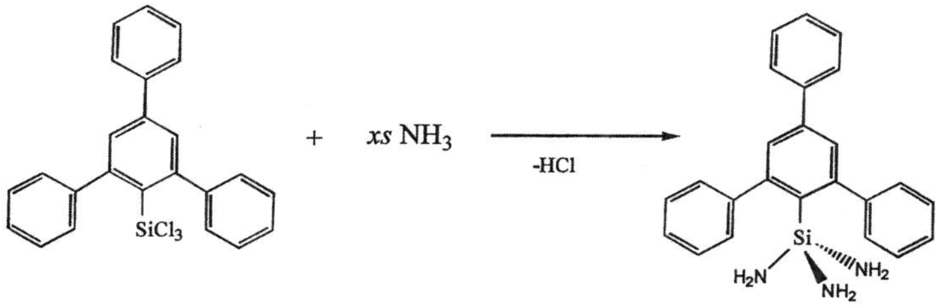

$$\text{(9)}$$

The reduction of $Cl_3SiC_6H_2$-2,4,6-Ph_3 with lithium metal in the presence of trimethylsilyl chloride led to the isolation of a silafluorene compound that was formed by an intramolecular ring closure.[100] Although the mechanism is unknown, it is believed that reduction affords a silyl radical, which may

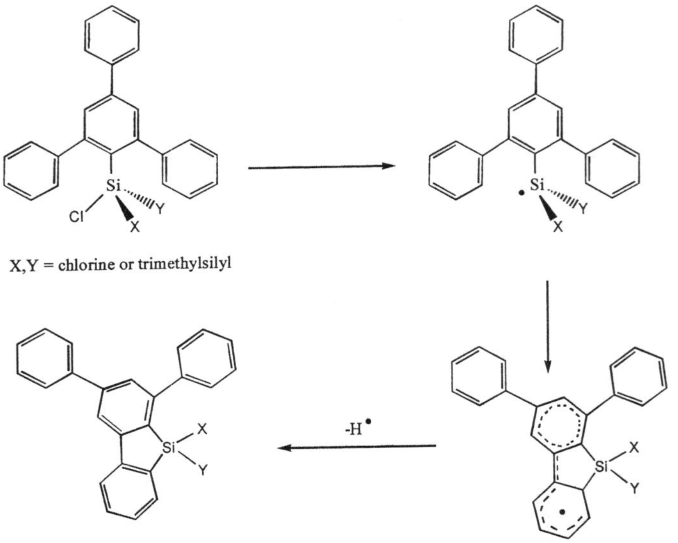

X,Y = chlorine or trimethylsilyl

SCHEME 7.   Reduction of $Cl_3SiC_6H_2$-2,4,6-Ph_3 to give a silafluorene derivative.[100]

then add to the *ortho*-phenyl ring to provide a delocalized intermediate. This loses a hydrogen atom to give the final product (Scheme 7).

Finally, it has been reported that the titanium-catalyzed ($TiCl_4$/Et_2AlCl) cyclotrimerization of 1-phenyl-2-(trimethylsilyl)acetylene affords the terphenyl silane 1,3,5-triphenyl-2,4,6-tris(trimethylsilyl)benzene in trace amounts.[101]

3. Germanium, Tin, and Lead Compounds

It is convenient to discuss the derivatives of the remaining heavier group 14 terphenyl elements together, since they have many features in common. This is particularly true for the divalent compounds.

The thermally stable monomeric diterphenyl species $M(C_6H_3$-2,6-$Mes_2)_2$ (where M = Ge, Sn, and Pb) were prepared[102] in moderate yields by the treatment of 2 equiv of the lithium aryl salt with $GeCl_2 \cdot$ dioxane, $SnCl_2$, or $PbCl_2$, respectively. The dimeric monoterphenyl metal halide compounds $\{M(Cl)C_6H_3$-2,6-$Mes_2\}_2$ (where M = Ge and Sn) were prepared similarly by using 1 equiv of the lithium salt. As depicted in Scheme 8, the compounds $\{Ge(Cl)C_6H_3$-2,6-$Mes_2\}_2$ and $\{Sn(Cl)C_6H_3$-2,6-$Mes_2\}_2$ may also be prepared via ligand exchange, from the diterphenyl derivatives and 1 equiv of MCl_2. Attempts to obtain the monoterphenyl lead halide compound have been

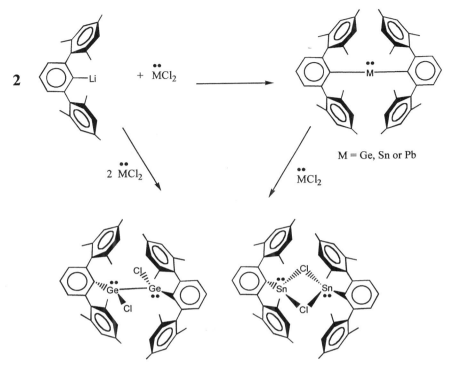

SCHEME 8.   Synthetic routes to $M(C_6H_3\text{-}2,6\text{-}Mes_2)_2$ (M = Ge, Sn or Pb) and $\{M(Cl)C_6H_3\text{-}2,6\text{-}Mes_2\}_2$ (M = Ge or Sn).[102]

unsuccessful to date. The synthesis of the bulkier $Sn(Cl)C_6H_3\text{-}2,6\text{-}Trip_2$ can also be accomplished in ca. 30% yield by ligand exchange.[103]

The X-ray crystal structures of the bisterphenyl derivatives are isomorphous and essentially isostructural. They exist as V-shaped, discrete monomers with the closest M–M (M = Ge, Sn or Pb) contacts being 8.409(1), 8.392(3), and 8.388(1) Å, respectively. The C–M–C bond angles are essentially invariant, having the values 114.2(2), 114.7(2), and 114.5(6)°, respectively. The angles are the widest reported for any σ-bonded MR_2 organometallic derivative with monodentate ligands.[104]

The terphenyl metal halide compound $\{Ge(Cl)C_6H_3\text{-}2,6\text{-}Mes_2\}_2$ (Fig. 21)[102] has a dimeric structure featuring Ge–Ge bonds and pyramidal geometry at each germanium, which is in contrast with the monomeric structure of Ge(Cl)Mes*.[104a] The difference may be attributed to the different steric properties of the Mes* and $-C_6H_3\text{-}2,6\text{-}Mes_2$ ligands.[72] The Ge–Ge bond distance of 2.443(2) Å is similar to that of a single bond and is longer

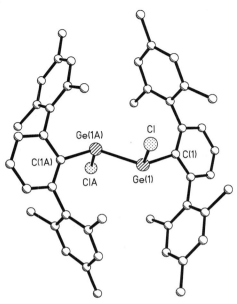

Fig. 21. Schematic drawing of the Ge–Ge bonded dimeric structure of $\{Ge(Cl)C_6H_3$-Mes$\}_2$.[102]

than the Ge–Ge distance in most other germanium dimers of the type $(GeR_2)_2$,[105] and the compound is dissociated to monomers in solution.

The tin analogue $\{Sn(Cl)C_6H_3$-2,6-Mes$_2\}_2$, also has a dimeric structure, but in this case the metals are bridged by chlorides to afford a Sn---Sn separation of 3.997(1) Å. A notable feature of the structure is the difference in length of the bridging Sn–Cl distances, (2.600(2) and 2.685(2) Å), which suggests that there may be a tendency to dissociate to monomers. This structure is similar to those observed in the related derivatives $\{Sn(Cl)N(SiMe_3)_2\}_2$,[106a] $\{Sn(Cl)TMP\}_2$[106a] (TMP = 2,2,6,6-tetramethyl-piperidino) and $[Sn(Cl)\{C(SiMe_2Ph)_3\}]_2$.[106b] The only other reported bis(terphenyl) compound is $Sn(C_6H_3$-2,4,6-Ph$_3)_2$[107,108] which has not been structurally characterized. It reacts with $[Fe_3(CO)_{12}]$ to afford the bridging stannanediyl complex $[Fe_2(CO)_8\{\mu\text{-}Sn(C_6H_3$-2,4,6-Ph$_3)_2\}]$.[108]

Reduction of $\{Ge(Cl)C_6H_3$-2,6-Mes$_2\}_2$ with 1 equiv of KC_8 (1 equiv per germanium) afforded the dark blue, neutral cyclotrigermenyl radical species $[GeC_6H_3$-2,6-Mes$_2]_3^{\cdot}$.[109] The X-ray crystal structure revealed a three-membered Ge_3 ring with an average Ge–Ge distance of 2.35(7) Å. In valence bond terms, the ring is expected to involve a double bond between two members with the unpaired electron on the remaining atom.

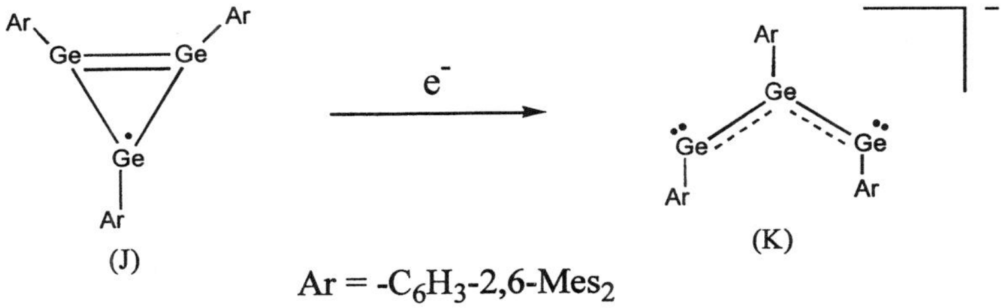

$$Ar = -C_6H_3\text{-}2,6\text{-}Mes_2$$

SCHEME 9.   Reduction of $\{Ge(Cl)C_6H_3\text{-}2,6\text{-}Mes_2\}_2$ to give the cyclic radical $(GeC_6H_3\text{-}2,6\text{-}Mes_2)_3^{\cdot}$ or the allyl anion analogue $(GeC_6H_3\text{-}2,6\text{-}Mes_2)_3^{-}$.[109]

Although the crystal structure had a disordered Ge_3 core, the EPR spectrum was consistent with the location of the unpaired electron on a single germanium.

When an excess of KC_8 (2 equiv per Ge) is used, the trigermanium anion species $\{GeC_6H_3\text{-}2,6\text{-}Mes_2\}_3^{-}$ is isolated in 23.5% yield as its potassium salt. Apparently, the addition of a further electron to the cyclic radical $\{GeC_6H_3\text{-}2,6\text{-}Mes_2\}_3^{\cdot}$ results in the opening of the Ge_3 ring to afford an analogue of the allyl anion (Scheme 9). The anion $\{GeC_6H_3\text{-}2,6\text{-}Mes_2\}_3^{-}$ has a Ge–Ge distance of 2.422(2) Å and a Ge–Ge–Ge angle of 159.19(10)°. The K^+ cation is associated with it through K^+–aromatic ring interactions.

As mentioned earlier, the complex $(\eta^5\text{-}C_5H_5)(CO)_2MoGeC_6H_3\text{-}2,6\text{-}Mes_2$, which was prepared according to Eq. (10), has a molybdenum–germanium triple bond.[110] The X-ray crystal structure (Fig. 22) shows that the Mo–Ge–C angle (172.2(2)°) is almost linear and the Mo–Ge bond length, 2.271(1) Å, is much shorter than the single bond lengths in $(\eta^5\text{-}C_5H_5)(CO)_2\text{-}Mo(GePh_3)\{C(OEt)Ph\}^{111}$ (2.658(2) Å) and in $(\eta^5\text{-}C_5H_5)(\eta^2\text{-}C_6H_{11})(CO)_2\text{-}Mo(GePh_3))^{112}$ (2.604(2) Å). The short Mo–Ge distance is consistent with the presence of a Mo–Ge triple bond between the 15-electron $(\eta^5\text{-}C_5H_5)(CO)_2Mo$ fragment and the $GeC_6H_3\text{-}2,6\text{-}Mes_2$ moiety.

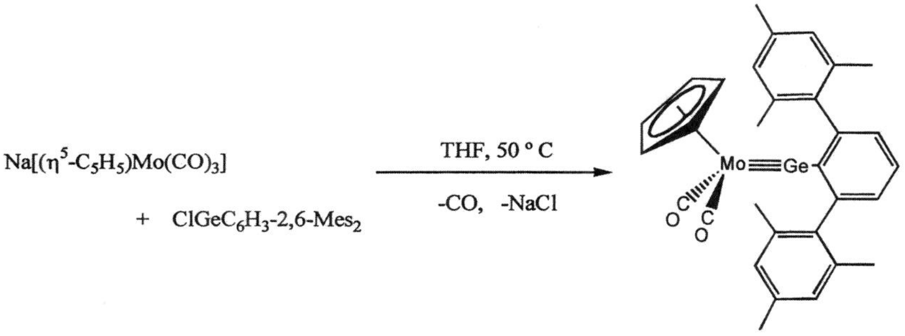

$$Na[(\eta^5\text{-}C_5H_5)Mo(CO)_3]$$

$$+ \quad ClGeC_6H_3\text{-}2,6\text{-}Mes_2 \quad \xrightarrow[\text{-CO, -NaCl}]{\text{THF, 50 °C}}$$

(10)

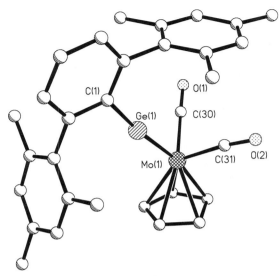

FIG. 22.   Schematic drawing of the complex $(\eta^5\text{-}C_5H_5)(CO)_2MoGeC_6H_3\text{-}2,6\text{-}Mes_2$,[110] which has a Mo–Ge triple bond.

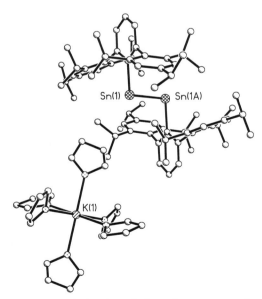

FIG. 23.   Computer-generated drawing of the solvent separated salt $[K(THF)_6][2,6\text{-}Trip_2H_3C_6SnSnC_6H_3\text{-}2,6\text{-}Trip_2]$.[103]

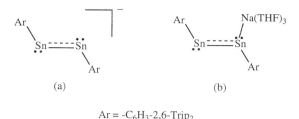

(a)   (b)

Ar = -C₆H₃-2,6-Trip₂

FIG. 24.   Schematic drawing of [2,6-Trip₂H₃C₆SnSnC₆H₃-2,6-Trip₂]⁻ as (a) a solvent-separated ion pair or as (b) its sodium salt.[103,115]

The reduction of the terphenyl tin halide $Sn(Cl)C_6H_3$-2,6-Trip₂ with KC₈ afforded the singly reduced anion $[SnC_6H_3$-2,6-Trip₂$]_2^{-\cdot}$.[103] The compound may be crystallized as either the solvent-separated ion pairs $[K(THF)_6][SnC_6H_3$-2,6-Trip₂$]_2$ or [K(dibenzo-18-crown-6)(THF)₃] $[SnC_6H_3$-2,6-Trip₂$]_2 \cdot 2THF$. The structure of $[K(THF)_6][SnC_6H_3$-2,6-Trip₂$]_2$ is represented in Fig. 23. It features an unpaired electron in a π-orbital that is formed by overlap of the tin p-orbitals, and there is a lone pair at each tin (Fig. 24a). This species, therefore, has a formal bond order of 1.5, and the narrow ca. 95° bond angle suggests little hybridization at the tin atoms. The Sn–Sn bonds have an average distance of 2.806(16) Å, which is close to the Sn–Sn single bond distance in gray tin (2.80 Å)[113], and slightly longer than the quasi-double bond length (2.764(2) Å) in $Sn_2[CH(SiMe_3)_2]_4$.[114] Attempts to isolate the neutral species by using exactly 1 equiv of KC₈ per $Sn(Cl)C_6H_3$-2,6-Trip₂ were unsuccessful. Apparently, it is very susceptible to reduction and the anion is always produced during the reaction. When sodium anthracenide is used as reductant with $Sn(Cl)C_6H_3$-2,6-Trip₂, the contact ion-pair compound $[Na(THF)_3][SnC_6H_3$-2,6-Trip₂$]_2$[115] is formed. The X-ray crystal structure reveals a similar *trans*-bent geometry to that of the solvent-separated potassium salts. The Na–Sn bond length is 3.240(7) Å (Fig. 24b) and the Sn–Sn bond distance, 2.811(1) Å, remains essentially unchanged. The C–Sn–Sn angles (av. 97.8°) are also comparable to those in the solvent-separated potassium salts.

Terphenyl ligands have also been used to stabilize a stannaselenone.[116] The deselenation of the tetraselenastannolane, Tbt(Ar)SnSe₄ (where Tbt = -C₆H₃-2,4,6-{CH(SiMe₃)₂}₃ and Ar = 2″-diisopropyl-*m*-terphenyl-2′-yl (Ditp)), with 3 equiv of Ph₃P affords the stannaselenone Tbt(Ar)SnSe in 84% yield (Scheme 10). If only 2 equiv of the phosphine reagent are used, the diselenastannirane Tbt(Ar)SnSe₂ is produced in 56% yield. The Sn–Se bond distance in Tbt(Ar)SnSe is 2.375(3) Å, which is shorter than reported Sn–Se single bond lengths (2.55–2.60 Å)[116] and is very similar to the Sn–Se distance (2.394(1) Å) in the terminal selenido complex [η^4-Me₈taa]SnSe, where Me₈taa = octamethyldibenzotetraaza[14]annulene di-

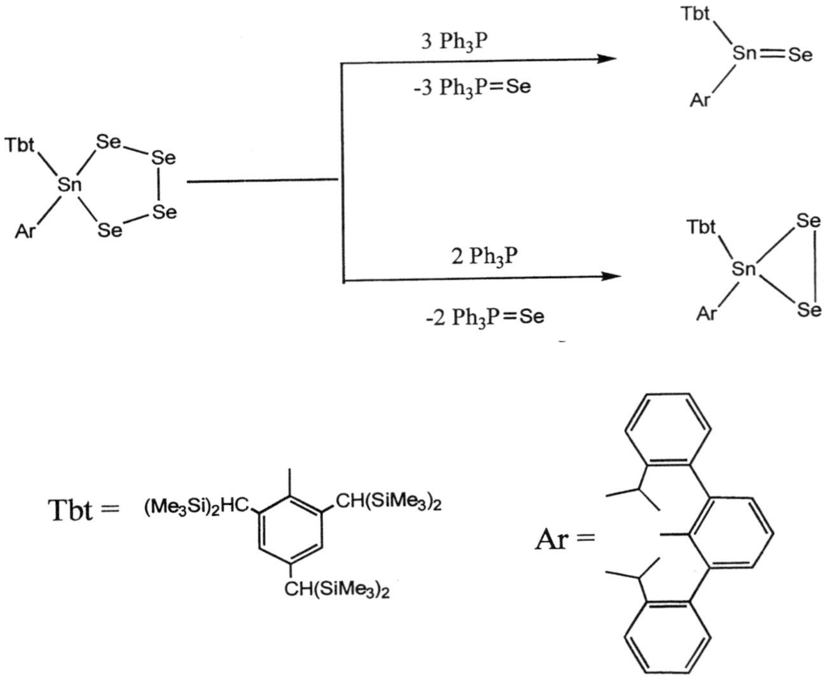

SCHEME 10. Synthesis of stannaselone and a diselenastannirane from a tetraselenastan-nolane.[116]

anion.[117] The average Sn–Se bond distance in Tbt(Ar)-SnSe$_2$ is 2.530(3) Å and is within the values reported for single bonds. The Se–Se distance (2.524(4) Å) is about 0.2 Å longer than typical Se–Se single bonds.[116]

C. Group 15 Terphenyl Derivatives

1. Nitrogen Compounds

The nitration of triphenylbenzene, 1,3,5-Ph$_3$C$_6$H$_3$, to give O$_2$NC$_6$H$_2$-2,4,6-Ph$_3$ in high yield (70%) was reported in 1957.[118] Reduction of this compound with SnCl$_2$ affords the amine H$_2$NC$_6$H$_2$-2,4,6-Ph$_3$. The crystal structure of the compound O$_2$NC$_6$H$_2$-2,4,6-Ph$_3$ has only been determined recently.[119] In the solid state there are two symmetry-independent molecules with similar torsion angles (64 and 65°) between the plane of the nitro group and that of the ring. A slight modification of the amine synthesis in which SnCl$_2$ was replaced by zinc granules has also been reported to afford the product in ca. 80% yield.[120] The reaction of 2,4,6-triphenylaniline with Na$_2$MoO$_4$ in DME and in the presence of Me$_3$SiCl and NEt$_3$ at 80°C for

3 days yields the oxoimido complex [MoCl$_2$O(NC$_6$H$_2$-2,4,6-Ph$_3$)(DME)].[120] The crystal structure of this complex shows that the molybdenum has distorted octahedral coordination with *cis*-oriented organoimide and oxide ligands. The Mo–N bond length (1.756(7) Å) and the Mo–N–C bond angle (172.2(7)°) are similar to those found in the complex [Mo(NPh)$_2$ (S$_2$CNEt$_2$)$_2$] (Mo–N 1.754(4) Å Mo–N–C 169.4(4)°), which also has a "bent" two-electron donor (1σ, 1π-interaction Mo–N 1.789(4) Å; Mo–N–C 139.4°).[120]

A range of thio-2,4,6-triphenylaniline compounds has been synthesized via the reaction with the arenesulfenyl chloride and NEt$_3$. Oxidation of these with PbO$_2$/K$_2$CO$_3$ leads to the generation of surprisingly stable radicals, which have been characterized by UV/vis, X-ray crystallography, and ESR spectroscopy.[121]

Benzene solutions of these radicals are greenish blue or yellowish green and absorb in the visible region at 513–535 and 626–644 nm. The solutions are quite stable and can be stored for months without decomposition. No reaction with oxygen was detected upon exposure of their solutions to the atmosphere for 5 h. At low temperatures no dimerization was observed, and a structural determination of [ArSNC$_6$H$_2$-2,4,6-Ph$_3$]$^-$ (Ar = 2,4-Cl$_2$C$_6$H$_2$) shows that it is monomeric. The Ar–S–N fragment is coplanar and subtends an angle of 18.7° to the central ring of the terphenyl group. The S–N bond distance is 1.637(5) Å. On the basis of the almost planar conformation, the unpaired electron is expected to delocalize over the whole molecular array. This is confirmed by the EPR spectrum, which shows a 1:1:1 triplet with relatively large hyperfine splittings (in the partially deuterated BrC$_6$H$_4$S–N–C$_6$H$_2$-2,4,6-C$_6$D$_5$) by the anilino *meta* (0.133–0.134 mT) and arylthioyl *ortho* protons (0.88 mT).

Further work on this system has led to a range of terphenyl thioaminyls.[122] As shown earlier, these thioaminyls can be oxidized using PbO$_2$ to give persistent radicals. Several of these species have been isolated in the crystalline form and structural analysis of both terphenyl derivatives shows an approximately planar π system, with approximate 21° torsion angles for both phenylthiyl moieties in 2-*t*-Bu-4,6-Ph$_2$H$_2$C$_6$NHSAr$_2$ 4-*t*-Bu-2,6-Ph$_2$H$_2$C$_6$NHSAr$_2$ (see Fig. 25). Extensive delocalization of the unpaired electron is therefore expected over the anilino and phenylthiyl groups, but not over the pendant phenyl rings, which have torsion angles >45° to the central phenyl ring. This expectation is confirmed by EPR spectroscopy, which shows a 1:1:1 triplet for both types of radical with splittings of 0.921–0.948 mT. Deuteration of the phenyl groups on the anilino benzene ring gives rise to further splitting of the nitrogen 1:1:1 triplet by the anilino *meta* (0.126–0.138 mT) and phenylthiyl *ortho* and *para* protons (0.077–0.096 mT).

Ar = 4-MeC$_6$H$_4$
Ar = 4-ClC$_6$H$_4$
Ar = 4-BrC$_6$H$_4$
Ar = 4-BrC$_6$D$_4$

Ar = 2,4-Cl$_2$C$_6$H$_3$
Ar = 3,5-Cl$_2$C$_6$H$_3$
Ar = 3-NO$_2$C$_6$H$_4$
Ar = 4-NO$_2$C$_6$H$_4$

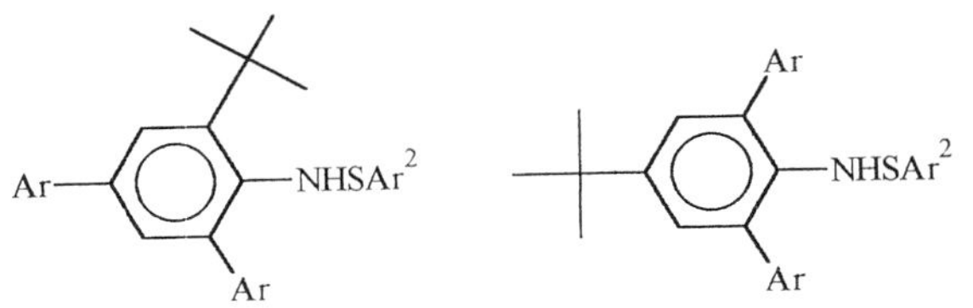

Ar = Ph, Ar^2 = 3-NO$_2$C$_6$H$_4$
Ar = Ph, Ar^2 = 4-NO$_2$C$_6$H$_4$
Ar = Ph, Ar^2 = 2,4-Cl$_2$C$_6$H$_3$
Ar = Ph, Ar^2 = 3,5-Cl$_2$C$_6$H$_3$
Ar = 4-ClC$_6$H$_4$, Ar^2 = 3-NO$_2$C$_6$H$_4$
Ar = 4-ClC$_6$H$_4$, Ar^2 = 4-NO$_2$C$_6$H$_4$
Ar = 4-ClC$_6$H$_4$, Ar^2 = 2,4-Cl$_2$C$_6$H$_3$

Ar = Ph, Ar^2 = 3-NO$_2$C$_6$H$_4$
Ar = Ph, Ar^2 = 4-NO$_2$C$_6$H$_4$
Ar = Ph, Ar^2 = 2,4-Cl$_2$C$_6$H$_3$
Ar = Ph, Ar^2 = 3,5-Cl$_2$C$_6$H$_3$
Ar = Ph, Ar^2 = 4-ClC$_6$H$_4$
Ar = Ph, Ar^2 = 4-BrC$_6$H$_4$
Ar = 4-ClC$_6$H$_4$, Ar^2 = 3-NO$_2$C$_6$H$_4$
Ar = 4-ClC$_6$H$_4$, Ar^2 = 4-NO$_2$C$_6$H$_4$
Ar = 4-ClC$_6$H$_4$, Ar^2 = 2,4-Cl$_2$C$_6$H$_3$
Ar = 4-ClC$_6$H$_4$, Ar^2 = 3,5-Cl$_2$C$_6$H$_4$

FIG. 25. Schematic drawing of terphenyl-substituted thiazyl radicals and thioaminyls.[121,122]

2. *Phosphorus Compounds*

The compounds 2,6-Mes$_2$H$_3$C$_6$PCl$_2$[123] and 2,6-Trip$_2$H$_3$C$_6$PCl$_2$[124] have been isolated in high yield ($>70\%$) via the reaction of the aryl lithium salt and PCl$_3$. The ^{31}P NMR spectra show characteristic resonances at 160.1 and 157.1 ppm, respectively. Although no complete set of structural data has been published for either compound, the P–C distance in 2,6-Mes$_2$H$_3$C$_6$PCl$_2$ was reported to be 1.842 Å.[125] These phosphines are useful precursors to a range of compounds. In the presence of water 2,6-Mes$_2$H$_3$C$_6$PCl$_2$ hydrolyzes to form the phosphinic acid 2,6-Mes$_2$-

$H_3C_6P(O)(OH)H$, which was characterized spectroscopically by FTIR (ν PH 2451; ν PO 1192, 963 cm^{-1}), ^1H NMR (O–H 7.89 ppm, br, P–H 6.36 ppm d); ^{31}P NMR (26.4 ppm d; J_{HP} 590 Hz), and by X-ray diffraction.

$$ (11) $$

Ar = Mes

In the solid state, the phosphinic acid is dimerized through hydrogen bonding, in spite of the bulky terphenyl substituent. This type of dimerization is also commonly observed in less bulky phosphinic and phosphonic acids.[126] The P–O distances, 1.508(2) and 1.521(2) Å, and the O–H and O–H$\cdots$O distances, 1.06(4) and 1.46(4) Å, are consistent with localized hydrogen bonding. The P–C distance (1.812(2) Å) is slightly shorter than the P–C bond distance in the parent phosphine. The terphenyl groups are in a *trans* conformation with respect to the bridging acid moiety.

The reduction of $2,6\text{-Mes}_2H_3C_6PCl_2$ by $LiAlH_4$ affords $2,6\text{-Mes}_2H_3C_6PH_2$ in 90% yield. The ^{31}P NMR spectrum features a triplet centered at -146.3 ppm with a J_{HP} of 210 Hz. Attempts at deprotonation with n-BuLi at low temperature only yield the salt $2,6\text{-Mes}_2H_3C_6P(Li)H$ (^{31}P NMR -137.8 ppm, J_{HP} 176 Hz). The formation of $ArPLi_2$ could not be induced, even with excess n-BuLi. In contrast, the less hindered $1,3,5\text{-Ph}_3H_3C_6PH_2$ (formed via the reaction between the Grignard reagent and PCl_3 at $-60°C$ followed by reduction with $LiAlH_4$ at $0°C$)[127] can be dilithiated with n-BuLi. *In situ* lithiation of the phosphine in the presence of dichloro(2,4,6-triisopropylphenyl)-*tert*-butylsilane affords the phosphasilene 1-(2,4,6-triphenylphenyl)-2-*tert*-butyl-2-(2,4,6-triisopropylphenyl)phosphasilene solely as the *E*-isomer. This compound has been characterized by ^{31}P NMR (86.7 ppm) and ^{29}Si NMR (180.0 ppm, $^1J_{PSi}$ 151 Hz).[127]

$$ (12) $$

$R^1 = {}^tBu$
$R^2 = 2,4,6\text{-triisoproplyphenyl}$
Ar = phenyl

The chloride, $2,6\text{-Mes}_2H_3C_6PCl_2$ can also be reduced by magnesium (with sonication) to give the orange diphosphene $2,6\text{-Mes}_2H_3C_6PPC_6H_3\text{-}2,6\text{-Mes}_2$ in good yield.[128] The ^{31}P NMR spectrum displays a singlet at 493.2 ppm, which agrees closely with the value reported for the analogous compound $\text{Mes*}P{=}P\text{Mes*}$ (494 ppm). In the solid state the diphosphene displays a *trans*-orientation of the terphenyl ligands. The P–C bond distances are 1.840(3) and 1.842(3) Å, which are similar to that found in the phosphinic acid (1.812(2) Å) and free phosphine (1.842 Å). The reported P–P

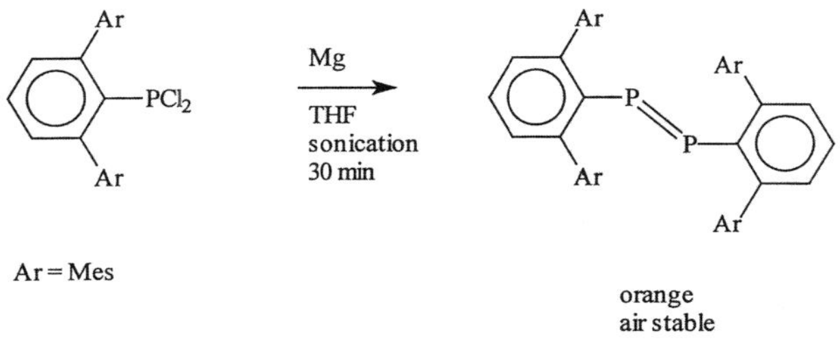

$$\text{(13)}$$

Ar = Mes

orange
air stable

distance of 1.985(2) Å is even shorter than that found in the first structurally characterized diphosphene $\text{Mes*}P{=}P\text{Mes*}$ (2.034(2) Å),[16] although C–P–P bond angles of 97.5(1) and 109.8(1)° and the C–P–P–C torsion angle of 177.1(2)° are comparable to those found in this compound (102.8(1) and 172.2(1)°, respectively). The analogous reduction of $2,6\text{-Trip}_2H_3C_6PCl_2$ with magnesium in THF does not lead to the corresponding diphosphene. Instead, interaction of the phosphorus with one of the isopropyl groups is observed. This results in the structure depicted schematically in Eq. (14). The P–C(ring) bond distances average 1.81 Å, the P–C(isopropyl) distance is 1.868(8) Å, and the C–P–C(isopropyl) angles are 103.9(9)° and 101.7(9)°. The reduction of $2,6\text{-Trip}_2H_3C_6PCl_2$ by potassium in hexane, however, leads to the orange diphosphene in

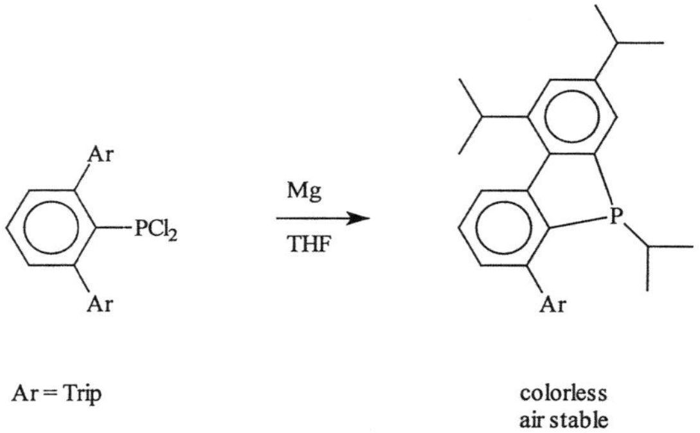

$$\text{(14)}$$

Ar = Trip

colorless
air stable

which only a trace amount of the *ortho*-metallated product can be detected by ^{1}H NMR spectroscopy.[124] The ^{31}P NMR spectrum of the diphosphene has a peak at 520.3 ppm, which is near the middle of the known diphosphene range.

The diphosphenes ArP=PAr (Ar = C_6H_2-2,6-Mes_2-4-CH_3, C_6H_3-2,6-dimethylphenyl) have also been synthesized and characterized by similar methods,[128] i.e., ^{1}H, ^{31}P NMR, HRMS, UV-vis. The ^{31}P NMR spectra display resonances at 492.7 and 494.4 ppm, which are almost identical to those found in the analogous species ArP=PAr (Ar = Mes*, 494 ppm; Ar = -C_6H_3-2,6-Mes_2, 493.2 ppm). The purple radical anions [ArPPAr]$^-$ can be generated by one-electron reversible reduction using cyclic voltammetry (0.1-M n-Bu_4NBF_4 in THF) or by chemical reduction in THF, using lithium, sodium, or magnesium metal. Reduction potentials for ArP=PAr (Ar = -C_6H_3-2,6-Mes_2, C_6H_2-2,6-Mes_2-4-CH_3, -C_6H_3-2,6-dimethylphenyl) are higher than those found for Mes*P=PMes* ($E_{1/2}$ = -2.08, -2.15, -2.08 vs -1.93, respectively). Ab initio calculations indicate that the difference does not arise from any orientational effects of the aromatic residues relative to the P=P bond. Rather, reduction is linked to the length (strength) of the P–P bond. In Mes*P=PMes* the P–P bond is 2.034(2) Å compared to 1.985(2) Å in ArP=PAr (Ar = C_6H_3-2,6-Mes_2). A stronger P–P bond would give rise to a higher energy LUMO, which would be more difficult to reduce. Calculations show that the LUMO is the P=P π* level, and the HOMO is the n_+ (except for Ar = C_6H_3-2,6-Mes_2, where π is slightly favored). It has also been reported that the HOMO for Mes*P=PMes* is the n_+ molecular orbital on the basis of density functional theory calculations.[129] The EPR spectra of the radical anions generated by reduction with lithium on magnesium in THF reveal the expected triplet centered at about g = 2.008 with an equivalent hyperfine coupling of 4.6 mT to each ^{31}P nucleus. The relatively small coupling constant implies that the electron occupies the P=P π* orbital. Anions generated by sodium reduction afford a more complicated spectrum indicating the presence of two species. At low temperatures the signals attributed to the uncharacterized minor species disappear. Also, the addition of 18-crown-6 to the solutions of the sodium anion removes these minor species. The reduction of 2,6-$Mes_2H_3C_6PPC_6H_3$-2,6-Mes_2 with potassium does not lead to the radical anion but to a species formulated as 2,6-$Mes_2H_3C_6P(H)K$ on the basis of ^{31}P NMR spectroscopy (-124.4 ppm, d, J_{HP} 177.4 Hz).

3. Arsenic Compounds

Although many sterically demanding groups have been used to stabilize low coordinate arsenic species,[130] terphenyl derivatives have only been reported relatively recently. The derivative 2,6-Mes_2-4-$MeH_2C_6AsCl_2$[131]

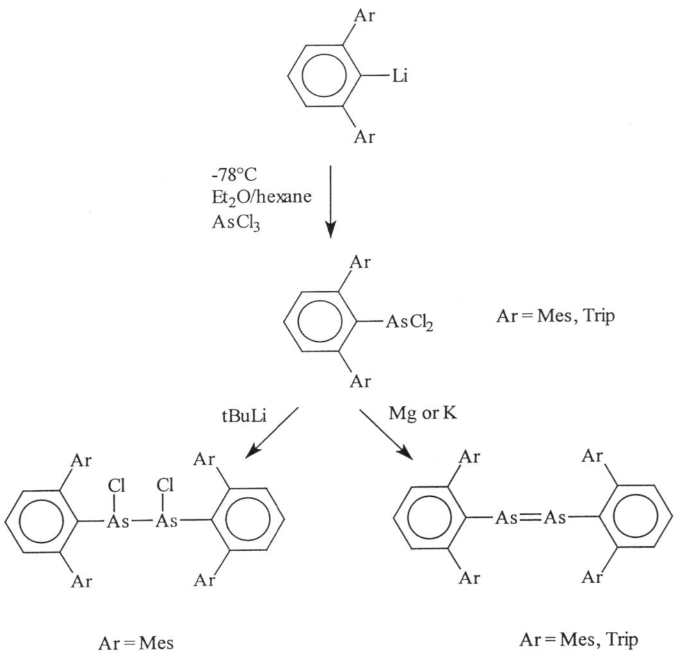

SCHEME 11.   Reduction of $Cl_2AsC_6H_3$-2,6-Mes_2 or $Cl_2AsC_6H_3$-2,6-$Trip_2$.[124,131]

was used to synthesize 2,6-Mes_2-4-$MeH_2C_6AsPC_6H_2$-2,6-Mes_2-4-Me, which was not structurally characterized owing to disorder problems. Current work also shows that compounds of the type 2,6-$Mes_2H_3C_6AsCl_2$[124] and 2,6-$Trip_2H_3C_6AsCl_2$[124] can be formed readily by the reaction of the lithium salt and $AsCl_3$ at low temperature in ca. 50% yield. These compounds have been characterized by ^{1}H and ^{13}C NMR, IR, melting point, and in the case of the latter, by X-ray crystallography.[124] Disorder of the -$AsCl_2$ moiety was observed, but this has been successfully modeled. The As–C and As–Cl bond distances are 2.00(avg.) and 2.15(avg.) Å.

Preliminary investigations of the reactivity of these terphenylarsenic chlorides have shown that reduction can lead to products having either single or double As–As bonds (Scheme 11). Reduction with *t*-BuLi gives the orange diarsane $\{As(Cl)C_6H_3$-2,6-$Mes_2\}_2$, which has a *trans* conformation of the terphenyl groups and 1.985(7) Å, with C–As–As angles of 105.9(2) and 99.5(2)°. The As–As bond length is 2.5068(12)Å and the As–C distances are 1.982(7). Interestingly, the chlorine atoms are eclipsed with As–Cl distances of 2.189(2) and 2.175(2) Å.

The compounds 2,6-$Trip_2H_3C_6AsCl_2$ and 2,6-$Mes_2H_3C_6AsCl_2$ can be reduced by magnesium and potassium, respectively, to give the halide-free diarsenes in yields of ca. 50% and ca. 20%.[124,130c] Their structures display the terphenyl ligands in the *trans* conformation. The As–As bond lengths (2.285(3) and 2.286(1) Å) are slightly longer than that calculated for HAs = AsH[132] and that measured in Mes*As = AsCH(SiMe$_3$)$_2$ (2.224(2) Å).[130b] In the -C$_6$H$_3$-2,6-Trip$_2$ derivative there are two different C–As–As angles of 96.4(2)° and 107.9(3)°. Unsymmetric C–As–As angles have also been seen in Mes*As = AsCH(SiMe$_3$)$_2$ (C–As–As 99.9(3)° and 93.6(3)°),[130b] where there are different substituents at each arsenic. In the -C$_6$H$_3$-2,6-Mes$_2$ derivative the corresponding angles are symmetric with C–As–As = 96.6(1)°.

The reaction of 2,6-Trip$_2$H$_3$C$_6$AsCl$_2$ with Li$_2$PMes has led to the unsymmetric multiply bonded species 2,6-Trip$_2$H$_3$C$_6$As = PMes.[130c] The ^{31}P NMR spectrum has a peak at 534.2 ppm that is quite similar to the ^{31}P shifts (575 and 533 ppm, respectively) seen in the related species Mes*P = AsCH(SiMe$_3$)$_2$ and Mes*As = PCH(SiMe$_3$)$_2$.[130] The As–P bond distance is 2.134(2) Å and the C–As–P, C–P–As angles are 101.5(2) and 96.8(2)°, respectively. Comparison with the only other structurally characterized P–As doubly bonded species Mes*P = AsCH(SiMe$_3$)$_2$ shows quite similar bond angles (101.4(9), 96.4(9)°) and an As–P bond distance of 2.125(1) Å.

4. *Antimony Compounds*

The terphenylantimony halides 2,6-Mes$_2$H$_3$C$_6$SbCl$_2$[124] and 2,6-Trip$_2$H$_3$C$_6$-SbCl$_2$[124] can be synthesized by a straightforward reaction of SbCl$_3$ with 1 equiv of the corresponding terphenyl lithium reagent at low temperature. Yields of ca. 50% were obtained and both compounds have been fully characterized. The X-ray structure of 2,6-Trip$_2$H$_3$C$_6$SbCl$_2$ shows disorder at the antimony and chlorine positions, however. The Sb–C and Sb–Cl bond lengths are ca. 2.22 and ca. 2.35 Å, respectively. The reduction of 2,6-Trip$_2$H$_3$C$_6$SbCl$_2$ was attempted with magnesium, KC$_8$, or potassium metal. No reduction was seen with magnesium in THF even with sonication. Partial reduction occurred with KC$_8$ upon stirring for 3 days, to afford an orange crystalline material consisting of a mixture of the distibene and the 1,2-dichlorodistibane in ca. 4:1 ratio. The Sb–Sb bond length in the distibene is 2.668(2) Å and in the distibane it is 2.892(8) Å. Full reduction can be achieved by reaction with potassium metal in hexane for 3 days to afford the green distibene in high (>80%) yield.[124] The 2.668(2) Å Sb–Sb doubly bonded distance may be compared with the 2.642(1) Å observed in the first distibene 2,4,6-{(Me$_2$Si)$_2$CH}$_3$H$_2$C$_6$SbSbC$_6$H$_2$-2,4,6-{CH(SiMe$_3$)$_2$}$_3$.[133] The synthesis of this molecule was accomplished via the selective deselenation of the cyclic species (SeSbC$_6$H$_2$-2,4,6-{CH(SiMe$_3$)$_2$})$_3$ with P(NMe$_2$)$_3$.

A stable unsymmetrical stibaphosphene has been synthesized via the reaction of 2,6-Trip$_2$H$_3$C$_6$SbCl$_2$ and Li$_2$PMes.[124] However, it is obtained in

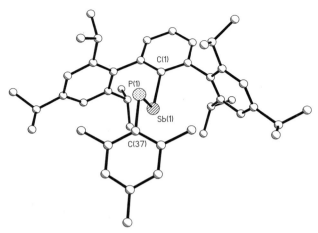

Fɪɢ. 26.   Schematic drawing of the unsymmetrical stibaphosphene MesPSbC$_6$H$_3$-2,6-Trip$_2$.[124]

low (1%) yield (the other products are Ar–Sb–Sb–Ar and (MesP)$_n$). The X-ray crystal structure (Fig. 26), shows that the aromatic ligands are oriented *trans* to each other with an Sb–P distance of 2.335(2) Å and C–Sb–P, C–P–Sb angles of 100.9(2)° and 95.7(3)°, respectively. The ^{31}P NMR spectrum has a signal at 543 ppm, which is 77 ppm upfield from the only other known Sb–P multiply bonded species, ArP=SbCH(SiMe$_3$)$_2$ (^{31}P NMR 620 ppm),[130] which was too unstable to be structurally characterized.

5. *Bismuth Compounds*

The reaction of LiC$_6$H$_2$-2,4,6-Ph$_3$ with BiCl$_3$ in THF in 2:1 or 3:1 stoichiometry yields ClBi(C$_6$H$_2$-2,4,6-Ph$_3$)$_2$ or Bi(C$_6$H$_2$-2,4,6-Ph$_3$)$_3$ in ca. 60% yield.[134] Ligand exchange reactions between BiCl$_3$ and either of these products in the appropriate ratio afford Cl$_2$BiC$_6$H$_2$-2,4,6-Ph$_3$ in 63% yield. More recently, this compound has been synthesized by the 1:1 reaction of the lithium reagent and BiCl$_3$ in low yield (20%).[135] Attempted reduction of ClBi(C$_6$H$_2$-2,4,6-Ph$_3$)$_2$ with Na/NH$_3$-, Mg/THF or with cobaltocene/toluene results in decomposition. The attempted reduction of Cl$_2$BiC$_6$H$_2$-2,4,6-Ph$_3$ with Na/NH$_3$ or Mg/THF also leads to decomposition, but reduction with cobaltocene/toluene produces BiAr$_3$.[135] It was proposed that this is due to a disproportionation reaction of a nonisolated, monovalent, unstable 2,4,6-triphenylphenylbismuth(I) species. The compounds Bi(C$_6$H$_2$-2,4,6-Ph$_3$)$_3$ and ClBi(C$_6$H$_2$-2,4,6-Ph$_3$)$_2$ crystallize as the solvates Bi(C$_6$H$_2$-2,4,6-Ph$_3$)$_3$ · 4THF and ClBi(C$_6$H$_2$-2,4,6-Ph$_2$)$_2$ · 2PhMe. The triaryl has pyramidal coordination at bismuth with interligand angles of 104.4(2), 103.7(4), and 109.4(4)°.

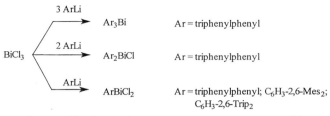

SCHEME 12.   Synthesis of terphenylbismuth derivatives.[135]

The terphenyl groups are arranged in a propeller-like conformation around the bismuth with average Bi–C distances of *ca.* 2.35 Å.[134] This is in the range of Bi–C distances in other derivatives with bulky substituents, e.g., $Bi[C_6H_2-2,4,6-(CF_3)_3]_3$ (2.36–2.38 Å)[136] but is longer than Bi–C bonds in less crowded triaryls, e.g., $Bi(C_6H_5)_3$ (Bi–C = ca. 2.24 Å).[137]

The monoaryl $Cl_2BiC_6H_2-2,4,6-Ph_3$ is dimeric in the solid state with Bi–Cl distances of 3.074(1) Å (bridging) and 2.532(1) Å (terminal).[135] The Bi–C bond length (2.266(4) Å) is somewhat shorter than that found in the bis and tris aryls, probably as a result of the reduced steric crowding. The intermolecular Bi–Bi distance, 4.316(1) Å, is shorter than the sum of the van der Waals radii, 4.68 Å. Also, this molecule displays a Menshutkin interaction[138] in which there is relatively close contact between bismuth and one of the pendant aryl groups resulting in a Bi–centroid distance of 3.53 Å (calculated 4.04 Å).

The reaction[124] of $BiCl_3$ with $LiC_6H_2-2,6-Mes_2$ or $LiC_6H_3-2,6-Trip_2$ affords the corresponding aryl bismuth dichloride in yields of >70%. Both compounds were spectroscopically characterized. The crystal structure of the former species has been determined. This molecule is dimeric in the solid state when crystallized from Et_2O. The bridging and the terminal Bi–Cl bond lengths are ca. 3.12 and 2.52 Å, respectively. The Bi–C distances are 2.289(2) and 2.208(5) Å. The Bi–Bi separation, 4.427 Å, is also shorter than the sum of the van der Waals radii. When crystallized from C_6H_6, a monomer is obtained that has Bi–C and Bi–Cl distances of 2.269(10), 2.502(3), and 2.531(3) Å, respectively. The bismuth atom displays a Menshutkin interaction[138] to one of the *orthoaryl* rings (Bi–ring centroid = 3.467(10) Å). The distortion of the external angles at the *ipso* carbon, which have the values C2–C1-Bil 130.9(1.6), C6–C1–Bil 110.3(1.5)°, is shown in Fig. 27. The reduction of $2,6-Trip_2H_3C_6BiCl_2$ with potassium in hexane affords a deep purple product.[124] An identical product is obtained from the reaction of $2,6-Trip_2H_3C_6BiCl_2$ with Li_2PMes or with H_2PMes in the presence of DBU. Although the crystal structure has not been obtained to date, UV-vis, spectroscopy,[1] H and ^{13}C NMR spectroscopy, and analytical

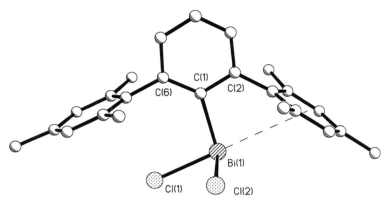

FIG. 27.   Schematic drawing of the structure of 2,6-Trip$_2$H$_3$C$_6$BiCl$_2$.[124]

data imply that this product is the dibismuthene 2,6-Trip$_2$H$_3$C$_6$BiBiC$_6$H$_3$2,6-Trip$_2$. Its structure is probably similar to that of 2,4,6-{(Me$_3$Si)$_2$CH}$_3$H$_2$C$_6$-BiBiC$_6$H$_2$-2,4,6-{CH(SiMe$_3$)$_2$}$_3$, which was the first example of a stable compound with a Bi=Bi double bond.[23]

D. *Group 16 Terphenyl Derivatives*

This section briefly summarizes synthetic routes to the terphenylchalcogenols, much of whose chemistry concerns their ligation to metals.[43–55] This burgeoning subject is beyond the scope of this review.

1. *Oxygen Compounds*

ortho-Diphenyl phenols were first prepared in 1900 by Hill *et al.*,[140] by the condensation reaction of sodium nitromalonic aldehyde with dibenzyl ketone to give 2,6-diphenyl-4-nitrophenol [Eq. (15)].

$$\text{Ph}\diagup\!\!\diagdown\text{Ph} + \text{Na}\left[\begin{array}{c}\text{O}\quad\text{O}\\\text{H}\diagdown\!\!\diagup\!\!\overset{\ominus}{\diagdown}\!\!\diagup\text{H}\\\text{NO}_2\end{array}\right]\xrightarrow[-\text{NaOH}]{-\text{H}_2\text{O}}\quad\begin{array}{c}\text{NO}_2\\\text{Ph}\diagup\!\!\diagdown\text{Ph}\\\text{OH}\end{array}\tag{15}$$

The synthesis of the unsubstituted 2,6-diphenylphenol was first reported in 1939. It was achieved by the treatment of dipheny ether with phenylisodium in benzene.[141] Scheme 13 represents a simplified mechanism for this reaction. Since then, several improved synthetic methods of syntheses, as

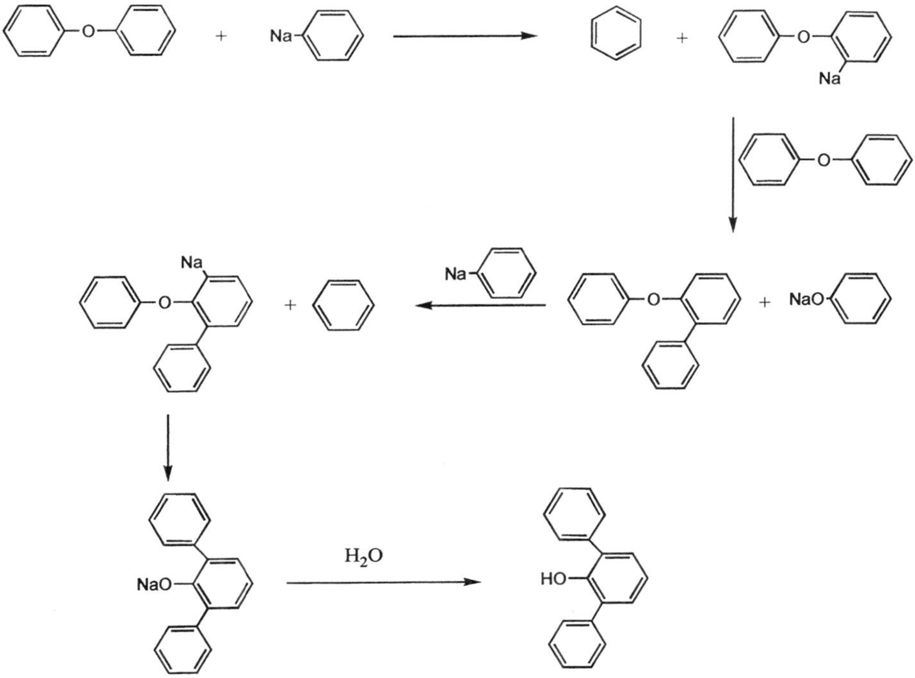

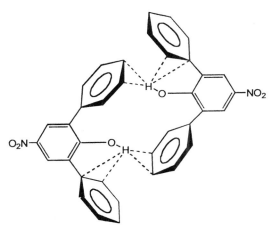

SCHEME 13.   Synthetic scheme for HOC_6H_3-2,6-Ph_2.[142]

FIG. 28.   Schematic drawing of the structure of 2,6-diphenyl-4-nitrophenol, which features O–H---π interactions.[144]

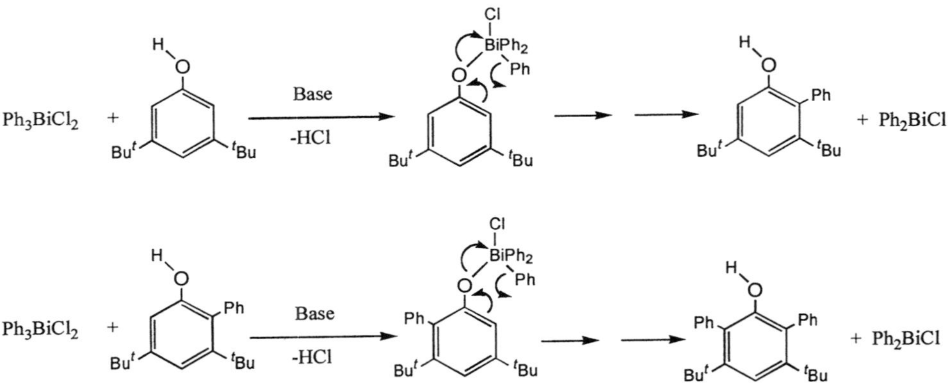

SCHEME 14. Synthetic scheme for an *ortho*-metallation resistant hindered 2,6-diphenyl phenol derivative.[142]

well as the syntheses of other terphenyl phenol derivatives, have appeared in the literature.[142]

In the solid state, 2,6-diphenylphenol has a monomeric structure with an intramolecular O–H---π interaction to one of the *ortho*-phenyl substituents.[143] Interestingly, the 2,6-diphenyl-4-nitrophenol derivative has a dimeric structure in the solid state with intermolecular, as well as intramolecular, O-H---π interactions between *ortho*-phenyl substituents.[144] The latter compound represented the first example of bifurcated OH---π bonding (Fig. 28).

More recently, the syntheses of the *t*-Bu substituted 2,6-diphenyl phenol derivatives have been reported.[142] They were synthesized through the regio-specific *ortho*-phenylation of a variety of 3,5-substituted phenols via ligand coupling (or reductive elimination) reactions with the use of pentavalent bismuth[145a] or aryl lead triacetate[145b] reagents. For example, the phenylation of 3, 5-di-*tert*-butylphenol with Ph_3BiCl_2 under basic conditions afforded the 2,6-diphenyl derivative, via a bismuth alkoxy intermediate, in 77% yield (Scheme 14).

These bulky terphenolate ligands have proved useful in the prevention of cyclometallation of the *ortho*-substituents[146] in early-transition-metal phenolates. The bulky *meta*-substituents prevent the rotation of the *ortho*-phenyl ring from becoming coplanar with the central phenoxide core, which is a requirement for C–H bond activation.

Finally, it should be mentioned that the lithium terphenolates may easily be formed by treatment of the phenols with *n*-butyllithium. These derivatives are important reagents for the introduction of the terphenolate ligand into metal complexes.[46] At present, there is only one published crystal structure of a terphenolate. This is the compound $\{LiOC_6H\text{-}2,6\text{-}Ph_2\text{-}3,5\text{-}t\text{-}Bu_2\}_3$, which has a trimeric structure featuring a Li_3O_3 ring with Li–O distances of 1.78(1) and 1.840(7) Å.[147]

2. Sulfur Compounds

Terphenylthiols can be prepared directly from the phenol or lithium derivatives. For example, Dilworth *et al.* have reported the synthesis of $HSC_6H_3\text{-}2,6\text{-}Ph_2$[148] by a multistep reaction (Newman–Kaines rearrangement) which converts the thiol to the phenol [Eq. (16)]. The arylthiols $HSC_6H_2\text{-}2,4,6\text{-}Ph_3$,[149] $HSC_6H_3\text{-}2,6\text{-}Mes_2$,[150] and $HSC_6H_3\text{-}2,6\text{-}Trip_2$[151]

(16)

were obtained in good yield by the treatment of the corresponding lithium salts with elemental sulfur, followed by either hydrolysis with dilute H_2SO_4 solution or treatment with $LiAlH_4$.

Other sulfur functionalities may be introduced by the reaction of the lithium derivatives with SO_2Cl_2 or SO_2 as electrophiles.[93] The reaction of $LiC_6H_3\text{-}2,6\text{-}Dmp_2$ with SO_2Cl_2 gave the sulfonyl chloride in 70% yield. The reaction with SO_2 gave $LiO_2SC_6H_3\text{-}2,6\text{-}Dmp_2$, which after treatment with HCl yielded the sulfinic acid $HO_2SC_6H_3\text{-}2,6\text{-}Dmp_2$ in 70% yield. Treatment of the sulfonyl chloride with 12-fold excess $LiAlH_4$ produces the thiol in 64% yield. Alternatively, reduction of the sulfonyl chloride with phosphorus or iodine in acetic anhydride yields the acetylated thiol in 80% yield, and in acetic acid gives the thiol in 51% yield.

Alkali metal derivatives of terphenyl chalcogenolate ligands, the majority of which are lithium–sulfur species, usually crystallize with Lewis base adducts of donor molecules such as diethyl ether, THF, or TMEDA.[152] The bulky terphenythiolate lithium derivatives of $\text{-}SC_6H_2\text{-}2,4,6\text{-}Ph_3$, $\text{-}SC_6H_3\text{-}2,6\text{-}Mes_2$, and $\text{-}SC_6H_3\text{-}2,6\text{-}Trip_2$ are among the few examples of unsolvated species to have been structurally characterized.[151,153] The lithium derivatives of the terphenyl thiols are obtained, most simply, by treatment of the thiol

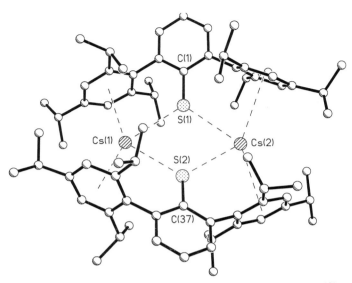

Fig. 29.   Computer-generated drawing of $(CsSC_6H_3\text{-}2,6\text{-}Trip_2)_2$.[154]

with *n*-butyllithium.[149,153] A comparison of the structures of these lithium derivatives clearly demonstrates a progression of steric requirements for the three terphenylthiolate ligands. The $-SC_6H_2\text{-}2,4,6\text{-}Ph_3$ salt has a tetrameric structure, whereas the $-SC_6H_3\text{-}2,6\text{-}Mes_2$ salt is trimeric.[151] These structures are in effect four- and three-rung ladders and are related to the infinite ladder structure seen in $[PhCH_2SLi \cdot NC_5H_5]_n$.[154] The more hindering $-SC_6H_3\text{-}2,6\text{-}Trip_2$ salt is dimeric with bridging thiolate ligands.[153a] All complexes display varying degrees of interaction between the lithiums and the *ortho*-substituent aromatic rings.

Direct reaction of the heavier alkali metals with the thiols produces the corresponding metal salts. The sodium, potassium, rubidium, and cesium salts of $-SC_6H_3\text{-}2,6\text{-}Trip_2$ have been prepared.[151] All have dimeric structures in the solid state, and like the lithium derivative, display varying degrees of interactions between the metal and the *ortho*-Trip aromatic rings. The X-ray crystal structure of the cesium derivative, $[CsSC_6H_3\text{-}2,6\text{-}Trip_2]_2$, is shown in Fig. 29.[153]

The rubidium aluminate complex $RbAlMe_2(SC_6H_3\text{-}2,6\text{-}Trip_2)_2$[155] has also been prepared by the reaction of the rubidium terphenylthiolate with $AlMe_3$ [Eq. (17)].

$$2AlMe_3 + 2RbSC_6H_3\text{-}2,6\text{-}Trip_2 \longrightarrow RbAlMe_4 + RbAlMe_2SC_6H_3\text{-}2,6\text{-}Trip_2 \quad (17)$$

The X-ray crystal structure shows that the two thiolate ligands bridge the

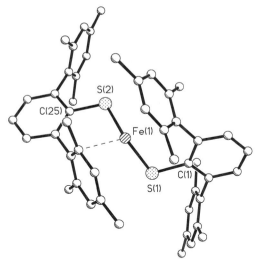

FIG. 30. Computer-generated drawing of $Fe(SC_6H_3\text{-}2,6\text{-}Mes_2)_2$.[53]

rubidium ion and the $AlMe_2$ moiety. The Rb^+ ion also interacts with two Trip rings in a way that is similar to the $(KSC_6H_3\text{-}2,6\text{-}Trip_2)_2$ structure.[151] The complex is a rare example of a structurally characterized species having both aluminum and a heavier alkali metal. Prior examples for rubidium include the crystal structure of the azide RbN_3AlMe_3[156] and a powder diffraction study of $Rb[AlMe_4]$.[157]

The group 2 metal terphenyl thiolates are limited to the magnesium complexes $\{Mg(SC_6H_2\text{-}2,4,6\text{-}Ph_3)_2\}_2$[158] $Mg(SC_6H_3\text{-}2,6\text{-}Mes_2)_2$,[153] and $Mg(SC_6H_3\text{-}2,6\text{-}Trip_2)_2$,[155] which were made by reaction of the thiol with dibutylmagnesium. The compounds $Mg(SC_6H_2\text{-}2,4,6\text{-}Ph_3)_2$ and $Mg(SC_6H_3\text{-}2,6\text{-}Trip_2)_2$ have been structurally characterized by X-ray crystallography.[158,155] The $\text{-}SC_6H_2\text{-}2,4,6\text{-}Ph_3$ species has a dimeric structure with bridging thiolate ligands. In contrast, the $\text{-}SC_6H_3\text{-}2,6\text{-}Trip_2$ species has an unique monomeric structure that also supports the proposed monomeric formulation for $Mg(SC_6H_3\text{-}2,6\text{-}Mes_2)_2$, based on ^1H NMR evidence in solution.[153]

As mentioned earlier, the bulky terphenyl thiolate ligand $\text{-}SC_6H_3\text{-}2,6\text{-}Mes_2$ has been shown to stabilize nominal two-coordination in transition metal complexes.[53,54] Figure 30 represents the X-ray crystal structure of the iron derivative, $Fe(SC_6H_3\text{-}2,6\text{-}Mes_2)_2$.[53]

3. Selenium Compounds

The selenols $HSeC_6H_3\text{-}2,6\text{-}Mes_2$[159] and $HSeC_6H_3\text{-}2,6\text{-}Trip_2$[155] can be synthesized in moderate yields by standard synthetic routes. They are made in a similar manner to the thiols by reaction of the lithium salt with 1 equiv of elemental selenium, followed by reduction with HBF_4. This affords the selenols $HSeC_6H_3$-2,6-Mes_2 and $HSeC_6H_3$-2,6-$Trip_2$ in 60 and 64% yields, respectively. The lithium salt $LiSeC_6H_3$-2,6-Mes_2 may be prepared by treatment of the selenol with *n*-butyllithium. The oxidation of the selenolate with an aqueous basic solution of $K_3Fe(CN)_6$ affords the diselenide in 35% yield.[159] In the solid state, the diselenide has a Se–Se–C angle of 102.5°, and a normal Se–Se bond length of 2.339(1) Å.[160] The torsion angle between the Se–C vectors (128.3°) is unusually large, falling outside the normal range (74–87°) for other diselenides,[161] and is most likely due to the steric interactions between the large terphenythiolate ligands.

Only a few group 1 and 2 metal derivatives of selenolates have been structurally characterized. They are prepared with the same methods used for the thiolates.[155,158] At present there are no crystal structures of lithium terphenyl selenolates. However, the potassium and rubidium salts, which are dimeric, have been structurally characterized.[155] They are isomorphous, both to each other and to the closely related thiolate analogues.[153a] Currently, there are no reported terphenylselenolates reported for the alkaline-earth metals.

4. *Tellurium Compounds*

Strähle and co-workers have reported the synthesis of the ditelluride $(TeC_6H_2$-2,4,6-$Ph_3)_2$.[162] It was prepared in 63% yield by treatment of 1 equiv of elemental tellurium with the Grignard reagent $BrMgC_6H_2$-2,4,6-Ph_3. The complex features a Te–Te bond distance of 2.2945(6) Å and a Te–Te–C angle of 95.1°, which are comparable to those in other ditellurides.[163] The prominent feature in this structure is the very small torsion angle of 66.1° between the Te–C vectors. This is much smaller than seen in other ditellurides and is attributed to crystal lattice packing effects. $(TeC_6H_2$-2,4,6-$Ph_3)_2$ reacts with Ph_3PAuBF_4 to give the complex $[(Ph_3PAu)_2TeC_6H_2$-2,4,6-$Ph_3]BF_4$ in 45% yield, which has a pyramidal $RTeAu_2$ moiety. This complex may then be oxidized with I_2 [Eq. (18)] to give the gold(III) dimeric complex $(I_2AuTeC_6H_2$-2,4,6-$Ph_3)$.

$$2[(Ph_3PAu)_2TeC_6H_2\text{-}2,4,6\text{-}Ph_3]^+ + 4I_2 \longrightarrow (I_2AuTeC_6H_2\text{-}2,4,6\text{-}Ph_3)_2 +$$
$$2PPh_3I_2 + 2Ph_3PAu^+ \quad (18)$$

E. *Group 17 Terphenyl Derivatives*

Terphenyl halides form a very important group of compounds since they are the most common starting point for further derivatives. The bromination

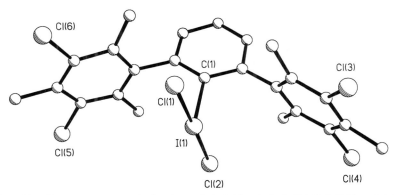

FIG. 31. Computer-generated drawing of 1-dichloroiodo-2,6-bis(3,5-dichloro-2,4,6-tri-methylphenyl)benzene.[168]

of triphenylbenzene, which involved the simple addition of Br_2 to 1,3,5-triphenylbenzene in CS_2 was first reported in 1935.[28] A more recent synthesis which involves the use of iron powder as a catalyst was published in 1994.[12] A high yield of 84% was reported. The solid state structure features a C–Br bond of 1.892(5) Å.[12] A more general synthetic method has been developed by Hart *et al.* for *in situ* terphenyl halide formation. There are two routes to the intermediate *m*-terphenyl Grignard, which can then be quenched with iodine. The first route (Scheme 1a) involves the addition of 3 equiv of aryl Grignard to 1-iodo-2,6-dichlorobenzene; the second method (Scheme 1b) involves the low temperature lithiation of 1,3-dichlorobenzene and the subsequent addition of 2 equiv of aryl Grignard reagent.[25,26] The latter route is now preferred for halogenated terphenyl species owing to the ready availability of the 1,3-dichlorobenzene, the requirement of just 2 equiv of aryl Grignard reagent, and the fact that it involves fewer steps (see Scheme 1b). A wide range of halogenated terphenyls has been synthesized via this method,[25,26,164,165] e.g., 1-bromo-2,6-diphenylbenzene, 1-bromo-3,5-diphenylbenzene, 1,4-dibromo-2,3,5,6-tetraphenylbenzene, 1,4-dibromo-2,3,5,6-tetra-*o*-tolylbenzene, 1-bromo-2,6-di-*p*-tolylbenzene, 1-bromo-2,6-(bromobenzyl)benzene, 1-bromo-2,6-(thiobenzyl)benzene, and their corresponding iodo derivatives.

For synthesis on a larger scale, it has been found that the workup described by Hart *et al.* is impractical, and an alternative has been published.[27] This differs in that aryl halide by-products are removed by distillation under reduced pressure. The residue is then refluxed in EtOH, and filtration effects the recovery of crude halogenated terphenyl product. Sublimation under reduced pressure affords the pure product in up to 75% yield.

The perfluoronated 1-bromo-terphenyl has been synthesized[166] and structurally characterized.[167] The synthesis was effected by the reaction of 2-lithiononafluorobiphenyl with (pentafluorophenyl)lithium, via a proposed aryne intermediate. Bromination was accomplished by Li–Br exchange with 2-bromononafluorobiphenyl. An unusual example of a hypervalent iodo-terphenyl has been synthesized via the reaction of 2,6-dimesitylphenyl-iodide with Cl_2 gas in $CHCl_3$ at 0°C affording, 1-dichloroiodo-2,6-bis(3,5-dichloro-2,4,6-trimethylphenyl)benzene in moderate yield (60%).[168] The solid state structure (Fig. 31) shows that the dichloroiodo moiety has T-shaped (linear Cl–I–Cl 178.3(1)°) geometry with a torsion angle to the central phenyl ring of 79°. The I–Cl distances are 2.469(4) and 2.491(4) Å. This compound undergoes facile decomposition in

$$n = 1,2 \tag{19}$$

CH_2Cl_2 solution [Eq. (19)], even at reduced temperatures, to give predominantly 1-bromo-2,6-bis(3,5-dichloro-2,4,6-trimethylphenyl)benzene with trace amounts of mono- and dichlorinated derivatives.

ACKNOWLEDGMENTS

We are grateful to the donors of the Petroleum Research Fund administered by the American Chemical Society and the National Science Foundation for financial support.

REFERENCES

(1) For example: (a) Wilberg, N. *Coord. Chem. Rev.* **1997**, 163, 217; (b) Westerhausen, M.; Digeser, M. H.; Wieneke, M.; Nöth, H.; Knizek, J. *Eur. J. Inorg. Chem.* **1998**, 517; Westerhausen, M.; Digeser, M.; Knizek, J.; Schwartz, W. *Inorg. Chem.* **1998**, *37*, 619; (c) Arnold, J. *Progr. Inorg. Chem.* **1995**, *43*, 353.

(2) Lappert, M. F. In Bassindale, A. R.; Gaspar, P. G., Eds.; *Frontiers in Organosilicon Chemistry;* Royal Society of Chemistry: Cambridge, 1991; p. 231.

(3) Eaborn, C.; Smith, J. D.; Izod, K. *J. Organomet. Chem.* **1995**, *500*, 89; Eaborn, C.; Smith, J. D. *Coord. Chem. Rev.* **1996**, *154*, 125.

(4) (a) Harris, D. H.; Lappert, M. F. *J. Organomet. Chem. Library* **1976**, *2*, 13; (b) Eller, P. G.; Bradley, D. C.; Hursthouse, M. B.; Meek, D. W. *Coord. Chem. Revs.* **1977**, *24*, 1; (c) Lappert, M. F.; Power, P. P.; Sanger, A. R.; Srivastava, R. C. *Metal and Metalloid Amides;* Ellis Horwood–Wiley: Chichester, 1979; (d) Cummins, C. C. *Progr. Inorg. Chem.* **1998**, *47*, 485.

(5) (a) Weese, K. J.; Bartlett, R. A.; Murray, B. D.; Olmstead, M. M.; Power, P. P. *Inorg. Chem.* **1987**, *26*, 2409; (b) Becker, G.; Hitchcock, P. B.; Lappert, M. F.; McKinnon, I. A. *J. Chem. Soc. Chem. Commun.* **1989**, 1312; (c) Chisholm, M. F.; Folting, K.; Haubrich, S. T.; Martin, J. D. *Inorg. Chim. Acta* **1993**, *213*, 17.

(6) (a) Bartlett, R. A.; Feng, X.; Olmstead, M. M.; Power, P. P.; Weese, K. J. *J. Am. Chem. Soc.* **1987**, *109*, 485; (b) Paetzold, P.; Pelzer, C.; Boese, R. *Chem. Ber.* **1988**, *121*, 151; (c) Bartlett, R. A.; Dias, H. V. R.; Feng, X.; Power, P. P. *J. Am. Chem. Soc.* **1989**, *111*, 1306.

(7) For example: Baker, R. T.; Ovenall, D. W.; Calabrese, J. C.; Westcott, S. A.; Taylor, N. J.; Williams, I. D.; Marder, T. B. *J. Am. Chem. Soc.* **1990**, *112*, 9399; Iverson, C. N.; Smith, M. R., III. *J. Am. Chem. Soc.* **1995**, *117*, 4403; Hartwig, J. F.; He, X. *Angew. Chem. Int. Ed. Engl.* **1996**, *35*, 315.

(8) Koschmieder, S. U.; Wilkinson, G. *Polyhedron* **1991**, *10*, 135.

(9) Klose, A.; Solari, E.; Floriani, C.; Chiesi-Villa, A.; Rizzoli, C.; Re, N. *J. Am. Chem. Soc.* **1994**, *116*, 9123.

(10) Nobel, D.; van Koten, G.; Spek, A. L. *Angew. Chem. Int. Ed. Engl.* **1989**, *28*, 208.

(11) Lingnau, R.; Strähle, J. *Angew. Chem. Int. Ed. Engl.* **1988**, *27*, 436.

(12) Haaland, A.; Rypdal, K.; Verne, H.-P.; Scherer, W. R. Thiel, W. R. *Angew. Chem. Int. Ed. Engl.* **1994**, *33*, 2443.

(13) (a) Muller, H.; Seidel, W.; Görls, H. *Angew. Chem. Int. Ed. Engl.* **1995**, *34*, 325; (b) Wehmschulte, R. J.; Power, P. P. *Organometallics* **1995**, *14*, 3264.

(14) He, X.; Olmstead, M. M.; Power, P. P. *J. Am. Chem. Soc.* **1992**, *114*, 9668.

(15) Yoshifuji, M.; Shima, I.; Inamoto, N. *Tetrahedron Lett.* **1979**, 3963.

(16) Yoshifuji, M.; Shima, N.; Inamoto, N.; Hirotsu, K.; Higuchi, T. *J. Am. Chem. Soc.* **1981**, *103*, 4587.

(17) Cetinkaya, B.; Gumrukcu, I.; Lappert, M. F.; Atwood, J. L.; Shakir, R. *J. Am. Chem. Soc.* **1980**, *102*, 2086.

(18) Cetinkaya, B.; Hitchcock, P. B.; Lappert, M. F.; Misra, M. C.; Thorne, A. J. *J. Chem. Soc. Chem. Commun.* **1984**, 148.

(19) Hitchcock, P. B.; Lappert, M. F.; Samways, B. J.; Weinberg, E. L. *J. Chem. Soc. Chem. Comun.* **1983**, 1492.

(20) (a) du-Mont, W. W.; Kubiniok, S.; Lange, L.; Pohl, S.; Saak, W.; Wagner, I. *Chem. Ber.* **1991**, *124*, 1315; (b) Ruhlandt-Senge, K.; Power, P. P. *Inorg. Chem.* **1991**, *30*, 3683.

(21) For example: Block, E.; Eswarakrishnan, V.; Gernon, M.; Ofori-Okai, G.; Sana, C.; Tang, K.; Zubieta, J. *J. Am. Chem. Soc.* **1989**, *111*, 658.

(22) For example: Tokitoh, N.; Manmaru, K.; Okazaki, R. *Organometallics* **1994**, *13*, 167.

(23) Tokitoh, N.; Arai, Y.; Okazaki, R.; Nagase, S. *Science* **1997**, *277*, 78.

(24) (a) Suslick, K. S.; Fox, M. M. *J. Am. Chem. Soc.* **1983**, *105*, 3507; (b) Cook, B. B.; Reinert, J. J.; Suslick, K. *J. Am. Chem. Soc.* **1986**, *108*, 7286.

(25) (a) Du, C.-J.; Hart, H.; Ng, D. K.-K. *J. Org. Chem.* **1986**, *51*, 3162; (b) Hanoda, K.; Hart, H.; Du, C.-S. *J. Org. Chem.* **1985**, *50*, 5521.

(26) Saednya, A.; Hart, H. *Synthesis* **1996**, 1455.

(27) Schiemenz, B.; Power, P. P. *Organometallics* **1996**, *15*, 958.

(28) Kohler, E. P.; Blanchard, L. W. *J. Am. Chem. Soc.* **1935**, *57*, 367.

(29) (a) Olmstead, M. M.; Power, P. P. *J. Organomet. Chem.* **1991**, *408*, 1; (b) Girolami, G. S.; Riehl, M. E.; Suslick, K. S.; Wilson, S. R. *Organometallics* **1992**, *11*, 3907.

(30) Hope, H.; Power, P. P. *J. Am. Chem. Soc.* **1983**, *105*, 5320.

(31) Beno, M. A.; Hope, H.; Olmstead, M. M.; Power, P. P. *Organometallics* **1985**, *4*, 2117.

(32) Bartlett, R. A.; Dias, H. V. R.; Power, P. P. *J. Organomet. Chem.* **1988**, *341*, 1.

(33) Ruhlandt-Senge, K.; Ellison, J. J.; Wehmschulte, R. J.; Pauer, F.; Power, P. P. *J. Am. Chem. Soc.* **1993**, *115*, 11353.

(34) Schiemenz, B.; Power, P. P. *Angew. Chem. Int. Ed. Engl.* **1996**, *35*, 2150.

(35) Niemeyer, M.; Power, P. P. *Organometallics* **1997**, *16*, 3258.

(36) Niemeyer, M.; Power, P. P. *Inorg. Chem.* **1997**, *36*, 4688.

(37) (a) Coates, G. E.; Glockling, F.; Hvek, N. D. *J. Chem. Soc.* **1952**, 4496; (b) Almenningen, A.; Haaland, A.; Nilson, J. E. *Acta Chim. Scand.* **1968**, *22*, 972; (c) Clark, A. H.; Haaland, A. *Acta Chim. Scand.* **1970**, *124*, 3024.

(38) Schomaker, V.; Stevenson, D. P. *J. Am. Chem. Soc.* **1941**, *63*, 27.

(39) Ellison, J. J.; Power, P. P. *J. Organomet. Chem.* **1997**, *526*, 263.

(40) Spek, A. L.; Voorbergen, P.; Schart, G.; Bloomberg, C.; Bickelhaupt, F. *J. Organomet. Chem.* **1974**, *77*, 147.

(41) Toney, J.; Stucky, G. D. *J. Chem. Soc. Chem. Commun.* **1967**, 1168.

(42) Hwang, C.-S.; Power, P. P., unpublished results, 1997.

(43) (a) Smith, G. D.; Fanwick, P. E.; Rothwell, I. P. *J. Am. Chem. Soc.* **1989**, *111*, 750; (b) Smith, G. D.; Visciglio, V. M.; Fanwick, P. E.; Rothwell, I. P. *Organometallics* **1992**, *11*, 1064; (c) Maringgle, W.; Meller, A.; Noltemeyer, M.; Sheldrick, G. M. *Z. Anorg. Allg. Chem.* **1986**, *536*, 24.

(44) (a) Fanwick, P. E.; Rothwell, I. P.; Kriley, C. E. *Polyhedron* **1996**, *15*, 2403; (b) Kerschner, J. L.; Rothwell, I. P.; Huffman, J. C.; Streib, W. E. *Organometallics* **1988**, *7*, 1871; (c) Kerschner, J. L.; Fanwick, P. E.; Rothwell, I. P.; Huffman, J. C. *Organometallics* **1989**, *8*, 1431; (d) Steffey, B. D.; Chamberlain, L. R.; Chesnut, R. W.; Chebi, D. E.; Fanwick, P. E.; Rothwell, I. P. *Organometallics* **1989**, *8*, 1419; (e) Chesnut, R. W.; Yu, J. S.; Fanwick, P. E.; Rothwell, I. P.; Huffman, J. C. *Polyhedron* **1990**, *9*, 1051; (f) Kriley, C. E.; Kerschner, J. L.; Fanwick, P. E.; Rothwell, I. P. *Organometallics* **1993**, *12*, 2051; (g) Chesnut, R. W.; Steffey, B. D.; Rothwell, I. P.; Huffman, J. C. *Polyhedron* **1989**, *7*, 753; (h) Kerschner, J. L.; Kriley, C. E.; Fanwick, P. E.; Rothwell, I. P. *Acta Crystallogr., Sect. C (Cr. Str. Comm.)* **1994**, *50*, 1193; (i) Kerschner, J. L.; Yu, J. S.; Fanwick, P. E.; Rothwell, I. P.; Huffman, J. C. *Organometallics* **1989**, *8*, 1414; (j) Lockwood, M. A.; Fanwick, P. E.; Rothwell, I. P. *J. Chem. Soc. Chem. Commun.* **1996**, 2013; (k) Kriley, C. E.; Fanwick, P. E.; Rothwell, I. P. *J. Am. Chem. Soc.* **1994**, *116*, 5225; (l) Kerschner, J. L.; Fanwick, P. E.; Rothwell, I. P. *J. Am. Chem. Soc.* **1987**, *109*, 5840.

(45) (a) Lockwood, M. A.; Fanwick, P. E.; Rothwell, I. P. *Polyhedron* **1995**, *14*, 3363; (b) Jaques, D.; Clark, J. R.; Chebi, D. E.; Fanwick, P. E.; Rothwell, I. P. *Acta Crystallogr., Sect. C (Cr. Str. Comm.)* **1994**, *50*, 898; (c) Lockwood, M. A.; Potyen, M. C.; Steffey, B. D.; Fanwick, P. E.; Rothwell, I. P. *Polyhedron* **1995**, *14*, 3293; (d) Kerschner, J. L.; Rothwell, I. P.; Huffman, J. C.; Streib, W. E. *Organometallics* **1988**, *7*, 1871; (e) Kerschner, J. L.; Fanwick, P. E. Rothwell, I. P. *J. Am. Chem. Soc.* **1988**, *110*, 8235; (f) Steffey, B. D.; Chesnut, R. W.; Kerschner, J. L.; Pellechia, P. J.; Fanwick, P. E.; Rothwell, I. P. *J. Am. Chem. Soc.* **1989**, *111*, 378; (g) Kerschner, J. L.; Fanwick, P. E.; Rothwell, I. P. *J. Am. Chem. Soc.* **1987**, *109*, 5840; (h) Kerschner, J. L.; Torres, E. M.; Fanwick, P. E.; Rothwell, I. P.; Huffman, J. C. *Organometallics* **1989**, *8*, 1424.

(46) (a) Hill, J. E.; Balaich, G.; Fanwick, P. E.; Rothwell, I. P. *Organometallics* **1993**, *12*, 2911; (b) Chesnut, R. W.; Fanwick, P. E.; Rothwell, I. P. *Inorg. Chem.* **1988**, *27*, 752; (c) Chamberlain, I. R.; Durfee, L. D.; Fanwick, P. E.; Kobriger, L.; Latesky, S. L.; McMullen, A. K.; Rothwell, I. P.; Folting, K.; Huffman, J. C.; Streib, W. E.; Wang, R. *J. Am. Chem. Soc.* **1987**, *109*, 390; (d) Dilworth, J. R.; Hanich, J.; Krestel, M.; Beck,

J.; Strahle, J. *J. Organomet. Chem.* **1986**, *315*, C9; (e) Fanwick, P. E.; Ogilvy, A. E.; Rothwell, I. P. *Organometallics* **1987**, *6*, 73; (f) Darensbourg, D. J.; Niezgoda, S. A.; Reibenspies, J. H.; Draper, J. D. *Inorg. Chem.* **1997**, *36*, 5686; (g) Chesnut, R. W.; Jacob, G. G.; Yu, J. S.; Fanwick, P. E.; Rothwell, I. P. *Organometallics* **1991**, *10*, 321; (h) Hill, J. E.; Fanwick, P. E.; Rothwell, I. P. *Inorg. Chem.* **1991**, *30*, 1143; (i) Hill, J. E.; Balaich, G. J.; Fanwick, P. E.; Rothwell, I. P. *Organometallics* **1991**, *10*, 3428; (j) Hill, J. E.; Nash, J. M.; Fanwick, P. E.; Rothwell, I. P. *Polyhedron* **1990**, *9*, 1617; (k) Torrez, E.; Profilet, R. D.; Fanwick, P. E.; Rothwell, I. P. *Acta Crystallogr., Sect. C* (*Cr. Str. Comm.*) **1994**, *50*, 902; (l) Kerschner, J. L.; Fanwick, P. E.; Rothwell, I. P.; Huffman, J. C. *Inorg. Chem.* **1989**, *28*, 780; (m) Hill, J. E.; Fanwick, P. E.; Rothwell, I. P. *Organometallics* **1992**, *11*, 1771; (n) Yu, J. S.; Fanwick, P. E.; Rothwell, I. P. *Acta Crystallogr., Sect. C* (*Cr. Str. Comm.*) **1992**, *48*, 1759; (o) Chebi, D. E.; Fanwick, P. E.; Rothwell, I. P. *Polyhedron* **1990**, *9*, 969; (p) Clark, J. R.; Fanwick, P. E.; Rothwell, I. P. *J. Chem. Soc., Chem. Commun.* **1995**, 553; (q) Yasuda, H.; Nakayama, Y.; Takei, K.; Nakamura, A.; Kai, Y.; Kanehisa, N. *J. Organomet. Chem.* **1994**, *473*, 105; (r) Hill, J. E.; Fanwick, P. E.; Rothwell, I. P. *Organometallics* **1992**, *11*, 1775; (s) Hill, J. E.; Fanwick, P. E.; Rothwell, I. P. *Organometallics* **1991**, *10*, 15; (t) Darensbourg, D. J.; Mueller, B. L.; Bischoff, C. J.; Chojnacki, S. S.; Reibenspies, J. H. *Inorg. Chem.* **1991**, *30*, 2418; (u) Baliach, G. J.; Hill, J. E.; Waratuke, S. A.; Fanwick, P. E.; Rothwell, I. P. *Organometallics* **1995**, *14*, 656; (v) Minhas, R. K.; Edema, J. J. H.; Gambarotta, S.; Meetsma, A. *J. Am. Chem. Soc.* **1993**, *115*, 6710; (w) Zambrano, C. H.; Profilet, R. D.; Hill, J. E.; Fanwick, P. E.; Rothwell, I. P. *Polyhedron* **1993**, *12*, 689; (x) Waratuke, S. A.; Johnson, E. S.; Thorn, M. G.; Fanwick, P. E.; Rothwell, I. P. *J. Chem. Soc., Chem. Commun.* **1996**, 2617; (y) Chamberlain, L. R.; Durfee, L. D.; Fanwick, P. E.; Kobriger, L. M.; Latesky, S. L.; McMullen, A. K.; Steffey, B. D.; Rothwell, I. P.; Folting, K.; Huffman, J. C. *J. Am. Chem. Soc.* **1987**, *109*, 6068; (z) Chesnut, R. W.; Durfee, L. D.; Fanwick, P. E.; Rothwell, I. P.; Folting, K.; Huffman, J. C. *Polyhedron* **1987**, *6*, 2019; (aa) Quignard, F.; Leconte, M.; Basset, J. M.; Leh-yeh Hsu; Alexander, J. J.; Shore, S. G. *Inorg. Chem.* **1987**, *26*, 4272; (bb) Hill, J. E.; Profilet, R. D.; Fanwick, P. E.; Rothwell, I. P. *Angew. Chem., Int. Ed. Engl.* **1990**, *29*, 664; (cc) Hill, J. E.; Fanwick, P. E.; Rothwell, I. P. *Organometallics* **1990**, *9*, 2211; (dd) Lockwood, M. A.; Potyen, M. C.; Steffey, B. D.; Fanwick, P. E.; Rothwell, I. P. *Polyhedron* **1995**, *14*, 3293; (ee) Lockwood, M. A.; Fanwick, P. E.; Eisenstein, O.; Rothwell, I. P. *J. Am. Chem. Soc.* **1996**, *118*, 2762; (ff) O'Donoghue, M. B.; Schrock, R. R.; LaPointe, A. M.; Davis, W. M. *Organometallics* **1996**, *15*, 1334; (gg) Riley, P. N.; Profilet, R. D.; Salberg, M. M.; Fanwick, P. E.; Rothwell, I. P. *Polyhedron* **1998**, *17*, 773.

(47) (a) Deacon, G. B.; Tiecheng Feng; Nickel, S.; Ogden, M. I.; White, A. H. *Aust. J. Chem.* **1992**, *45*, 671; (b) Deacon, G. B.; Nickel, S.; MacKinnon, P.; Tiekink, E. R. T. *Aust. J. Chem.* **1990**, *43*, 1245; (c) Deacon, G. B.; Nickel, S.; Tiekink, E. R. T. *J. Organomet. Chem.* **1991**, *409*, C1; (d) Deacon, G. B.; Gatehouse, B. M.; Shen, Q.; Ward, G. N.; Tiekink *Polyhedron* **1993**, *12*, 1289; (e) Deacon, G. B.; Tiecheng Feng; Skelton, B. W.; White, A. H. *Aust. J. Chem.* **1995**, *48*, 741; (f) Clark, D. L.; Deacon, G. B.; Tiecheng Feng; Hollis, R. V.; Scott, B. L.; Skelton, B. W.; Watkin, J. G.; White, A. H. *J. Chem. Soc., Chem. Commun.* **1996**, 1729.

(48) (a) Webster, M.; Browning, D. J.; Corker, J. M. *Acta Crystallogr., Sect. C* (*Cr. Str. Comm.*) **1996**, *52*, 2439; (b) Smith, G. D.; Fanwick, P. E., Rothwell, I. P. *Inorg. Chem.* **1990**, *29*, 3221; (c) Maruoka, K.; Imoto, H.; Saito, S.; Yamamoto, H. *J. Am. Chem. Soc.* **1994**, *116*, 4131; (d) Smith, G. D.; Fanwick, P. E.; Rothwell, I. P. *Inorg. Chem.* **1989**, *28*, 618.

(49) Darensbourg, D. J.; Holtcamp, M. W. *Macromolecules* **1995**, *28*, 7577.

(50) Dilworth, J. R.; Hu, J. *Adv. Inorg. Chem.* **1994**, *40*, 411.
(51) Bishop, P. T.; Dilworth, J. R.; Nicholson, T.; Zubieta, J. *J. Chem. Soc. Dalton Trans.* **1991**, 385.
(52) Buyuktas, B. S.; Olmstead, M. M.; Power, P. P. *J. Chem. Soc. Chem. Commun.*, in press.
(53) Ellison, J. J.; Ruhlandt-Senge, K.; Power, P. P. *Angew. Chem. Int. Ed. Engl.* **1994**, *33*, 1178.
(54) Ellison, J. J.; Power, P. P. *Inorg. Chem.* **1994**, *33*, 4231.
(55) Ellison, J. J.; Ruhlandt-Senge, K.; Hope, H.; Power, P. P. *Inorg. Chem.* **1995**, *34*, 49.
(56) Simons, R. S.; Power, P. P. *J. Am. Chem. Soc.* **1996**, *118*.
(57) He, X.; Olmstead, M. M.; Power, P. P. *J. Am. Chem. Soc.* **1992**, *114*, 9668.
(58) Theopold, K. H.; Silvestre, J.; Byrne, E. K.; Richeson, D. S. *Organometallics* **1989**, *8*, 2001.
(59) Huffman, J. C.; Nugent, W. A.; Kochi, J. K. *Inorg. Chem.* **1980**, *19*, 2749.
(60) Li, X.-W.; Pennington, W. T.; Robinson, G. H. *J. Am. Chem. Soc.* **1995**, *117*, 7578.
(61) Li, X.-W.; Xie, J. M.; Schreiner, P. R.; Gripper, R. C.; Crittendon, R. C.; Campana, C. F.; Schaefer III, H.-F.; Robinson, G. H. *Organometallics* **1996**, *15*, 3798.
(62) Su, J. R.; Li, X.-W.; Crittendon, R. C.; Robinson, G. H. *J. Am. Chem. Soc.* **1997**, *119*, 547.
(63) Su, J. R.; Li, X.-W.; Crittendon, R. C.; Campana, C. F.; Robinson, G. H. *Organometallics* **1997**, *16*, 4511.
(64) Haubrich, S. T.; Power, P. P. *J. Am. Chem. Soc.* **1998**, *120*, 2202.
(65) Niemeyer, M.; Power, P. P. *Angew. Chem. Int. Ed. Engl.* **1998**, *37*, 1277.
(66) Hope, H.; Pestana, D. C.; Power, P. P. *Angew. Chem. Int. Ed. Engl.* **1991**, *30*, 691.
(67) Grigsby, W. J.; Power, P. P. *J. Am. Chem. Soc.* **1996**, *118*, 7981.
(68) Brotherton, R. J.; McCloskey, A. L.; Boone, J. L.; Manasevit, H. M. *J. Am. Chem. Soc.* **1960**, *82*, 6245.
(69) (a) Moezzi, A.; Olmstead, M. M.; Power, P. P. *J. Am. Chem. Soc.* **1992**, *114*, 2715; (b) Moezzi, R. A.; Bartlett, R. A.; Power, P. P. *Angew. Chem. Int. Ed. Engl.* **1992**, *31*, 1082; (c) Power, P. P. *Inorg. Chim. Acta* **1992**, *198–200*, 443.
(70) Grigsby, W. J.; Power, P. P. *Chem. Eur. J.* **1997**, *3*, 368.
(71) Wehmschulte, R. J.; Grigsby, W. J.; Schiemenz, B.; Bartlett, R. A.; Power, P. P. *Inorg. Chem.* **1996**, *35*, 6694.
(72) Wehmschulte, R. J.; Power, P. P. *Inorg. Chem.* **1994**, *33*, 5611.
(73) Wehmschulte, R. J.; Power, P. P. *Inorg. Chem.* **1996**, *35*, 3262 **1998**, *22*, 1125.
(74) (a) Wehmschulte, R. J.; Power, P. P. *New J. Chem.,* **1998**, *22*, 1125. (b) Wehmschulte, R. J.; Power, P. P. *Inorg. Chem.,* **1998**, *37*, 6908.
(75) Li, X.-W.; Pennington, W. T.; Robinson, G. H. *Organometallics* **1995**, *14*, 2109.
(76) Crittendon, R. C.; Li, X.-W.; Su, J. R.; Robinson, G. H. *Organometallics* **1997**, *16*, 2443.
(77) Meller, A.; Pusch, S.; Pohl, E.; Häming, L.; Herbst-Irmer, R. *Chem. Ber.* **1993**, *126*, 2255.
(78) Li, X.-W.; Robinson, G. L.; Pennington, W. T. *Main Group Chem.* **1996**, *1*, 301.
(79) Cowley, A. H.; Isom, H. S.; Decken, A. *Organometallics* **1995**, *14*, 2589.
(80) Rahbarnoohi, H.; Heeg, M. J.; Oliver, J. P. *Organometallics* **1994**, *13*, 2123.
(81) Robinson, G. H.; Li, X.-W.; Pennington, W. T. *J. Organomet. Chem.* **1995**, *501*, 399.
(82) (a) Petrie, M. A.; Power, P. P.; Dias, H. V. R.; Ruhlandt-Senge, K.; Waggoner, K. M.; Wehmschulte, R. J. *Organometallics,* **1993**, *12*, 1086; (b) Schultz, S.; Pusch, S.; Potul, E.; Dielkus, S.; Herbst-Irmer, R.; Meller, A.; Roesky, H. W. *Inorg. Chem.* **1993**, *32*, 3343.
(83) Uhl, W.; Layh, M.; Hildenbrand, T. *J. Organomet. Chem.* **1989**, *364*, 289.
(84) He, X.; Bartlett, R. A.; Olmstead, M. M.; Ruhlandt-Senge, K.; Sturgeon, B. E.; Power, P. P. *Angew. Chem. Int. Ed. Engl.* **1993**, *32*, 717.
(85) Linti, G.; Frey, R.; Schmidt, M. *Z. Naturforsch. B: Chem. Sci.* **1994**, *49B*, 958.
(86) Klinkhammer, K. W. *Angew. Chem. Int. Ed. Engl.* **1997**, *36*, 2320.

(87) Xie, J. M.; Grev, R. S.; Gu, J. D.; Schaefer III, H. F.; Schleyer, P. v. R.; Su, J. R.; Li, X.-W.; Robinson, G. H. *J. Am. Chem. Soc.,* **1998**, *120*, 773.

(88) Cotton, F. A.; Cowley, A. H.; Feng, X. *J. Am. Chem. Soc.* **1998**, *120*, 1795.

(89) Allen, T. L.; Fink, W. H.; Power, P. P., submitted for publication.

(90) Cotton, F. A.; Feng, X. *J. Organometallics* **1998**, *17*, 128.

(91) Weiss, J.; Stetzkamp, D.; Nuber, B.; Fischer, R. A.; Boehme, C.; Frenking, G. *Angew. Chem. Int. Ed. Engl.* **1997**, *36*, 70.

(92) Martin, L. R.; Einstein, F. W. B.; Pomeroy, R. K. *Inorg. Chem.* **1985**, *24*, 2777.

(93) (a) Lüning, U; Wangnick, C.; Peters, K.; von Schnering, H. G. *Chem. Ber.* **1991**, *124*, 397; (b) Lüning, U.; Baumgartner, H. *Synlett.* **1993**, 571; (c) Lüning, U.; Baumgartner, H.; Manthey, C.; Meynhardt, B. *J. Org. Chem.* **1996**, *61*, 7922; (d) Lüning, U. J. *Mater. Chem.* **1997**, *7*, 175.

(94) Chen, C. T.; Chadha, R.; Siegel, J. S.; Hardcastle, K. *Tetrahedron Lett.* **1995**, *36*, 8403.

(95) (a) Callot, H. J.; Metz, F. *Tetrahedron* **1985**, 4495; (b) Callot, H. J; Albrecht-Gary, A. M.; Joubbeh, M. A.; Metz, B.; Metz, F. *Inorg. Chem.* **1989**, *28*, 3633.

(96) Dimroth, K.; Neubauer, G. *Chem. Ber.* **1959**, *92*, 2042.

(97) (a) Goto, K.; Holler, M.; Okazaki, R. *Tetrahedron Lett.* **1996**, *37*, 18; (b) Goto, K.; Holler, M.; Okazali, R. *J. Am. Chem. Soc.* **1997**, *119*, 1460; (c) Goto, K.; Okazaki, R. *Liebigs Ann./Recueil,* **1997**, 2393.

(98) Simons, R. S.; Haubrich, S. T.; Mork, B. V.; Power, P. P. *Main Group Chem.,* **1998**, *2*, 275.

(99) Ruhlandt-Senge, K.; Bartlett, R. A.; Olmstead, M. M.; Power, P. P. *Angew. Chem. Int. Ed. Engl.* **1993**, *32*, 425.

(100) Millevolte, A. J.; van den Winkel, Y.; Powell, D. R.; West, R. *Organometallics* **1997**, *16*, 5375.

(101) Schmid, G.; Thewalt, U.; Sedmera, P.; Hanus, V.; Mach, K. *J. Organomet. Chem.* **1994**, *466*, 125.

(102) Simons, R. S.; Pu, L.; Olmstead, M. M.; Power, P. P. *Organometallics* **1997**, *16*, 1920.

(103) Olmstead, M. M.; Simons, R. S.; Power, P. P. *J. Am. Chem. Soc.* **1997**, *119*, 11705.

(104) (a) See for Ge: Jutzi, P.; Schmidt, H.; Neumann, B.; Stammler, H. G. *Organometallics* **1996**, *15*, 741; (b) see for Sn: Weidenbruch, M.; Schlaefke, J.; Schäfer, A.; Peters, K.; von Schnering, H.G.; Marsmann, H. *Angew. Chem. Int. Ed. Engl.* **1994**, *33*, 1846; (c) see for Sn: Grutzmacher, H.; Pritzkow, H.; Edelmann, F. T. *Organometallics* **1991**, *10*, 23; (d) see for Sn: Lay, H.; Pritzkow, H.; Grutzmacher, H. *J. Chem. Soc. Chem. Commun.* **1992**, 260; (e) see for Pb: Brooker, S.; Buijink, J. K.; Edelmann, F. T. *Organometallics* **1991**, *10*, 25; (f) see for Sn and Pb: Klinkhammer, K.W.; Schwartz, W. *Angew. Chem. Int. Ed. Engl.* **1995**, *34*, 1334.

(105) For example: (a) Hitchcock, P. B.; Lappert, M. F.; Miles, S. J.; Thorne, A. J. *J Chem. Soc. Chem. Commun.* **1984**, 480; (b) Kira, M.; Iwamoto, T.; Maruyama, T.; Kabuto, C.; Sakurai, H. *Organometallics* **1996**, *15*, 7783; (c) Snow, J. T.; Murakami, S.; Masamune, S.; Williams, D. J. *Tetrahedron Lett.* **1984**, *25*, 4191.

(106) (a) Chorley, R. W.; Hitchcock, P. B.; Jolly, B. S.; Lappert, M. F.; Lawless, G. A. *J. Chem. Soc. Chem. Commun.* **1991**, 1302; (b) Eaborn, C.; Izod, K.; Hitchcock, P. B.; Sözerli, S. E.; Smith, J. D. *J. Chem. Soc. Chem. Commun.* **1995**, 1829.

(107) Cardin, C. J.; Cardin, D. J.; Lawless, G. A.; Power, M. B. *Proc. R. Ir. Acad.* **1989**, *89*, 399.

(108) Cardin, C. J.; Cardin, D. J.; Convery, M. A.; Devereux, M. M.; Twamley, B.; Silver, J. *J. Chem. Soc. Dalton Trans.* **1996**, 1145.

(109) Olmstead, M. M.; Pu, L.; Simons, R. S.; Power, P. P. *J. Chem. Soc. Chem. Commun.,* **1997**, 1595.

(110) Simons, R. S.; Power, P. P. *J. Am. Chem. Soc.* **1996**, *118*, 11966.

(111) Chan, L. Y. Y.; Dean, W. K.; Graham, W. A. G. *Inorg. Chem.* **1977**, *16*, 1067.

(112) Carré, F.; Colomer, E.; Corriu, R. J. P.; Vioux, A. *Organometallics* **1984**, *3*, 970.
(113) Wells, A. F. *Structural Inorganic Chemistry,* 5th ed.; Clarendon: Oxford, p. 1279.
(114) Goldberg, D. E.; Harris, D. H.; Lappert, M. F.; Thomas, K. M. *J. Chem. Soc. Chem. Commun.* **1976**, 261.
(115) Pu, L.; Haubrich, S. H.; Power, P. P. *J. Organomet. Chem.,* in press.
(116) Saito, M.; Tokitoh, N.; Okazaki, R. *J. Am. Chem. Soc.* **1997**, *119*, 11124.
(117) Kuchta, M. C.; Parkin, G. *J. Am. Chem. Soc.* **1994**, *116*, 8372.
(118) Dimroth, K.; Brauninger, G.; Neubauer, G. *Chem. Ber.* **1957**, *90*, 1634.
(119) Anulewicz, R.; Pniewska, B.; Milart, P. *Acta Crystallogr., Sec. C,* **1994**, *50*, 1516.
(120) Clark, G. R.; Nielson, A. J.; Rickard, C. E. F. *J. Chem. Soc., Dalton Trans.* **1996**, 4265.
(121) Miura, Y.; Tanaka, A.; Hirotsu, K. *J. Org. Chem.* **1991**, *56*, 6638.
(122) Miura, Y.; Momoki, M.; Fuchikami, T.; Teki, Y.; Itoh, K.; Mizutani, H. *J. Org. Chem.* **1996**, *61*, 4300.
(123) Urnezius, E.; Prostasiewicz, J. D. *Main Group Chemistry,* **1996**, *1*, 369.
(124) Twamley, B.; Power, P. P. Unpublished results.
(125) Urnezius, E.; Prostasiewicz, J. D. Unpublished results, cited in Ref. 123.
(126) (a) Corbridge, D. E. C. *The Structural Chemistry of Phosphorus*; Elsevier Scientific: New York, 1974; (b) Corbridge, D. E. C. *Phosphorus, An Outline of its Chemistry, Biochemistry and Technology,* 4th Ed.; Elsevier Scientific: New York, 1990.
(127) Winkel, Y. v. d.; Bastiaans, H. M. M.; Bickelhaupt, F. *J. Organomet. Chem.* **1991**, *405*, 183.
(128) Shah, S.; Burdette, S. C.; Swavey, S.; Urbach, F. L.; Protasiewicz, J. D. *Organometallics* **1997**, *16*, 3395.
(129) Cowley. A. H.; Decken, A.; Norman, N. C.; Kruger, C.; Lutz, F.; Jacobsen, H.; Ziegler, T. *J. Am. Chem. Soc.* **1997**, *119*, 3389.
(130) (a) Cowley, A. H.; Kilduff, J. E.; Lasch, J. G.; Mehotra, S. K.; Norman, N. C.; Pakulski, M.; Whittlesey, B. R.; Atwood, J. A.; Hunter, W. E. *Inorg. Chem.* **1984**, *23*, 2582; (b) Cowley, A. H.; Lasch, J. G.; Norman, N. C.; Pakulski, M. *J. Am. Chem. Soc.* **1983**, *105*, 5506; (c) Twamley, B. T.; Power, P. P. *J. Chem. Soc. Chem. Commun.* **1998**, 1623.
(131) Tsuji, K.; Fuji, Y.; Sasaki, S.; Yoshifuji, M. *Chem. Lett.* **1997**, 855.
(132) Nagase, S.; Suzuki, S.; Kurakake, T. *J. Chem. Soc. Chem. Commun.* **1990**, 1724.
(133) Tokitoh, N.; Arai, Y.; Sasamori, T.; Okazaki, R.; Nagase, S.; Uekusa, H.; Ohashi, Y. *J. Am. Chem. Soc.* **1998**, *120*, 433.
(134) Li, X.-W.; Lorberth, J.; Massa, W.; Wocadlo, S. *J. Organomet. Chem.* **1995**, *485*, 141.
(135) Avtomonov, E. V.; Li, X.-W.; Lorberth, J. *J. Organoment. Chem.,* **1997**, *530*, 71.
(136) Whitmire, K. H.; Labahn, D.; Roesky, H. W.; Noltemeyer, M.; Sheldrick, G. M. *J. Organomet. Chem.* **1991**, *402*, 55.
(137) Hawley, D. M.; Ferguson, G. *J. Chem. Soc. (A)* **1968**, 2059.
(138) Mootz, D.; Handler, V. *Z. Anorg. Allg. Chem.,* **1986**, *533*, 23 and references cited therein.
(139) Menshutkin, B. N. *Zh. Russ. Fiz. Khim. Ova.,* **1911**, *43*, 1298, 1785; Smith W.; Davis, G. W. *J. Chem. Soc.* **1882**, *41*, 411.
(140) Hill, H. B. *Amer. Chem. J.* **1900**, *24*, 5; Charles, E.; Jones, S.; Kenner, J. *J. Chem. Soc.* **1931**, 1842; Kenner, J.; Morton, G. F. *J. Chem. Soc.* **1934**, 679.
(141) Lütringhaus, A.; Säät, V. *Liebigs. Ann. Chem.* **1939**, *542*, 241.
(142) (a) Lütringhaus, A.; Ambros, D. *Chem. Ber.* **1956**, *89*, 463; (b) Betts, A. W.; Davey, W. *J. Chem. Soc.* **1961**, 3340; (c) Hay, A. S.; Clark, R. F. *Macromolecules,* **1970**, *3*, 533; (d) McManus, M. J.; Berchtold, G. A. *J. Org. Chem.* **1986**, *51*, 2784; (e) Abramov, V. S. *Zh. Obhsch. Khim.* **1952**, *22*, 647. *ibid* **1953**, *47*, 5351; (f) Dana, D. E.; Hay, A. S. *Synthesis,* **1982**, 164; (g) Tafeenko, V. A.; Bogdan, T. V.; Aslanov, L. A. *Zh. Strukt. Khim.* **1994**, *35*, 155.

(143) Nakatsu, K.; Yoshioka, H.; Kunimoto, K.; Kinugasa, T.; Ueji, S. *Acta. Cryst. B* **1978,** *34,* 2347.

(144) Ueji, S.; Nakatsu, K.; Yoshioka, H.; Kinoshita, T. *Tetrahedron Lett.* **1982,** *23,* 1173.

(145) (a) Barton, D. H. R.; Bhatnagar, N. Y.; Blazejewski, J.; Charpiot, B.; Finet, J.; Lester, D. J.; Motherwell, W. B.; Papoula, M. T. B.; Stanforth, S. P. *J. Chem. Soc. Perkin Trans. I* **1985,** 2657; (b) Barton, D. H. R.; Donnelly, D. M. X.; Guiry, P. J.; Reibeuspies, J. H. *J. Chem. Soc., Chem. Commun.* **1990,** 1110.

(146) Vilardo, J. S.; Lockwood, M. A.; Hanson, L. G.; Clark, J. R.; Parkin, B. C.; Fanwick, P. E.; Rothwell, I. P. *J. Chem. Soc. Dalton Trans.* **1997,** 3353.

(147) Vilardo, J. S.; Fanwick, P. E.; Rothwell, I. P. *Polyhedron* **1998,** *17,* 769.

(148) Bishop, P. T.; Dilworth, J. R.; Nicholson, T.; Zubieta, J. *J. Chem. Soc. Dalton Trans.* **1991,** 385. (See also Newman, M. S.; Kaines, H. A. *J. Org. Chem.* **1966,** *31,* 3980, and ref. 50.

(149) Ruhlandt-Senge, K.; Power, P. P. *Bull. Soc. Chim. Fr.* **1992,** *129,* 594.

(150) Ellison, J. J.; Ruhlandt-Senge, K.; Power, P. P. *Angew. Chem. Int. Ed. Engl.* **1994,** *33,* 1178.

(151) Niemeyer, M.; Power, P. P. *Inorg. Chem.* **1996,** *15,* 7264.

(152) (a) Beswick, M. A.; Wright, D. S. *Comprehensive Organometallic Chemistry II,* Vol. 1; Pergamon: New York, 1995; Ch. 1; (b) Pauer, F.; Power, P. P. In *Lithium Chemistry: A Theoretical and Experimental Overview*; Sapse, A. M.; Schleyer, P. V. R., Eds.; Wiley: New York, 1995; Ch. 9.

(153) Ellison, J. J.; Power, P. P. *Inorg. Chem.* **1994,** *33,* 4231.

(154) Banister, A. J.; Clegg, W.; Gill, W. R. *J. Chem. Soc., Chem. Commun.* **1987,** 850.

(155) Niemeyer, M.; Power, P. P. *Inorg. Chim. Acta* **1997,** *263,* 201.

(156) Wolfrum, R.; Sauermann, G. *J. Organomet. Chem.* **1969,** *18,* 27.

(157) Atwood, J. L.; Cummings, J. M. *J. Cryst. Mol. Struct.* **1977,** *7,* 257.

(158) Ruhlandt-Senge, K. *Inorg. Chem.* **1995,** *34,* 3499.

(159) Ellison, J. J.; Ruhlandt-Senge, K.; Hope, H. H.; Power, P. P. *Inorg. Chem.* **1995,** *34,* 49.

(160) Huheey, J. E. *Inorganic Chemistry,* 3rd ed.; Harper and Row: New York, 1983; p. 258.

(161) Back, T. G.; Codding, P. W. *Can. J. Chem.* **1983,** *61,* 2749.

(162) Schultz Lang, E.; Maichle-Mössmer, C.; Strähle J. *Z. Anorg. Allg. Chem.* **1994,** *620,* 1678.

(163) (a) Kruse, F. H.; Marsh, R. E.; McCullough, J. D. *Acta Crystallogr.* **1957,** *10,* 201; (b) Llabres, G.; Dideberg, O.; Dupont, L. *Acta Crystallogr.* **1972,** *B29,* 2438; (c) Van den Bossche, G.; Spirlet, M. R.; Dideberg, O.; Dupont, L. *Acta Crystallogr.* **1979,** *B35,* 1727; (d) Ludlow, S.; McCarthy, A. E. *J. Organomet. Chem.* **1981,** *219,* 169; (e) Van den Bossche, G.; Spirlet, M. R.; Dideberg, O.; Dupont, L. *Acta Crystallogr.* **1984,** *C40,* 1011; (f) du Mont, W. W.; Lange, L.; Karsch, H. H.; Peters, K.; Peters, E. M.; von Schnering, H. G. *Chem. Ber.* **1988,** *121,* 11; (g) Edelmann, A.; Brooker, S.; Bertel, N.; Noltmeyer, M.; Roesky, H. W.; Sheldrick, G. M.; Edelmann, F. T. *Z. Naturforsch.* **1992,** *47b,* 305; (h) Junk, T.; Irgolic, K. J.; Meyers, E. A. *Acta Crystallogr.* **1993,** *C49,* 975.

(164) Hart, H.; Rajakumar, P. *Tetrahedron* **1995,** *51,* 1313.

(165) Vinod, T. K.; Rajakumar, P.; Hart, H. *Tetrahedron* **1995,** *51,* 2267.

(166) Cohen, S. C.; Tomlinson, A. J.; Wiles, M. R.; Massey, A. G. *J. Organomet. Chem.,* **1968,** *11,* 385.

(167) Bowen Jones, J.; Brown, D. S. *Acta Crystallogr.,* **1980,** *B36,* 3189.

(168) Protasiewicz, J. D. *J. Chem. Soc., Chem. Commun.* **1995,** 1115.

ADVANCES IN ORGANOMETALLIC CHEMISTRY, VOL. 44

Hydrogen Donor Abilities of the Group 14 Hydrides

CHRYSSOSTOMOS CHATGILIALOGLU

I.Co.C.E.A.
Consiglio Nazionalle delle Ricerche
40129 Bologna, Italy

MARTIN NEWCOMB

Department of Chemistry
Wayne State University
Detroit, Michigan 48202

I

INTRODUCTION

The reactions of atoms or radicals with silicon hydrides, germanium hydrides, and tin hydrides are the key steps in formation of the metal-centered radicals [Eq. (1)]. Silyl radicals play a strategic role in diverse areas of science, from the production of silicon-containing ceramics to applications in polymers and organic synthesis.[1] Tin hydrides have been widely applied in synthesis in radical chain reactions that were well established decades ago.[2,3] Germanium hydrides have been less commonly employed but provide some attractive features for organic synthesis.

67

$$R\cdot \ + \ R_3M\text{–}H \ \longrightarrow \ R\text{–}H \ + \ R_3M\cdot$$

$$M = Si, Ge, Sn$$

(1)

This review focuses on the kinetics of reactions of the silicon, germanium, and tin hydrides with radicals. In the past two decades, progress in determining the absolute kinetics of radical reactions in general has been rapid. The quantitation of kinetics of radical reactions involving the Group 14 metal hydrides in condensed phase has been particularly noteworthy, progressing from a few absolute rate constants available before 1980 to a considerable body of data we summarize here.

The review is divided into sections according to the type of metal hydride for convenience in discussing the information systematically. At one extreme, kinetic studies have been performed with many types of silicon hydrides, and much of the data can be interpreted in terms of the electronic properties of the silanes imparted by substituents. At the other extreme, kinetic studies of tin hydrides are limited to a few stannanes, but the rate constants of reactions of a wide range of radical types with the archetypal tin hydride, tributylstannane, are available. Kinetic isotope effects for the various hydrides are collected in a short section, and this is followed by a section that compares the kinetics of reactions of silicon, germanium, and tin hydrides.

II

BOND DISSOCIATION ENERGIES AND KINETIC METHODS

Throughout this work, we will refer to the heats of reactions and the methods used for the kinetic determinations. We have collected background information for discussion in this section. A list of important bond dissociation energies is followed by brief descriptions of the more important kinetic methods used for determining radical reaction rate constants discussed in this review.

A. Bond Dissociation Energies

Knowledge of homolytic bond dissociation energies (BDEs) is critically important for understanding radical chemistry. The bond energies of organic compounds have been reviewed extensively, but we will use recommended R–H BDE values for organic compounds given in a recent excellent

TABLE I

BOND DISSOCIATION ENERGIES (BDEs) IN kcal/mol[a]

Molecule	DH^b		Group 14 hydride[c]	D_{rel}	DH	Ref.[d]
CH_4	104.9	± 0.1	Me_3Si-H		95.0 ± 0.5	6
CH_3CH_2-H	101.1	± 0.4	Et_3Si-H	0	95.1	7
$(CH_3)_2CH-H$	98.6	± 0.4	$Me_3SiSi(Me_2)-H$	−4.8	90.3	7
$(CH_3)_3C-H$	96.5	± 0.4	$(MeS)_3Si-H$	−7.6	87.5	9
$H_2C=CHCH_2-H$	88.2	± 2.1	$(Me_3Si)_3Si-H$	−11.1	84.0	7
$PhCH_2-H$	88.5	± 1.5	Bu_3Ge-H	−6.5	88.6	8
$c-C_3H_6$	106[e]		Ph_3Ge-H	−9.0	86.1	8
$H_2C=CH_2$	111.2	± 0.8	Bu_3Sn-H	−16.5	78.6	10
$Ph-H$	111.2	± 0.8				
$HC(O)CH_2-H$	94.3	± 2.2				
$NCCH_3$	94.8	± 2.1				
$HOCH_2-H$	96.06	± 0.15				
$HOC(O)-H$	>89.5					

[a] The values listed are X–H bond dissociation enthalpies; DH_{298}.

[b] Values from Ref. 4 unless noted.

[c] See text for method.

[d] Reference for experimental D_{rel} values; with the exception of Me_3SiH, all DH_{298} values for group 14 hydrides were calculated in this work.

[e] Value from Ref. 5 adjusted to fit scale of Ref. 4.

compilation (Table I).[4] It is noteworthy that, for saturated hydrocarbons, the simple alkyl C–H bond energies in that work differ little from those in a classic listing of BDE values.[5] Most of the BDE values for simple organic compounds are known with precisions of 0.5–2 kcal/mol.

Accurate and precise bond energies of the group 14 hydrides are more difficult to find. On the basis of the kinetics of the reaction $Me_3Si + HBr = Me_3SiH + Br$, Marshall and co-workers obtained the bond dissociation enthalpy $D°(Me_3Si-H)$ with high precision (Table I).[6] Absolute bond energies for other substituted silicon, germanium, and tin hydrides are poor with uncertainties that derive mainly from the experimental procedures. However, studies involving photoacoustic calorimetry[7-10] have resulted in apparently precise relative bond energies for a representative series of group 14 hydrides that can be placed on an absolute scale with some confidence.

The photoacoustic calorimetry technique employs photolysis by laser pulses of a mixture containing di-*tert*-butyl peroxide, an appropriate metal hydride, and solvent. Photolysis of the peroxide gives t-BuO· radicals that abstract a hydrogen atom from the hydride, and the measured photoacoustic signal is proportional to the overall reaction enthalpy. After calibration, and using literature values for the heats of formation of some species, $D°(R_3M-H)$ can be derived. However, studies have shown that the effect of ignoring the small volume changes that occur in the reactions leads to an underestimation of the bond dissociation enthalpy by 2.1 kcal/mol.[11] Furthermore, unknown correction terms may be needed to derive values that are basically bond dissociation enthalpies in the gas phase. The relative data obtained by this technique should be reliable, however, because the uncertainties due to the solvent are canceled, and the relative errors should be ±1 kcal/mol.

Table I includes the relative bond dissociation enthalpies obtained for some group 14 hydrides by photoacoustic calorimetry.[7-10] The data demonstrate that, for the trialkyl-substituted series, the bond strengths decrease by 6.5 and 16.5 kcal/mol on going from silane to germane and to stannane, respectively. The silicon–hydrogen bonds can be dramatically weakened by successive substitution of the Me_3Si group at the Si–H functionality. A substantial decrease in the bond strength is also observed by replacing alkyl with methylthio groups.

The relative bond energies from the photoacoustic calorimetry studies can be placed on an absolute scale in the following manner. On the basis of thermodynamic data, an approximate value of $D°(Me_3SiSiMe_2-H) =$ 90.3 kcal/mol can be calculated.[12,13] Using this value, we have converted the relative bond energy values to absolute $D°$ values that are also listed in Table I. The agreement in the Si–H BDE values for Me_3SiH and Et_3SiH is gratifying and suggests that our approach for estimating the group 14 BDE values is reliable.

B. *Radical Kinetic Methods*

Rate constants can be determined directly or by indirect techniques. The latter approach involves competition between the reaction with an unknown rate constant and another reaction (the basis reaction) with a known rate constant. Indirect methods can be highly precise, but at some point the kinetics must be placed on an absolute scale by comparison to directly measured rate constants. Therefore, indirect methods incorporate the absolute errors of the calibrated basis reaction as well as the random errors of the competition study and those of any preceding competition studies used in calibration of the basis reaction.

Most radicals react in self- or cross-termination reactions with another radical in diffusion-limited processes. Therefore, most of the useful radical–molecule reactions are very fast. The lifetimes of the radicals are typically in the microsecond range, and photochemical generation of radicals is usually required for direct kinetic studies. Continuous or pulsed irradiation from a high-intensity UV lamp or laser can be employed. Pulse radiolysis is often used for studies in aqueous solutions.

Steady-State ESR Spectroscopy. In this method, the ESR spectrum is monitored, usually with constant sample irradiation. Radical concentrations reach a steady state, and the rate constants for termination can be determined from the radical concentrations if the rate constants for radical formation are known. Estimates of the rate of formation of radicals are often a source of large errors. The technique often is applied at low temperatures so that diffusion is slowed and adequate radical concentrations are achieved, and extrapolation of the results to ambient temperatures can give large errors.

When radical **A·** reacts to form product radical **B·** with an appropriate rate constant, the absolute concentrations of each radical can be determined in a steady-state ESR experiment. This ratio and a measured or calculated rate of destruction of **A·** and **B·** by diffusion-controlled radical–radical reactions can be used to calculate the rate constant for formation of **B·** from **A·**.

Kinetic ESR Spectroscopy.[14] In principle, any method that permits quantitation of the concentrations of radicals with time can be applied for kinetic studies. This method employs an ESR spectrometer to monitor radicals following generation by a pulsed energy source. The sensitivity of the ESR and ease of quantitation of radical concentrations coupled with the ability to identify radicals unequivocally by ESR are major attractions of the method. For fast reactions, light can be pulsed with a chopper, and ESR signal intensity versus time following the pulse can be averaged. In recent applications, lasers have been employed as the light source.

Laser Flash Photolysis (LFP). Conventional absorbance spectroscopy became a premier method for measuring radical kinetics in condensed phase when coupled with flash photolysis from high-energy lasers with short pulse durations. The drawbacks for kinetic absorbance spectroscopy involved poor sensitivity due to small concentrations of radicals and problems with radical identification and quantitation. Relatively high concentrations of radicals are achieved by LFP, and accumulated spectroscopic information for a number of radical types is now available.[15] In direct LFP, formation or decay of signals from the radical are monitored, usually by UV-vis spectroscopy. The most accurate results are obtained when observed rate constants are in the range of 10^5–10^7 s^{-1}. A lower limit of about 1×10^4 s^{-1} usually exists because radical–radical reactions and reactions of radicals with residual oxygen in the degassed solvent occur with pseudo-first-order rate constants approaching this value.

In optimal situations, LFP rate constants for first-order processes can be measured with precisions as good as 1% at the 95% confidence interval. Bimolecular processes are conveniently studied by LFP under pseudo-first-order conditions, and precisions in the rate constants typically are better than 10%. Under pseudo-first-order conditions, the observed rate constant for formation or decay of a radical will be $k_{obs} = k_0 + k_X[X]$, where X is the species reacting with the radical in a bimolecular process. The reaction is conducted with varying concentrations of X, and a plot of k_{obs} vs $[X]$ has a slope of k_X.

LFP-Probe Method. In cases where the radicals of interest do not contain a useful chromophore, the LFP technique can be modified by incorporation of a probe radical reaction that gives a product with a chromophore. The probe reaction can be unimolecular or bimolecular, a constant concentration of probe reagent is employed in the latter case. Formation of the detectable species occurs with an observed first-order or pseudo-first-order rate constant equal to k_0. In the presence of another reagent X that reacts with the original radical, the rate constant for formation of detectable species is $k_{obs} = k_0 + k_X[X]$, and the bimolecular rate constant is determined (as before) by conducting the reaction at varying concentrations of X. Note that the LFP-probe technique is a direct method even though the reactant or product of interest is not monitored.

Competition Kinetic Method. Indirect, competition kinetic methods have become increasingly popular as the number of calibrated radical reactions has increased and the precision of the rate constants has improved.[16] Any radical reaction with a known rate constant can serve as a basis reaction in a competition kinetic study.

Scheme 1 illustrates the design of an experiment that could be used to determine the rate constant for H-atom abstraction from a group 14 hydride. Radical $A \cdot$ reacts with the hydride to give product $A–H$. In competition with this reaction, radical $A \cdot$ gives radical $B \cdot$ in a unimolecular or bimolecular reaction with a known rate constant, and product radical $B \cdot$ also reacts with the hydride, giving $B–H$. The rate constant for reaction of $A \cdot$ with the metal hydride can be determined from the product distribution, the known rate constant for conversion of $A \cdot$ to $B \cdot$, and the concentrations

$$A\cdot \; + \; R_3M–H \; \xrightarrow{\; k_H \;} \; A–H \; + \; R_3M\cdot$$

$$A\cdot \; + \; B–X \; \xrightarrow{\; k_X \;} \; A–X \; + \; B\cdot$$

$$B\cdot \; + \; R_3M–H \; \longrightarrow \; B–H \; + \; R_3M\cdot$$

SCHEME 1.

of reagents. Kinetic expressions for a number of different competition experiments have been catalogued.[16]

Radical Clocks. Calibrated unimolecular radical reactions (cyclizations, fragmentations, or rearrangements) applied as basis reactions in competition kinetic studies are often referred to as "radical clocks."[16,17] The application of a radical clock is attractive because the experiment is simplified in terms of the number of reagents necessary and the kinetic invariance provided by the unimolecular reaction. A radical clock reaction competing with a bimolecular process conducted under pseudo-first-order conditions (excess reagent) is especially simple to analyze. Another advantage of radical clocks is that alkyl radical cyclizations and ring openings have been shown to be insensitive to solvent polarity effects.[18–21] However, the kinetics of heteroatom-substituted radical reactions such as decarbonylation of acyl radicals and fragmentations of ester-substituted radicals are affected by changes in solvent polarity.[20–22]

A major source of error in any indirect method is inaccuracy of the basis rate constants. Errors can result from determinations of rate constants by a sequence of several indirect studies or by an unanticipated solvent effect on the kinetics of a basis reaction. An error can also result in calibration of a radical clock if the requisite assumption that the clock radical will react with a rate constant equal to that of a simple model radical is not correct. Nevertheless, indirect methods in general, and radical clock studies in particular, have been the workhorse of radical kinetic determinations.

LFP-Clock Method. In this method, rate constants for the radical clock reactions are measured directly by LFP, and the clocks are used in conventional competition kinetic studies for the determination of second-order rate constants. The advantages are that the clock can be calibrated with good accuracy and precision in the solvent of interest, and light-absorbing reagents can be studied in the competition reactions. The method is especially useful when limited kinetic information is available for a class of radicals.

III

REACTIONS OF RADICALS WITH SILICON HYDRIDES

Although we will deal with organic radicals in solution, it is worth mentioning that the reactivity of atoms and small organic radicals with silanes in the gas phase has been studied extensively. For example, the bond dissociation energies of a variety of Si–H bonds are based on the reaction of iodine or bromine with the corresponding silanes.[1]

A. *Carbon-Centered Radicals*

The reaction of carbon-centered radicals with silicon hydrides is of great importance in chemical transformations under reducing conditions where an appropriate silane is either the reducing agent or the mediator for the formation of new bonds.[23]

The kinetic data for reactions of carbon-centered radicals with various silanes and the silanthrane derivatives **1–6** are numerous as shown in

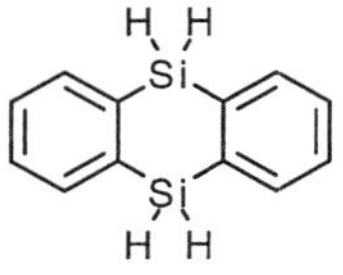

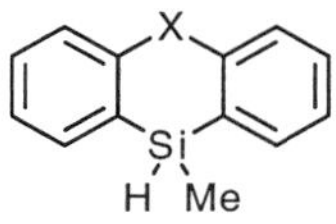

2: X = SiMe$_2$
3: X = SiMe(SiMe$_3$)
4: X = SnMe$_2$
5: X = O
6: X = S

Table II. Most of the data was obtained from radical clock studies. The neophyl radical rearrangement[24] [Eq. (2)] was used for the majority of the kinetic data in Table II, but the ring expansion rearrangement reactions[25–27] of radicals **7** and **8**, cyclizations of 5-hexenyl type radicals,

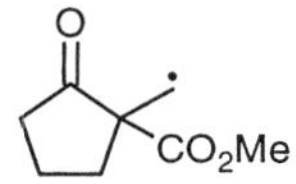

7 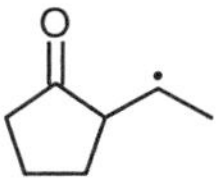   **8**

and other radical rearrangements[16] and trapping reactions also were employed.[28–36] The rate constants for the hydrogen atom abstraction from (Me$_3$Si)$_3$SiH by acyl radicals were measured by using competing decarbonylation reactions as the radical clocks [Eq. (3)]; the reference data was the Arrhenius parameters for the α-cleavage reaction of the propanoyl radical in the gas phase.[37,38] It is noteworthy that the hydrogen donation to acyl radicals R$_3$CC($\cdot$)O is essentially independent of the number of alkyl substituents R; that is, acyl radical precursors to 1°, 2°, and 3° alkyl radicals reacted with R$_3$MH with the same rate constants. Hydrogen atom abstraction from Ph$_2$SiH$_2$ or Ph$_3$SiH and chlorine atom abstraction from CCl$_4$ are the competing bimolecular processes that were used for determining rate constants for phenyl radicals[39]; the reference rate constants were obtained by LFP.[40]

(2)

(continues)

TABLE II

Rate Constants for Reactions of Carbon-Centered Radicals with Silicon Hydrides

Silane	Radical	Solvent	Rate constant (M^{-1} s^{-1}) (temp in °C as subscript)	Rate expression (log k) ($\theta = 2.3RT$ kcal/mol)	Ref.
Et_3SiH	$PhC(Me)_2CH_2\cdot$	$m\text{-}(t\text{-Bu})_2C_6H_4$	$k_{27} = 6.4 \times 10^2$	$8.7\text{-}8.0/\theta^a$	28
	$CH_2{=}CHCH_2C(\cdot)Me_2$	C_6H_6	$k_{50} = 3 \times 10^{3b}$		30
$PhSiH_3$	$PhC(Me)_2CH_2\cdot$	$m\text{-}(t\text{-Bu})_2C_6H_4$	$k_{110} = 2.9 \times 10^{4a}$		29
$PhSi(H)Me_2$	$CH_3(CH_2)_{14}CH_2\cdot$	n-Heptane	$k_{90} = 1.1 \times 10^{4c}$		31
$2\text{-MeOC}_6H_4Si(H)Me_2$	$PhC(Me)_2CH_2\cdot$	C_6H_6	$k_{80} = 1.1 \times 10^{4a}$		32
Ph_2SiH_2	$PhC(Me)_2CH_2\cdot$	$m\text{-}(t\text{-Bu})_2C_6H_4$	$k_{110} = 5.6 \times 10^{4a}$		29
	$C_6H_5\cdot$	CCl_4	$k_{60} = 3.4 \times 10^{7d}$		39, 40
1	$RCH_2\cdot$ (**7**)	C_6H_6	$k_{80} = 2.1 \times 10^{6e}$		27
	$R_2CH\cdot$ (**8**)	C_6H_6	$k_{100} = 2.6 \times 10^{5e}$		33
$Ph_2Si(H)Me$	$CH_3(CH_2)_{14}CH_2\cdot$	n-Heptane	$k_{90} = 1.9 \times 10^{4c}$		31
2	$PhC(Me)_2CH_2\cdot$	C_6H_6	$k_{80} = 0.8 \times 10^{4a}$		32
3	$PhC(Me)_2CH_2\cdot$	C_6H_6	$k_{80} = 4.5 \times 10^{4a}$		32
4	$PhC(Me)_2CH_2\cdot$	C_6H_6	$k_{80} = 4.0 \times 10^{4a}$		32
5	$PhC(Me)_2CH_2\cdot$	C_6H_6	$k_{80} = 1.7 \times 10^{4a}$		32
6	$PhC(Me)_2CH_2\cdot$	C_6H_6	$k_{80} = 2.2 \times 10^{4a}$		32
	$PhC(Me)_2CH_2\cdot$	C_6H_6	$k_{80} = 8.7 \times 10^{4a}$		32
Ph_3SiH	$PhC(Me)_2CH_2\cdot$	$m\text{-}(t\text{-Bu})_2C_6H_4$	$k_{110} = 4.6 \times 10^{4a}$	$8.7\text{-}7.0/\theta^f$	29, 31
	$CH_2{=}CHCH_2C(\cdot)Me_2$	C_6H_6	$k_{50} = 9 \times 10^{3b}$		30
	$C_6H_5\cdot$	CCl_4	$k_{60} = 2.1 \times 10^{7d}$		39, 40

TABLE II (*continued*)

Silane	Radical	Solvent	Rate constant ($M^{-1} s^{-1}$) (temp in °C as subscript)	Rate expression (log k) ($\theta = 2.3RT$ kcal/mol)	Ref.
$(MeS)_3SiH$	$RCH_2\cdot$ (7)	C_6H_6	$k_{80} = 3.9 \times 10^5$ [e]		27
$(i\text{-}PrS)_3SiH$	$RCH_2\cdot$ (7)	C_6H_6	$k_{80} = 3.7 \times 10^5$ [e]		27
$Me_3SiSi(H)Me_2$	$PhC(Me)_2CH_2\cdot$	$m\text{-}(t\text{-}Bu)_2C_6H_4$	$k_{27} = 3.5 \times 10^3$	$9.0-7.5/\theta$ [a]	29
$Me_3SiSiMe_2Si(H)Me_2$	$PhC(Me)_2CH_2\cdot$	$m\text{-}(t\text{-}Bu)_2C_6H_4$	$k_{27} = 5.5 \times 10^3$	$9.0-7.2/\theta$ [a]	29
$(Me_3Si)_2Si(H)Me$	$PhC(Me)_2CH_2\cdot$	$t\text{-}Bu\text{-}C_6H_5$	$k_{27} = 3.2 \times 10^4$	$8.9-6.0/\theta$ [a]	34
$H(PhSiH)_nH$	$RCH_2\cdot$ [g]	C_6H_6	$k_{85} \approx 10^5$ [h]		35
$(Me_3Si)_3SiH$	$CH_2=CH(CH_2)_4CH_2\cdot$	$n\text{-}Octane$	$k_{27} = 3.8 \times 10^5$	$8.9-4.5/\theta$ [i]	36
	$CH_2=CH(CH_2)_4C(\cdot)HMe$	$n\text{-}Tetradecane$	$k_{27} = 1.4 \times 10^5$	$8.3-4.3/\theta$ [i]	36
	$CH_2=CH(CH_2)_4C(\cdot)Me_2$	$n\text{-}Tetradecane$	$k_{27} = 2.6 \times 10^5$	$7.9-3.4/\theta$ [i]	36
	$R_3CC(\cdot)O$	$Me\text{-}C_6H_5$	$k_{27} = 1.8 \times 10^4$	$8.2-5.4/\theta$ [k]	37, 38
	Ph·		$k_{20} = 3 \times 10^8$ [l]		36

[a] Depends on literature values for neophyl rearrangement.

[b] Approximate value; depends on the rate constant for reaction of the radical with *N*-hydroxypyridine-2-thione ester.

[c] Depends on literature value for primary alkyl radical recombination.

[d] Calculated using the relative kinetic data with Cl_4C at 60°C and the absolute rate constant for reaction of $C_6H_5\cdot$ with Cl_4C.

[e] Depends on literature values for ring expansion.

[f] A rate constant of 3.0×10^4 M^{-1} s^{-1} at 90°C has been determined by an independent method (Ref. 31). The Arrhenius expression is derived by assuming the log A and taking both experimental values.

[g] 5-Hexenyl or neophyl radical.

[h] Lower (5×10^4 M^{-1} s^{-1}) and higher (6×10^5 M^{-1} s^{-1}) limit values for each SiH moiety of $H(PhSiH)_nH$ have been obtained.

[i] Depends on literature values for 5-hexenyl-type rearrangements.

[k] R = alkyl or H; depends on literature Arrhenius parameters for decarbonylation of the propanoyl radical in the gas phase.

[l] Derived in this work; see text.

$$R_3C \overset{\overset{\displaystyle O}{\|}}{\cdot} \longrightarrow R \overset{\overset{\displaystyle R}{|}}{\underset{R}{\cdot}} + C\equiv O \qquad (3)$$

R = H or alkyl

The Arrhenius expression for the reaction of the *o*-(allyloxy)phenyl radical (**9**) with (Me$_3$Si)$_3$SiH relative to this unimolecular rearrangement [Eq. (4)] has been measured, *viz.*, $\log(k_C/k_H)$ (M) $= 2.6 - 1.6/\theta$.[36] When the competition study was performed, however, reliable absolute rate constants for the cyclization of radical **9** to radical **10** were not available, although a

$$\qquad (4)$$

9   1 0

relative Arrhenius function for cyclization of **9** and reaction with Bu$_3$SnH was known.[41] In Section V, we will discuss reactions of aryl radicals with Bu$_3$SnH, but the important points for the analysis of the silane kinetics are the following. Even without calibration of radical clock **9**, one can subtract the relative Arrhenius function for reaction of Bu$_3$SnH with **9** from that for reaction of (Me$_3$Si)$_3$SiH with **9** to obtain a relative rate expression for reactions of the two H-atom transfer reagents with the aryl radical. Specifically, $\log(k_{SnH}/k_{SiH}) = 0.5 - 0.1/2.3RT$ (kcal/mol), or Bu$_3$SnH reacts 2.7 times faster with aryl radical **9** at ambient temperature than does (Me$_3$Si)$_3$SiH. If one assumes that this relationship holds for any aryl radical, then the rate constant that has been determined[42] for reaction of Bu$_3$SnH with "Ph·" radical (see later discussion) can be used to give an approximate rate constant for reaction of (Me$_3$Si)$_3$SiH with the phenyl radical of 3×10^8 M^{-1} s^{-1} at ambient temperature.

The kinetic data for halogenated carbon-centered radicals with silicon hydrides are also numerous, as shown in Table III. Gas phase reactions of F$_3$C· with Me$_3$SiH have been reported.[43,44] The rate constants for the trichloromethyl radical were obtained by two different competitive approaches. For Me$_3$SiH and Et$_3$SiH, reaction of cyclohexane with the Cl$_3$C· radical was used as the reference; the relative rate constants were combined with previously available Arrhenius parameters in the liquid phase of the reference systems.[45] For the other silanes, the relative rates of hydrogen atom abstractions by the Cl$_3$C· radical[46] were combined with the rate constant for Et$_3$SiH.[45] The rate constant at 77°C for the reaction of Cl$_3$SiH with the Cl$_3$C· radical in CCl$_4$ obtained from the preceding method is identical to that calculated from the Arrhenius parameters for the same reaction in the gas phase.[47] The kinetic data for perfluoro-*n*-alkyl radicals

TABLE III

RATE CONSTANTS FOR REACTIONS OF HALOGENATED CARBON-CENTERED RADICALS WITH SILICON HYDRIDES

Silane	Radical	Solvent	Rate constant (M^{-1} s^{-1}) (temp in °C as subscript)	Rate expression ($\theta = 2.3RT$ kcal/mol)	Ref.
Me₃SiH	F₃C·	Gas	$k_{27} = 1.7 \times 10^5$	$9.3-5.6/\theta^a$	43, 44
	Cl₃C·	CCl₄/c-C₆H₁₂	$k_{27} = 1.5 \times 10^2$	$8.5-8.7/\theta^b$	45
Et₃SiH	n-C₈F₁₇·	Et₃SiH	$k_{30} = 7.5 \times 10^{5\,c,d}$		48, 49
	n-C₄F₉·	1,3-(CF₃)₂C₆H₄	$k_{25} = 5.0 \times 10^{5\,c}$		50
	(CF₃)₂CF·	Et₃SiH	$k_{26} = 3.6 \times 10^{6\,e}$		52
	(CF₃)₃C·	Et₃SiH	$k_{26} = 2.4 \times 10^{8\,e}$		52
	Cl₃C·	CCl₄/c-C₆H₁₂	$k_{27} = 5.0 \times 10^2$	$8.6-8.1/\theta^b$	45
(CF₃CH₂CH₂)₃SiH	n-C₄F₉·	(Not specified)	$k_{25} = 2.1 \times 10^{5\,c}$		51
t-BuSi(H)Me₂	n-C₄F₉·	(Not specified)	$k_{25} = 4.9 \times 10^{4\,c}$		51
PhSi(H)Me₂	Cl₃C·	CCl₄	$k_{77} = 2.9 \times 10^{3\,f}$		46
Ph₂SiH₂	Cl₃C·	CCl₄	$k_{77} = 2.9 \times 10^{3\,f}$		46
Ph₂Si(H)Me	Cl₃C·	CCl₄	$k_{77} = 3.5 \times 10^{3\,f}$		46
Ph₃SiH	Cl₃C·	CCl₄	$k_{77} = 5.1 \times 10^{3\,f}$		46
Ph₃SiSi(H)PhMe	Cl₃C·	CCl₄	$k_{77} = 2.5 \times 10^{4\,f}$		46
(Me₃Si)Si(H)Me₂	n-C₄F₉·	(Not specified)	$k_{25} = 3.1 \times 10^{6\,c}$		51
(Me₃Si)₂Si(H)Me	n-C₈F₁₇·	C₆D₆	$k_{30} = 1.6 \times 10^{7\,c}$		48, 49
(Me₃Si)₃SiH	n-C₈F₁₇·	C₆D₆	$k_{30} = 5.1 \times 10^{7\,c}$		48, 49
Cl₃SiH	F₃C·	Gas	$k_{27} = 1.9 \times 10^4$	$8.8-6.2/\theta^a$	44
	Cl₃C·	CCl₄	$k_{77} = 3.5 \times 10^{2\,f}$	$8.6-9.7/\theta^{a,g}$	46, 47

[a] Depends on literature Arrhenius parameters for the X₃C· radical recombination.

[b] Depends on literature Arrhenius parameters for Cl₃C· radical with cyclohexane.

[c] Depends on the rate constant for the addition of perfluoro-n-alkyl radical to 1-hexene. For reactions of C₈F₁₇·, the original literature report stated that the radical was perfluoroheptyl; we are grateful to Prof. W. R. Dolbier, Jr., for the correction.

[d] Reference 50 claims this value is suspect because of experimental difficulties.

[e] Depends on the rate constant for the addition of the radical to pentafluorostyrene.

[f] Calculated using the relative kinetic data with Et₃SiH and the rate constant of Cl₃C· with Et₃SiH from this table.

[g] Arrhenius expression in gas phase.

were obtained by competition of the appropriate silane with the addition to 1-hexene,[48–51] whereas the rate constants of perfluoroisopropyl and per-fluoro-*tert*-butyl radicals toward Et_3SiH were obtained by competition with the addition to pentafluorostyrene.[52]

Mechanistic studies have shown that the attack of primary alkyl radicals on Et_3SiH occurs in about 60% of the cases at the SiH moiety and in 40% at the ethyl groups at 403 K.[28] H-atom abstraction from the ethyl groups in reaction of a perfluoroalkyl radical with Et_3SiH amounted to about 4% of the total reactions.[49] In the case of $(Me_3Si)_3SiH$, the attack on the Si–H bond and the methyl groups occurs with about 95 : 5 regioselectivity favoring abstraction of the Si–H.[53] Such side-chain reactions do not occur with tin hydrides; see Section VI.

The trends in reactivity for H-atom abstraction are the following: (1) For a primary alkyl radical, the rate constants increase along the series $Et_3SiH < Ph_3SiH < (MeS)_3SiH < (Me_3Si)_3SiH$ with the expected intermediate values for silanes having mixed substituents. Similar behavior is envisaged with other carbon-centered radicals. (2) For triethylsilane, the rate constants decrease along the series $(CF_3)_3C\cdot > (CF_3)_2CF\cdot > R_fCF_2\cdot > RCH_2\cdot > R_3C\cdot > Cl_3C\cdot$ and cover six orders of magnitude. However, trends in which the rate constants decrease along the series $R_fCF_2\cdot > C_6H_5\cdot > RCH_2\cdot \geq R_2CH\cdot \geq R_3C\cdot > RC(\cdot)O$ are also delineated for other silanes. The preexponential factors all lie in the expected range for a bimolecular reaction, and the majority fit within the range $8.5 < \log A < 9.0$. The anticipated decrease of the A factor with increasing steric hindrance of the attacking radical is also observed. The activation energy is clearly the major factor in determining the radical–silane reactivity. The trends just outlined can be entirely attributed to more favorable thermodynamic factors along the series. However, polarized transition states, such as those of canonical structures **11** and **12**, have been suggested to explain the reactivity of the $Cl_3C\cdot$ radical[54] and perfluoroalkyl radicals,[52] respectively.

$$\overset{\ominus}{Cl_3C}\cdots\overset{\oplus}{H}\cdots\cdot SiR_3 \qquad\qquad \overset{\delta-}{R_f}\cdots\overset{\delta+}{H}\cdots\cdot SiR_3$$

11   **12**

Charge development in the transition states for reactions of $Cl_3C\cdot$ radicals with *m*- and *p*-substituted phenylsilanes **13** was evaluated in Hammett studies; the relative rate constants correlate with σ values with $\rho \approx -0.5$.[46,54] Although a canonical structure such as **11** could provide an explanation of this behavior, the relatively small magnitude of ρ values could simply reflect differences in bond strengths of the substrates in the series,[55] i.e., if electron-withdrawing substituents also strengthen the silicon–hydrogen bond in **13**,

$R = R' = H$
$R = H, R' = Me$
$R = R' = Me$
$R = Me, R' = \alpha\text{-Np}$

13

then a Hammett correlation with a negative ρ value is expected irrespective of the abstracting species.

Phenyl substitution on the silicon, whether single or multiple, has only a small effect on the rate constant for H-atom abstraction from a silane, in contrast with the behavior observed with the carbon analogues. For example, the rate constants increase by a factor of 1–2 for each substitution of a methyl group by a phenyl group along the series $Me_3SiH < PhSi(H)Me_2 < Ph_2Si(H)Me < Ph_3SiH$ for reactions with $RCH_2\cdot$ and $Cl_3C\cdot$ radicals, the effects being cumulative. On the other hand, the rate constants for H-atom transfer increase along the series $PhSiH_3 < Ph_3SiH < Ph_2SiH_2$ for reaction with the neophyl radical; however, taking into account the statistical number of hydrogens abstracted changes the order to that expected: $PhSiH_3 < Ph_2SiH_2 < Ph_3SiH$. The lack of resonance stabilization by phenyl groups on the silicon-centered radical has been attributed to the larger size of silicon (compared to carbon) and to the pyramidal nature of the radical center.[56] The silanthrane derivatives **1–6** are more reactive than the corresponding diphenylsilanes (Table I). For example, the 5,10-dihydrosilanthrene (**1**) is an order of magnitude more reactive than Ph_2SiH_2 toward primary alkyl radicals (taking into account the statistical number of hydrogens abstracted). The enhancement in the reactivity of the silanthrane derivatives is probably due to stabilization of the silyl radical induced either by a transannular interaction of the vicinal Si substituent or by the quasiplanar arrangement of the radical center.[27,32,33,57] Ph_3SiD was used as a reference for the deuterium donation in order to obtain relative rate constants for the hydrogen transfer to the benzyl radical from 23 compounds; the values range from 0.12 for mesitylene to 37 for 9,10-dihydroanthracene.[58]

Table II shows that for a particular radical the rate constants increase substantially with successive substitutions of alkyl or phenyl groups on silicon by thiyl or silyl groups. In particular, for primary alkyl radicals for which more data are available, replacement of a methyl group with a Me_3Si group results in a rate increase of an order of magnitude, and the effect is cumulative. Similar behavior is observed for the $Cl_3C\cdot$ radical and the $n\text{-}C_8F_{17}\cdot$ radical (Table III). These results are in good agreement with the thermodynamic data for the respective silanes, which show a weakening of the Si–H bond strength upon replacement of an alkyl substituent with a thiyl or silyl group. The rate constants for reactions of primary, secondary, and tertiary alkyl radicals with $(Me_3Si)_3SiH$ are quite similar in the range of temperatures that are useful for chemical transformation in the liquid phase. This is due to a compensation of entropic and enthalpic effects through this series of alkyl radicals (Table II).

Intramolecular hydrogen abstraction by primary alkyl radicals from the Si–H moiety has been reported as a key step in several unimolecular chain transfer reactions.[59,60] In particular, the 1,5-hydrogen transfer of radicals **14–17** [Eq. (5)], generated from the corresponding iodides, was studied in

$$(5)$$

14: R = t-Bu, X = O, Y = H
15: R = t-Bu, X = O, Y = Ph
16: R = t-Bu, X = CH$_2$, Y = H
17: R = Ph, X = CH$_2$, Y = H

competition with the addition of the radicals to allyltributylstannane, and approximate rate constants for the hydrogen transfer were obtained. Values at 80°C are 0.4×10^4, 1×10^4, 1×10^4, and 2×10^4 for **14–17**, respectively. Using the rate constants for reactions of the neophyl radical with the appropriate silanes as standards, effective molarities of 1–2 M are calculated.

B. *Nitrogen-Centered Radicals*

The kinetic data for reactions of nitrogen-centered radicals with silicon hydrides is limited to rate constants for piperidinyl radical **18** (Table IV) by using ESR spectroscopy.[61] The two remarkable features of the data are

18

the low preexponential factors and the order of reactivity of phenyl-substituted derivatives in which the rate constants increase along the series $Ph_3SiH < Ph_2SiH_2 < PhSiH_3$ (taking into account the statistical number of reactive hydrogen atoms in each silane). This behavior, which is opposite to that observed in the analogous reactions of alkyl and alkoxyl radicals (see later discussion), was explained in terms of steric effects in the transition state that demand a very specific orientation of the abstracting aminyl radical and the hydrogen–silicon bond in the silane.[61]

TABLE IV

RATE CONSTANTS FOR REACTIONS OF NITROGEN-CENTERED RADICALS WITH SILICON HYDRIDES

Silane	Radical	Solvent	Rate constant (M^{-1} s^{-1}) (temp in °C as subscript)	Rate expression ($\theta = 2.3RT$ kcal/mol)	Ref.
Et_3SiH	**18**	t-Bu-C_6H_5	$k_{25} = 2.7$	$5.2-6.5/\theta$	61
$PhSiH_3$	**18**	t-Bu-C_6H_5	$k_{25} = 119.2$	$6.9-6.6/\theta$	61
Ph_2SiH_2	**18**	t-Bu-C_6H_5	$k_{25} = 32.1$	$5.8-5.8/\theta$	61
Ph_3SiH	**18**	t-Bu-C_6H_5	$k_{25} = 5.3$	$5.7-6.7/\theta$	61
$Me_3SiSi(H)Me_2$	**18**	t-Bu-C_6H_5	$k_{25} = 10.0$	$4.8-4.4/\theta$	61
$(Me_3Si)_3SiH$	**18**	t-Bu-C_6H_5	$k_{25} = 34.4$	$4.8-5.2/\theta$	61

C. *Oxygen-Centered Radicals*

The reaction of thermally and photochemically generated *tert*-butoxyl radicals with trisubstituted silanes [Eqs. (6) and (7)] has been used extensively for the generation of silyl radicals in ESR studies, in time-resolved optical techniques, and in organic synthesis. Absolute rate constants for reaction (7) were measured directly by LFP techniques,[56,62,63] whereas the gas phase kinetic values for reactions of Me$_3$SiH were obtained by competition with decomposition of the *tert*-butoxyl radical.[64,65]

$$t\text{-BuO–OBu-}t \xrightarrow{\ h\nu \text{ or } \Delta\ } 2\ t\text{-BuO•} \qquad (6)$$

$$t\text{-BuO•} + R^1R^2R^3SiH \longrightarrow t\text{-BuOH} + R^1R^2R^3Si• \qquad (7)$$

Rate constants and activation parameters for reaction (7) in the liquid phase are collected in Table V. These values reflect the overall (or molecular) reactivity of the substrates regardless of the site or mechanism of the reaction. Mechanistic studies have shown that reaction of the *t*-BuO· radical with Et$_3$SiH at 300 K involves H-atom abstraction from both the Si–H moiety and the ethyl groups with about 80 : 20 regioselectivity favoring Si–H abstraction,[56] whereas the reaction of the alkoxyl radical with (Me$_3$Si)$_3$SiH at 300 K occurs with 95 : 5 selectivity favoring Si–H abstraction over abstraction of hydrogen from the methyl groups.[66] The Arrhenius parameters for the gas-phase reaction of Me$_3$SiH with the alkoxyl radical are also included in Table V. An initial discrepancy between the activation energies for the reactions of *t*-BuO· radicals with Et$_3$SiH[56] and with Me$_3$SiH,[64] e.g., 2.6 vs 3.7 kcal/mol, was attributed to small temperature changes or to inaccuracies in the reference reaction in the gas phase experiments.[67] However, the gas-phase reaction was later studied under different experimental conditions, and the revised activation energy was found to be 2.1 kcal/mol.[65]

The rate constants for reaction of *t*-BuO· radical with silanes increase along the series Et$_3$SiH < Ph$_3$SiH < (MeS)$_3$SiH < (Me$_3$Si)$_3$SiH with the expected intermediate values for silanes having mixed substituents. In particular, the rate constants decrease along the two series PhSiH$_3$ < Ph$_2$SiH$_2$ < Ph$_3$SiH and Me$_3$SiH < Me$_3$SiSi(H)Me$_2$ < (Me$_3$Si)$_2$Si(H)Me < (Me$_3$Si)$_3$SiH. The available preexponential factors lie in the expected range and, therefore, the activation energy is expected to be the major factor in determining the reactivity between the *t*-BuO· radical and the silanes. As in the case of carbon-centered radicals, the trends just outlined can be entirely attributed to more favorable thermodynamic factors along the series.

Absolute rate constants for the reaction of Et$_3$SiH with aroyloxyl radicals, *p*-X-C$_6$H$_4$C(O)O·, where X = MeO, Me, H, and Cl, were measured by

TABLE V
RATE CONSTANTS FOR REACTIONS OF OXYGEN-CENTERED RADICALS WITH SILICON HYDRIDES

Silane	Radical	Solvent	Rate constant (M^{-1} s^{-1}) (temp in °C as subscript)	Rate expression ($\theta = 2.3RT$ kcal/mol)	Ref.
n-$C_5H_{11}SiH_3$	t-BuO·	$(t\text{-BuO})_2/C_6H_6$[a]	$k_{27} = 1.1 \times 10^7$		56
Me_3SiH	t-BuO·	Gas	$k_{27} = 9.3 \times 10^6$	$8.5–2.1/\theta$[c]	65
Et_3SiH	t-BuO·	$(t\text{-BuO})_2/C_6H_6$[a]	$k_{27} = 5.7 \times 10^6$	$8.7–2.6/\theta$	56
	$C_2H_5C(O)O·$	CCl_4	$k_{24} = 5.6 \times 10^6$		69
	4-Me-$C_6H_4C(O)O·$	CCl_4	$k_{24} = 7.4 \times 10^6$		69
	4-MeO-$C_6H_4C(O)O·$	CCl_4	$k_{24} = 4.8 \times 10^6$		68, 69
	4-Cl-$C_6H_4C(O)O·$	CCl_4	$k_{24} = 3.8 \times 10^6$		69
n-Bu_3SiH	$(CF_3)_2NO·$	$CFCl_3$ or $CF_2ClCFCl_2$	$k_{27} = 0.78$	$5.5–7.7/\theta$	72
t-$BuSi(H)Me_2$	$PhC(Me)_2OO·$	$PhCHMe_2$	$k_{73} = 0.10^d$		70
$PhSiH_3$	t-BuO·	$(t\text{-BuO})_2/C_6H_6$[a]	$k_{27} = 7.5 \times 10^6$		56
	$PhC(Me)_2OO·$	$PhCHMe_2$	$k_{73} = 0.90^d$		70
$PhSi(H)Me_2$	t-BuO·	$(t\text{-BuO})_2/C_6H_6$[a]	$k_{27} = 6.6 \times 10^6$		56
	$PhC(Me)_2OO·$	$PhCHMe_2$	$k_{73} = 0.21^d$		70
Ph_2SiH_2	t-BuO·	$(t\text{-BuO})_2/C_6H_6$[a]	$k_{27} = 1.3 \times 10^7$		62
	$PhC(Me)_2OO·$	$PhCHMe_2$	$k_{73} = 0.84^d$		70
1	$PhC(Me)_2OO·$	$PhCHMe_2$	$k_{73} = 7.2^d$		70
$Ph_2Si(H)Me$	$PhC(Me)_2OO·$	$PhCHMe_2$	$k_{73} = 0.43^d$		70
Ph_3SiH	t-BuO·	$(t\text{-BuO})_2/C_6H_6$[a]	$k_{27} = 1.1 \times 10^7$		62
	$PhC(Me)_2OO·$	$PhCHMe_2$	$k_{73} = 0.67^d$		70
$Me_3SiSi(H)Me_2$	t-BuO·	$(t\text{-BuO})_2/isooctane$[b]	$k_{24} = 1.7 \times 10^7$		63
$(Me_3Si)_2Si(H)Me$	t-BuO·	t-Bu-C_6H_5	$k_{24} = 6.2 \times 10^{7e}$		34
$(Me_3Si)_3SiH$	t-BuO·	$(t\text{-BuO})_2/isooctane$[b]	$k_{24} = 1.1 \times 10^8$		66
	$PhC(Me)_2OO·$	$PhCHMe_2$	$k_{73} = 66.3^d$		70, 71
$H(PhSiH)_nH$	$PhC(Me)_2OO·$	$PhCHMe_2$	$k_{73} = 24.7^{d,g}$		70
$(MeS)_3SiH$	t-BuO·	$(t\text{-BuO})_2/isooctane$[b]	$k_{27} = 4.4 \times 10^7$		9
$(i\text{-PrS})_3SiH$	t-BuO·	$(t\text{-BuO})_2/isooctane$[b]	$k_{27} = 4.5 \times 10^7$		9
Cl_3SiH	t-BuO·	$(t\text{-BuO})_2/C_6H_6$[a]	$k_{27} = 4.0 \times 10^{7f}$		56

[a] 2:1(v/v) Di-tert-butyl peroxide/benzene.

[b] 1:4 (v/v) Di-tert-butyl peroxide/isooctane.

[c] Measurements were carried out in the temperature range 130–185°C; depends on literature Arrhenius parameters for the unimolecular decomposition of tert-butoxyl radicals.

[d] Depends on literature value for cumylperoxyl radical recombination.

[e] Average value of competitive studies between $Me_3SiSi(H)Me_2/(Me_3Si)_2Si(H)Me$ and $Me_3SiSi(H)Me_2/(Me_3Si)_3SiH$.

[f] Approximate value due to low reproducibility (see original work).

[g] Value for each SiH moiety.

laser flash photolysis and were found to be in the range of $3.8–7.4 \times 10^6$ M^{-1} s^{-1} at 24°C (Table V).[68,69] The rate constants for reaction of the t-BuO· and PhC(O)O· radicals with Et_3SiH are identical within experimental error. Furthermore, the rate constants were correlated by the Hammett equation using σ^+ substituent constants, and a ρ value of $+(0.68 \pm 0.56)$ was obtained.

It is noteworthy that the absolute rate constants for the reaction of the benzophenone triplet with Et_3SiH, n-$C_5H_{11}SiH_3$, $PhSiH_3$, and Cl_3SiH have been measured by LFP,[56] and comparison of the kinetic data with corresponding data for reactions of t-BuO· radicals shows that these two transient species have a rather similar reactivity toward silanes. Furthermore, the xanthate and the p-methoxyacetophenone triplets were found to be more and less reactive, respectively, than the benzophenone triplet with Et_3SiH.[56] Similar behavior of excited states in reactions with tin hydrides is discussed in Section V.

The kinetics of the reactions of the cumylperoxyl radical with a variety of silanes have been measured by using inhibited hydrocarbon oxidation methodology (Table V).[70,71] The trends in reactivity for the cumylperoxyl radical are the same as those observed for the t-BuO· radical, although the reactions are about seven orders of magnitude slower. Specifically, the rate constants increase along the two series $PhSi(H)Me_2 < Ph_2Si(H)Me <$ Ph_3SiH and $PhSiH_3 < Ph_2SiH_2 < Ph_3SiH$ when taking into account the statistical number of reactive hydrogens. Furthermore, the rate constants increase by about 1 and 2 orders of magnitude, respectively, in comparing Ph_3SiH to 5,10-dihydrosilanthrene (**1**) and $(Me_3Si)_3SiH$.

The Arrhenius parameters for the reaction of the persistent $(CF_3)_2NO·$ radical with n-Bu_3SiH also were determined by ESR spectroscopy (Table V).[72] It has been suggested that the unusually low preexponential factor is due to geometric constraints on the transition state. The similar reactivities of $(CF_3)_2NO·$ and $PhCMe_2OO·$ radicals toward n-Bu_3SiH and Ph_3SiH, respectively, are expected because both the thermochemistries and spin distributions for these two radicals are rather similar.[72,73]

IV

REACTIONS OF RADICALS WITH GERMANIUM HYDRIDES

As with silicon hydrides, the reaction of atoms or radicals with germanium hydrides is the key step for the majority of reactions forming germyl radicals. However, kinetic data for the reactions of organic radicals with germanium hydrides in solution are limited to carbon- and oxygen-centered radicals.

A. *Carbon-Centered Radicals*

The kinetic data for these reactions are numerous, as shown in Table VI. Most of values were obtained by radical clock methods. The ring expansion of radical **7** has been employed as the clock in a study that provided much of the data in Table VI.[74] Cyclizations of 5-hexenyl-type radicals also have been used as clocks,[75-77] and other competition reactions have been used.[78] Hydrogen atom abstraction from n-Bu$_3$GeH by primary alkyl radicals containing a trimethylsilyl group in the α-, β-, or γ-position were obtained by the indirect method in competition with alkyl radical recombination.[79]

7

19

Attempts were made to study reactions of σ-type radicals (phenyl, vinyl, cyclopropyl) with Bu$_3$GeH and Bu$_3$SnH by LFP.[41] However, these studies employed diaroyl or diacyl peroxides as radical precursors with the assumption that the aroyloxyl and acyloxyl radicals produced by photolysis would decarboxylate "instantly" on the time scale of the studies. It is now known that decarboxylation of the benzoyloxyl radical (PhCO$_2$·) is relatively slow,[69] and it is likely that the measured kinetics were for reactions of a mixture of radicals. Slow decarboxylation of the vinylacyloxyl radical also might have resulted in errors, but the decarboxylation of the cyclopropyl-acyloxyl radical probably was fast enough that the kinetics were for the reaction of the cyclopropyl radical. The problems with the LFP approach for σ-type radicals and the eventual determination of a rate constant for reaction of the "Ph·" radical with Bu$_3$SnH are discussed in more detail in Section V.

The rate constant for reaction of the methyl radical with Bu$_3$GeH is based on the absolute rate constant for reaction of the CH$_3$· radical with Bu$_3$SnH and on a pair of competition experiments involving the reaction of CH$_3$I with a mixture of Bu$_3$SnH and Bu$_3$SnD and another mixture containing Bu$_3$SnD and Bu$_3$GeH.[75] A rate constant of 1.4×10^5 M^{-1} s^{-1} was determined for reaction of the CH$_3$· radical with Bu$_3$GeH by use of the rotating sector method,[80] but diffusional rate constants apparently were slightly underestimated in that study on the basis of the results with tin hydrides (Section V), and this would result in an underestimation of the rate constant for reaction of the germanium hydride. Because the rate constant for reaction of the CH$_3$· radical found in the rotating sector study

TABLE VI
RATE CONSTANTS FOR REACTIONS, OF CARBON-CENTERED RADICALS WITH GERMANIUM HYDRIDES

Germane	Radical	Solvent	Rate constant (M^{-1} s^{-1}) (temp in °C as subscript)	Rate expression ($\theta = 2.3RT$ kcal/mol)	Ref.
PhCH$_2$(Et)GeH$_2$	RCH$_2$· (7)	C$_6$H$_6$	k_{80} = 1.3 × 10^{6a}		74
n-Bu$_3$GeH	CH$_3$·	c-C$_6$H$_{12}$	k_{25} = 5 × 10^{5b}		78
	CH$_2$=CH(CH$_2$)$_4$CH$_2$·	n-Octane	k_{80} = 3.4 × 10^5	8.4–4.7/θ^c	75, 76
	RCH$_2$· (7)	C$_6$H$_6$	k_{80} = 3.8 × 10^{5a}		74
	Me$_3$SiCH$_2$CH$_2$CH$_2$·	C$_6$D$_6$	k_{80} = 1.0 × 10^{5d}		79
	Me$_3$SiCH$_2$CH$_2$·	C$_6$D$_6$	k_{33} = 8.8 × 10^{4d}		79
	Me$_3$SiCH$_2$·	C$_6$D$_6$	k_{33} = 6.3 × 10^{5d}		79
	CH$_2$=CH(CH$_2$)$_4$C(·)HMe	n-Nonane	k_{80} = 7.6 × 10^4	8.3–5.5/θ^c	76
	C$_6$H$_5$·	C$_6$H$_6$	k_{29} = 2.6 × 10^{8e}		41
	Me$_2$C=CH·	C$_6$H$_6$	k_{27} = 3.5 × 10^{7f}		41
	c-C$_3$H$_5$·	C$_6$H$_6$	k_{30} = 1.3 × 10^{7g}		41
	n-C$_8$F$_{17}$·^{h}	C$_6$D$_6$	k_{30} = 1.5 × 10^7		81
	CCl$_3$·	c-C$_6$H$_{12}$	k_{25} = 5 × 10^{5i}		78
(PhCH$_2$)$_3$GeH	RCH$_2$· (7)	C$_6$H$_6$	k_{80} = 3.0 × 10^{6a}		74
mesGeH$_3$	RCH$_2$· (7)	C$_6$H$_6$	k_{80} = 2.8 × 10^{6a}		74
mes$_2$GeH$_2$	RCH$_2$· (7)	C$_6$H$_6$	k_{80} = 2.1 × 10^{6a}		74
19	RCH$_2$· (7)	C$_6$H$_6$	k_{80} = 1.9 × 10^{7a}		74
Ph$_3$GeH	RCH$_2$· (7)	C$_6$H$_6$	k_{80} = 3.8 × 10^{6a}		74
(Me$_3$Si)$_3$GeH	CH$_2$=CH(CH$_2$)$_4$CH$_2$·	C$_6$H$_6$	k_{50} = 1.9 × 10^{7c,j}		74, 77

[a] Depends on literature values for ring expansion.

[b] This value is based on the absolute rate constant for the reaction of the CH$_3$· radical with n-Bu$_3$SnH and a pair of competitive experiments involving the reaction of CH$_3$I with n-Bu$_3$SnH/n-Bu$_3$SnD and n-Bu$_3$SnD/n-Bu$_3$GeH. A rate constant of 1.4 × 10^5 M^{-1} s^{-1} has been measured by the rotating sector method (Ref. 78).

[c] Depends on literature values for 5-hexenyl-type rearrangements.

[d] Depends on literature values for alkyl radical recombination.

[e] The value is likely to be for a mixture of the Ph· radical and the PhC(O)O· radical; see text.

[f] The value might be for a mixture of the vinyl radical and its acyloxyl precursor; see text.

[g] The value might be for a mixture of the cyclopropyl radical and its acyloxyl precursor; see text.

[h] Depends on the rate constant for the addition of the radical to 1-hexene. The original report stated that the perfluoroheptyl radical was studied; we thank Prof. W. R. Dolbier, Jr., for the correction.

[i] The original value obtained from the rotating sector application was multiplied by a factor of 4 (see text).

[j] An approximate rate constant of 3 × 10^6 M^{-1} s^{-1} at 25°C has been determined by the same procedure (Ref. 77).

was about a factor of 4 smaller than that determined in the competition study, we have adjusted the rate constant for reaction of the $Cl_3C\cdot$ radical by the same factor; the kinetic value listed in Table VI for $Cl_3C\cdot$ radical is not the same as that in the original report.[80] The kinetic data for reactions of other halogenated carbon-centered radicals with germanium hydrides is limited to the reaction of the perfluoro-n-octyl radical with Bu_3GeH; the value was determined by competition against addition of the perfluoroalkyl radical to 1-hexene.[81]

The trends in reactivity for the H-atom transfer reaction are the following: (1) For the primary alkyl radical **7**, the rate constants increase along the series Bu_3GeH < Ph_3GeH < $(Me_3Si)_3GeH$. (2) For Bu_3GeH, the rate constants decrease along the series $R_fCF_2\cdot$ > $Me_3SiCH_2\cdot$ > $RCH_2CH_2\cdot$ $\approx$ $Cl_3C\cdot$ $\approx$ $R_3C\cdot$. The available preexponential factors lie in the expected range and suggest that the activation energy is the major factor in determining the radical–germane reactivity. Although the trends just outlined can be entirely attributed to more favorable thermodynamic factors along the series,[74] a polarized transition state, similar to the canonical structure **20**, was suggested to explain the reactivity of the $Me_3SiCH_2\cdot$ radical.[79]

$$Me_3SiCH_2 \overset{\delta^-}{----} \overset{\cdot}{H} \overset{\delta^+}{---} GeBu_3$$

20

21

Aryl substitution on germanium, whether single or multiple, has only a small effect on the rate constants for hydrogen atom transfer, whereas the rate constant increases substantially with substitution of an alkyl group on Ge by a silyl group, much as observed with the silanes. A strong substituent effect also was observed for germane **19**.

B. *Oxygen-Centered Radicals*

Rate constants for reactions of germanium hydrides with oxygen-centered radicals are collected in Table VII. The reaction of photogenerated *tert*-butoxyl radicals with germanium hydrides has often been the method of choice for germyl radical formation for ESR and time-resolved spectroscopic studies. The absolute rate constants for such reactions were measured directly by means of LFP.[8,56] The kinetic values for reaction of the t-BuO$\cdot$ radical with Bu_3GeH and Ph_3GeH are identical. The rate constants for reaction of the t-BuO$\cdot$ radical with phenyl-substituted germanes increase along the series Ph_3GeH < Ph_2GeH_2 < $PhGeH_3$, but when one accounts for the statistical number of hydrogens in each germane, the reactivity is

TABLE VII
Rate Constants for Reactions of Oxygen-Centered Radicals with Germanium Hydrides

Germane	Radical	Solvent	Rate constant (M^{-1} s^{-1}) (temp in °C as subscript)	Rate expression ($\theta = 2.3RT$ kcal/mol)	Ref.
Me_3GeH	t-BuO·	$(t$-BuO$)_2/C_6H_6$[a]	$k_{22} = 6.7 \times 10^7$		8
n-Bu$_3$GeH	t-BuO·	$(t$-BuO$)_2/C_6H_6$[a]	$k_{27} = 9.2 \times 10^7$		56
	PhC(Me)$_2$OO·	PhCHMe$_2$	$k_{73} = 18.6^b$		70
	$(CF_3)_2$NO·	$CFCl_3$ or $CF_2ClCFCl_2$	$k_{27} = 67.5$	6.2–6.0/θ	72
$PhGeH_3$	t-BuO·	$(t$-BuO$)_2/C_6H_6$[a]	$k_{22} = 4.4 \times 10^8$		8
Ph_2GeH_2	t-BuO·	$(t$-BuO$)_2/C_6H_6$[a]	$k_{22} = 1.8 \times 10^8$		8
Ph_3GeH	t-BuO·	$(t$-BuO$)_2/C_6H_6$[a]	$k_{27} = 9.2 \times 10^7$		56
	TEMPO (21)	C_6H_6	$k_{60} = 1.8 \times 10^{-4}$		61
$(Me_3Si)_3GeH$	PhC(Me)$_2$OO·	PhCHMe$_2$	$k_{73} = 4.7 \times 10^{3c}$		70

[a] 2:1(v/v) di-*tert*-butyl peroxide/benzene.
[b] Depends on the literature value for the cumylperoxyl radical recombination.
[c] Depends on the literature value for reaction of the cumylperoxyl radical with cumene.

the same for the series. These kinetic data suggest that rate constants for the reaction of the t-BuO$\cdot$ radical with Ge–H moieties are independent of the substituent (whether an H, alkyl, or phenyl group) and are close to $9 \times 10^7 \ M^{-1} \ s^{-1}$.

The kinetics for the reaction of the cumylperoxyl radical with Bu$_3$GeH and (Me$_3$Si)$_3$GeH have been measured by using inhibited hydrocarbon oxidation methodology.[70] The Me$_3$Si-substituted derivative was found to be much more reactive, in agreement with expectations.

Kinetic data for the reaction of the persistent nitroxides such as TEMPO (**21**) and the (CF$_3$)$_2$NO$\cdot$ radical with Ph$_3$GeH and Bu$_3$GeH, respectively, were determined by ESR spectroscopy.[61,72] The fluorinated nitroxide was more than four orders of magnitude more reactive than TEMPO. On the other hand, the similar reactivities of the (CF$_3$)$_2$NO$\cdot$ and PhCMe$_2$OO$\cdot$ radical toward Bu$_3$GeH are expected because both the thermochemistries and spin densities for these two radicals are rather similar.[72,73]

V
REACTIONS OF RADICALS WITH TIN HYDRIDES

The reduction of alkyl halides by tin hydrides in radical chain reactions was well established in the 1960s,[2] and the essence of this reaction is an important component of the modern synthetic chemist's repertoire. As with silicon and germanium hydrides, tin hydrides react with radicals by H-atom transfer to give a stannyl radical. Unlike the silanes, where rate constants are available for a variety of silicon hydrides, few kinetic studies with tin hydrides have been performed. Rate constants for reactions of tributyltin hydride (Bu$_3$SnH) with a number of radicals are known, and rate constants for reactions of Me$_3$SnH and Ph$_3$SnH with a few radicals have been measured or can be calculated from results in the literature. Bu$_3$SnH holds a prominent position in the absolute kinetic scale of radical chemistry for two reasons. Many rate constants for Bu$_3$SnH reactions with radicals were determined directly by LFP techniques or via the LFP-clock combined method, and these kinetic values apparently are both precise and accurate. In addition, most of the absolute rate constants for unimolecular radical clock reactions are known from competition kinetics studies using Bu$_3$SnH, and the resulting radical clock kinetic values have been integrated into a wide range of bimolecular radical reactions via competition kinetics.[16]

A. *Carbon-Centered Radicals*

Because of the apparent importance of tin hydrides in synthetic sequences, reactions of R$_3$SnH with alkyl radicals were among the first radical reactions whose kinetics were calibrated on an absolute scale. Ingold's group studied these reactions by the rotating sector technique and determined several rate constants at ambient temperature that are similar to presently accepted values.[80] Absolute kinetics of reactions of Bu₃SnH with several alkyl radicals were measured directly by LFP by Ingold's group in the single most important radical kinetic study performed,[82] because these kinetic values have been integrated into a number of radical clocks and then into bimolecular reactions of the clocks with other trapping agents.[16] A later LFP study extended the results to other radicals.[41]

Table VIII contains rate constants for reactions of tin hydrides with carbon-centered radicals. A striking feature of Table VIII in comparison to other tables in this work is the high percentage of reactions for which Arrhenius parameters were determined by direct LFP or the LFP-clock method. These results are expected to be among the most accurate listed in this work. Scores of radical clocks have been studied with Bu₃SnH, but the objectives of those studies were to determine rate constants for the clocks using tin hydride trapping as the calibrated basis reaction.

For the series of alkyl radicals CH₃·, RCH₂·, R₂CH·, R₃C·, there is a regular decrease in LFP-measured rate constants for reactions with Bu₃SnH. Reductions in the exothermicities of the reactions with more highly substituted alkyl radicals are likely to be a factor in the kinetics. In addition, the results suggest that entropic effects are at play in the kinetics. The BDE of a C–H bond in cyclopropane is greater than that in a normal alkane, and the cyclopropyl radical reacts faster with Bu₃SnH.[41]

In unpublished work[83] that extends previous studies,[84] the rate constants for cyclization of the primary alkyl radical clock **22** [Eq. (8)] in THF were measured by LFP with high precision. Use of these recently measured rate constants for cyclization of **22** and the previous results for reaction of **22** with Bu₃SnH in THF[84] gives a rate constant at 22°C slightly (about 20%) smaller than that from the LFP studies in the mixed hydrocarbon–peroxide solvent. Cyclization of radical clock **22** in PhCF₃ solvent also was calibrated by LFP, and a rate constant for reaction of Bu₃SnH with **22** at 80°C in this solvent is available.[85] The value in PhCF₃ at 80°C is also slightly (about 20%) smaller than that calculated from the LFP results for a primary alkyl radical at this temperature. Taken together, the results suggest that solvent effects in the reactions of Bu₃SnH with alkyl radicals are minimal.

$$\tag{8}$$

22

TABLE VIII
Rate Constants for Reactions of Carbon-Centered Radicals with Tin Hydrides

Radical	Solvent, method	Rate expression (log k) ($\theta = 2.3RT$ kcal/mol)	Rate constant, $(Ms)^{-1}$ (at 20°C unless noted)	Ref.
		Bu₃SnH		
CH₃·	Isooctane/ROOR,[a] LFP[b]	$(9.39 \pm 0.28)-(3.23 \pm 0.34)/\theta$	9.5×10^6	82
CH₃CH₂·	Isooctane/ROOR,[a] LFP[b]	$(9.14 \pm 0.42)-(3.80 \pm 0.57)/\theta$	2.0×10^6	82
CH₃CH₂CH₂CH₂·	Isooctane/ROOR,[a] LFP[b]	$(9.06 \pm 0.31)-(3.65 \pm 0.41)/\theta$	2.2×10^6	82
RCH₂CH₂· (recommended)	Isooctane/ROOR[a,c]	$(9.1 \pm 0.2)-(3.7 \pm 0.3)/\theta$	2.2×10^6	41
RCH₂CH₂· (22)	THF, LFP-clock[d]		1.7×10^6 (22°C)[e]	83, 84
RCH₂CH₂· (22)	PhCF₃, LFP-clock[d]		5×10^6 (80°C)	85
(CH₃)₃CCH₂·	Isooctane, LFP[b]		3.0×10^6	41
(CH₃)₂CH·	Isooctane/ROOR,[a] LFP[b]	$(8.5 \pm 0.2)-(2.7 \pm 0.2)/\theta$	1.3×10^6	82
c-C₆H₁₁·	Isooctane/ROOR,[a] LFP[b]	$(8.71 \pm 0.37)-(3.47 \pm 0.49)/\theta$	1.9×10^6	82
(CH₃)₃C·	Isooctane/ROOR,[a] LFP[b]	$(9.24 \pm 0.78)-(3.97 \pm 1.15)/\theta$	1.7×10^6	82
PhCH₂·	Isooctane/ROOR,[a] LEP[b]	$(8.43 \pm 0.14)-(2.95 \pm 0.19)/\theta$	$\leq3 \times 10^5$	82
PhCH₂·	Cyclohexane, competition[f]		3.0×10^5	86
PhCH₂· (Bu₃SnD)[g]	Hexane, competition[f]		1.3×10^4	86
Ph₂CR· (23) (Et₃SnH)[h]	Octane, competition[h]		ca. 100	87
c-C₆H₅·	Benzene, LFP[b,i]	$(9.3 \pm 0.5)-(1.9 \pm 0.6)/\theta$	7.6×10^7	41
Me₂C=CH·	Pentane, LFP[b,i]	$(9.7 \pm 0.3)-(1.6 \pm 0.4)/\theta$	3.2×10^8	41
C₆H₅·[j]	THF or cyclohexane, competition[j]		7.8×10^8 (ambient)	42
ROCH₂· (24)	(Benzene),[k] clock[l]	$9.1-5.0/\theta$	2.3×10^5	88
RCH(·)OCH₃ (29)	THF, LFP[b]		$(3.5 \pm 0.8) \times 10^5$ (25°)	19
RCH(·)OCH₃ (29)	THF, LFP-clock[d]	$(8.4 \pm 0.6)-(3.8 \pm 0.9)/\theta$	3.6×10^5	19
RC(O)OCH₃· (25)	Benzene, clock[l]	$(9.3 \pm 0.7)-(4.3 \pm 0.7)/\theta$	1.2×10^6	88
RCH(·)CO₂Et (30)	THF, LFP-clock[d]	$(8.15 \pm 0.6)-(2.2 \pm 0.6)/\theta$	3.2×10^6	89
RCH(·)CONEt₂ (28, 31)	THF, LFP-clock[d]	$(7.6 \pm 0.3)-(1.5 \pm 0.4)/\theta$	3.0×10^6	90
R₂C(·)CO₂Et (26)	THF, LFP-clock[d]	$(8.5 \pm 0.9)-(4.3 \pm 1.2)/\theta$	1.9×10^5	89
R₂C(·)CN (27)	THF, LFP-clock[d]	$(6.8 \pm 0.8)-(1.7 \pm 0.9)/\theta$	3.4×10^5	89
Me₃SiCH₂·	Isooctane or benzene, LFP[b]	$(10.2 \pm 0.5)-(3.90 \pm 0.62)/\theta$	1.9×10^7	79
Me₃Si(CH₂)₂CH₂·	tert-Butylbenzene, LFP[b]	$(8.4 \pm 0.7)-(2.81 \pm 0.95)/\theta$	2.0×10^6	79
R₃CC(·)=O (32)[m]	Toluene, clock[l]	$8.2-3.5/\theta$	3.9×10^5	37, 38
CH₃CH₂C(·)=O (Bu₃SnD)[g]	Hexane, LFP[b]		3.0×10^5 (23°C)	91
ROC(·)=O (33)	THF, LFP-clock[d]		1.7×10^5 (2°C)	92
n-C₈F₁₇·[n]	Benzene, competition[o]		2.0×10^8 (25°C)	93

TABLE VIII (continued)

Radical	Solvent, method	Rate expression (log k) ($\theta = 2.3RT$ kcal/mol)	Rate constant, $(Ms)^{-1}$ (at 20°C unless noted)	Ref.
Me₃SnH				
$(CH_3)_3C\cdot$	Cyclohexane		7×10^5[b,p]	80
Ph₃SnH				
$RCH_2\cdot$		$3.4 \times Bu_3SnH$ rate constants for $RCH_2\cdot$ (70–130°C)[p]		
$R_2CH\cdot$		$4.3 \times Bu_3SnH$ rate constants for $R_2CH\cdot$ (100°C)[p]		
$R_3C\cdot$		$3.7 \times Bu_3SnH$ rate constants for $R_3C\cdot$ (100°C)[p]		
$(CH_3)_3C\cdot$	Cyclohexane		9×10^6[p]	80
$Me_3COOC(Me)_2CH_2\cdot$	Benzene	$5 \times Bu_3SnH$ rate constant (25°C)[p]		95
Ph₂RSnH[q]				
$RCH_2\cdot$	*tert*-Butylbenzene, clock[l]	$(8.9 \text{ to } 9.0)-(3.0 \text{ to } 3.2)/\theta$	$(3 \text{ to } 5) \times 10^6$	96
$(n\text{-}C_6F_{13}CH_2CH_2)_3SnH$				
$RCH_2CH_2\cdot$ (**22**)	$PhCF_3$, LFP-clock[d]	$(8.9 \pm 0.2)-(3.1 \pm 0.3)/\theta$	3.8×10^6	85

[a] The solvent was a mixture of isooctane and di-*tert*-butyl peroxide.
[b] Direct LFP
[c] Recommended value from combination of results for ethyl and butyl radicals.
[d] Direct LFP calibration of radical clock and indirect calibration of tin hydride.
[e] Calculated from results of Refs. 83 and 84.
[f] Depends on rate constants for radical self-termination reactions.
[g] Rate constants for reaction of Bu_3SnD.
[h] Rate constant for reaction of Et_3SnH; depends on thermochemical assumptions for the clock reaction; see text.
[i] The rate constants could be for a mixture of radicals; see text.
[j] Average results for aryl radicals.
[k] Solvent not stated; benzene is assumed based on related studies.
[l] Radical clock method.
[m] Similar results obtained for precursors to 1°, 2°, and 3° radicals.
[n] The radical stated in the original work was perfluoroheptyl; we thank Prof. W. R. Dolbier for informing us of the correction.
[o] Depends on rate constants for addition to styrenes.
[p] The rate constant was derived in this work; see text.
[q] R = Me, Et, Bu, *i*-Pr, *c*-C_6H_{11}, or Me_3SiCH_2; see original reference for specific values.

The benzyl radical reacted too slowly with Bu$_3$SnH to permit direct LFP kinetic studies,[82] but rate constants for reaction of PhCH$_2$· with Bu$_3$SnH were determined by competition kinetics.[86] A rate constant for reaction of the diphenylalkyl radical 23 with Et$_3$SnH has been deduced from competition kinetic studies against rearrangement of 23 to its ring-opened isomer [Eq. (9)] in a procedure that required assumptions about the thermochemistry of the ring-opening reaction.[87] The rate constants for H-atom transfer from tin hydride to phenyl-substituted carbon radicals follow the exothermicity of the reactions.

$$\text{(9)}$$

23

Rate constants for reactions of Bu$_3$SnH (and Bu$_3$GeH) with σ-type carbon-centered radicals have proven to be problematical. An initial attempt to measure the rate constants for reaction of the phenyl radical with Bu$_3$SnH involved LFP studies employing dibenzoyl peroxide as the radical source,[41] but it was later realized that decarboxylation of the benzoyloxyl radical (PhCO$_2$·) to the phenyl radical was slow (see later discussion). Eventually, rate constants for reaction of aryl radicals with Bu$_3$SnH and Bu$_3$SnD at ambient temperature were determined by a combination of methods that included competition reactions and absolute rate constants for several aryl radical reactions, with the assumption that any aryl radical would react with about the same rate constants.[42]

Rate constants for reaction of a vinyl radical (2,2-dimethylethenyl) with Bu$_3$SnH (and Bu$_3$GeH) also were reported from LFP studies in which the diacyl peroxide was the radical source,[41] but these values might be in error for the same reason as before. Specifically, the C–H BDE for ethylene is the same as that for benzene, and it is possible that decarboxylation of the vinylacyloxyl radical (Me$_2$C=CHCO$_2$·) is slow and that the measured rate constants are for reaction of metal hydrides with the acyloxyl radical or a mixture of radicals. This conjecture is supported by the facts (1) that the rate constants and Arrhenius parameters for reaction of Me$_2$C=CH· with Bu$_3$SnH are quite similar to those found in the study with the PhCO$_2$· that, we will argue later, is likely to be a mixture of rate constants for reaction of the aroyloxyl radical and the phenyl radical with the tin hydride, and (2) the vinyl radical might be expected to react with a rate constant similar to that of Ar·, but the value found in the LFP study that employed the diacyl peroxide was smaller than that for Ar·.

We note that the rate constants for reaction of the cyclopropyl radical with

Bu$_3$SnH (and Bu$_3$GeH) also involved photolysis of the diacyl peroxide.[41] Despite the fact that a reactive σ-radical is formed, we believe the decarboxylation of the first-formed cyclopropylacyloxyl radical was likely to be "instant" on the time scale of the LFP study. The C–H BDE of cyclopropane is about 5 kcal/mol less than those of benzene and ethylene, and the decarboxylation of c-C$_3$H$_5$CO$_2$· would be expected to be faster than that of PhCO$_2$·, a radical with a lifetime of 0.5 μs at ambient temperature.[69] If the cyclopropylacyloxyl decarboxylation is five times faster than benzoyloxyl decarboxylation, then the cyclopropyl radical was by far the predominant species present in the LFP study.[41] Further evidence that the study actually measured reactions of the cyclopropyl radical is found in the rate constants for the Bu$_3$SnH reactions, which are smaller than those found for reactions of oxygen-centered radicals with the tin hydride.

Rate constants for reactions of Bu$_3$SnH with some α-substituted carbon-centered radicals have been determined. These values were obtained by initially calibrating a substituted radical clock on an absolute kinetic scale and then using the clock in competition kinetic studies with Bu$_3$SnH. Radical clocks **24** and **25** were calibrated by kinetic ESR spectroscopy,[88] whereas rate constants for clocks **26–31** were measured directly by LFP.[19,89,90] For one case, reaction of Bu$_3$SnH with radical **29**, a rate constant was measured directly by LFP using the cyclization of **29** as the probe reaction.[19]

24   **25**

26: R = Me, X = C(O)OEt   **29:** X = OMe
27: R = Me, X = CN   **30:** X = C(O)OEt
28: R = H, X = C(O)NEt$_2$ **31:** X = CONEt$_2$

The rate constants for reaction of Bu$_3$SnH with the primary α-alkoxy radical **24** and the secondary α-alkoxy radical **29** are in reasonably good agreement. However, one would not expect the primary radical to react less rapidly than the secondary radical. The kinetic ESR method used to calibrate **24** involved a competition method wherein the cyclization reactions competed with diffusion-controlled radical termination reactions, and diffusional rate constants were determined to obtain the absolute rate constants for the clock reactions.[88] The LFP calibrations of radical clocks

26–31 were direct kinetic measurements, and the accuracy of the method is apparent by comparison of the directly measured LFP rate constant at 25°C for reaction of **29** with the tin hydride and the rate constant calculated from the combined LFP-clock approach. We suggest that the values for reaction of Bu$_3$SnH with the primary clock **24** might be less reliable than those obtained from **29**.

For the primary and secondary α-alkoxy radicals **24** and **29**, the rate constants for reaction with Bu$_3$SnH are about an order of magnitude smaller than those for reactions of the tin hydride with alkyl radicals, whereas for the secondary α-ester radical **30** and α-amide radicals **28** and **31**, the tin hydride reaction rate constants are similar to those of alkyl radicals. Because the reductions in C–H BDE due to alkoxy, ester, and amide groups are comparable, the exothermicities of the H-atom transfer reactions will be similar for these types of radicals and cannot be the major factor resulting in the difference in rates. Alternatively, some polarization in the transition states for the H-atom transfer reactions would explain the kinetic results. The "electron-rich" tin hydride reacts more rapidly with the electron-deficient α-ester and α-amide radicals than with the electron-rich α-alkoxy radicals.

The tertiary α-ester (**26**) and α-cyano (**27**) radicals react about an order of magnitude less rapidly with Bu$_3$SnH than do tertiary alkyl radicals. On the basis of the results with secondary radicals **28–31**, the kinetic effect is unlikely to be due to electronics. The radical clocks **26** and **27** also cyclize considerably less rapidly than a secondary radical counterpart (**26** with R = H) or their tertiary alkyl radical analogue (i.e., **26** with R = X = CH$_3$), and the "slow" cyclization rates for **26** and **27** were ascribed to an enforced planarity in ester- and cyano-substituted radicals that, in the case of tertiary species, results in a steric interaction in the transition states for cyclization.[89] It is possible that a steric effect due to an enforced planar tertiary radical center also is involved in the kinetic effect on the tin hydride reaction rate constants.

An electronic effect on the kinetics of Bu$_3$SnH reactions with silicon-containing alkyl radicals is apparent from the LFP-measured rate constants.[79] The α-silyl radical Me$_3$SiCH$_2\cdot$ reacts with Bu$_3$SnH an order of magnitude faster than does a primary alkyl radical. The kinetic acceleration was ascribed to an α-effect of silicon in stabilizing negative charge development in the transition state for the H-atom transfer reaction, similar to that shown in structure **20** in Section IV. The γ-silyl-substituted radical Me$_3$Si(CH$_2$)$_2$CH$_2\cdot$ reacts with Bu$_3$SnH with rate constants nearly identical to those of an alkyl radical.

Radical clock competition kinetic studies of reactions of Bu$_3$SnH with acyl radicals have been reported. Relative rate constants for reactions of

Bu$_3$SnH in competition with decarbonylation of acyl radicals **32** [Eq. (10)] were determined in the same studies that determined rate constants for reactions of (Me$_3$Si)$_3$SiH with the acyl radicals.[37,38] The results from the three radical clocks were similar and are taken to be representative of any acyl radical. A rate constant for reaction of Bu$_3$SnD with an acyl radical was determined by LFP with IR detection.[91] The reaction of a related radical, alkoxycarbonyl radical **33**, with Bu$_3$SnH was studied in competition with decarboxylation [Eq. (11)].[92] Rate constants for decarboxylation of **33** were measured directly by LFP methods, but the radical clock study with Bu$_3$SnH could only be studied at one temperature; from the limited data that could be obtained, radical **33** reacts with Bu$_3$SnH with about the same rate constants as the acyl radicals **32**. The rate constants for acyl radical reactions with Bu$_3$SnH are smaller than those for reactions of alkyl radicals, consistent with the reduced C–H BDE of a formyl group and the nucleophilic character of the acyl radical.

$$R_3C \overset{O}{\underset{}{\|}} \cdot \longrightarrow R \overset{R}{\underset{R}{\cdot}} + C\equiv O \qquad (10)$$

32: R = H or alkyl

$$ \longrightarrow + CO_2 \qquad (11)$$

33

Several reactions of halogen-substituted carbon-centered radicals with silanes have been studied, but limited kinetic information is available for reactions of halogen-substituted radicals with tin hydrides. A rate constant for reaction of the perfluorooctyl radical with Bu$_3$SnH was determined by competition against addition of this radical to styrenes, reactions that were calibrated directly by LFP methods.[93] At ambient temperature, the n-C$_8$F$_{17}$· radical reacts with tin hydride two orders of magnitude faster than does an alkyl radical, consistent with the electron-deficient nature of the perfluoroalkyl radical and the electron-rich character of the tin hydride. Similar behavior was noted previously for reactions of silanes with perhaloalkyl radicals.

Few kinetic studies of reactions of alkyl radicals with tin hydrides other than Bu$_3$SnH have been reported. Studies of the reactions of the *tert*-butyl radical with Me$_3$SnH and Ph$_3$SnH were performed by the rotating sector method,[80] but an error in absolute values exists in that method as judged by differences in rate constants for reactions of Bu$_3$SnH with alkyl radicals determined by rotating sector[80] and LFP.[82] For reactions of the *tert*-butyl radical with Me$_3$SnH, Ph$_3$SnH, and Bu$_3$SnH, the same diffusional rate constants were assumed.[80] Therefore, we have used the *relative* rate constants for these three stannanes[80] and the absolute rate constant for reaction of Bu$_3$SnH with the *tert*-butyl radical[82] to calculate absolute rate constants. The kinetic results for reactions of Me$_3$SnH and Ph$_3$SnH with (CH$_3$)$_3$C· at 25°C listed in Table VIII were derived here and are not the values originally listed.[80]

A similar approach can be used to calculate rate constants for reaction of Ph$_3$SnH with 1°, 2°, and 3° radicals at higher temperatures. Walling and Cioffari reported relative rate constants in the range of 70–130°C for the cyclizations of five 5-hexenyl radicals in competition with trapping by both Bu$_3$SnH and Ph$_3$SnH.[94] These results provide relative rate constants for reactions of the two tin hydrides, and one can combine them with the absolute rate constants[82] for Bu$_3$SnH reactions with the appropriate radical type (1°, 2°, and 3°) to calculate the rate constants for reactions of Ph$_3$SnH with these classes of radicals.

The preceding approach can also be used for estimating a crude rate constant for reaction of Ph$_3$SnH with an α-oxy radical. Relative rate constants for the unimolecular cyclization reaction of radical **34** [Eq. (12)] and trapping by Bu$_3$SnH and Ph$_3$SnH were reported.[95] The absolute rate constant for the Bu$_3$SnH trapping reaction is not known, but it will probably be within the range of those for reaction of the tin hydride with α-alkoxy and with alkyl radicals (ca. 1×10^6 M^{-1} s^{-1} at 25°C). A rate constant for reaction of Ph$_3$SnH with **34** might be estimated from the competition studies and an assumed kinetic value for Bu$_3$SnH, but the important point is that Ph$_3$SnH reacted five times faster with radical **34** than did Bu$_3$SnH.

$$\begin{array}{ccc} \text{(structure)} & \longrightarrow & \text{(structure)} + \text{Me}_3\text{CO·} \end{array} \qquad (12)$$

3 4

For the limited examples available, the order of reactivity with alkyl radicals is Me$_3$SnH < Bu$_3$SnH < Ph$_3$SnH, a trend that is similar to that observed with silanes and germanes. One might assume that room temperature kinetics for reactions of this series of stannanes with alkyl radicals could be estimated with reasonable accuracy according to the following guidelines: Me$_3$SnH will react about 0.4 times as fast as Bu$_3$SnH, and Ph$_3$SnH will react about 4–5 times as fast as Bu$_3$SnH. In the discussions of silanes, where rate constants for a range of aryl silanes were considered, and of germanes, we noted the small kinetic effect of replacing an alkyl group on Si or Ge with a phenyl group. Even from the limited information available with Ph_3SnH, it is clear that substitution of a phenyl for an alkyl group on the tin atom has a similar small accelerating effect.

A study[96] of several diphenylalkylstannanes, RPh_2SnH (R = Me, Et, Bu, i-Pr, c-C_6H_{11}, Me_3SiCH_2), found rate constants for reactions with the 5-hexenyl radical clock in a narrow range of 8×10^6 to 12×10^6 M^{-1} s^{-1} at 80°C. As one might expect, these values are between those of Bu_3SnH and Ph_3SnH (6×10^6 and 21×10^6 M^{-1} s^{-1}, respectively, at 80°C). The small changes in rate constants for the series were due to slight changes in the energies of activation and correlated nicely with steric parameters.

A "fluorous" tin hydride reagent, $(n$-$C_6F_{13}CH_2CH_2)_3SnH$, developed for use in fluorous-phase methodology, was calibrated in $PhCF_3$ solvent against the primary alkyl radical clock **22** for which rate constants were determined by LFP in the same solvent.[85] The fluorous tin hydride reacts with the primary alkyl radical about twice as fast as does Bu_3SnH. This kinetic acceleration might be due to slight electron withdrawal by the remote perfluorohexyl groups that decreases the "nucleophilicity" of the tin hydride and accelerates reactions with electron-rich radicals.

B. *Oxygen-Centered Radicals*

The application of radical methodology for formation of heteroatom–carbon bonds is limited in comparison to applications for carbon-carbon bond forming reactions, and kinetic data for reactions of heteroatom-centered radicals with tin hydrides are also limited (Table IX). Because of its ease of formation and high reactivity, the *tert*-butoxyl radical has been employed in a number of kinetic studies. Rate constants for reaction of t-BuO· with Bu_3SnH were measured by LFP by Scaiano using two techniques that gave similar results, direct LFP with detection of the $Bu_3Sn·$ radical and the LFP-probe method wherein t-BuO· reacted with the tin hydride and Ph_2CHOH and the $Ph_2C(·)OH$ radical was monitored.[97] LFP studies of t-BuO· were later extended to include Ph_3SnH.[62]

On the basis of the discovery that decarboxylation of the benzoyloxy radical is not fast, one might be tempted to assign the LFP-measured rate constants originally ascribed to reaction of Bu_3SnH (and Bu_3GeH) with Ph· to the reaction of the metal hydrides with $PhCO_2·$.[41] This is not recommended, however. The decarboxylation of $PhCO_2·$ occurs with a rate constant of about 2×10^6 s^{-1} at ambient temperature, or the lifetime of this radical is about 0.5 μs.[69] Although detailed LFP results from the "Ph·" study are not available in the original report,[41] such studies usually involve monitoring transient formation or decay (in this case growth of the $Bu_3Sn·$

TABLE IX
Rate Constants for Reactions of Oxygen- and Nitrogen-Centered Radicals with Tin Hydrides

Stannane	Radical	Solvent, method	Rate constant (M^{-1} s^{-1})	Rate expression ($\theta = 2.3RT$ kcal/mol)	Ref.
Oxygen-centered radicals					
Bu_3SnH	$(CH_3)_3CO\cdot$	ROOR-isooctane, LFP	2.0×10^8 (22°C)		97
Ph_3SnH	$(CH_3)_3CO\cdot$	ROOR-benzene, LFP	4.2×10^8 (27°C)		62
Bu_3SnH	$4\text{-}(MeO)\text{-}C_6H_4CO_2\cdot$	C_6F_6, LFP	1.7×10^8 (24°C)		68
Bu_3SnH	$PhC(Me)_2OO\cdot$	$PhCHMe_2$, competition[a]	1.6×10^3 (73°C)		70
Ph_3SnH	$PhC(Me)_2OO\cdot$	$PhCHMe_2$, competition[a]	4.3×10^3 (73°C)		71
Bu_3SnH	$(CF_3)_2NO\cdot$	$CFCl_3$ or $CF_2ClCFCl_2$, ESR	490	$5.9\text{–}4.3/\theta$	72
Bu_3SnH	TEMPO (**21**)	Benzene, ESR	0.025 (60°C)		61
Bu_3SnH	$^3(Me_2CO)^*$	Acetonitrile, LFP	5.4×10^8 (ambient)		100
Bu_3SnH	$^3(Ph_2CO)^*$	(Not specified), LFP	2.9×10^8 (22°C)		97
Bu_3SnH	$^1(Me_2CO)^*$	Me_4Sn, LFP	1.0×10^9 (ambient)		100
Bu_3SnH	$^1(CH_3C(O)C(O)CH_3)^*$	Cyclohexane, LFP	2.7×10^7 (ambient)		100
Nitrogen-centered radicals					
Bu_3SnH	$R_2N\cdot$ (**36**)	THF, LFP-clock	4.3×10^5 (20°C)	$9.11\text{–}4.66/\theta$	101
Bu_3SnH	$R_2C{=}N\cdot$ (**37**)	Toluene, LFP-clock	ca. 3×10^3 (25°C)		102
Bu_3SnH	$RC(O)N(\cdot)R'$ (**38**)	THF, LFP-clock	1.3×10^9 (20°C)	$10.26\text{–}1.55/\theta$	103
Ph_3SnH	$R_2NH\cdot+$ (**39**)	Acetonitrile LFP	2.4×10^8 (22°C)		104
Bu_3SnH	$^3(RN{=}NR)^*$ (**40**)	Benzene, LFP	3.5×10^7 (ambient)		100
Bu_3SnH	$^1(RN{=}NR)^*$ (**40**)	Benzene, LFP	9.8×10^8 (ambient)		100
Bu_3SnH	$^1(RN{=}NR)^*$ (**41**)	Acetonitrile, LFP	6.2×10^7 (ambient)		105

[a] Depends on literature values for cumylperoxyl radical recombination.

signal was followed) over about 1–2 μs. If this was the case, the observed kinetics were for a mixture of reactions of the $PhCO_2\cdot$ and $Ph\cdot$ radicals with the tin hydride. The $4\text{-}(CH_3O)C_6H_4CO_2\cdot$ radical (**35**) decarboxylates about an order of magnitude less rapidly than does the $PhCO_2\cdot$ radical,[69] and a rate constant for reaction of this radical with Bu_3SnH[68] is more secure. Support for the conjecture that the original LFP kinetic study involved a combination of radical reactions with the tin hydride is found in the fact that the original rate constant at ambient temperature (6 × 10^8 s^{-1})[41] lies between the rate constants for reactions of Ph· and **35** with the tin hydride.

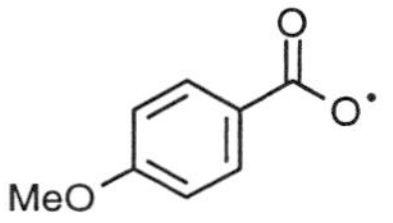

3 5   **21 (TEMPO)**

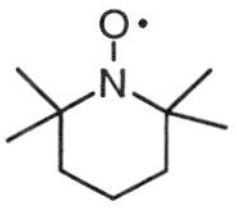

Reactions of oxygen-centered radicals with tin hydride that give compounds with weaker O–H bond energies are considerably slower than the above reactions as expected. Thus, the cumylperoxyl radical, $PhC(Me)_2OO\cdot$, reacted with Bu_3SnH and Ph_3SnH with rate constants that are five orders of magnitude smaller than those for reactions with the $t\text{-BuO}\cdot$ radical.[70,71] The nitroxyl radical $(CF_3)_2NO\cdot$ reacted slowly with Bu_3SnH,[72] and the very stable TEMPO radical (**21**) reacted quite slowly with Bu_3SnH.[61]

The excited states of carbonyl compounds are often considered to be similar to alkoxyl radicals because of the unpaired radical on oxygen. In the initial study of the kinetics of reaction of the $t\text{-BuO}\cdot$ radical with Bu_3SnH, Scaiano also reported a rate constant for reaction of the benzophenone triplet with Bu_3SnH measured directly by LFP.[97] This rate constant was similar to that obtained for reaction of the alkoxyl radical, as might be expected, and the formation of the $Bu_3Sn\cdot$ radical indicated that the triplet "quenching" reaction of the tin hydride involved H-atom transfer. However, the rate constant for reaction of benzophenone triplet with the tin hydride was considerably smaller than rate constants previously assigned for reactions of the acetone and acetophenone excited states on the basis of quenching studies.[98,99] Scaiano revisited this seeming enigma with direct LFP studies of reactions of both singlet and triplet excited states of acetone and a diazo compound (see later discussion). The singlet and triplet excited states of acetone were found to be highly reactive with Bu_3SnH, as was the singlet excited state of butane-2,3-dione (biacetyl).[100]

C. Nitrogen-Centered Radicals

Limited kinetic results are available for nitrogen-centered radicals reacting with tin hydrides (Table IX). Most of the rate constants were determined by the LFP-clock method. The LFP-calibrated clocks were radicals **36–38**, each of which cyclizes. Thus, rate constants for H-atom transfer from Bu₃SnH to the dialkylaminyl radical **36**,[101] the iminyl radical **37**,[102] and the amidyl radical **38**[103] are available. A rate constant at ambient temperature was determined directly by the LFP-probe method for reaction of Ph₃SnH with the dialkylaminium cation radical **39**, which fragments to give the diphenylmethyl radical; Ph₃SnH was used in the study because the solubility of Bu₃SnH in acetonitrile was not sufficient for measurable trapping.[104]

The limited kinetic data for reactions of tin hydride with nitrogen-centered radicals apparently demonstrates the combined effects of the enthalpies of the reactions and polarization in the transition states for H-atom transfer. The aminyl and iminyl radicals are electron-rich, and the N–H bonds formed are relatively weak; these radicals react relatively slowly with tin hydride. On the other hand, the electrophilic amidyl and aminium cation radicals form strong N–H bonds and react rapidly with the tin hydride reagents.

Diazo compounds are photoreduced to hydrazines in the presence of silanes and stannanes in a process that apparently involves initial H-atom transfer to an excited state of the diazo compound.[105] LFP studies of reactions of Bu₃SnH with excited states of diazo compounds **40** and **41** found fast reactions. The singlet diazo species reacted with Bu₃SnH about as fast as singlet carbonyl compounds, whereas the triplet diazo species reacted with tin hydride somewhat slower than triplet excited states of carbonyl compounds.[100,105]

VI
KINETIC ISOTOPE EFFECTS

Primary kinetic isotope effects (KIE) in reactions of the silicon, germanium, and tin hydrides must be relatively small. The SiH stretching frequencies of a wide range of silanes average 2100 cm^{-1}, and the Sn–H stretching frequency of Bu$_3$SnH is only 1815 cm^{-1}.[106] Complete loss of these vibrations in the transition states of linear H-atom abstraction reactions at ambient temperatures would give maximum values of k_H/k_D = 3.8 and 3.1, respectively, for silanes and stannanes. The values for stannanes should be further reduced because of the high exothermicity of the reactions resulting from the low bond energy of R$_3$SnH.

KIE values have been determined from independent kinetic studies of reactions of R$_3$MH and R$_3$MD and from competition studies with mixtures of the hydride and deuteride (or tritiide). Given the high inherent precision of a competing reaction, this method would be expected to be preferred. However, we noted in Section III that low amounts of H-atom abstraction occur from the alkyl groups in reactions of Et$_3$SiH and (Me$_3$Si)$_3$SiH, and the extent of the alkyl-chain reaction must be factored into the analyses. In the case of reactions of stannanes, H-atom abstraction from the alkyl groups is not a problem. It has been shown that insignificant amounts of H are abstracted in the reaction of Bu$_3$SnD with the highly reactive phenyl radical,[42] and one can assume that this is also the case for less exothermic reactions.

Primary KIE in reactions of carbon-centered radicals with a variety of silanes are well documented. The deuterium KIE for reaction of the neophyl radical with Et$_3$SiH[28] and with silane **2**[32] were 2.2 (130°C) and 3.4 (80°C), respectively. The deuterium KIE for the reactions of primary alkyl radicals with Me$_3$SiSi(H)Me$_2$,[63] (Me$_3$Si)$_2$Si(H)Me,[34] and (Me$_3$Si)$_3$SiH[36] have been reported as 3.0 (120°C), 2.3 (90°C), and 2.3 (80°C), respectively, although these values are not adjusted for alkyl-group reactions. The k_H/k_D values for reaction of the benzyl radical with Ph$_3$SiH[58] was 3.9 (170°C), and that for reaction of the Cl$_3$C· with α-NpPhMeSiH[46] was 3.4 (77°C). The highly reactive t-BuO· radical reacted with Me$_3$SiSiMe$_2$H[63] with k_H/k_D = 1.7 (21°C).

The deuterium KIE values are generally in the range expected for linear three-center hydrogen transfer reactions,[44,107] and they track nicely with the rate constants for the reactions with the faster, more exothermic reactions displaying smaller KIEs. The large KIE value for reaction of the benzyl radical is noteworthy in that it exceeds the theoretical maximum for the classical model in a manner apparently similar to that seen with tin hydride (see below).

The deuterium KIEs for reactions of alkyl radicals with Bu$_3$SnH at 27°C determined by LFP were 2.3 ($CH_3\cdot$), 1.9 ($CH_3CH_2\cdot$), and 2.3 ($n\text{-}C_4H_9\cdot$).[82] From rotating sector studies, the deuterium KIE for reaction of $(CH_3)_3C\cdot$ at 25°C with Bu_3SnH was 2.7.[80] The trend is consistent with decreasing exothermicity of the reactions as the radicals become more highly substituted. In an extensive study of tritium KIEs in tin hydride reactions, Kozuka and Lewis[108] found k_H/k_T values for reactions of hexyl and cyclohexyl radicals with Bu_3SnH at 25°C of 3.07 and 2.95, which are equivalent to k_H/k_D = 2.2 and 2.1, respectively, using the relationship[109] $(k_H/k_T) = (k_H/k_D)^{1.442}$. At 80°C, the k_H/k_T values for a series of 1°, 2°, and 3° radicals were quite similar (2.65, 2.72, and 2.53, respectively), in contrast to the k_H/k_D trend noted earlier, whereas tritium KIE for reactions of Ph_3SnH with the same series of radicals at 80°C gave values of 2.55, 2.30, and 2.14, respectively.[108] Comparison of the tritium KIE values for Bu_3SnH with those for Ph_3SnH indicates a reduced Sn–H bond energy in Ph_3SnH, consistent with the greater reactivity of this tin hydride in comparison to Bu_3SnH.

The relatively stable benzyl radical reacted with Bu_3SnH with k_H/k_D = 2.3 at 20°C as determined by independent competition kinetic studies with Bu_3SnH and Bu_3SnD,[86] whereas the k_H/k_T value for benzyl reacting with Bu_3SnH at 21°C was 5.69,[108] equivalent to k_H/k_D = 3.3. The disagreement in the two determinations probably reflects errors in the absolute values for the former determination because it is unlikely that the H-atom abstraction occurred from the butyl chains in the competition reaction that gave the tritium KIE (see above). The large tritium KIE values for reactions of benzyl radical with Bu_3SnH, obtained at various temperatures,[108] exceed the maximum values of classical theoretical models, and mirror the behavior seen in reaction of $PhCH_2\cdot$ with Ph_3SiH.

Small KIE values were found in reactions of highly reactive radicals with tin hydride as expected. In the study that provided rate constants for aryl radical reactions with Bu_3SnH, the phenyl radical was allowed to react with Bu_3SnD in THF-d_0 and with Bu_3SnH in THF-d_8; the deuterium KIE for reaction of Ph· with the tin hydride calculated from these results was k_H/k_D = 1.6 at 31°C.[42] The highly reactive t-BuO· radical had a KIE of k_H/k_D = (1.23 ± 0.15) for reaction with Bu_3SnH and Bu_3SnD at 22°C as determined by LFP studies.[97] Similar small KIE values of k_H/k_D for reaction with Bu_3SnH were found for the singlet excited states of acetone (k_H/k_D = 1.28) and diazo compound **40** (k_H/k_D = 1.27).[100]

VII

COMPARISONS OF REACTIVITIES OF GROUP 14 HYDRIDES

In Table X a number of rate constants taken from the preceding sections are listed for comparison. The measured rate constants for the reactions

TABLE X

COMPARISON OF RATE CONSTANTS FOR THE REACTIONS OF RADICALS WITH VARIOUS GROUP 14 HYDRIDES[a]

Radical	Et$_3$SiH	Bu$_3$GeH	(TMS)$_3$SiH	Bu$_3$SnH
Ph·			3×10^8	7.8×10^8
t-BuO·	4.6×10^{6b}	9.2×10^7	1.1×10^8	2.0×10^8
R$_f$CF$_2$·	5.0×10^5	1.5×10^7	5.1×10^7	2.0×10^8
Me$_3$SiCH$_2$·		6.3×10^5		2.2×10^7
RCH$_2$·	3.2×10^{2b}	9.5×10^4	3.9×10^5	2.5×10^6
RC(O)·			1.8×10^4	3.9×10^5
PhC(Me)$_2$OO·	0.1^c	19	66	1.6×10^3

[a] Rate constants (M^{-1} s^{-1}) at room temperature; peroxyl rate constants at 73°C.
[b] Corrected value referring to attack at the SiH moiety.
[c] Rate constant for reaction of t-BuSi(Me)$_2$H.

of *tert*-butoxyl and primary alkyl radicals with Et$_3$SiH have been corrected in order to provide values referring to SiH moieties (Section III). For a particular radical, the rate constants increase along the series Et$_3$SiH < Bu$_3$GeH < (Me$_3$Si)$_3$SiH < Bu$_3$SnH which are in good agreement with the thermodynamic data for the group 14 hydrides listed in Table I. For example, the relative rate constants of primary alkyl radicals with Et$_3$SiH, Bu$_3$GeH, (Me$_3$Si)$_3$SiH, and Bu$_3$SnH are 1:250:1000:6579, whereas the exothermicities of the reactions are 3.1, 9.6, 10.7, and 19.6 kcal/mol, respectively. For a particular group 14 hydride, the rate constant decreases along the series Ph· > t-BuO· > R$_f$CF$_2$· > Me$_3$SiCH$_2$· > RCH$_2$· > RC(O)· > PhCMe$_2$OO·. The anticipated decrease in the difference of the rate constants for the group 14 hydrides as the reaction becomes more exothermic and faster is also observed.

The available preexponential factors all lie in the expected range for bimolecular reactions, and the majority fit within the range 8.5 < log A < 9.0 M^{-1} s^{-1}. An anticipated decrease of the A factor with the increasing steric hindrance of the attacking radical is also observed. Although the activation energy is clearly the major factor in determining the radical–hydride reactivity, the trends cannot be entirely attributed to the more favorable thermodynamic factors along the series.

Polarized transition states have been often suggested to explain anomalous reactivity. For example, the Me$_3$SiCH$_2$· radical is appreciably more reactive than *n*-butyl radical toward Bu$_3$GeH and Bu$_3$SnH (Table X), although the BDE for Me$_3$SiCH$_2$–H is expected[79] to be 2–3 kcal/mol less than the corresponding primary C–H BDE. This trend in reactivity is attributed to the ability of the silicon to stabilize developing negative charge in the transition state for the H-atom transfer reaction. The reactions of $RCH(\cdot)OCH_3$ and $RCH(\cdot)CO_2Et$ with Bu_3SnH provide other examples in which polarized transition states have been invoked. In comparison with the reaction of a secondary alkyl radical with Bu_3SnH, the rate constants for $RCH(\cdot)OCH_3$ and $RCH(\cdot)CO_2Et$ are one order of magnitude slower and slightly faster, respectively, although in both cases the exothermicities of the reactions with the substituted radicals are ca. 5 kcal/mol less than that for reaction of a secondary alkyl radical with the tin hydride. It has been suggested that the reduced exothermicities of the reactions of the secondary α-ester radical $RCH(\cdot)CO_2Et$ and analogous secondary α-amide radicals with tin hydride are outweighed by rate-accelerating effects due to polar contributions stabilizing the transition states.[19,90]

The trends in reactivity of phenyl-substituted group 14 hydrides deserve some comment. For a particular radical, the rate constants increase along the expected series: $Ph_3SiH < Ph_3GeH < Ph_3SnH$. For example, the rate constants for reactions of primary alkyl radicals increase by a factor of 5.6 on going from Et_3SiH to Ph_3SiH, of 4.6 on going from Bu_3GeH to Ph_3GeH, and of 3.4 on going from Bu_3SnH to Ph_3SnH. Similar behavior can be envisaged for other alkyl radicals or the *tert*-butoxyl radical. Also, "partial" phenyl substitution has a small effect on the rate constant along the series of a particular heteroatom. However, the rate constants slightly increase along the series $PhMH_3 < Ph_2MH_2 < Ph_3MH$ when taking into account the statistical number of hydrogens abstracted. The lack of resonance stabilization by phenyl groups on the metal-centered radical has been attributed to the larger size of silicon, germanium, and tin (compared to carbon) and to the pyramidal nature of the radical center.

The enormous increase in reactivity imparted by Me_3Si substituents is not limited only to silicon hydrides. That is, the rate constants for reactions of primary alkyl radicals increase by a factor of 1200 on going from Et_3SiH to $(Me_3Si)_3SiH$, and by a factor of 110 on going from Bu_3GeH to $(Me_3Si)_3GeH$. Similarly, the rate constants for reactions of peroxyl radicals increase by a factor of 660 on going from $t\text{-}BuSi(H)Me_2$ to $(Me_3Si)_3SiH$, and by a factor of 250 on going from Bu_3GeH to $(Me_3Si)_3GeH$.

VIII

CONCLUSION

This collection of kinetic data for reactions of the group 14 hydrides with various radicals should provide essential information to chemists in both the synthetic and physical communities. As noted in the introduction, the kinetic data for these reactions obtained in the past two decades has been extensive, and we believe much of it is quite accurate and precise. Whereas one might have been concerned about the order of magnitude for the kinetics of many of these reactions 20 years ago, the uncertainties of most of the kinetics listed here are probably less than a factor of 2. For kinetics measured by direct LFP and LFP-clock methods, the precisions are typically within 10–25% at the 95% confidence level, and the absolute errors are unlikely to be much greater.

We have focused on experimental values in this work, but it is important to note two areas of group 14 hydride kinetics that are not discussed but in which we expect to see considerable progress in the near future. One is the kinetics of reactions of radicals centered on atoms of the third row and above, and the other is computational estimates of rate constants.

There are several examples in which group 16 radicals other than oxygen-centered radicals abstract hydrogen from group 14 hydrides. The synthesis of 2-functionalized allyl tris(trimethylsilyl)silanes is based on radical chain reactions in which the key step is the reaction either of $PhS\cdot$ or $PhSO_2\cdot$ with $(Me_3Si)_3SiH$.[110] The reduction of halides and xanthates to their corresponding hydrocarbons[111] as well as the hydrosilylation of alkenes either intermolecularly[112] or intramolecularly[113] by unreactive silanes were successful in the presence of alkylthiols. These reactions succeed because the thiol overcomes the slow hydrogen donation rate of Et_3SiH by acting as the initial H-atom donor to the alkyl radical, and the thiyl radical thus formed reacts rapidly with the silane to regenerate thiol.[111] Evidence that a variety of thiyl radicals,[114] as well as the $PhSO_2\cdot$ radical[115] and the $PhSe\cdot$ radical,[116] abstract hydrogen from Bu_3SnH has also been reported. Some of these reactions have been demonstrated to be fast on the time scale of seconds because catalytic amounts of a specific transfer agent are rapidly replenished by a sacrificial group 14 donor,[111,116] but, to our knowledge, there exist no reliable kinetic data for these classes of reactions. Because of the increasing utility of some of these reactions in organic synthesis, we expect that the kinetics of several classes will be the subject of study in the near future.

Furthermore, a number of reactions in which silyl radicals abstract hydrogen from silicon hydrides have been discovered. An example is shown in Eq. (13), which has been suggested as one of the propagation steps in the autoxidation of $(Me_3Si)_3SiH$.[117] Similar reactions have been observed in the oxidation of $(Me_3Si)_2Si(H)Me$ and $H(PhSiH)_nH$.[118] The identity reaction, i.e., the reaction of a silyl radical with the corresponding hydride, has been suggested in the cis–trans isomerization shown in Eq. (14).[57] To our knowledge, there also are no reliable kinetic data for these classes of reactions.

$$(Me_3Si)_3SiH + Me_3SiSi(\bullet)(OSiMe_3)_2 \longrightarrow$$

$$(Me_3Si)_3Si\bullet + Me_3SiSi(H)(OSiMe_3)_2$$

(13)

(14)

A number of approaches based on calculations have been applied in order to obtain information on the factors that influence the reactivity of radical reactions. In particular, for the hydrogen atom transfer [Eq. (15)] the applied methodologies have spanned from high-level ab initio to empirical calculations.

$$X\text{–}H + Y\bullet \longrightarrow X\bullet + H\text{–}Y$$

(15)

Zavitsas introduced a nonparametric model for calculating energies of activation for hydrogen abstraction involving group 14 hydrides, and the overall average deviation between experimental and calculated values is astonishingly small (less than 1 kcal/mol).[119] For the reaction in Eq. (15), the method takes into account the importance of X–Y antibonding in addition to the enthalpy of reaction. Roberts proposed an empirical algorithm that relates the activation energies to the four molecules involved in reaction (15). However, this empirical approach failed to reproduce the experimental values for the majority of the known identity reactions (i.e., where X = Y), and it was necessary to advance specific rationalizations in order to justify the deviations from the experimental activation energies.[120] Denisov applied another empirical method based on a parabolic model in some radical reactions involving silicon hydrides.[121]

On the other hand, high-level computational methods are limited, for obvious reasons, to very simple systems.[122] Calculations are likely to have limited accuracy due to basis set effects, relativistic contributions, and spin orbit corrections, especially in the case of tin hydrides, but these concerns can be addressed. Given the computational economy of density functional theories and the excellent behavior of the hybrid-DFT B3LYP[123] already demonstrated for calculations of radical energies,[124] we anticipate good progress in the theoretical approach. We hope that this collection serves as a reference for computational work that we are certain will be forthcoming.

It is with great pleasure that we dedicate this review to Dr. Keith U. Ingold, a mentor and friend and leader in the field of radical kinetics, on the occasion of his 70th birthday.

REFERENCES

(1) Chatgilialoglu, C. *Chem. Rev.* **1995**, *95*, 1229.
(2) Kuivila, H. G. *Acc. Chem. Res.* **1968**, *1*, 290.
(3) Giese, B. *Silicon, Germanium, Tin, Lead Compd.* **1986**, *9*, 99.
(4) Berkowitz, J.; Ellison, G. B.; Gutman, D. *J. Phys. Chem.* **1994**, *98*, 2744.
(5) McMillen, D. F.; Golden, D. M. *Ann. Rev. Phys. Chem.* **1982**, *33*, 493.
(6) Ding, L.; Marshall, P. *J. Am. Chem. Soc.* **1992**, *114*, 5754; Goumri, A.; Yuan, W.-J.; Marshall, P. *J. Am. Chem. Soc.* **1993**, *115*, 2539; Kalinovski, I. J.; Gutman, D.; Krasnoperov, L. N.; Goumri, A.; Yuan, W.-J.; Marshall, P. *J. Phys. Chem.* **1994**, *98*, 9551.
(7) Kanabus-Kaminska, J.; Hawari, J. A.; Griller, D.; Chatgilialoglu, C. *J. Am. Chem. Soc.* **1987**, *109*, 5267.
(8) Clark, K. B.; Griller, D. *Organometallics* **1991**, *10*, 746.
(9) Chatgilialoglu, C.; Guerra, M.; Guerrini, A.; Seconi, G.; Clark, K. B.; Griller, D.; Kanabus-Kaminska, J.; Martinho-Simões, J. A. *J. Org. Chem.* **1992**, *57*, 2427.
(10) Burkey, T. J.; Majewski, M.; Griller, D. *J. Am. Chem. Soc.* **1986**, *108*, 2218.
(11) Clark, K. B.; Wayner, D. D. M.; Demirdji, S. H.; Koch, T. H. *J. Am. Chem. Soc.* **1993**, *115*, 2447.
(12) Bullock, W. J.; Walsh, R.; King, K. D. *J. Phys. Chem.* **1994**, *98*, 2595.
(13) O'Neal, H. E.; Ring, M. A.; Richardson, W. H.; Liceiardi, G. F. *Organometallics* **1989**, *8*, 1968.
(14) Griller, D.; Ingold, K. U. *Acc. Chem. Res.* **1980**, *13*, 193.
(15) Chatgilialoglu, C. In *Handbook of Organic Photochemistry*; Scaiano, J. C., Ed.; CRC Press: Boca Raton, FL, 1989; Vol. 2; pp. 3–11.
(16) Newcomb, M. *Tetrahedron* **1993**, *49*, 1151.
(17) Griller, D.; Ingold, K. U. *Acc. Chem. Res.* **1980**, *13*, 317.
(18) Beckwith, A. L. J.; Bowry, V. W.; Ingold, K. U. *J. Am. Chem. Soc.* **1992**, *114*, 4983.
(19) Johnson, C. C.; Horner, J. H.; Tronche, C.; Newcomb, M. *J. Am. Chem. Soc.* **1995**, *117*, 1684.
(20) Newcomb, M.; Horner, J. H.; Emanuel, C. J. *J. Am. Chem. Soc.* **1997**, *119*, 7147.
(21) Horner, J. H.; Tanaka, N.; Newcomb, M. *J. Am. Chem. Soc.* **1998**, *120*, 10379.
(22) Tsentalovich, Y. P.; Fischer, H. *J. Chem. Soc., Perkin Trans. 2* **1994**, 729.
(23) Chatgilialoglu, C. *Acc. Chem. Res.* **1992**, *25*, 188; Chatgilialoglu, C.; Ferreri, C.; Gimisis, T. In *The Chemistry of Organic Silicon Compounds, Vol 2*; Rappoport, S.; Apeloig, Y., Eds.; Wiley: London, 1998; Part 2; pp 1539–1579.
(24) Franz, J. A.; Barrows, R. D.; Camaioni, D. M. *J. Am. Chem. Soc.* **1984**, *106*, 3964.
(25) Chatgilialoglu, C.; Ferreri, C.; Sommazzi, A. *J. Am. Chem. Soc.* **1996**, *118*, 7223.
(26) Chatgilialoglu, C.; Ferreri, C.; Lucarini, M.; Venturini, A.; Zavitsas, A. A. *Chem. Eur. J.* **1997**, *3*, 376.
(27) Chatgilialoglu, C.; Timokhin, V. I.; Ballestri, M. *J. Org. Chem.* **1998**, *63*, 1327.
(28) Chatgilialoglu, C.; Ferreri, C.; Lucarini, M. *J. Org. Chem.* **1993**, *58*, 249.
(29) Ballestri, M.; Chatgilialoglu, C.; Guerra, M.; Guerrini, A.; Lucarini, M.; Seconi, G. *J. Chem. Soc., Perkin Trans. 2* **1993**, 421.
(30) Newcomb, M.; Park, S.-U. *J. Am. Chem. Soc.* **1986**, *108*, 4132.
(31) Lesage, M.; Martinho-Simões, J. A. M.; Griller, D. *J. Org. Chem.* **1990**, *55*, 5413.
(32) Oba, M.; Kawahara, Y.; Yamada, R.; Mizuta, H.; Nishiyama, K. *J. Chem. Soc., Perkin Trans. 2* **1996**, 1843.
(33) Gimisis, T.; Ballestri, M.; Ferreri, C.; Chatgilialoglu, C.; Boukherroub, R.; Manuel, G. *Tetrahedron Lett.* **1995**, *36*, 3897.

(34) Chatgilialoglu, C.; Guerrini, A.; Lucarini, M. *J. Org. Chem.* **1992**, *57*, 3405.
(35) Chatgilialoglu, C.; Ferreri, C.; Vecchi, D.; Lucarini, M.; Pedulli, G. F. *J. Organometal. Chem.* **1997**, *546*, 475.
(36) Chatgilialoglu, C.; Dickhaut, J.; Giese, B. *J. Org. Chem.* **1991**, *56*, 6399.
(37) Chatgilialoglu, C.; Lucarini, M. *Tetrahedron Lett.* **1995**, *36*, 1299.
(38) Chatgilialoglu, C.; Ferreri, C.; Lucarini, M.; Pedrielli, P.; Pedulli, G. F. *Organometallics* **1995**, *14*, 2672.
(39) Bridger, R. F.; Russell, G. A. *J. Am. Chem. Soc.* **1963**, *85*, 3754.
(40) Weldon, D.; Holland, S.; Scaiano, J. C. *J. Org. Chem.* **1996**, *61*, 8544.
(41) Johnston, L. J.; Lusztyk, J.; Wayner, D. D. M.; Abeywickreyma, A. N.; Beckwith, A. L. J.; Scaiano, J. C.; Ingold, K. U. *J. Am. Chem. Soc.* **1985**, *107*, 4594.
(42) Garden, S. J.; Avila, D. V.; Beckwith, A. L. J.; Bowry, V. W.; Ingold, K. U.; Lusztyk, J. *J. Org. Chem.* **1996**, *61*, 805.
(43) Morris, E. R.; Thynne, J. C. J. *Trans. Faraday Soc.* **1970**, *66*, 183.
(44) Arthur, N. L.; Bell, T. N. *Rev. Chem. Intermed.* **1978**, *1*, 37.
(45) Baruch, G.; Horowitz, A. *J. Phys. Chem.* **1980**, *87*, 2535.
(46) Sommer, L. H.; Ulland, L. A. *J. Am. Chem. Soc.* **1972**, *94*, 3903; Nagai, Y.; Matsumoto, H.; Hayashi, M.; Tajima, E.; Watanabe, H. *Bull. Chem. Soc. Jpn.* **1971**, *44*, 3113.
(47) Rice, J. A.; Treacy, J. J.; Sidebottom, H. W. *Int. J. Chem. Kinet.* **1984**, *16*, 1505.
(48) Rong, X. X.; Pan, H. Q.; Dolbier, W. R.; Smart, B. E. *J. Am. Chem. Soc.* **1994**, *116*, 4521.
(49) Dolbier, W. R.; Rong, X. X. *J. Fluorine Chem.* **1995**, *72*, 235.
(50) Delest, B.; Shtarev, A. B.; Dolbier, W. R., Jr. *Tetrahedron,* in press.
(51) Shtarev, A. B.; Dolbier, W. R. Unpublished results.
(52) Dolbier, W. R.; Li, A. R. *J. Chem. Soc. Perkin Trans. 2* **1998**, 79.
(53) Ferreri, C.; Ballestri, M.; Chatgilialoglu, C. Unpublished results.
(54) Nagai, Y.; Matsumoto, H.; Hayashi, M.; Tajima, E.; Ohtsuki, M.; Sekikawa, N. *J. Organometal. Chem.* **1971**, *29*, 209.
(55) Zavitsas, A. A.; Fogel, G.; Halwagi, K. E.; Legotte, P. A. D. *J. Am. Chem. Soc.* **1983**, *105*, 6960; Zavitsas, A. A.; Pinto, J. A. P. *J. Am. Chem. Soc.* **1972**, *94*, 7390.
(56) Chatgilialoglu, C.; Scaiano, J. C.; Ingold, K. U. *Organometallics* **1982**, *1*, 466.
(57) Kyushin, S.; Shinnai, T.; Kubota, T.; Matsumoto, H. *Organometallics* **1997**, *16*, 3800.
(58) Bockrath, B.; Bittner, E.; McGrew, J. *J. Am. Chem. Soc.* **1984**, *106*, 135.
(59) Curran, D. P.; Xu, J. Y.; Lazzarini, E. *J. Chem. Soc., Perkin Trans. 1* **1995**, 3049.
(60) Curran, D. P.; Xu, J. Y.; Lazzarini, E. *J. Am. Chem. Soc.* **1995**, *117*, 6603.
(61) Lucarini, M.; Marchesi, E.; Pedulli, G. F.; Chatgilialoglu, C. *J. Org. Chem.* **1998**, *63*, 1687.
(62) Chatgilialoglu, C.; Ingold, K. U.; Lusztyk, J.; Nazran, A. S.; Scaiano, J. C. *Organometallics* **1983**, *2*, 1332.
(63) Lusztyk, J.; Maillard, B.; Ingold, K. U. *J. Org. Chem.* **1986**, *51*, 2457.
(64) Park, C. R.; Song, S. A.; Lee, Y. E.; Choo, K. Y. *J. Am. Chem. Soc.* **1982**, *104*, 6445.
(65) Lee, Y. E.; Choo, K. Y. *Int. J. Chem. Kinet.* **1986**, *18*, 267.
(66) Chatgilialoglu, C.; Rossini, S. *Bull. Soc. Chim. France* **1988**, 298.
(67) Chatgilialoglu, C. *Gazz. Chem. It.* **1986**, *116*, 511.
(68) Chateauneuf, J.; Lusztyk, J.; Ingold, K. U. *J. Am. Chem. Soc.* **1988**, *110*, 2877.
(69) Chateauneuf, J.; Lusztyk, J.; Ingold, K. U. *J. Am. Chem. Soc.* **1988**, *110*, 2886.
(70) Chatgilialoglu, C.; Timokhin, V. I.; Zaborovskiy, A. B.; Lutsyk, D. S.; Pyrstansky, R. E. Submitted for publication.
(71) Chatgilialoglu, C.; Timokhin, V. I.; Lutsyk, D. S.; Prystansky, R. E. *Chem. Commun.* **1999**, 405.
(72) Doba, T.; Ingold, K. U. *J. Am. Chem. Soc.* **1984**, *106*, 3958.
(73) Chatgilialoglu, C.; Malatesta, V.; Ingold, K. U. *J. Phys. Chem.* **1980**, *84*, 3597.

(74) Chatgilialoglu, C.; Ballestri, M.; Escudíe, J.; Pailhous, I. *Organometallics* **1999**, *18*.

(75) Luszytk, J.; Maillard, B.; Lindsay, D. A.; Ingold, K. U. *J. Am. Chem. Soc.* **1983**, *105*, 3578.

(76) Lusztyk, J.; Maillard, B.; Deycard, S.; Lindsay, D. A.; Ingold, K. U. *J. Org. Chem.* **1987**, *52*, 3509.

(77) Chatgilialoglu, C.; Ballestri, M. *Organometallics* **1995**, *14*, 5017.

(78) Carlsson, D. J.; Ingold, K. U.; Bray, L. C. *Int. J. Chem. Kinet.* **1969**, *1*, 315.

(79) Wilt, J. W.; Lusztyk, J.; Peeran, M.; Ingold, K. U. *J. Am. Chem. Soc.* **1988**, *110*, 281.

(80) Carlsson, D. J.; Ingold, K. U. *J. Am. Chem. Soc.* **1968**, *90*, 7047.

(81) Dolbier, W. R.; Rong, X. X.; Smart, B. E.; Yang, Z. Y. *J. Org. Chem.* **1996**, *61*, 4824.

(82) Chatgilialoglu, C.; Ingold, K. U.; Scaiano, J. C. *J. Am. Chem. Soc.* **1981**, *103*, 7739.

(83) Newcomb, M.; Choi, S.-Y.; Horner, J. H. Unpublished results.

(84) Ha, C.; Horner, J. H.; Newcomb, M.; Varick, T. R.; Arnold, B. R.; Lusztyk, J. *J. Org. Chem.* **1993**, *58*, 1194.

(85) Horner, J. H.; Martinez, F. N.; Newcomb, M.; Hadida, S.; Curran, D. P. *Tetrahedron Lett.* **1997**, *38*, 2783.

(86) Franz, J. A.; Suleman, N. K.; Alnajjar, M. S. *J. Org. Chem.* **1986**, *51*, 19.

(87) Halgren, T. A.; Roberts, J. D.; Horner, J. H.; Martinez, F. N.; Newcomb, M. Submitted.

(88) Beckwith, A. L. J.; Glover, S. A. *Aust. J. Chem.* **1987**, *40*, 157.

(89) Newcomb, M.; Horner, J. H.; Filipkowski, M. A.; Ha, C.; Park, S. U. *J. Am. Chem. Soc.* **1995**, *117*, 3674.

(90) Musa, O. M.; Choi, S. Y.; Horner, J. H.; Newcomb, M. *J. Org. Chem.* **1998**, *63*, 786.

(91) Brown, C. E.; Neville, A. G.; Rayner, D. M.; Ingold, K. U.; Lusztyk, J. *Aust. J. Chem.* **1995**, *48*, 363.

(92) Simakov, P. A.; Martinez, F. N.; Horner, J. H.; Newcomb, M. *J. Org. Chem.* **1998**, *63*, 1226.

(93) Avila, D. V.; Ingold, K. U.; Lusztyk, J.; Dolbier, W. R.; Pan, H. Q.; Muir, M. *J. Am. Chem. Soc.* **1994**, *116*, 99.

(94) Walling, C.; Cioffari, A. *J. Am. Chem. Soc.* **1972**, *94*, 6059.

(95) Bloodworth, A. J.; Davies, A. G.; Griffin, I. M.; Muggleton, B.; Roberts, B. P. *J. Am. Chem. Soc.* **1974**, *96*, 7599.

(96) Dakternieks, D.; Henry, D. J.; Schiesser, C. H. *J. Phys. Org. Chem.* **1999**, *12*, 233.

(97) Scaiano, J. C. *J. Am. Chem. Soc.* **1980**, *102*, 5399.

(98) Wagner, P. J. *J. Am. Chem. Soc.* **1967**, *89*, 2503.

(99) Wagner, P. J.; Kelso, P. A.; Zepp, R. G. *J. Am. Chem. Soc.* **1972**, *94*, 7480.

(100) Nau, W. M.; Cozens, F. L.; Scaiano, J. C. *J. Am. Chem. Soc.* **1996**, *118*, 2275.

(101) Musa, O. M.; Horner, J. H.; Shahin, H.; Newcomb, M. *J. Am. Chem. Soc.* **1996**, *118*, 3862.

(102) Le Tadic-Biadatti, M. H.; Callier-Dublanchet, A. C.; Horner, J. H.; Quiclet-Sire, B.; Zard, S. Z.; Newcomb, M. *J. Org. Chem.* **1997**, *62*, 559.

(103) Horner, J. H.; Musa, O. M.; Bouvier, A.; Newcomb, M. *J. Am. Chem. Soc.* **1998**, *120*, 7738.

(104) Horner, J. H.; Martinez, F. N.; Musa, O. M.; Newcomb, M.; Shahin, H. E. *J. Am. Chem. Soc.* **1995**, *117*, 11124.

(105) Adam, W.; Moorthy, J. N.; Nau, W. M.; Scaiano, J. C. *J. Org. Chem.* **1997**, *62*, 8082.

(106) Kuivila, H. G. *J. Organometal. Chem.* **1964**, *1*, 47.

(107) Strong, H. L.; Borwnawell, M. L.; Filippo, J. S., Jr. *J. Am. Chem. Soc.* **1983**, *105*, 6526.

(108) Kozuka, S.; Lewis, E. S. *J. Am. Chem. Soc.* **1976**, *98*, 2254.

(109) Swain, C. G.; Stivers, E. C.; Reuver, J. F., Jr.; Schaad, L. J. *J. Am. Chem. Soc.* **1958**, *80*, 5885.

(110) Chatgilialoglu, C.; Ballestri, M.; Vecchi, D.; Curran, D. P. *Tetrahedron Lett.* **1996**, *37*, 6383.
(111) Cole, S. J.; Kirwan, J. N.; Roberts, B. P.; Willis, C. R. *J. Chem. Soc., Perkin Trans. 1* **1991**, 103.
(112) Smadja, W.; Zahouily, M.; Journet, M.; Malacria, M. *Tetrahedron Lett.* **1991**, *32*, 3683; Dang, H. S.; Roberts, B. P. *Tetrahedron Lett.* **1995**, *36*, 2875; Haque, M. B.; Roberts, B. P. *Tetrahedron Lett.* **1996**, *37*, 9123.
(113) Cai, Y. D.; Roberts, B. P. *J. Chem. Soc., Perkin Trans. 1* **1998**, 467.
(114) Ueno, Y.; Miyano, T.; Okawara, M. *Tetrahedron Lett.* **1982**, *23*, 443.
(115) Baldwin, J. E.; Adlington, R. M.; Birch, D. J.; Crawford, J. A.; Sweeney, J. B. *J. Chem. Soc., Chem. Commun.* **1986**, 1339; Baldwin, J. E.; Adlington, R. M.; Lowe, C.; O'Niel, I. A.; Sanders, G. L.; Schofield, C. J.; Sweeney, J. B. *J. Chem. Soc., Chem. Commun.* **1988**, 1030.
(116) Crich, D.; Yao, Q. W. *J. Org. Chem.* **1995**, *60*, 84.
(117) Chatgilialoglu, C.; Guarini, A.; Guerrini, A.; Seconi, G. *J. Org. Chem.* **1992**, *57*, 2207.
(118) Chatgilialoglu, C.; Guerrini, A.; Lucarini, M.; Pedulli, G. F.; Carrozza, P.; Da Roit, G.; Borzata, V.; Lucchini, V. *Organometallics* **1998**, *17*, 2169.
(119) Zavitsas, A. A.; Chatgilialoglu, C. *J. Am. Chem. Soc.* **1995**, *117*, 10645; Zavitsas, A. A. *J. Chem. Soc., Perkin Trans. 2* **1996**, 391; Zavitsas, A. A. *J. Chem. Soc., Perkin Trans. 2* **1998**, 499.
(120) Roberts, B. P.; Steel, A. J. *J. Chem. Soc., Perkin Trans. 2* **1994**, 2155; Roberts, B. P. *J. Chem. Soc., Perkin Trans. 2* **1996**, 2719.
(121) Denisov, E. T. *Russ. Chem. Rev.* **1997**, *66*, 859.
(122) Dakternieks, D.; Henry, D. J.; Schiesser, C. H. *J. Chem. Soc., Perkin Trans. 2* **1998**, 591; Dakternieks, D.; Henry, D. J.; Schiesser, C. H. *Organometallics* **1998**, *17*, 1079.
(123) Becke, A. D. *J. Chem. Phys.* **1993**, *98*, 5648.
(124) Fox, T.; Kollman, P. A. *J. Phys. Chem.* **1996**, *100*, 2950.

ADVANCES IN ORGANOMETALLIC CHEMISTRY, VOL. 44

Doubly Bonded Derivatives of Germanium

JEAN ESCUDIE and HENRI RANAIVONJATOVO

Hétérochimie Fondamentale et Appliquée, UPRES A 5069, Université P. Sabatier, 31062 Toulouse Cédex 4, France

I

INTRODUCTION

Over the past 15 years, dramatic and exciting progress has been made in the organometallic chemistry of group 14 elements: Many species once thought to exist only as reactive intermediates, such as doubly bonded compounds $M=M'$ (M = Si, Ge, Sn; M' = C, M, N, P, As, O, S, Se, Te) have been synthesized and isolated owing to steric and electronic stabilization. Among them, many heavy homologues of alkenes, imines, ketones, etc., are now known, but the study of doubly bonded germanium compounds $>Ge=M'$ is still a growing area. The synthesis of a new class of low-coordinate germanium compounds, the germaallenes $>Ge=C=X$ (X = C, P), heavy homologues of allenes and heterocumulenes, has been reported. By contrast, derivatives with a germanium triply bonded to a group 14 or 15 element are still unknown.

Many reviews have been devoted to low-coordinate germanium derivatives; the first reviews were written in 1982 by Satgé[1] and in 1984 by Wiberg[2] on transient derivatives. Other reviews describing some aspects of the synthesis, the physical data, and the reactivity of transient or stable $>Ge=X$ compounds have been published later by some of us,[3-5] Rivière et al.,[6] Baines and Stibbs,[7] Masamune et al.,[8] and Grev[9] for theoretical studies. For a general review on main group element analogues of carbenes, olefins and small rings, see also Ref. 10.

The purpose of this article is to review the papers published between January 1995 and October 1998 on $>Ge=M'$ derivatives (M' = group 14, 15, or 16 elements). The early work in this field has not been summarized.

We have divided the paper in chapters devoted successively to $>Ge=C<$, $>Ge=Ge<$, $>Ge=Si<$, $>Ge=Sn<$, $>Ge=N-$, $>Ge=P-$, $>Ge=O$, $>Ge=S$ (Se, Te), and $>Ge=C=X$. In every chapter we will describe the new routes to such doubly bonded derivatives, their physical data together with theoretical calculations, and their reactivity. Only the doubly bonded compounds between germanium and main group elements are described. Thus, the derivatives in which the germanium is multiply bonded to a transition metal are not reported here.

II

GERMENES ($>Ge=C<$)

A. *Synthesis*

The previously reported stable germenes[11-15] have been synthesized by the three routes displayed in Scheme 1. Other routes were possible for

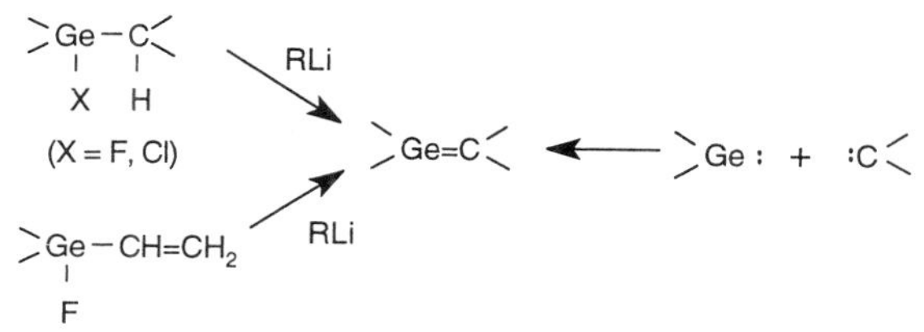

SCHEME 1.

transient germenes[1-3]; however, the range of methods available for synthesizing germenes was limited. During the past 4 years, novel methods have been described and routes previously employed have been used again to obtain both transient and stable germenes.

1. Reactions Involving a Germylene

a. *Coupling reaction between a germylene and CS_2*. A new method allowed the synthesis by Okazaki *et al.* of the stable compound 5^{16}: it is a reaction between carbon disulfide and the overcrowded diarylgermylene 1 which gives, via thiagermirane thione 3 and germathiocarbonyl ylide 4, the first germaketene dithioacetal 5 as an air-sensitive orange compound (Scheme 2). The starting 1 was obtained from the corresponding dibromogermane and lithium naphthalenide (i)[16] or generated *in situ* by thermolysis of 2,3-diphenylgermirene 2 (ii),[17] whereas the sequential nucleophilic substitution of GeI_2 with TbtLi and TipLi did not allow the preparation of 1 because of the lack of reproducibility of this method.[18] 5 is stabilized owing to the large steric hindrance due to the Tip and especially the huge Tbt groups, which makes the germene unit inert towards methanol even in

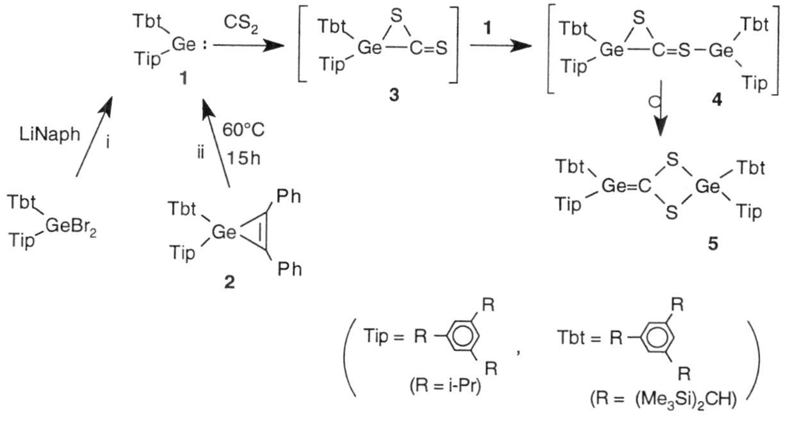

SCHEME 2.

refluxing benzene, in dramatic contrast with other germenes, which react very easily with alcohols.

b. *Coupling reaction between a germylene and a carbene.*   Schumann *et al.* have synthesized the stable derivative 6[19] with tricoordinated Ge and C by reaction of the bis(amino)germylene 7 with the transient carbene 8 [Eq. (1)]. Similar compounds with tin or lead instead of germanium have also been prepared by a similar route.[19]

$$
\begin{array}{ccc}
\text{i-Pr}_2\text{N} & & \text{i-Pr}_2\text{N} \\
\diagdown & \xrightarrow{[(\text{Me}_3\text{Si})_2\text{N}]_2\text{Ge} : \ 7} & \diagdown \\
\text{i-Pr}_2\text{N} & & \text{i-Pr}_2\text{N}
\end{array}
$$

8   6   (1)

with $=\text{Ge}\diagup^{\text{N(SiMe}_3)_2}_{\diagdown\text{N(SiMe}_3)_2}$

2. Peterson-Type Reactions

A novel route using a Peterson-type reaction for synthesizing the stable germene 10 has been reported by Apeloig *et al.*[20] [Eq. (2)]. 10 displays a new type of substitution pattern, with two silyl substituents attached to germanium. A similar reaction performed from germyllithium 11 afforded in this case the transient germene 12,[21] which was characterized by dimerization and trapping reactions. The analogous stable adamantylidenesilene[20] has been obtained by the same route. Note that many transient silenes have been synthesized by similar Peterson-type reactions (for reviews on silenes, see Refs. 22–24).

$$
\begin{array}{ll}
\text{RMe}_2\text{Si} \diagdown & \\
\text{Me}_3\text{Si} - \text{GeLi.3THF} & \xrightarrow[-\text{Me}_3\text{SiOLi}]{\text{Ad=O}} \quad \text{Ad=Ge} \diagup^{\text{SiMe}_2\text{R}}_{\diagdown\text{SiMe}_3} \\
\text{Me}_3\text{Si} \diagup &
\end{array}
$$

$$
\begin{array}{lll}
\text{R} = \text{t-Bu} \ \textbf{9} & & \text{R} = \text{t-Bu} \ \textbf{10} \\
\text{Me} \ \textbf{11} & \text{Ad} = \text{2-adamantylidene} & \text{Me} \ \textbf{12}
\end{array}
$$

(2)

Some new methods have been used to prepare exclusively transient germenes, which were identified by trapping reactions.

3. [2 + 2] Cycloreversion of Four-Membered Rings

A [4] → [2 + 2] photocycloreversion of 13 gave 14,[25] which was characterized in solution by UV and by its chemical reactivity (see Section II,C,5). Dimethylgermene 15[26] was obtained by a thermal cycloreversion from germacyclobutanes 16 and particularly 17, which appeared as a more selective precursor than 16, and trapped and identified in an Ar matrix at 12 K [Eq. (3)].

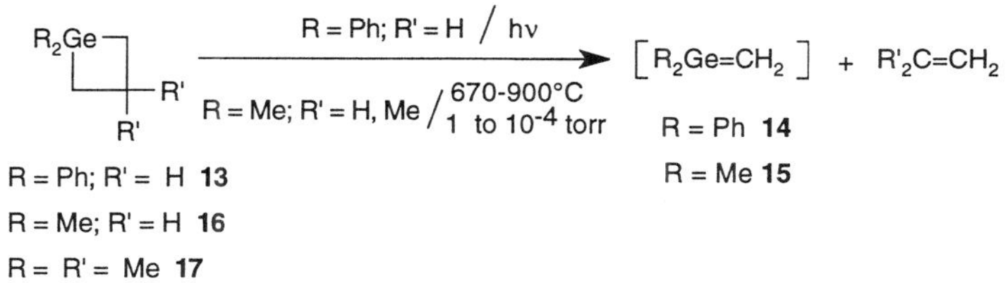

$$R_2Ge \xrightarrow[\substack{R = Me;\ R' = H,\ Me\ /\ \substack{670\text{-}900°C \\ 1\ \text{to}\ 10^{-4}\ \text{torr}}}]{R = Ph;\ R' = H\ /\ h\nu} \left[R_2Ge{=}CH_2 \right] + R'_2C{=}CH_2$$

$$
\begin{array}{ll}
R = Ph;\ R' = H\ \mathbf{13} & R = Ph\ \mathbf{14} \\
R = Me;\ R' = H\ \mathbf{16} & R = Me\ \mathbf{15} \\
R = R' = Me\ \mathbf{17} &
\end{array}
$$

(3)

15 was also obtained by Khabashesku *et al.*[26] by a thermal cycloreversion of another type of four-membered ring, the 1-germa-3-thietane **18**, previously synthesized by Barrau *et al.*,[27] who proved the ability of this heterocycle to give dimethylgermene[28] (Scheme 3). **15**, physicochemically characterized at 12 K, was obtained with secondary products due to a partial isomerization under the studied conditions [see Section II,C,1, Eq. (12)] and also to other routes of decomposition of **18**, i and ii, occurring in a minor ratio.[26]

Thermolysis at 180°C of dithiaphosphadigerminane **19**[29] leads to transient 2-thia-1,3-digermetane **20**, which gives a $[4] \rightarrow [2 + 2]$ decomposition with formation of both dimethylgermene **15** and dimethylgermanethione **21** [Eq. (4)]. Note that a silene and a silanethione were also simultaneously obtained by West *et al.* in a rather similar decomposition by photolysis of a 3-thia-1,2-disiletane,[30] a four-membered ring with a Si–Si–S–C linkage.

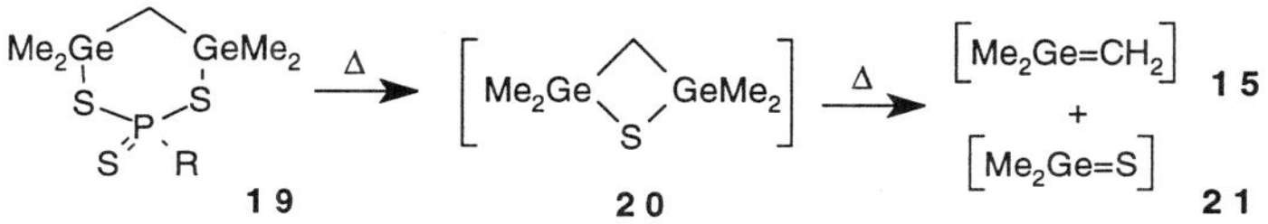

$$
\underset{\mathbf{19}}{\overset{\displaystyle Me_2Ge \diagup\diagdown GeMe_2}{\underset{\displaystyle S{\diagdown}P{\diagup}S}{}\ \underset{S'\ R}{}}}
\ \xrightarrow{\Delta}\
\underset{\mathbf{20}}{\left[Me_2Ge\diagdown_{S}\diagup GeMe_2 \right]}
\ \xrightarrow{\Delta}\
\begin{array}{c}\left[Me_2Ge{=}CH_2 \right] \\ + \\ \left[Me_2Ge{=}S \right]\end{array}
\begin{array}{c}\mathbf{15} \\ \\ \mathbf{21}\end{array}
$$

(4)

4. Retro Diels–Alder reactions

The germene $Me_2Ge{=}C(SiMe_3)_2$ **22** was obtained by thermolysis of **23**[31] (E = Ge) [Eq. (5)]. A very complete mechanistic study of the Diels–Alder and ene-reactions has been performed using **22** (see Section II,C,8). The corresponding silene (E = Si) has also been obtained by the same route.[31]

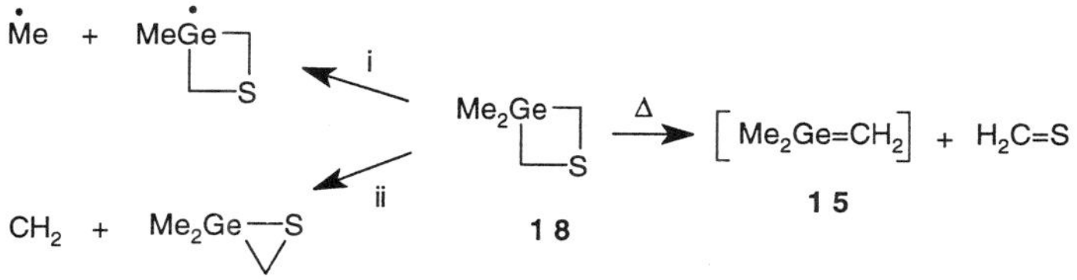

SCHEME 3.

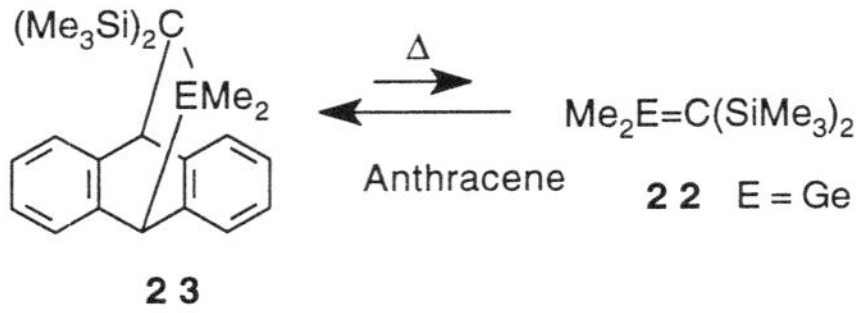

$$(5)$$

5. Defluorosilylation or Defluorolithiation

a. *Defluorosilylation.* A defluorosilylation of **25** at 100°C, initiated by small amounts (0.1 equiv) of a base such as triethylamine, gave **24**[32] [Eq. (6)]. This reaction allowed the synthesis of transient germenes with chlorine or fluorine on germanium. Only the elimination of Me$_3$SiF was observed, and never the elimination of Me$_3$SiCl.[32]

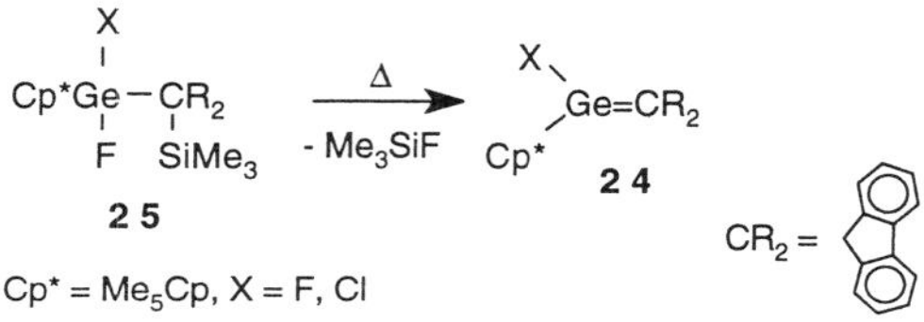

$$(6)$$

b. *Defluorolithiation.* In the absence of trapping reagents, the thermolysis of **26** (prepared by Wiberg *et al.*[33]) gave many unidentified products.[34] By contrast, the thermolysis of **26** at 100°C in C$_6$H$_6$ in the presence of trapping reagents gave adducts that prove the intermediate formation of transient germene **27**[34] obtained after a rearrangement (Scheme 4). Trapping reactions indicate that the equilibrium between **27** and **28** lies toward **27**. The analogous silene was obtained by a similar reaction when germanium was replaced by silicon.[35]

6. Rearrangements from a Germylene or a Germole

Photolysis of diazidogermacyclopentane **29** gave the germylene **30** and the germole **33** as main products,[36] but also the two germadienes **31** and

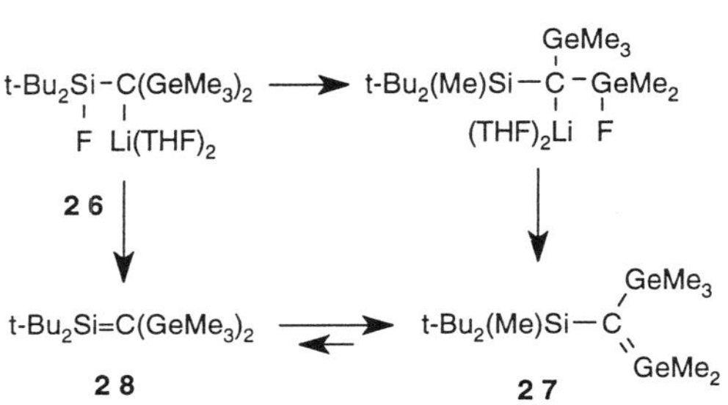

SCHEME 4.

32 as minor products [Eq. (7)]. **31** and **32** were characterized by UV and IR. A similar reaction was observed for the silicon analogue.[37]

$$(7)$$

B. *Physical Properties and Theoretical Studies*

1. *Structure Determinations and ^{13}C NMR*

5,[16] isolated as air-sensitive orange crystals, presents the shortest Ge=C bond length [1.771(16) Å][16] ever determined. The other bond distances previously reported for germenes **34** and **35** were respectively 1.827(4)[11a] and 1.803(4) Å.[38] **5** presents a trigonal planar geometry of the germene unit (the sums of the bond angles around Ge and C are respectively 359.7 and 360°) with a torsion angle found to be 4°[16] (6° in **35**[38] and 36° in **34**[11a]). The observed bond length is consistent with the calculated values for the parent germene $H_2Ge=CH_2$ (generally 1.76–1.81 Å[26,39–42] and 1.717 by MNDO[43]).

Calculations predict the Ge=C bond should be elongated by a fluorine substituent on carbon (by 0.01–0.02 Å)[42] and shortened by a fluorine substituent on germanium (by 0.01–0.02 Å).[42] This last result is in agreement with the calculations[44] performed on $Me_2Si=CH_2$, $F(Me)Si=CH_2$, and $F_2Si=CH_2$, in which the replacement of a methyl group by a fluorine increases the π-bond energy from 39 to 45 kcal/mol and even 50 in the difluorinated silene.

In **6** [Eq. (8)] the distance between germanium and the carbene-C atom [2.085(3) Å][19] is much longer than in "true" germenes and is even longer than in the classical Ge–C single bond. The sum of angles on germanium (303°) is very far from the 360° necessary for a planar germanium. Moreover, the ^{13}C NMR signals in **6** are respectively 145.45 ppm for the carbon bonded to the germanium and 145.62 for the two other carbons of the ring, and fall in the range typical for cyclopropenylium compounds. Thus, as said by the authors,[19] **6** is best described by the ylidic resonance structure **6a**, corresponding to a complex between a germylene and a carbene, rather than by the germene structure **6b**.

$$(8)$$

2. *Charges*

In the series of germenes **36–40**, the calculated charges on germanium and carbon differ greatly, depending on the substituents.[42] The presence of the electron-withdrawing fluorine on germanium increases the positive charge at germanium and thus the degree of $Ge=C$ bond ionicity, explaining the shortening of this bond. The reverse is observed in germene $H_2Ge=CHF$.[42]

$H_2Ge=CH_2$	$Me(H)Ge=CH_2$	$Me_2Ge=CH_2$	$F(H)Ge=CH_2$	$H_2Ge=CF_2$
3 6	**3 7**	**3 8**	**3 9**	**4 0**

Note that when they are substituted by the same groups the charge difference is greater in silenes than in germenes.[39]

Between 1,4-digermabutadiene $H_2Ge=CH-CH=GeH_2$ **41** and 2,3-digermabutadiene $H_2C=GeH-GeH=CH_2$ **42**, the total net charges exhibit a reversal in the intensities, respectively: Ge, +0.40; C, −0.29 for **41** and Ge, +0.29; C, −0.40 for **42**.[45]

3. *Ge=C Force Constants and Bond Orders*

These have been calculated for germenes substituted by alkyl groups, hydrogen, or fluorine on Ge and C, **36–40**.[42] The $Ge=C$ force constants are very close in **36–38** (4.74–4.79 mdyn/Å) but larger in the 1-fluorogermene **39** (4.94–4.99 mdyn/Å) and lower in 2-fluorogermene **40** (4.50–4.65 mdyn/Å). The force constant is much higher than in the Ge–C single bond (e.g., 2.72 mdyn/Å in $MeGeH_3$[46]) and proves the additional π-bonding between Ge and C.

The $Ge=C$ bond orders vary in the same direction[42] (1.57 in **38**, 1.62–1.65 in **39**, and 1.51–1.55 in **40**). It was previously estimated at 1.50 by MNDO in $Mes_2Ge=CR_2$ **35**.[38] The bond order in $Me_2Ge=CH_2$ (1.57) is slightly lower than the $Si=C$ bond order in $Me_2Si=CH_2$ (1.62[47] and 1.66[48]). This is consistent with the relative strength of the $Ge=C$ (31[49] and 32–33[41] kcal/mol) and $Si=C$ π-bonds (35–36 kcal/mol[49]) and provides an additional indication that the former is somewhat weaker than the latter.

4. *Infrared*

The observed value of Ge=C vibration for $Me_2Ge=CH_2$ (847.3 cm^{-1})[26] in argon matrix at 12 K is in good agreement with ab initio and density functional theory calculations that predict this band in the range 840–849 cm^{-1}.[26,42] This assignment was also supported by the calculated and observed $^{70}Ge/^{72}Ge/^{74}Ge/^{76}Ge$ isotopic splitting.[26] Calculations performed on germenes **36–40** showed a slight increase (2–12 cm^{-1}) from **36** to **38**; the fluorine substituent causes a more significant increasing shift (28–31 cm^{-1}) in **39** but a decreasing shift in **40** (49–52 cm^{-1}; see Ref. 42).

The Ge=C stretching modes are predicted at 854 and 869 cm^{-1}[45] in 1,4-digermabutadiene **41** *s-trans* and at 884 and 894 cm^{-1} in the 2,3-digermabutadiene **42** *s-trans* (882 and 893 cm^{-1} in the *s-cis*).[50] Similar values were found for cyclic 1-germadiene **31** (838 cm^{-1})[36] and 2-germadiene **32** (834 cm^{-1}),[36] close to those predicted on the bases of restricted Hartree–Fock calculations, respectively 830 and 843 cm^{-1}.[36]

5. *Ultraviolet/Visible Spectroscopy*

The Ge=C λ_{max} was unknown for a long time because of the difficulty in determining the spectrum in dilute solution without hydrolysis. The π–π^* transition has been determined and varies in a rather large range between 262 nm (**32**[36]), 301 nm (**31**[36]), 325 nm ($Ph_2Ge=CH_2$[25]), and 396 nm (ε: 2600) for germene **5**, which is an orange compound.[16]

6. *Ionization Energy*

Calculations gave for germene $Me_2Ge=CH_2$ values of 7.79–7.84 eV,[26] slightly depending on the method used. Unfortunately, the experimental determination of IE has not been possible because the steady-state concentration is not sufficient.[26]

7. *Geometry and Bonding in Germadienes*

The two germadienes **31** and **32**,[36] which have also been evidenced by IR and UV, and the two digermabutadienes **41**[45] and **42**[50] have been theoretically studied. As in butadiene, the 1,4-digermabutadiene **41** has a preferred planar *s-trans* form with two equivalent *gauche* conformers lying 3 kcal/mol higher in energy; the *cis*-planar conformer is a saddle point relating the two *gauche* forms.[45] In 2,3-digermabutadiene **42**, the long Ge–Ge central bond (compared to the C–C central bond in **41**) removes the steric hindrance that was causing the isomer of butadiene to distort into a nonplanar *gauche* form. Thus, **42** has two stable planar conformers close in energy corresponding to C_{2h} *s-trans* and C_{2v} *s-cis* forms; the *s-cis* isomer is disfavored by 0.4 kcal/mol.[50]

$$H_2Ge=CH-CH=GeH_2 \qquad H_2C=GeH-GeH=CH_2$$

3 1   **3 2**   **4 1**   **4 2**

Various approaches based on geometry (for example, 2% shortening of the central Ge–Ge bond in **42**, and 6% of the C–C central bond in **41**, whereas the shortening is less than 4% in butadiene) or energy criteria suggest that the π-conjugation is about half that in butadiene in **42**,[50] but, in contrast, one and one half times as large in **41**.[45] Calculations performed in digermadienes **31** and **32** yielded planar geometries[36] with central Ge–C (in **32**) and C–C (in **31**) bond lengths shorter than generally found, respectively 1.916 and 1.490 Å. This proves the conjugation of the two π-bonds. However, the degree of shortening of these bonds is considerably smaller than in silicon analogues, and thus indicates that π-conjugation is less effective in the germadienes[36] than in the siladienes.[37] An interesting feature in **41**[45] is the important weight of the diradical form **43**, which is expected to favor 1,4-coupling to germacyclobutene **44** [Eq. (9)]. A strong exothermicity of 44.3 kcal/mol is calculated for this ring closure,[45] in marked contrast with the experimental values (endothermicity of 9–11 kcal/mol) found in the butadiene-to-cyclobutene rearrangement.[51–53]

Because of the weight of **43**, the system **41/43** undergoes an easier rotation about its formally Ge=C double bond than about its formally C–C single bond. The diradical character of **41** should favor any addition with radical intermediates and polymerization reactions.[45]

$$\tag{9}$$

C. *Reactivity*

The reactivity of transient or stable germenes has already been extensively explored,[1–4,7] but some known reactions have been performed again in order to determine their mechanism. Various new reactions have also been described.

1. *Thermolysis*

The thermal decomposition of **5**[54] in the presence of DMB leads to germacyclopentene **45** and CS_2. It constitutes a new reaction mode of a germene that leads to the corresponding germylene [Eq. (10)]. Such a type of reaction is in sharp contrast to the thermolysis of the carbon analogue **46**, giving the thioketene **47** and the corresponding thioketone[55] [Eq. (11)].

$$ (10) $$

$$ (11) $$

Another type of thermal rearrangement has been observed with dimethylgermene **15**[26] leading to germylene **48** and then, probably via a germirane intermediate, to germylene **49**[26] [Eq. (12)].

$$ (12) $$

2. *Photochemical Rearrangement*

Photochemical rearrangements of transient germadienes **31**[36] and **32**[36] have been evidenced from the corresponding germole or germylene (see Section II,A,6).

3. *Dimerization*

Generally, the transient germenes display a head-to-tail dimerization. This is the case for $Me_2Ge=CH_2$[26] and $Ph_2Ge=CH_2$[25] described in this paper. Calculations[42] predict a virtually zero barrier for this type of dimerization for $H_2Ge=CH_2$, $Me(H)Ge=CH_2$, $Me_2Ge=CH_2$, and $F(H)Ge=CH_2$. It is the most common pathway, consistent with the significant polarization of the $Ge^{\delta+}=C^{\delta-}$ bond: A strongly polarized bond appears to be a major factor favoring head-to-tail dimerization.

In marked contrast, disilylgermene **12** gives only the head-to-head dimer **50**[21] [Eq. (13)]. The same mode of dimerization was reported for the analogous silene $(Me_3Si)_2Si=Ad$.[56]

Head-to-head dimerization is a less common mode. It occurs particularly in siloxysilenes $(R_3Si)_2Si=C(R)OSiR_3$ in which the double bond is not significantly polarized (for reviews on silenes, see Refs. 22–24), probably by a process initiated by the silicon–silicon bond formation. This process would involve the formation of a 1,4-diradical species whose radical centers are located on carbon.[24] A similar process could be postulated for **12**. Such head-to-head dimerization has also been predicted for $H_2Ge=CHF$.[42] The latter case is very likely controlled by the electron-withdrawing nature of the fluorine substituent, similar to the effect of the $OSiMe_3$ group at carbon on the dimerization of silenes $(R_3Si)_2Si=C(R)OSiMe_3$.[22–24]

Although the corresponding 1,2-disilacyclobutane **51** reverts thermally to the corresponding silene,[21, 57] heating of **50** at 50°C leads only to the isolation of germanium-containing oligomers via a digermene intermediate.[21,57] Attempts to trap a germene failed [Eq. (14)]. In contrast to the thermolysis, the photolysis of **50** at 254 nm proceeds in close analogy to that of **51**, giving the previously unknown digermene $(Me_3Si)_2Ge=Ge(SiMe_3)_2$[57] (see Section III,A,6).

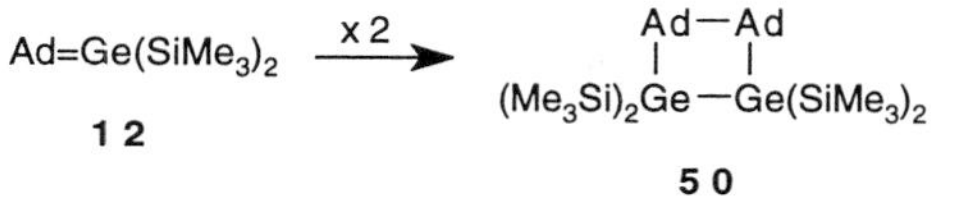

$$Ad=Ge(SiMe_3)_2 \xrightarrow{\times 2} \begin{matrix} Ad-Ad \\ | \quad | \\ (Me_3Si)_2Ge-Ge(SiMe_3)_2 \end{matrix} \tag{13}$$

$$\begin{matrix} \textbf{12} \end{matrix} \qquad \qquad \textbf{50}$$

$$\begin{matrix} Ad-Ad \\ | \quad | \\ (Me_3Si)_2M-M(SiMe_3)_2 \end{matrix} \xrightarrow[\Delta]{} \begin{cases} M=Si \quad (Me_3Si)_2Si=Ad \\ \\ M=Ge \quad Ge\ oligomers \end{cases} \tag{14}$$

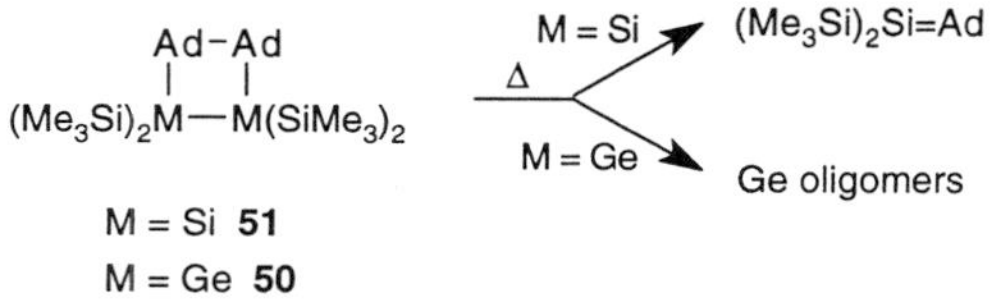

$$M = Si\ \textbf{51}$$
$$M = Ge\ \textbf{50}$$

4. *Oxygen*

When oxygen was bubbled into a THF solution of **5** in the presence of methanol, the orange color characteristic of **5** disappeared immediately.[54] The formation of products **52**, **53**, and **54** was interpreted in terms of an initial [2 + 2] cycloaddition of molecular oxygen to **5**, leading to a transient dioxagermetane followed by a [2 + 2] decomposition with formation of germanone **55**[54] (Scheme 5).

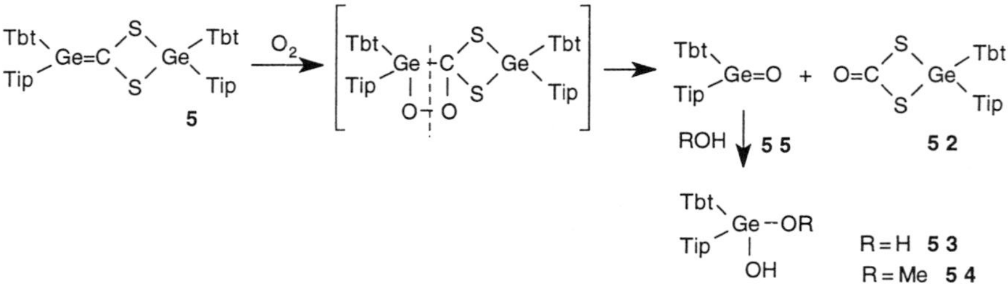

SCHEME 5.

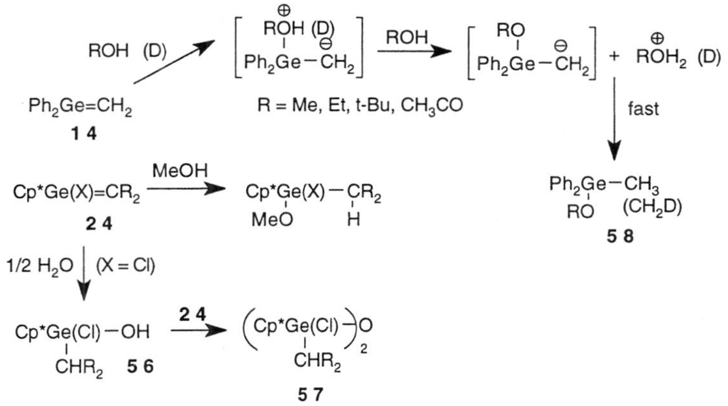

SCHEME 6.

5. Alcohols and Water

Addition of protonic reagents (alcohols, water, organic acids . . .) has been studied previously.[1-4,7] It is an easy way to prove the formation of transient germenes 14[25] and 24[32] (Scheme 6). The digermoxane 57[32] is probably formed by reaction between the germanol 56 and a second equivalent of germene 24, as previously proved in a similar reaction from fluorenylidenegermene.[58] Kinetic studies on the formation of alkoxymethyldiphenylgermanes 58 show that the alcoholysis proceeds by a mechanism involving the initial formation of a germene–alcohol complex[25] followed by an intermolecular base-catalyzed process involving a second molecule of alcohol (Scheme 6).

6. Chloroform

Fluoro- and chlorogermenes 24a and 24b are evidenced by a trapping reaction with the transient formation of 59, which loses dichlorocarbene to give 60[32] [Eq. (15)]. A similar reaction process with addition of the CH bond of chloroform to the Ge=N double bond of a germanimine Ge=N followed by loss of CCl_2 has also been reported.[59]

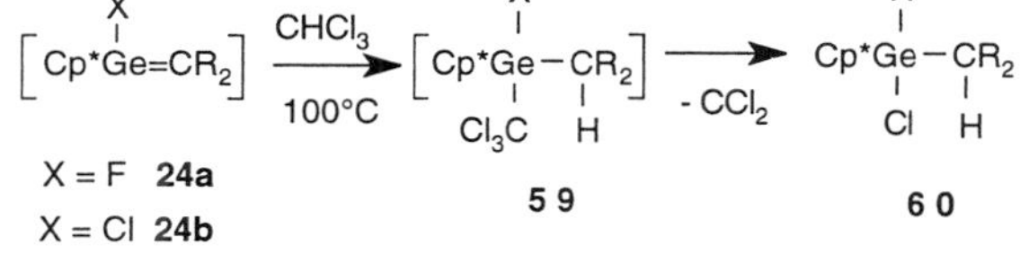

$$X = F \quad 24a$$
$$X = Cl \quad 24b$$

(15)

7. Insertion into a Ge–S bond

As we have seen previously (see Section II,A,3), the four-membered ring heterocycle **20** is a precursor of germene **15** that has been characterized by its dimerization and by insertion into the Ge–S bond of **20**[29] (Scheme 7). However, the formation of **63** could also arise by insertion of **21** into a Ge–C bond of the 1,3-digermacyclobutane **62**, and **64** could be formed from germene **15** and a four-membered ring $(Me_2Ge-S)_2$ intermediate.[29] The six-membered ring derivative **61** has also been observed.

8. Alkenes and 1,3-Dienes

Depending on the trapping reagents, various reactions have been observed: an ene reaction between propene, isobutene, and **22**[31] or **27**,[34] a [2 + 4] cycloaddition between butadiene and **12**,[21] **22**,[31] and **27**,[34] or both an ene reaction and a [2 + 4] cycloaddition between 2,3-dimethylbutadiene and **22**[31] or **27**,[34] and between piperylene or hexadiene and **22**.[31] Some of these reactions are summarized in Scheme 8. By contrast, **27** does not react with the C=C double bond of an alkene such as $CH_2=CH-OMe$.[34]

These studies proved that the reactions of **22** with butadienes and propenes take place both regioselectively and stereoselectively and are accelerated by electron-donating groups on propenes and butadienes (e.g., 2-methylpropene in relation to propene) and retarded by increasing bulkiness of substituents in 1,4- or 1,3-positions. As in the case of alkenes and silenes, the reactions of **22** occur in a concerted way and are HOMO (dienes or enes)–LUMO (dienophiles or enophiles) controlled.[31] However, some small differences are observed between germene **22** and the analogous

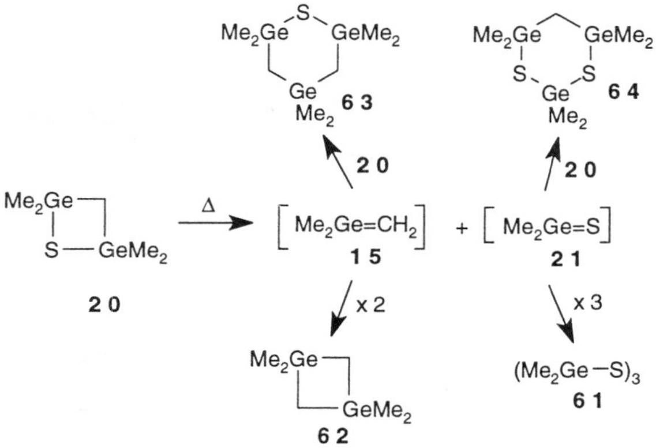

SCHEME 7.

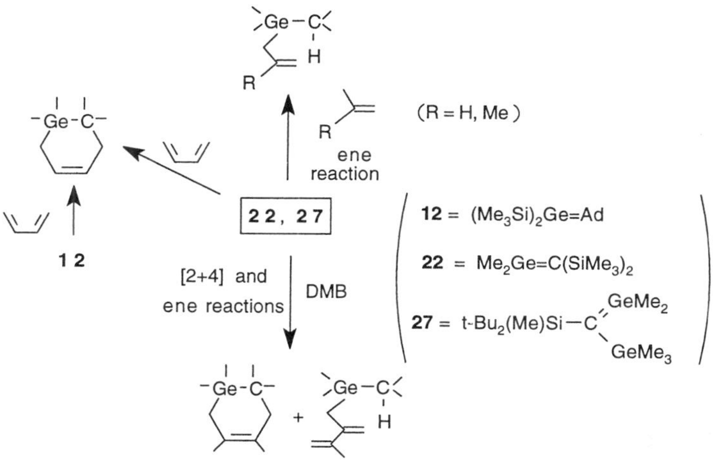

SCHEME 8.

silene $Me_2Si=C(SiMe_3)_2$ **65**: For example, with a diene/ene mixture, **22** appears as a better enophile, probably due to a π/π^* energy difference and a double bond polarity higher in **65** than in **22**.

III

DIGERMENES (>Ge=Ge<)

A. Synthesis

The digermenes were the first stable doubly bonded germanium species to be prepared,[3-5a,7,8] mainly by the routes displayed in Scheme 9. These routes have still been used during the past 4 years, but new ones have also been discovered. Depending on the method used and on the substituents on germanium, stable or transient digermenes have been obtained.

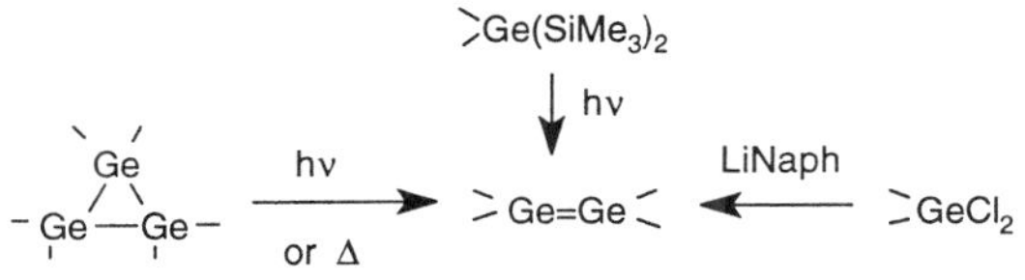

SCHEME 9.

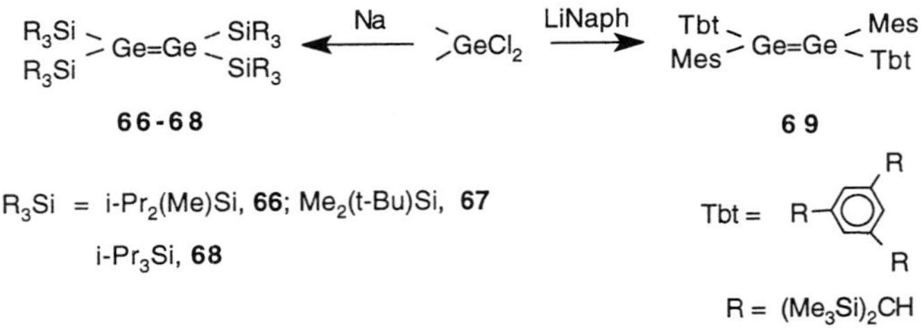

SCHEME 10.

1. From Dichlorogermanes

Four new stable digermenes **66–69** were synthesized from dichloroger-manes and sodium in toluene or lithium naphthalenide (Scheme 10). **66** and **68** have been characterized by X-ray[61] (see Section III,B,1). **66–68**[61] keep their digermene structure in solution, in contrast to **69**[60] (probably the E isomer due to the steric hindrance in the Z isomer), for which an interconversion with the corresponding germylene was observed.

2. Coupling Reaction between Germylenes

The new stable digermene **70** was obtained by Weidenbruch et al.[62] by a reaction between silylene **71** and germylene **72**. Surprisingly, this reaction does not give the expected germasilene **73** but the digermene **70**, by a germasilene → silylgermylene rearrangement followed by a dimerization. However, an alternative mechanism could occur, involving the direct inser-tion of the silylene **71** into one Ge–N bond of **72** without formation of the germasilene intermediate **73**[62] (Scheme 11). Only the overcrowded Z form

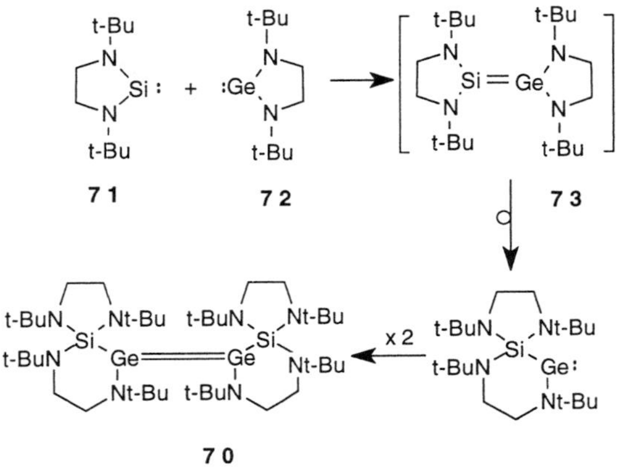

SCHEME 11.

was obtained, rather than the less sterically stressed *E* configuration. Four molecules of **70** cocrystallize with one molecule each of silylene **71** and germylene **72**.[62]

A similar reaction[62] between silylene **71** and stannylene [(Me$_3$Si)$_2$N]$_2$Sn does not afford the distannene analogous to digermene **70**, but a hydridodisilylstannane.[62]

3. *From GeCl$_2$ · Dioxane*

The stable digermene **74**[63] (which displays a nearly planar environment of both germanium atoms and dissociates into two germylenes in solution[63]) and cyclotrigermenes **75, 76**[64] were obtained by reaction of GeCl$_2$ · dioxane with, respectively, a Grignard reagent and *t*-Bu$_3$SiNa or *t*-Bu$_3$GeLi (Scheme 12).

The reaction mechanism to produce **75, 76** is unknown. The use of bulky silyl and germyl anions is crucial to stabilize the ring, since theoretical calculations indicate that electropositive substituents such as silyl and germyl groups lead to a remarkable relief of the strain of three-membered rings containing the group 14 elements[65,66] and because only substitution reactions occur in the reaction of GeCl$_2$ · dioxane with aryllithium to form germylenes.[67,68] However, with another bulky silyllithium such as (Me$_3$Si)$_3$-SiLi · (THF)$_3$, a disilagermirane[69] was obtained instead of a cyclotrigermene analogous to **75**.

4. *Photolysis of a Trigermirane*

Hexaarylcyclotrigermanes are known to be precursors of digermenes photochemically and thermally.[3–5a,7,8] Hexa-*tert*-butylcyclotrigermane[70] behaves rather differently, reacting thermally by cleavage of just one Ge–Ge bond to give the ring-expanded thia-,[71] selena-,[71] and telluratrigermetanes[72] with heavier chalcogens or open-chain dihalogenotrigermanes with halogens.[73,74] But, photochemically, it gives simultaneously the corresponding

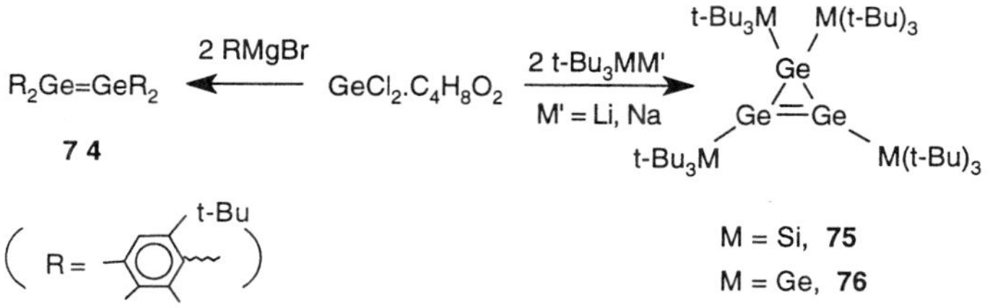

SCHEME 12.

germylene **77** and digermene **78**,[73,74] which is not stable and was characterized by trapping reactions [Eq. (16)].

$$(t\text{-}Bu_2Ge)_3 \xrightarrow[25°C]{hv} \quad t\text{-}Bu_2Ge \colon \; + \; t\text{-}Bu_2Ge=Ge(t\text{-}Bu)_2 \qquad (16)$$
$$\qquad\qquad\qquad\qquad\qquad\quad \mathbf{7\,7} \qquad\qquad \mathbf{7\,8}$$

5. *Photolysis of a Disilagermirane*

The photolysis of the peralkylated disilagermirane **79**[75] gives mainly the disilene **82** and the germylene **80**, which dimerizes to the corresponding digermene **81** (route a). The intermediate formation of germylene **80** was proved by trapping with DMB. The formation of germasilene **83** and silylene **84** (route b) was only a very minor pathway[75] (Scheme 13). The mode of decomposition observed (mainly loss of germylene) is the same as that previously reported by Baines *et al.*, thermally[76] or photochemically,[77] from hexamesitylsiladigermirane. Of course, in this case, because of the presence of one Si and two Ge atoms in the starting heterocycle, a germasilene was formed.

6. *[2 + 2] Decomposition of a 1,2-Digermacyclobutane*

Transient tetrasilyldigermene **85**[57] was clearly obtained by photolysis of **50** and trapped by butadiene (Scheme 14). By contrast, thermolysis of **50** at 60°C, even in the presence of trapping reagents, affords only oligomers, probably because the digermene **85** polymerizes at this temperature before it can be trapped.[57]

7. *Dehalogenation of a Dichlorodigermane*

Transient digermene **86** was postulated as possible intermediate in the reaction of 1,2-dichlorodigermacyclobutane **87** with Mg/MgBr₂ in the presence of an alkene, leading to the bicyclic compound **88**[78] [Eq. (17)]. However, an electron transfer reaction with the intermediate formation of a germyl radical is also a possibility.[78] Similarly, transient 1,2-digermacyclobu-

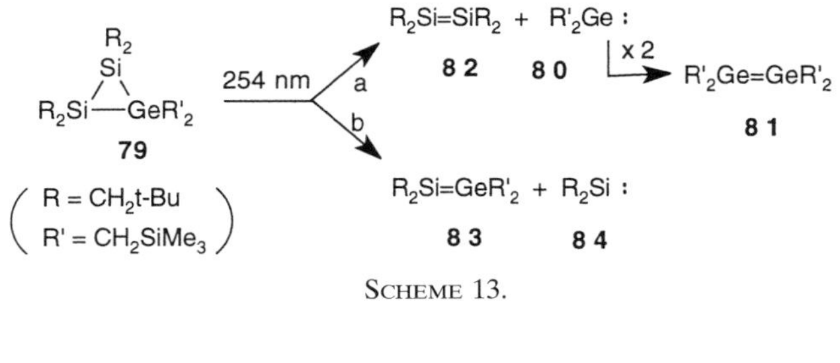

SCHEME 13.

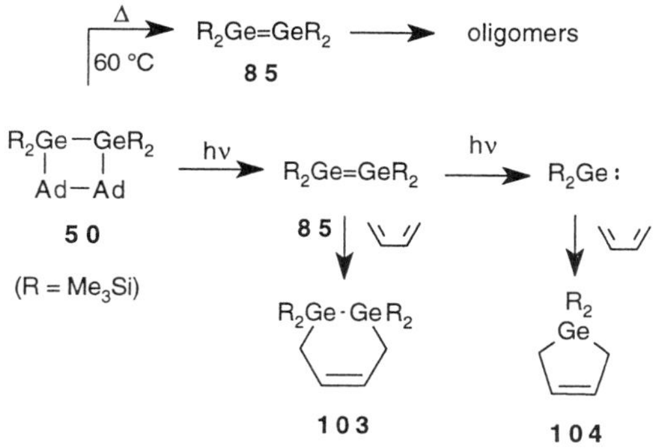

$$\xrightarrow[60\,°C]{\Delta} R_2Ge=GeR_2 \longrightarrow \text{oligomers}$$
$$\mathbf{85}$$

$$\underset{\mathbf{50}}{\overset{R_2Ge-GeR_2}{\underset{Ad-Ad}{\mid\quad\mid}}} \xrightarrow{h\nu} \underset{\mathbf{85}}{R_2Ge=GeR_2} \xrightarrow{h\nu} R_2Ge:$$

$$(R = Me_3Si)$$

$$R_2Ge\cdot GeR_2 \qquad \underset{Ge}{R_2}$$
$$\mathbf{103} \qquad\qquad \mathbf{104}$$

SCHEME 14.

tadiene **89** was postulated as intermediate in the reaction of **90** with Mg/MgBr$_2$.[79]

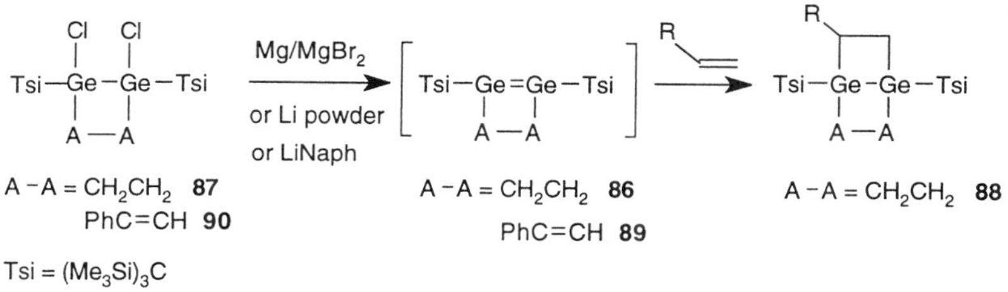

$$\underset{\substack{A-A = CH_2CH_2 \ \ \mathbf{87}\\ PhC=CH \ \ \mathbf{90}\\ Tsi = (Me_3Si)_3C}}{\overset{Cl\quad Cl}{\underset{A-A}{Tsi-Ge-Ge-Tsi}}} \xrightarrow[\substack{\text{or Li powder}\\ \text{or LiNaph}}]{Mg/MgBr_2} \left[\underset{\substack{A-A = CH_2CH_2 \ \ \mathbf{86}\\ PhC=CH \ \ \mathbf{89}}}{\overset{}{\underset{A-A}{Tsi-Ge=Ge-Tsi}}}\right] \longrightarrow \underset{\substack{A-A = CH_2CH_2 \ \ \mathbf{88}}}{\overset{R}{\underset{A-A}{Tsi-Ge-Ge-Tsi}}}$$

$$(17)$$

8. From a Digermyne Intermediate

The formation of transient 1,2-digermacyclobutadiene **91**[79] was postulated in the reaction of TsiGeCl · LiCl · 3THF[78] with Mg in the presence of diphenylacetylene. This reaction could involve the formation of the digermyne intermediate **92** generated from **93** formed by dimerization of the germylene anion radical **94** [Eq. (18)].

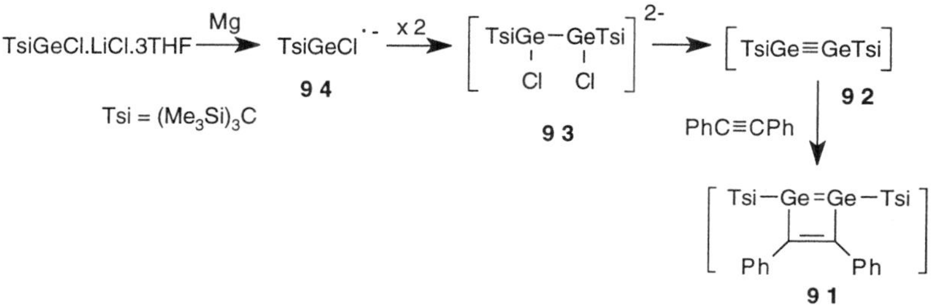

$$\text{TsiGeCl.LiCl.3THF} \xrightarrow{Mg} \underset{\mathbf{94}}{TsiGeCl^{\cdot-}} \xrightarrow{\times 2} \left[\underset{\mathbf{93}}{\overset{TsiGe-GeTsi}{\underset{Cl\quad Cl}{\mid\quad\mid}}}\right]^{2-} \longrightarrow \left[\underset{\mathbf{92}}{TsiGe\equiv GeTsi}\right]$$

$$Tsi = (Me_3Si)_3C \qquad\qquad PhC\equiv CPh \downarrow$$

$$\left[\underset{\mathbf{91}}{\overset{Tsi-Ge=Ge-Tsi}{\underset{Ph\qquad Ph}{\diagup\quad\diagdown}}}\right]$$

$$(18)$$

9. Cyclotrigermenyl Radical and Trigermenyl Allyl Anion

Treatment of the germylene **95**[80] with slightly less than 1 equiv of KC_8 (prepared from graphite and potassium) affords the radical **96** as dark blue crystals.[81] Reduction with a further equivalent of KC_8 gives the trigermenyl anion $K(GeR)_3$ **97**[81] [Eq. (19)]. Both **96** and **97** have been characterized by X-ray[81] (see Section III,B,1).

$$(19)$$

B. Physical Properties and Theoretical Studies

1. Structure Determinations (see Table I)

A very long $Ge=Ge$ bond length (2.454 Å) is observed for **70**,[62] considerably longer than found usually,[3–5a,7,8,82] and close to the value for a normal Ge–Ge single bond. Also unusual are the *trans* bent angles (41.3° and 42.3°), which exceed all previously reported bending angles (maximum of 36°). However, they lie in the range of the theoretically calculated *trans* bent angles (35–47°) for the parent compound $H_2Ge=GeH_2$[83,84] (and references cited therein).

Much shorter $Ge=Ge$ double bonds are reported for **74**[63] and the tetrasilyldigermenes **66**[61] and **68**,[61] in the expected range. In both cases a nearly planar environment of the germanium atoms was observed. Weidenbruch *et al.* explain this planar arrangement by the choice of the *ortho*-alkyl groups.[63] In the case of tetrasilyldigermenes **66** and **68**, the planar geometry should arise from the presence of silyl groups on germanium: Theoretical calculations have shown that electropositive substituents favor a planar arrangement, whereas electronegative or π-electron-donating substituents induce a larger bending deformation in $Ge=Ge$ compounds.[85]

Cyclotrigermenes **75**[64] and **76**[64] also have a short intracyclic $Ge=Ge$ double bond with a completely trigonal planar geometry.

In the cyclotrigermenyl radical **96** the average $Ge=Ge$ distance is 2.35(7) Å.[81] In valence bond terms the ring is expected to involve a double bond between two Ge with the unpaired electron located on the third Ge atom. Thus, the three following forms **96a**, **96b**, and **96c** can be written.

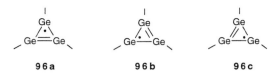

96a   96b   96c

In the allyl anion **97**,[81] the bond length [2.422(2) Å] is marginally shorter than a Ge–Ge single bond: the lengthening of the bond in relation to the Ge=Ge double bond is probably partly due to the increased interelectronic repulsion owing to the negative charges.[81]

By contrast, in the formally aromatic 2π-electron system of **98**[86] [Eq. (20)] the average Ge–Ge bond distance is 2.326(4) Å (**98a**) and 2.335(2) Å (**98b**), thus 0.1 Å shorter than in the anion **97**.[81] The three germanium atoms of this free germyl cation form an equilateral triangle.

$$
\begin{array}{ccc}
\underset{\substack{RGe=GeR}}{\overset{R_2}{\diagup Ge \diagdown}} & \xrightarrow{Ph_3C^+ \, BAr_4^-} & \underset{\substack{RGe-GeR}}{\overset{R}{\diagup \overset{\oplus}{Ge} \diagdown}} \, BAr_4^- \quad \xrightarrow{t\text{-}Bu_3GeNa} \\
\end{array}
$$

R = t-Bu$_3$Si **75**   R = t-Bu$_3$Si   Ar = Ph **98a**
t-Bu$_3$Ge **76**   C$_6$F$_5$ **98b**
                                              3,5-(CF$_3$)$_2$C$_6$H$_3$ **98c**

R = t-Bu$_3$Ge   Ar = C$_6$F$_5$ **98d**
                   3,5-(CF$_3$)$_2$C$_6$H$_3$ **98e**

$$
\underset{\substack{R-Ge=Ge-R \\ R=(t\text{-}Bu)_3Si}}{\overset{R \diagdown \quad \diagup Ge(t\text{-}Bu)_3}{Ge}} \qquad (20)
$$

Density functional theory has been used to understand and predict structures and bonding in main group compounds with multiple bonds.[87] The calculated bond length in Me$_2$Ge=GeMe$_2$ is predicted to be 2.313 Å.[87] Other calculations have also been made to analyze the bonding in H$_2$X=XH$_2$ (X = C, Si, Ge, Sn, Pb).[84] As expected from previous work, *trans* bent structures (46.3° for the C_{2h} H$_2$Ge=GeH$_2$) are minima with a considerable elongation of the Ge=Ge double bond relatively to planar structures (2.362 Å instead of 2.258 Å).[84] The dissociation energy H$_2$Ge=GeH$_2$ → 2 GeH$_2$ was found to be 32.7 kcal/mol.[84]

2. EPR Study

The EPR study of the cyclotrigermenyl radical **96** at room temperature displays a major signal at $g = 2.0069$ with a relative low hyperfine coupling to a single ^{73}Ge (16 G),[81] consistent with the location of the unpaired electron in an orbital of π-symmetry, indicating planar or near-planar geometry at the germanium center.

Photolysis of digermene **74** in benzene yielded the persistent radical [R$_3$Ge]· ,[88] g: 2.0049(2), a(^{73}Ge): 85(4) G. Thus, the photochemical behavior of **74** is similar to that of Bis$_2$Ge=GeBis$_2$ (Bis = (Me$_3$Si)$_2$CH).[89] Like its isoleptic stannyl radical, $[R_3Ge] \cdot$ was exceptionally thermostable: Even after irradiation has ceased and upon heating at 70°C for 6 h, there was no color change, nor any significant decrease of the signal strength. It is noteworthy that this was the first observation of a ^{13}C hyperfine coupling [10(1) G] made for a group 14 metal-centered radical.

3. UV Study

The absorption in the electronic spectrum lies in the normal range (λ_{max} between 413 and 475 nm) (see Table I) for acyclic digermenes.[60–63,75] A hypsochromic shift was observed for cyclotrigermene 75 (326 nm[64]).

A remarkable temperature dependence of the UV-vis spectra was observed for tetrasilyldigermene 68 in solution (red at room temperature, yellow at 77 K)[61]: Mechanistic studies have shown that this phenomenon is due to an equilibrium between two different conformations around the $Ge{=}Ge$ bond, for example, the bent structure and a twisted structure.

Contrary to the case of 68,[61] the thermochromism observed for the digermene 69[60] (blue at room temperature, yellow at low temperature) is due to an interconversion between the digermene and the corresponding germylene. This is the first spectroscopic observation of an equilibrium between a digermene and a germylene. The dissociation energy (14.7 kcal/mol)[60] is much smaller than the previously calculated value of 25–37 kcal/mol for $H_2Ge{=}GeH_2$,[41,90] indicating that the $Ge{=}Ge$ double bond is considerably weakened by the great steric repulsion caused by the huge Tbt groups. Note, however, that the disilene bearing the same groups as 69 exists only as a disilene and not as two silylenes below 50°C,[91–93] proving the subtle effect of the size of the substituents and probably of other factors on the equilibrium.

C. Reactivity

1. Formation of a Germyl Cation

Treatment of cyclotrigermenes 75 and 76 with various trityl tetraarylborates in benzene gives air- and moisture-sensitive crystals of the cyclotrigermenium salts 98a–e[86] [Eq. (20)]. Compounds 98 are stable in the absence of air. Addition of t-Bu$_3$GeNa to 98a gives a new cyclotrigermene. As proved by the X-ray structure determination, the cyclotrigermenium ions 98a,b are free germyl cations. Because of the steric congestion, 75 did not react with ethanol or diazomethane.[64] Thermolysis of 75 results in its decomposition,[64] but there was no evidence for a cycloelimination producing the triply bonded digermyne $RGe{\equiv}GeR$ and germylene R_2Ge.

TABLE I

X-Ray and UV-Visible Data

$R_2Ge=GeR_2$	$dGe=Ge$ (Å)	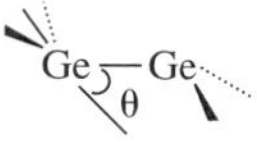 Ge—Ge θ	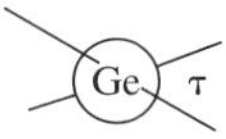 Ge τ	λ_{max} (nm) (ε)	Ref.
R = 2-t-Bu-4,5,6-Me₃C₆H **74**	2.2521(8)	7.9 10.4	20.4	440 (7,500)	63
R = i-Pr₂MeSi **66**	2.268(1) 2.266(1)	5.9 7.1	0.0 0.0	413 (16,100) 361 (2,900)	61
R = t-BuMe₂Si **67**				421 (7,000)	61
R = i-Pr₃Si **68**	2.298(1)	16.4	0.0	472 (2,000)	61
R = Me₃SiCH₂ **81**				420	75
R₂ = Tbt(Mes) E-**69**				439 (2,000)	60
R₂ = [structure] **70**	2.454(2)	41.3 42.3	22.3	475	62
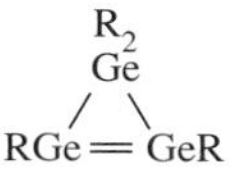 [structure] R = t-Bu₃Si **75**	2.239(4)			326 (1,300)	64
RGe=GeR R = t-Bu₃Ge **76**	2.239(5)			221 (55,000)	64
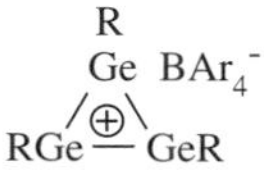 [structure] Ge BAr₄⁻ **98a** Ar = Ph	2.326(4)			276	86a
RGe—GeR **98c** (R = t-Bu₃Si) Ar = 3,5-(CF₃)₂Ph	2.335(2)			350	86b
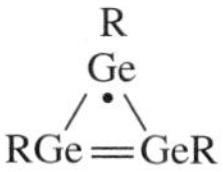 [structure] **96** (R = 2,6-Mes₂Ph)	2.35(7)				81
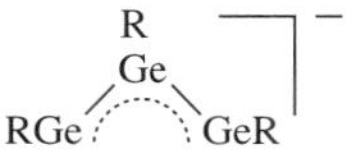 [structure] **97** R = 2,6-dimesitylphenyl	2.422(2)			266 (7,000)	81

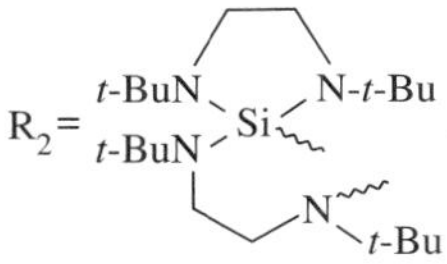

2. [2 + 2] Cycloadditions

a. *Phenylacetylene.* Tetra-*tert*-butyldigermene **78**,[73] photochemically formed from the corresponding cyclotrigermane, gives a [2 + 2] cycloaddition with phenylacetylene to form **99**[73] [Eq. (21)]. This is the first reaction of a tetraalkyldigermene with PhC≡CH, although similar cycloadditions were known from tetraaryldigermenes.[94,95] Note that a completely different reaction was reported between phenylacetylene and germanimines[96,97] or germaphosphenes,[98] which have a polarized double bond [Eq. (22)].

$$\text{t-Bu}_2\text{Ge}=\text{Ge(t-Bu)}_2 \xrightarrow[\text{hv}]{\text{PhC}\equiv\text{CH}} \begin{array}{c} \text{t-Bu}_2\text{Ge}-\text{Ge(t-Bu)}_2 \\ | \qquad | \\ \text{PhC}=\text{CH} \end{array} \qquad (21)$$

7 8   **9 9**

$$\text{Mes}_2\text{Ge}=\text{E-R} \xrightarrow{\text{PhC}\equiv\text{CH}} \begin{array}{c} \text{Mes}_2\text{Ge}-\text{E-R} \\ | \qquad | \\ \text{PhC}\equiv\text{C} \quad \text{H} \end{array} \qquad (22)$$

E = N, P

b. *Alkenes.* A [2 + 2] cycloaddition between transient 1,2-digerma-cyclobutadiene **89** and styrene leads to the bicyclic compound **100**[79] [Eq. (23)]. Together with **100**, the dihydro compound **101** was also obtained (ratio **100/101**: 64/36[79]), probably by abstraction of two hydrogen atoms from the solvent (THF). This is the first example of a cycloaddition reaction between a doubly bonded germanium compound and an alkene. A very similar reaction was reported from ethylene, styrene, and propene with the digermene intermediate **86**.[78] However, the authors say that other mechanisms can afford **88** and **102** directly from **87** without the intermediate formation of **86**.[78]

$$\left[\begin{array}{c} \text{Tsi}-\text{Ge}=\text{Ge}-\text{Tsi} \\ | \qquad | \\ \text{A}-\text{A} \end{array} \right] \xrightarrow[\text{(excess)}]{\text{R}} \begin{array}{c} \text{Tsi}-\text{Ge}-\text{Ge}-\text{Tsi} \\ | \qquad | \\ \text{A}-\text{A} \end{array} + \begin{array}{c} \text{Tsi}-\text{Ge}-\text{Ge}-\text{Tsi} \\ | \qquad | \\ \text{A}-\text{A} \end{array} \qquad (23)$$

A–A = CH$_2$CH$_2$ **86**   R = H, Me, Ph **8 8**   **1 0 2**

A–A = PhC=CH **8 9**   R = Ph **100**   **1 0 1**

Tsi = (Me$_3$Si)$_3$C

3. *[2 + 4] Cycloadditions*

a. *1,3-Dienes.* Digermenes were supposed for a long time not to react with dienes, but a [2 + 4] cycloaddition between tetrasilyldigermene **85** and butadiene has been reported[57] in the photolysis of 1,2-digermacyclobu-tane **50** in the presence of butadiene. **103** and **104** were formed in the ratio 5/1. It was suggested that the primary product of the photolysis is the digermene **85**, which under further irradiation dissociates into two germylenes[57] (Scheme 14).

The bicyclic derivative **105** was formed by a [2 + 4] cycloaddition between 2-methyl-1,3-butadiene and the hypothetical **86** [Eq. (24)].[78]

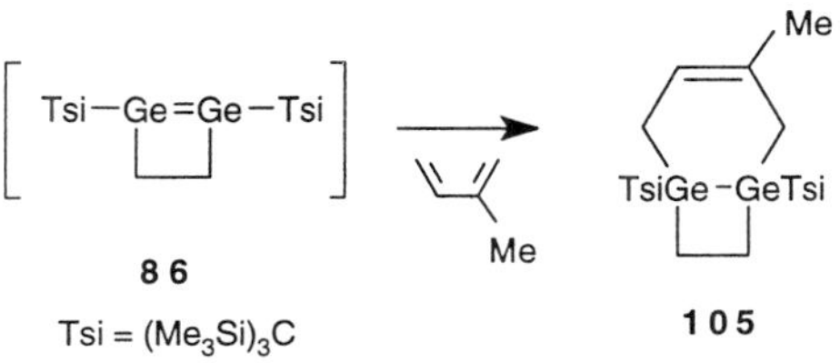

$$\text{(24)}$$

86

Tsi = $(Me_3Si)_3C$

105

b. *Quinones and Diketones.* A [2 + 4] cycloaddition was described between **78** and 3,5-di-*tert*-butyl-1,2-benzoquinone[74] affording **107** [Eq. (25)]. Trapping of germylene formed in the photolysis to give **106** was also observed.[74] The ratio between **106** and **107** was strongly dependent on the ratio of the reactants.

Attempts to trap the digermene **74** in solution with benzil afforded only the cycloadduct **109** because of the [1 + 4] cycloaddition with germylene **108**[63] [Eq. (26)]. Another proof of the formation of germylene **108** in solution from **74** was given by cryoscopic studies.[63]

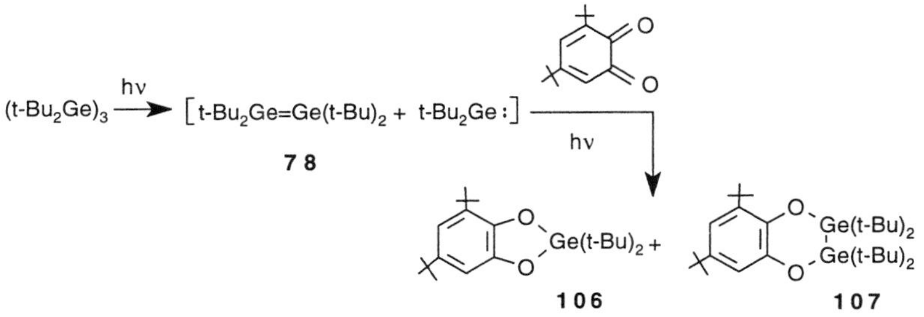

$$\text{(25)}$$

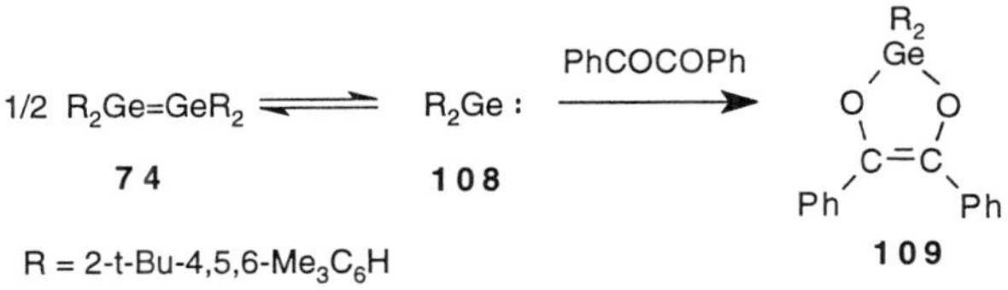

78   **106**   **107**

$$\text{(26)}$$

74   **108**   **109**

R = 2-t-Bu-4,5,6-Me_3C_6H

4. Oxygen

Oxygen reacts with digermenes **70**[62] and **74**[63] to afford in the two cases the corresponding 2,4-digerma-1,3-dioxetanes **110**[63] and **111**[62] (Scheme 15). However, two completely different mechanisms were involved in these reactions. **70** reacts with oxygen to give **111**[62] with retention of configuration, probably through the 1,2-dioxetane **112** that is initially formed and that then undergoes a spontaneous rearrangement.[62] Such a mechanism has previously been reported in the reaction of tetraaryldigermenes with oxygen, leading in this case to a stable 1,2-dioxetane[99] and also between disi-

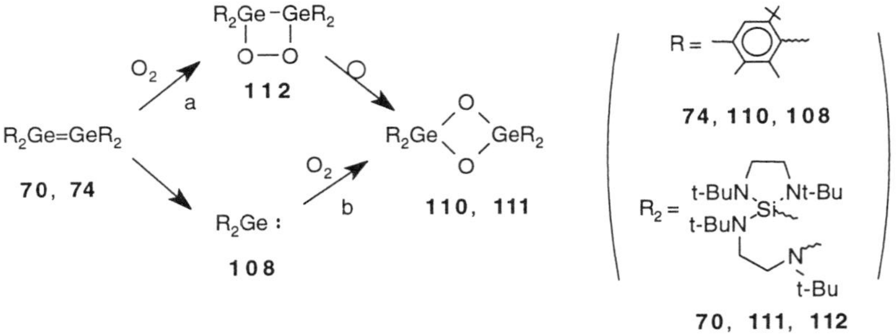

SCHEME 15.

lenes and oxygen.[100] By contrast, **110**[63] was obtained by mechanism b, involving probably a dimerization of a germanone intermediate formed from germylene **108** that exists in solution.

The absolute rate constant for quenching tetramesityldigermene with oxygen has been determined[101] by nanosecond laser flash photolysis from $(Mes_2Ge)_3$. The value determined $(2.3 \pm 0.4) \times 10^5$ M^{-1} s^{-1}) is about three orders of magnitude smaller than the values previously reported for $Me_2Ge=GeMe_2$, respectively 1.8×10^{8}[102] and 2.8×10^{8}.[102] This difference can, of course, be attributed to the stabilizing influence, both steric and electronic, of the mesityl groups. DFT pseudopotential calculations have been performed about the reaction between $H_2X=XH_2$ and oxygen or for the dimerization of $H_2X=O$[84] (X = C, Si, Ge, Sn, Pb) leading to 2,4-dimetalla-1,3-dioxetanes. The $H_2X=XH_2$ oxygenation energy decreases as expected from silicon to lead and is very exothermic for X = Ge. The Ge $\cdots$ Ge through-ring distance in these heterocycles is very short (2.668 Å) but the Ge$=$O distance is normal (1.848 Å).[84] No evidence for a Ge $\cdots$ Ge bond was found by natural bond orbital analysis, and such compounds appear as normal molecules. Note that in the case of their silicon analogues, the nature of bonding (existence or not of a possible cross-ring bonding between the silicon atoms) is now debated in the literature (for a review on disilenes, see Ref. 103 and references cited therein).

5. Decomposition of Digermene into Two Germylenes

As said previously, germylenes are obtained from digermenes **69** and **74** in solution[60,63] and from tetrasilyldigermene **85** by photolysis.[57] A similar phenomenon has already been reported,[3,4,7,8] this phenomenon depending on the substituents on germanium.

6. Reaction with Grignard Reagents,[104] Styrene, and Acrylonitrile

Such reactions have been performed on $Mes_2Ge=GeMes_2$ and also on the germasilene $Mes_2Ge=SiMes_2$ by Baines et al. (see Section IV,B).

IV

GERMASILENES (>Ge=Si<)

A. *Synthesis*

The chemistry and physicochemical characteristics of homonuclear doubly bonded compounds of the group 14 elements such as disilenes, digermenes, and distannenes, had already been extensively studied before the first mixed compound, the tetramesitylgermasilene $Mes_2Ge=SiMes_2$ **113**, was reported.[76,77] **113** is stable in solution only at low temperature[77] and a 1,2-shift of a mesityl group from germanium to silicon occurs easily, giving the silylgermylene **114** [Eq. (27)].

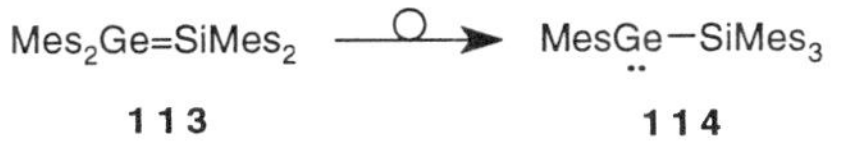

$$Mes_2Ge=SiMes_2 \xrightarrow{\quad \text{O} \quad} MesGe-SiMes_3$$

$$\text{113} \qquad\qquad\qquad\qquad \text{114}$$

(27)

To try to avoid such a 1,2-shift and to have a more stable germasilene, the two mesityl groups on silicon have been replaced by two *tert*-butyl groups.[105] Thus, the photolysis of siladigermirane **115** gave the germasilene **116**[105] and the germylene **117**. By contrast, thermolysis of **115** at 105°C for 22 h was unsuccessful. **116** appears to be stable at room temperature in the absence of light, but a photochemical cleavage is observed under continued photolysis to give the corresponding germylene and silylene[105] (Scheme 16).

Attempts to produce a germasilene by coupling a germylene and a silylene failed,[62] although the germasilene was possibly formed as intermediate (see Scheme 11).

B. *Reactivity*

Trapping of **116** with methanol or water in toluene affords regioselectively **118**[105] (Scheme 16). Note that from $Mes_2Ge=SiMes_2$ **113** the ratio of Mes_2-$Si(OH)Ge(H)Mes_2$ and $Mes_2Si(H)Ge(OH)Mes_2$ was 7/1.[106]

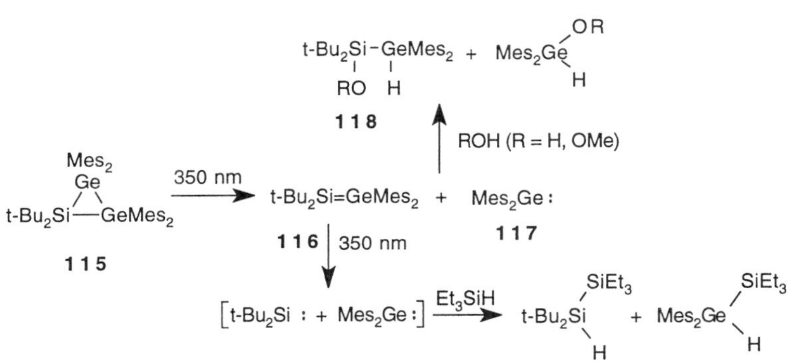

SCHEME 16.

The reactivity of **113** toward various reagents has already been studied by Baines *et al.*[76,77,105–107] This study was continued using electron-rich alkenes and electron-poor alkenes, in the presence of Et_3SiH to trap the germylene formed in the reaction (Scheme 17).[108–110] Thus, with 1-methoxy-butadiene or styrene, formal [2 + 2] cycloadditions occur.[109] With vinyl ether,[108] vinyl acetate,[108] allyltrimethylsilane,[110] *trans*-1-phenyl-2-vinylcyclo-propane,[110] or 2,3-dimethylbutadiene,[110] the 1,2-mesityl shift to give the silylgermylene **114** occurs at a faster rate than the cycloaddition and only the trapping of **114** has been observed. With acrylonitrile both types of reactions are observed: [2 + 2] cycloaddition giving **121** and trapping of germylene and silylgermylene leading to **119** and **120** in a 2/3/1 ratio.[109] Mechanistic studies performed on these reactions prove they occur via a radical pathway.[110]

An interesting comparison about the reactivity of disilene and digermene analogues toward these reagents has been made: The disilene reacts as the germasilene, giving formal [2 + 2] adducts with styrene and acryloni-trile,[108,109] whereas the digermene $Mes_2Ge=GeMes_2$ rearranges to a germ-ylgermylene at a faster rate than addition.[108,109]

As expected, the reactivity of crotonaldehyde and methyl vinyl ketone (MVK) is dominated by the reaction with the carbonyl moiety, leading

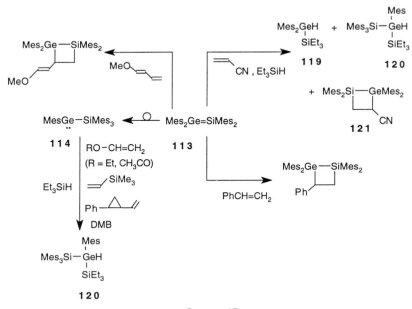

Scheme 17.

respectively to 122^{109} (exclusive formal [2 + 2] cycloaddition with regio-chemistry not definitively determined) and to **123**, **124**, and **125** in formal [2 + 2] and [2 + 4] cycloadditions in the ratio $8/1/4^{109}$ (Scheme 18). No product derived from the reaction between MVK and Mes$_2$Ge was detected. The reactivity of Mes$_2$Ge=SiMes$_2$ toward MVK falls between those of disilenes^{111} ([2 + 2] cycloadditions]) and germenes^{112} ([2 + 4] cycloadditions) and is consistent with the slight polarity of the Ge=Si double bond.

Methylmagnesium bromide (or iodide) add to Mes$_2$Ge=GeMes$_2$ and regioselectively to Mes$_2$Ge=SiMes$_2$ under mild conditions to produce a germylmagnesium halide species^{113} [Eq. (28)]. MeMgX adds more rapidly to the digermene compared to the germasilene, presumably because of the greater polarizability of the Ge=Ge double bond.

$$Mes_2M=GeMes_2 \xrightarrow[-70\,°C]{MeMgI} \underset{\underset{Me\ \ MgI}{|\ \ \ \ \ |}}{Mes_2M-GeMes_2} \xrightarrow{H^+} \underset{\underset{Me\ \ H}{|\ \ \ \ |}}{Mes_2M-GeMes_2} \qquad (28)$$

M = Si, Ge

When the reaction is performed at 110°C with an excess of Grignard reagent, an unprecedented methyl–mesityl ligand exchange is observed leading to Mes$_2$(Me)M–Ge(H)Me$_2$ (M = Si, Ge). A mechanism involving the α-elimination of MesMgX followed by the addition of MeMgX to the intermediate germylene has been proposed^{113} [Eq. (29)].

$$Mes_2M=GeMes_2 \xrightarrow[\substack{X = Br \quad M = Si \\ = I \quad \ \ = Ge}]{\substack{1)\ MeMgX \\ 2)\ NH_4Cl}} \underset{\underset{Me\ H}{|\ \ \ \ |}}{Mes_2M-GeMe_2} \qquad (29)$$

M = Si, Ge

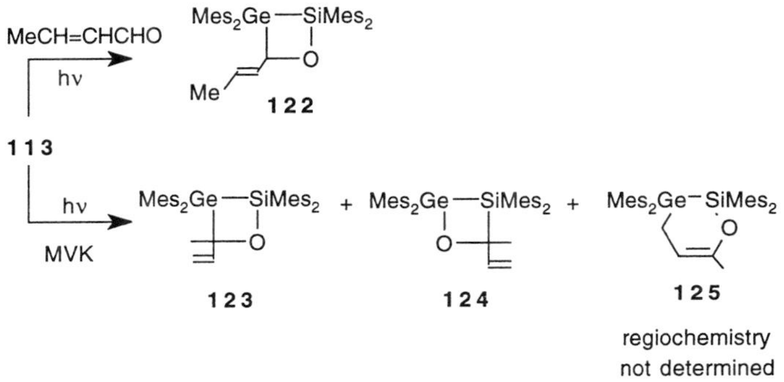

SCHEME 18.

V

STANNAGERMENE (>Sn=Ge<)

A. Synthesis

Stannagermene **126**, the first compound with a tin–germanium double bond, was synthesized by dehydrofluorination of the corresponding fluoro-stannylgermane **127** by *tert*-butyllithium in a 30/70 mixture of Et_2O/tolu-ene[114] [Eq. (30)]. The ^{119}Sn NMR showed that **128** was immediately formed at $-80°C$ (δ: 124.9 ppm, ^1JSnF: 1650.5 Hz). The stannagermene **126** was formed at about $-20°C$. Solutions of **126** are orange-red and very sensitive to air and moisture.[114]

B. Physical Properties

126 was characterized by a low-field chemical shift in^{119}Sn NMR ($+360$ ppm), in the expected range for a doubly bonded tin substituted by two Tip groups (e.g., $Tip_2Sn=SnTip_2$: 427 ppm,[115] $Tip_2Sn=PAr$ (Ar = 2,4,6-tri-*tert*-butylphenyl): 499.5 ppm,[116a] $Tip_2Sn=CR_2$ (CR_2 = fluorenylidene): 288 ppm[116b]).

$$Mes_2GeH_2 \xrightarrow[\text{2) Tip}_2\text{SnF}_2]{\text{1) t-BuLi}} \underset{\substack{| \quad | \\ F \quad H \\ \mathbf{127}}}{Tip_2Sn-GeMes_2} \xrightarrow[\text{- 80°C}]{\text{t-BuLi}} \left[\underset{\substack{| \quad | \\ F \quad Li}}{Tip_2Sn-GeMes_2} \right]$$

$$\text{(Tip = 2,4,6-iPr}_3\text{C}_6\text{H}_2)$$

$$\Bigg\downarrow_{\text{- LiF}} \quad \mathbf{128}$$

$$Tip_2Sn=GeMes_2$$

$$\mathbf{126}$$

(30)

C. Reactivity

126 was mainly characterized by trapping reactions: At $-20°C$ water and methanol add quantitatively to the Sn=Ge double bond to give **129**.[114] The reaction is regiospecific. One of the reasons is the polarity $Sn^{\delta+}Ge^{\delta-}$ of the Sn=Ge double bond, although this polarity is probably very low. Benzaldehyde gives with **126** a [2 + 4] cycloaddition to form **130** (Scheme 19). In this case also the reaction was regiospecific with oxygen bonded to Sn (^2JSnOC: 26.6 Hz, characteristic of a ^2JSnC).[114]

In the absence of trapping reagent, **126** decomposes at about room temperature to afford the distannagermirane **131**[114] ($\delta^{119}Sn$: -361.6 ppm,

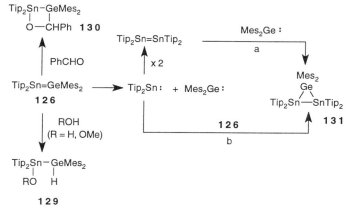

SCHEME 19.

$^1J^{117}Sn^{119}Sn$: 1440 Hz). The mechanism has not yet been elucidated, but an initial dismutation of **126** into germylene Mes_2Ge and stannylene Tip_2Sn can be postulated. Thus, the two routes a and b seem possible: addition of Tip_2Sn to the $Sn=Ge$ double bond or coupling of two stannylenes (a similar reaction has already been reported[115,117]) followed by the addition of a germylene to the $Sn=Sn$ double bond (Scheme 19). A head-to-head or head-to-tail dimerization of **126** followed by a loss of Mes_2Ge seems less probable.

In $M_{14}=M'_{14}$ compounds (M, M' = Si, Ge, Sn), only $>Si=Sn<$ derivatives have not yet been physicochemically characterized (a transient compound of this type has been postulated by Lappert *et al.* as possible intermediate in the reaction between a silylene and a stannylene leading to a silylstannylene[118]).

VI

GERMANIMINES ($>Ge=N-$)

The chemistry of germanium–nitrogen compounds was reviewed by M. Rivière-Baudet in 1995 with emphasis on her own contribution to this field and including her group's results on double-bond germanium–nitrogen species.[119] In a more recent review, J. Escudié and co-workers summarized the synthetic routes to germanimines and germaphosphenes (along with the homologous metallaimines $>M=N-$, metallaphosphenes $>M=P-$, and metallaarsenes $>M=As-$ (M = Si, Sn)) and compiled the stable ones together with their characteristic physical properties.[5b]

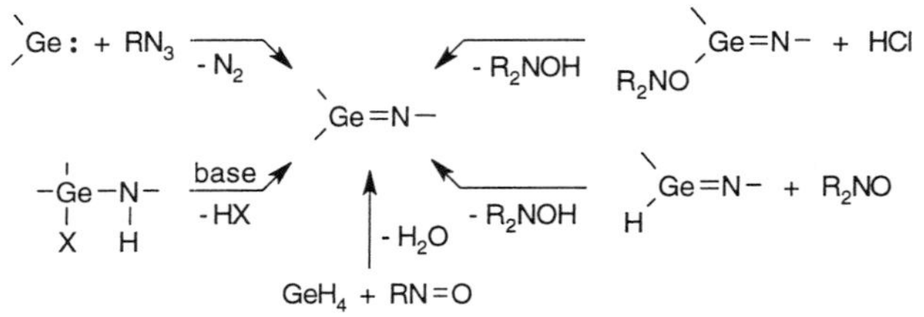

SCHEME 20.

A. Synthesis

Synthetic approaches to the hitherto described stable germanimines are summarized in Scheme 20 (for reviews see Refs. 3, 4, 5b, 6, 7, and 119) and new routes are described next.

1. Reaction between a Germylene and Diazomalonate Revisited

The reaction of the electron-rich germylene bis[bis(trimethylsilyl) amino]germane II with dimethyl 2-diazomalonate was reported in 1987 to provide the first example of stable germanimines.[120] The obviously persistent and thermodynamically stable reaction product was, on the basis of trapping experiments and in accordance with ^{1}H and ^{13}C NMR data, best considered to be the germanimine $[(Me_3Si)_2N]_2Ge=N-N=C(COOMe)_2$ **132**, obtained as the transoid isomer and converted upon storage in solution over a period of several days to the cisoid conformer **133**[120] (Scheme 21). About 10 years later, the product of this reaction between $Ge[N(SiMe_3)_2]_2$ and $N_2C(COOMe)_2$ was isolated in 95% yield. Instead of the previously presumed transoid germanimine **132**, it turned out to be the $2H$-1,3,4,2-oxadiazagermanine-5-carboxylate derivative **134**, as confirmed by an X-ray diffraction analysis.[121] The cisoid structure **133** for the second product of the reaction (formed upon storage in solution of the initial **134**) was also

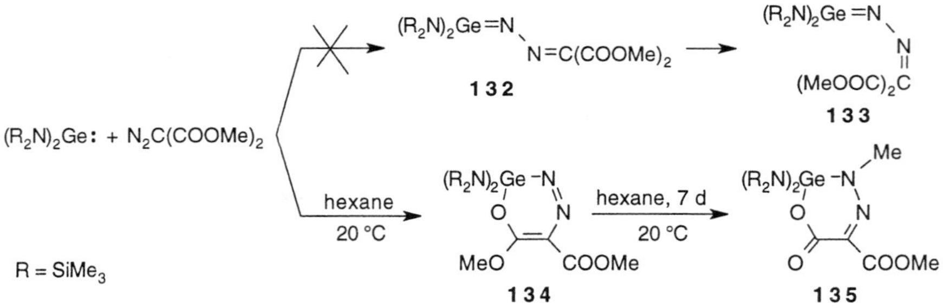

SCHEME 21.

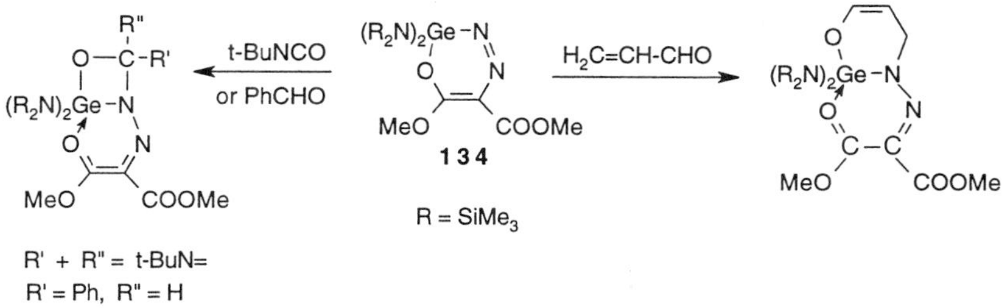

SCHEME 22.

proved incorrect. Indeed, an X-ray structure study established this second isomer to result from a methanide transfer from the methoxy group in position 6 to the ring nitrogen atom in position 3 to give **135**[122] (Scheme 21).

However, although **134** is not a germanimine, it did react with *tert*-butyl isocyanate, benzaldehyde, and acrolein to afford the heterocycles that would be expected from [2 + 2] or [2 + 4] cycloadditions with a moiety containing a Ge=N double bond[121] (Scheme 22). Whether a ring opening of **134** is induced by the formation of a donor bond from the reaction partner (thus including a five-coordinate transition state of the germanium atom) or by an equilibrium reaction proceeding via minimal amounts of the cisoid germanimine isomer was not settled.[121]

2. Dehydrohalogenation Reactions

Dehydrohalogenation reactions appeared to be a convenient route to double bond germanium nitrogen species and were commonly carried out with an organolithium compound as a base.[3,4,5b,6,7,96,97,123] Accordingly, the synthesis of two moderately hindered stable germanimines $Mes_2Ge=NR$ **136** and **137** has been reported.[59] Stabilization in these cases is achieved

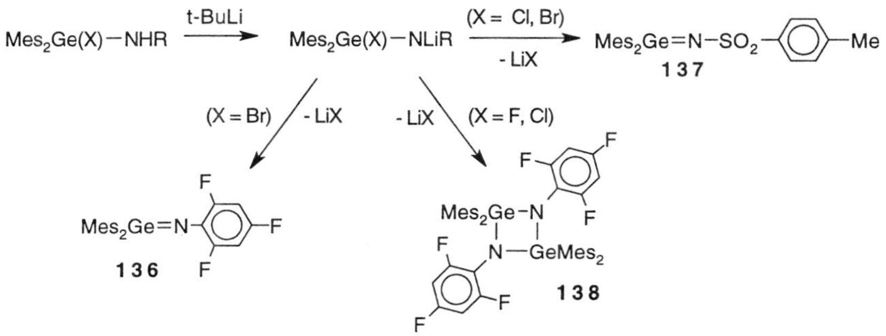

SCHEME 23.

$$Mes_2GeCl_2 + H_2NCN \xrightarrow[- 2 \, Et_3N.HCl]{2 \, Et_3N} (Mes_2Ge-NCN)_3 \xleftarrow{20 \, °C} Mes_2GeCl_2 + Li_2NCN$$

$$\mathbf{139}$$

SCHEME 24.

owing to the electron-withdrawing effects of the substituents on the nitrogen atom. The outcome of the treatment of $Mes_2Ge(X)NHR$ with *tert*-butyllithium ($X = F$, Cl, Br; $R = 2,4,6$-trifluorophenyl) depended on the nature of the halogen, leading to the expected germanimine **136** for $X = Br$, or to the cyclodigermazane $(Mes_2GeNR)_2$ **138** for $X = F$ or Cl, probably through coupling of two molecules of $Mes_2Ge(X)NLiR$ (Scheme 23).[59a,b]

A similar intermolecular process was assumed to take place in the dehydrohalogenation of dimesityldichlorogermane with cyanamide in the presence of Et_3N to yield $(Mes_2GeNCN)_3$ **139**, the trimer of the hypothetical germanimine $Mes_2Ge{=}N{-}C{\equiv}N$, as the major product. **139** was also obtained from Mes_2GeCl_2 and Li_2NCN at 20°C (Scheme 24).[124]

3. *Reaction between a Germylene and an Azide*

Coupling of a germylene with an aryl or silyl azide represents a well-established source of stable $Ge{=}N$ double bond derivatives.[125–130] The reaction of germylenes **140** and **141** with trimethylsilylazide was therefore performed to prepare the new stable base-stabilized germanimines **142**[131] and **143**,[132] respectively [Eq. (31)]. For the first time, a fluxional germanium nitrogen double bond, which originates from a reversible silatropic migration of a trimethylsilyl group, has been evidenced by a dynamic NMR study of the base-stabilized chiral germanimine **142**[131] [Eq. (32)].

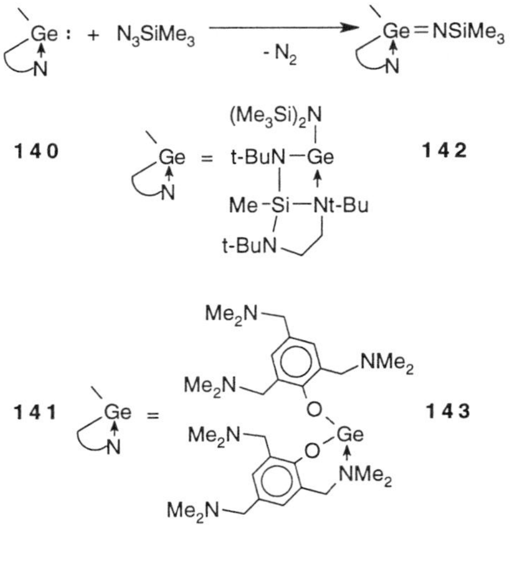

$$(31)$$

$$
\begin{array}{c}
\text{Me}_3\text{Si}\diagdown\quad\diagup\text{SiMe}_3 \\
\text{N} \\
\text{t-BuN}-\text{Ge}=\text{N}-\text{SiMe}_3 \\
\mid\quad\uparrow \\
\text{Me}-\text{Si}-\text{Nt-Bu} \\
\diagup \\
\text{t-BuN}
\end{array}
\rightleftharpoons
\begin{array}{c}
\text{Me}_3\text{Si}\diagdown \\
\text{N}\diagdown\diagup\text{SiMe}_3 \\
\text{t-BuN}-\text{Ge}-\text{N}-\text{SiMe}_3 \\
\mid\quad\uparrow \\
\text{Me}-\text{Si}-\text{Nt-Bu} \\
\diagup \\
\text{t-BuN}
\end{array}
\qquad (32)
$$

4. Rearrangement

An analogous [1,3]-sigmatropic migration of a trimethylsilyl group from a nitrogen to an oxygen atom giving a germanone intermediate has been postulated in the oxidation of the germylene **140** with trimethylamine N-oxide to lead to a transient germanimine, which further reacted with another Me_3NO molecule to form the dialkoxydiaminogermane **144**[131] [Eq. (33)].

$$
\begin{array}{c}
(\text{Me}_3\text{Si})_2\text{N} \\
\mid \\
\text{t-BuN}-\text{Ge}: \\
\mid\quad\uparrow \\
\text{Me}-\text{Si}-\text{Nt-Bu} \\
\diagup \\
\text{t-BuN} \\
\mathbf{140}
\end{array}
\xrightarrow[-\text{Me}_3\text{N}]{\text{Me}_3\text{NO}}
\left[
\begin{array}{c}
\text{Me}_3\text{Si}\diagdown\quad\diagup\text{SiMe}_3 \\
\text{N} \\
\text{t-BuN}-\text{Ge}=\text{O} \\
\mid\quad\uparrow \\
\text{Me}-\text{Si}-\text{Nt-Bu} \\
\diagup \\
\text{t-BuN}
\end{array}
\right]
\longrightarrow
\left[
\begin{array}{c}
\text{Me}_3\text{Si}\diagdown \\
\text{N} \\
\| \\
\text{t-BuN}-\text{Ge}-\text{OSiMe}_3 \\
\mid\quad\uparrow \\
\text{Me}-\text{Si}-\text{Nt-Bu} \\
\diagup \\
\text{t-BuN}
\end{array}
\right]
$$

$$
\Big\downarrow \text{Me}_3\text{NO} \qquad (33)
$$

$$
\begin{array}{c}
\text{H}\quad\text{OSiMe}_3 \\
\mid\quad\mid \\
\text{Me}_3\text{Si}-\text{N}-\text{Ge}-\text{O}-\text{CH}_2\text{N}(\text{CH}_3)_2 \\
\mid \\
\text{t-BuN}\qquad \mathbf{144} \\
\mid \\
\text{Me-Si} \\
\diagup\quad\diagdown \\
\text{t-BuN}\qquad\text{Nt-Bu}
\end{array}
$$

5. Miscellaneous Reactions

The preliminary communications on F_3C- and $(\text{F}_3\text{C})_2\text{NO}$-substituted germanimines **145**[133a,b] were gathered together in a more detailed article.[133c] Condensation reactions of GeH_4 with trifluoronitrosomethane and O-nitrosobis(trifluoromethyl)hydroxylamine afforded the non-"sterically protected" germanimines $\text{H}_2\text{Ge}=\text{NR}$ **145** [R = CF_3, $(\text{CF}_3)_2\text{NO}$], whose surprising relative stability appears to be conferred by the highly electronegative trifluoromethyl and bis(trifluoromethyl)nitroxy groups (Scheme 25). On standing at room temperature, these germanimines suffered polymerization. The H_2Ge moities underwent stepwise substitution reactions with bis(trifluoromethyl)nitroxyl, a powerful hydrogen abstractor and radical scavenger, to yield the correponding colorless liquid germanimines $[(\text{F}_3\text{C})_2\text{NO}]_n\text{H}_{2-n}\text{Ge}=\text{NR}$ **146**. The cleavage of the $(\text{F}_3\text{C})_2\text{NO}-\text{Ge}$ bond

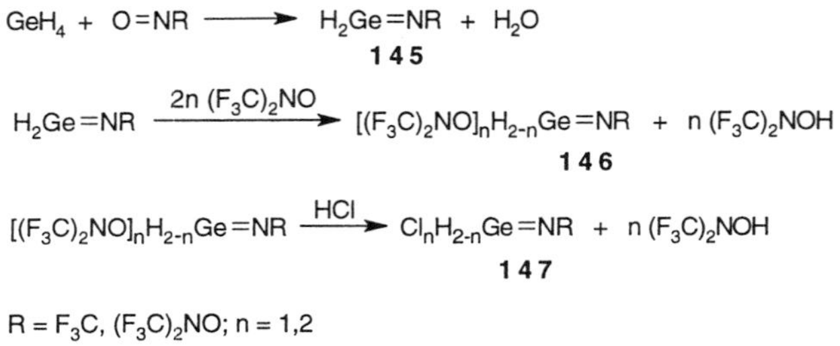

$$GeH_4 + O{=}NR \longrightarrow H_2Ge{=}NR + H_2O$$
$$\textbf{145}$$

$$H_2Ge{=}NR \xrightarrow{2n\ (F_3C)_2NO} [(F_3C)_2NO]_nH_{2-n}Ge{=}NR + n\ (F_3C)_2NOH$$
$$\textbf{146}$$

$$[(F_3C)_2NO]_nH_{2-n}Ge{=}NR \xrightarrow{HCl} Cl_nH_{2-n}Ge{=}NR + n\ (F_3C)_2NOH$$
$$\textbf{147}$$

$$R = F_3C, (F_3C)_2NO;\ n = 1,2$$

SCHEME 25.

with anhydrous hydrogen chloride afforded the corresponding chloro-germanimines $Cl_nH_{2-n}Ge{=}NR$ **147**.

B. Physical Properties

Germanimines in the preceding series[133] were isolated in 25–64% yield at low temperature by trap-to-trap fractionation under high vacuum. Molecular weight determination by Regnault's method gave satisfactory results. The presence of the $Ge{=}N$ double bond in **143**[132] and **145–147**[133] was indicated by IR absorption peaks in the 1000–1090 cm^{-1} region assigned to the stretching vibrational mode and confirmed for germanimines **145** by a regioselective addition of HI across the unsaturation.

The base-stabilized germanimine **142** formed pairs of enantiomers in the solid state, as could be proved by a single crystal X-ray diffraction analysis. The $Ge{=}N$ double bond distance of 1.701(5) Å corresponds to a shortening of 8.5% in relation to the standard Ge–N single bond length. This distance agrees well both with the computed bond length for $H_2Ge{=}NH$ (1.695 Å at the SCF level and 1.727 Å at the CI level)[39] and with the previously reported values (1.681–1.704 Å) for other structurally characterized germanimines.[5b,6,7] Steric interaction combined with the tendency of the nitrogen atom to form sp hybrids could account for the $Ge{=}N{-}Si$ angle widening (154.9°). A competition between the N-atoms of the N$_2$Si ring to form a N–Ge bond took place on heating the NMR sample.[131]

C. Reactivity

Reactivity of stable germanium–nitrogen double bond compounds has been widely documented in previous reviews.[3,4,6,7,119]

The presence of the $Ge{=}N$ double bond in germanimines **136**,[59] **137**,[59] and **143**[132] was chemically confirmed (Scheme 26). Water and alcohols (MeOH, *t*-BuOH) readily added across the Ge=N double bond with a regioselective formation of a germanium–oxygen bond.[59b,c] In the reaction of $Mes_2Ge=NSO_2PhMe$ **137** with 3,5-di-*tert*-butylorthodiphenol, only the dimesitylgermadioxolane **148** and *p*-toluenesulfonamide were character-ized.[59b] Chloroform behaved as a protic reagent leading to the transient adduct $Mes_2Ge(CCl_3)NHR$ **149**, which lost dichlorocarbene to yield the corresponding chlorogermylamine **150**.[59] A pseudo-Wittig reaction with benzaldehyde gave tetramesityldigermadioxetane **151** and the expected im-ine **152**, probably through decomposition of the initial [2 + 2] cycloadduct **153**.[59c] A [2 + 3] cycloaddition occurred between *N-tert*-butylphenylnitrone and the germanimine **136** to provide the five-membered heterocycle **154**,[59a,b] whereas no reaction was observed with **137**[59c] (Scheme 26).

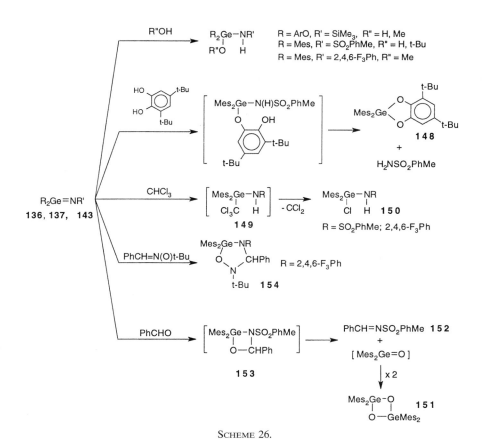

SCHEME 26.

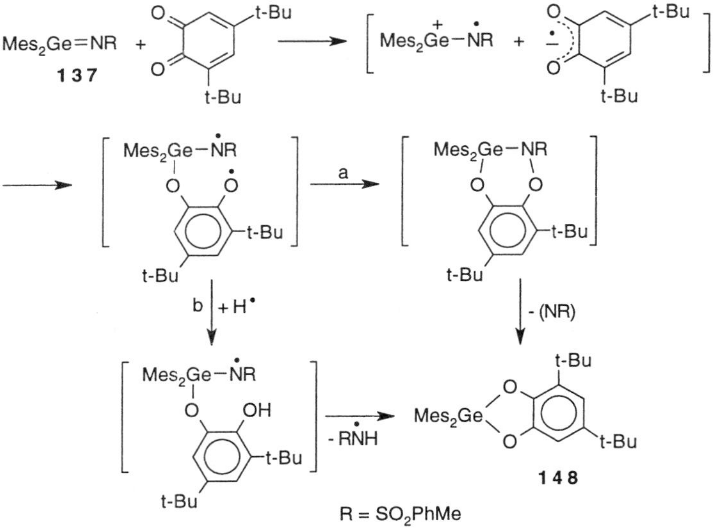

SCHEME 27.

In the reaction of $Mes_2Ge=NSO_2PhMe$ **137** with 3,5-di-*tert*-butyl-*o*-benzoquinone, a single electron transfer process was evidenced by an ESR study, as well as by the characterization of isobutene. The initially formed *o*-semiquinonic species could afford the dimesitylgermadioxolane **148** according to two different pathways[59c] (Scheme 27).

VII

GERMAPHOSPHENES (>Ge=P-)

A. *Synthesis*

Synthetic routes to stable germaphosphenes have previously been reviewed.[3,4,5b,7] The most general method consists in dehydrohalogenation of halogermylphosphanes. Bases such as DBU or $Me_3P=CH_2$ have marginally been employed and reported to lead to germaphosphene in low yields. Preference for fluorine as the halogen thwarts side reactions, and best yields are attained when experiments are carried out with *tert*-butyllithium because of its low nucleophilicity but high basicity.[134,135] Closely related to this method is the reaction between difluorodimesitylgermane and 2,4,6-triisopropylphenyldilithiophosphide to synthesize the corresponding stable germaphosphene[136] (Scheme 28).

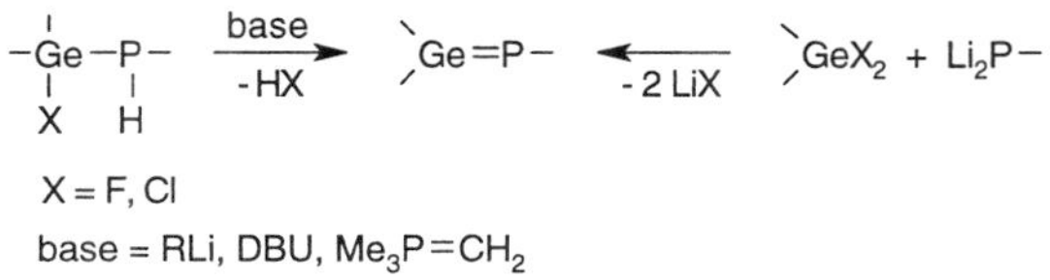

$$-\overset{|}{\underset{X}{Ge}}-\overset{|}{\underset{H}{P}}- \quad \xrightarrow[\text{- HX}]{\text{base}} \quad \diagup\hspace{-2pt}Ge{=}P{-} \quad \xleftarrow{\text{- 2 LiX}} \quad \diagup\hspace{-2pt}GeX_2 + Li_2P{-}$$

X = F, Cl

base = RLi, DBU, $Me_3P{=}CH_2$

SCHEME 28.

B. *Physical Properties*

Germaphosphenes thus prepared have been characterized by a low-field shift in ^{31}P NMR spectroscopy (δ ^{31}P: 145–175 ppm).[134–136] The Raman spectra exhibited strong lines at 503 and 501.5 cm^{-1}, respectively, for Mes_2-Ge=PAr[137] and (E)-Mes(t-Bu)Ge=PAr[135] (Mes = 2,4,6-trimethylphenyl; Ar = 2,4,6-tri-*tert*-butylphenyl). In the solid state, these two compounds displayed a nearly planar arrangement around the Ge=P unit, with a Ge=P double bond length of 2.139(3) Å and 2.143(4) Å, respectively, corresponding to a shortening of 8–9% in relation to the standard Ge–P single bond distance, in accordance with the sum of covalent radii of sp^2 hybridized germanium and phosphorus atoms (2.12 Å).[135,137]

C. *Reactivity*

The reactivity of stable germaphosphenes has been previously investigated and reviewed.[3,4,7] The chemical behavior of $Mes_2Ge{=}PAr$ **155** toward orthoquinones (tetrachloro-o-benzoquinone, 3,5-di-*tert*-butyl-o-benzoquinone, and 1,2-naphthoquinone) and α-diketones (benzil and biacetyl) has been examined. The reaction with o-quinones involved probably radical intermediates, and in each case only the less encumbered isomer of the corresponding [2 + 4] cycloadduct **156** was formed. Benzil led in a very similar manner to a 1,4,2,3-dioxagermaphosphin-5-ene **157**, whereas, as was observed with acetone, an ene-reaction occurred with biacetyl to afford **158**[138] (Scheme 29).

VIII

GERMANONES (>Ge=O)

A. *Synthesis and Reactivity*

Even though much effort has been directed toward the study of multiple bonding to germanium, compounds that exhibit Ge=O double bonding are almost completely limited to reactive intermediates.[1,3]

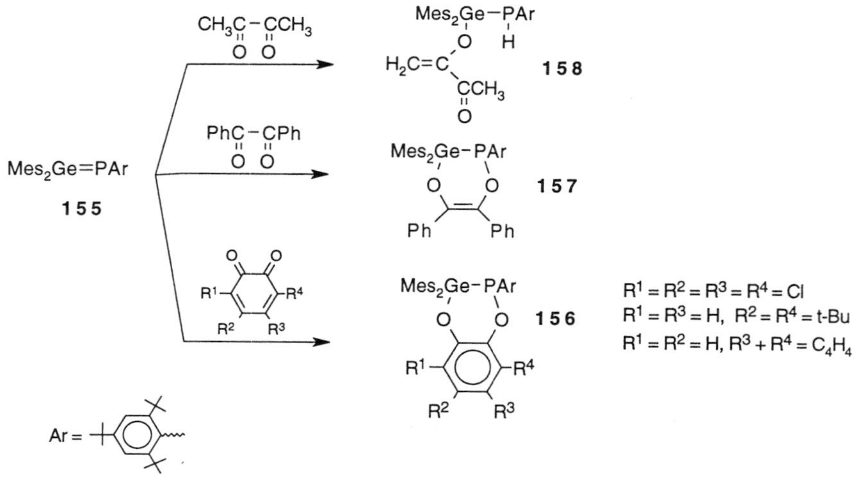

SCHEME 29.

1. Oxidation of Germylenes

Formation of the germaindanol **159**[139] by oxidation of the germylene $Ar_2Ge:$ (Ar = 2,4,6-tri-*tert*-butylphenyl) with trimethylamine *N*-oxide supported the early formation of the germanone $Ar_2Ge=O$. Obviously, the steric demand of the Ar group is not sufficient to stabilize the double-bonded germanium species, which could be neither isolated nor trapped and rapidly suffered an insertion of the $Ge=O$ unit into a CH bond[139] [Eq. (34)]. Similar behavior was observed from $Ar_2Ge=S$.[140]

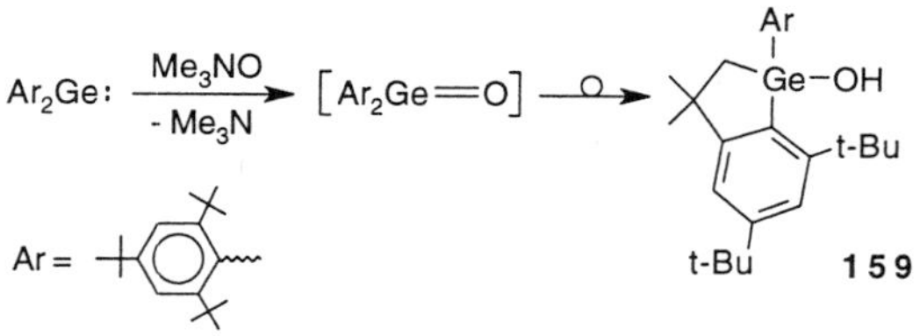

Intramolecular Lewis base complexation of the germanium atom also did not appear to be a valuable alternative to overcome the lability of these elusive intermediates, since the germanones expected from oxidation of the base-stabilized germylenes **140**[131] (with Me_3NO) and **141**[132] (with O_2, DMSO, or pyridine oxide) either rearrange as already mentioned [Section VI,A,4, Eq. (33)][131] or dimerize to digermadioxetane **160**[132] [Eq. (35)].

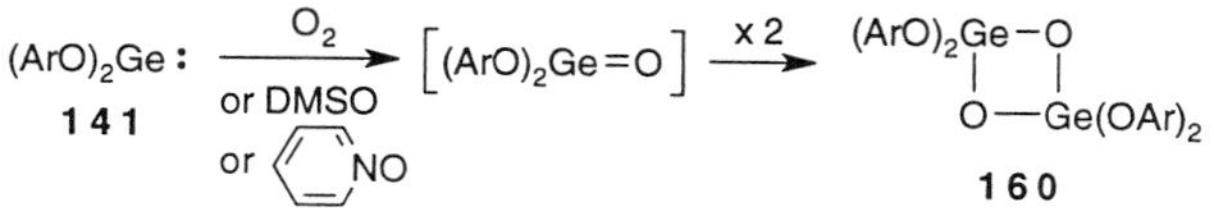

$$(ArO)_2Ge: \xrightarrow[\substack{\text{or DMSO} \\ \text{or } \includegraphics{NO}}]{O_2} \left[(ArO)_2Ge=O\right] \xrightarrow{\times 2} \substack{(ArO)_2Ge-O \\ | \quad | \\ O-Ge(OAr)_2}$$
$$\mathbf{141} \qquad\qquad\qquad\qquad\qquad\qquad\qquad \mathbf{160}$$

(35)

Ar = 2,4,6-tris[(dimethylamino)methyl]phenyl

Formation of the first germanone stable in solution, Tbt(Tip)Ge=O **55**,[141] was reported in 1995 from oxidation of the overcrowded germylene Tbt(Tip)Ge: with tribenzylamine *N*-oxide as an oxygen source. Although the germanone **55** was not spectroscopically characterized, a conclusive trapping experiment carried out 10 min after the disappearance of the blue color of the starting germylene demonstrated its existence at least for this lapse of time. For instance, the expected [2 + 3] cycloadduct **161** from a Ge=O moiety with MesCNO was obtained upon treatment with mesitonitrile. On standing for a longer period (10 h) in solution in the absence of trapping reagents, two diastereoisomeric intramolecular cyclization products **162** were formed[141] (Scheme 30).

2. Thermal Decomposition of Heterocycles

The large ring strain in the kinetically stabilized 1,2,4-oxazagermete **163**, as reflected by the large deviation of the CGeO bond angle from the tetrahedral value as well as by a long Ge–O bond length, should enhance the potential utility of such a small-ring derivative as a precursor of low coordinate germanium compounds. Thus, as expected, generation of intermediate germanone Tbt(Tip)Ge=O **55** upon heating a C_6D_6 solution of **163** was clearly indicated by the formation of the intramolecular cyclization products **162** and **164**.[145] Surprisingly, an ene reaction took place with 2,3-

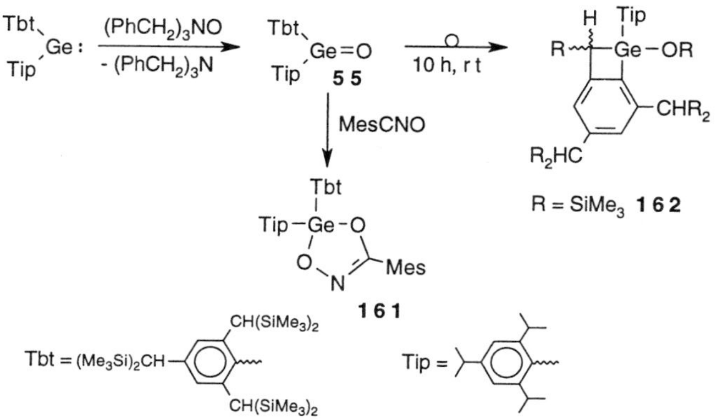

SCHEME 30.

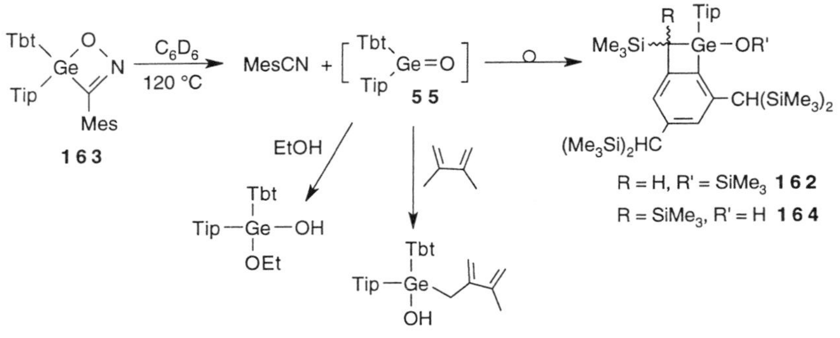

SCHEME 31.

dimethylbutadiene, whereas [4 + 2] cycloadditions are usually observed with heavier chalcogen double bond species $Tbt(Tip)Ge=E$ (E = S, Se, Te).[142–144] Trapping experiments of **55** with ethanol showed the addition of EtOH across the double bond system to occur more rapidly than the intramolecular cyclization process[145] (Scheme 31).

55 was also evidenced in the fragmentation of the 1,2-dioxaspirogermet-ane obtained by oxidation of germene **5**[54] (see Section II,C,4).

B. *Theoretical Studies on Germachalcogenones* $>Ge=E$ *(E = O, S, Se, Te)*

The relative stabilities of $R_2Ge=X$, *trans*-RGe–XR, and *cis*-RGe–XR isomers (X = O, R = H, Me; X = S, Se, R = H) were investigated by ab initio and DFT pseudopotential calculations.[84a,146] In fair agreement with the earlier computational studies,[39,147–149] the carbene-like structures lie very close in energy and the stability order of the *cis* and *trans* isomers depends on the basis set employed. The HGeOH isomers are favored over the strongly polarized double bond form $H_2Ge=O$. Methyl substitution preferably stabilizes the $Ge=O$ double bond isomer. Thus, the *cis*- and *trans*-germylenes HGeOH are located ~20 to 30 kcal/mol (depending on the calculation levels) lower than the parent germanone $H_2Ge=O$ and the energy differences are reduced to ~10 kcal/mol between dimethylgermanone and its divalent isomers. The energy difference between germylenes HGeXH and germachalcogenone H_2GeX isomers also decreases down group 16. Germanethione $H_2Ge=S$ is only 3–5 kcal/mol less stable than its *cis* and *trans* isomers, whereas MP4(SDTQ) and MP2 level calculations with the TZP($2df$, $2pd$) basis set predict germaneselone $H_2Ge=Se$ to be the global minimum isomer. A relatively high barrier (46.4 kcal/mol) was found for the $H_2Ge=Se \rightarrow$ *trans*-HGeSeH isomerization. Hydrogenation and cyclodi- and trimerization reactions of $H_2Ge=O$ are computed to be exothermic.[84]

IX

GERMANETHIONES (>Ge=S), GERMANESELONES (>Ge=Se), AND GERMANETELLONES (>Ge=Te)

A. *Synthesis*

More significant advances have been encountered in the field of the heavier homologous germanethiones, germaneselones, and germanetellones, as illustrated by the successful isolation and structural characterization of stable >Ge=E species (E = S; Se; Te).[4,7]

1. *Thermal Decomposition of 4-, 5-, and 6-Membered Ring Heterocycles with a Ge–Chalcogen Bond*

Desulfurization or deselenization of tetrathia- or selenagermolanes Tbt(Tip)GeE$_4$ (E = S; Se) using trivalent phosphorus derivatives afforded the first stable free germanethione Tbt(Tip)Ge=S **165**[142] and germaneselone Tbt(Tip)Ge=Se **166**[143] (Scheme 32). The outstanding stability of these "heavy ketones" follows from the steric demand of the combination of the overcrowded Tbt and Tip groups. Tokitoh and Okazaki focused on the potential of their dechalcogenation methodology to generate heavier congeners of carbonyl compounds >M=E (M = Si, Ge, Sn, Pb; E = S, Se, Te) and highlighted the scope of their research group's results, which cover the whole group 14 metals series, in an overview.[150]

A thermal retrocycloaddition of 5,5-diphenyl-1,2,4,3-trithiagermolanes **167** generated diarylgermanethiones Tbt(R)Ge=S (R = Mes, Tip), which under such severe reaction conditions (120°C in a sealed tube) were able to undergo [2 + 4] and [2 + 2] cycloadditions with 2,3-dimethyl-1,3-butadiene and dimethyl acetylenedicarboxylate, respectively. Formation of 1-thia-2-germacyclopentenethiones **169** was rationalized in terms of a further decomposition of the highly strained [2 + 2] cycloadducts thiagermetes **168**. Self-dimerization of the less hindered germanethione Tbt(Mes)Ge=S was also observed (Scheme 33).[151]

Fragmentation of the 4-membered ring thiadigermetane **20** into the tran-

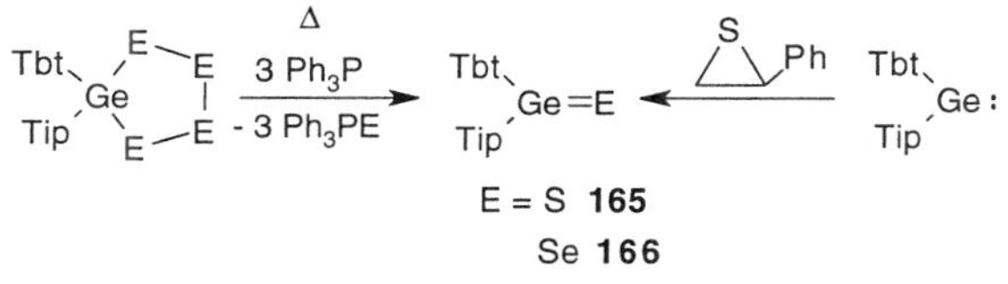

E = S **165**
Se **166**

SCHEME 32.

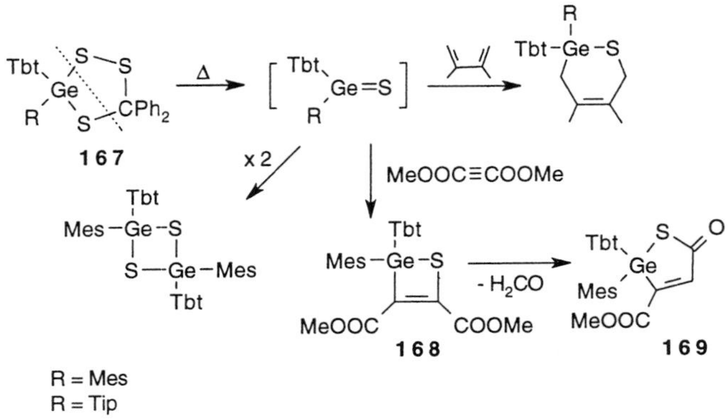

SCHEME 33.

sient germanethione $Me_2Ge=S$ **21** has already been mentioned (see Section II,A,3).[29]

2. Reactions between Germylenes and Chalcogens

Reaction of germylenes with elemental chalcogens (sulfur, selenium, tellurium) also allowed the synthesis of stable $>Ge=E$ species. This was first exemplified by the isolation of terminal chalcogenido complexes of germanium **170**[152] and **171–173**[153] (Scheme 34). Stabilization in these cases is achieved by intramolecular complexation of the metal atom by Lewis bases.

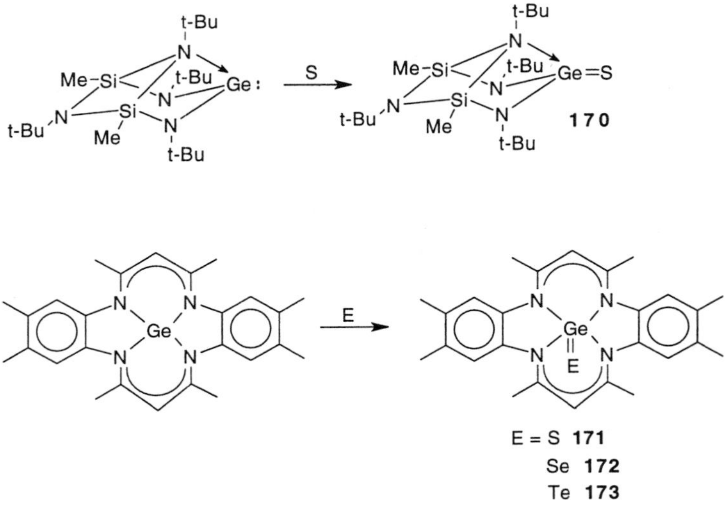

SCHEME 34.

More recently, other base-stabilized germylenes were employed in preparing novel tetra- or pentacoordinate germanechalcogenones **174–176**,[154] **177**,[131] **178**,[132,155] **179**,[132] and **180–183**[156] (Scheme 35).

The first tricoordinate germanetellones **184** and **185** were obtained and stabilized according to the same synthetic route, by taking advantage of

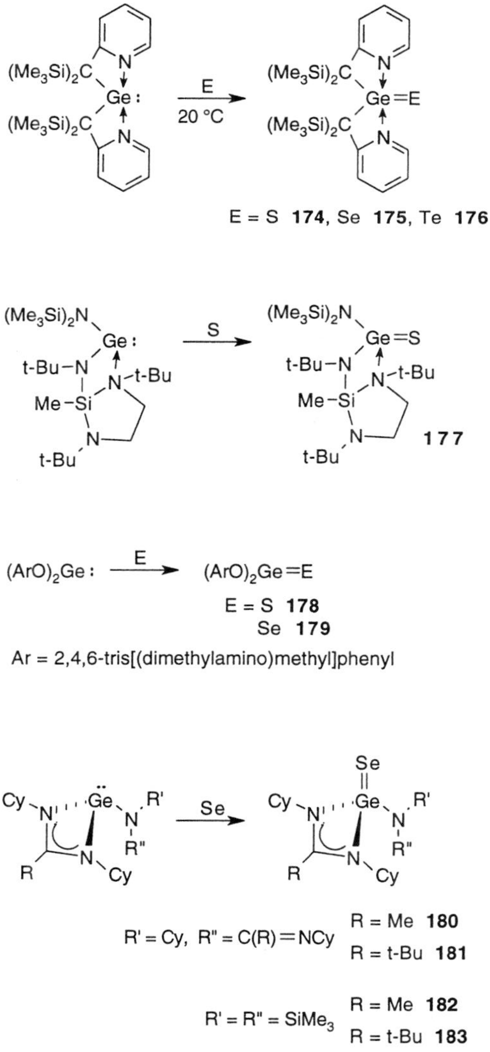

E = S **174**, Se **175**, Te **176**

177

(ArO)$_2$Ge : $\xrightarrow{\text{E}}$ (ArO)$_2$Ge=E

E = S **178**
Se **179**

Ar = 2,4,6-tris[(dimethylamino)methyl]phenyl

R' = Cy, R" = C(R)=NCy

R = Me **180**
R = t-Bu **181**

R' = R" = SiMe$_3$

R = Me **182**
R = t-Bu **183**

Scheme 35.

the efficient steric protection of the substituents. Hence, direct telluration of the severely congested germylenes Tbt(Tip)Ge: and Tbt(Dis)Ge:, generated by a thermal cycloreversion of the corresponding germirenes **2** or **186,** occurred when the three-membered ring Ge derivatives were allowed to react with elemental tellurium in a sealed tube at 90°C[144] [Eq. (36)].

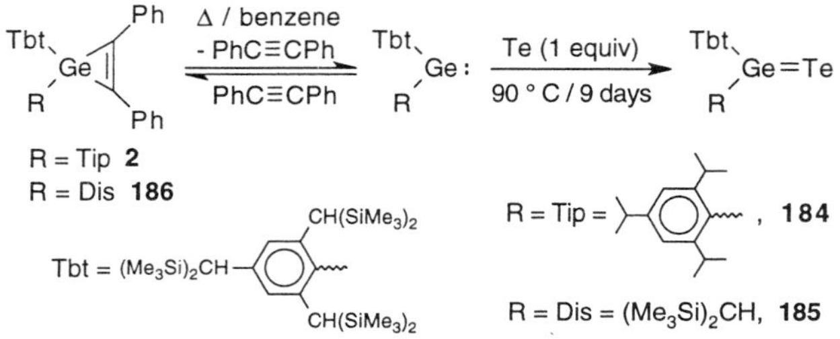

$$R = Tip \; \mathbf{2}$$
$$R = Dis \; \mathbf{186}$$

$$Tbt = (Me_3Si)_2CH$$

$$R = Tip = \quad , \; \mathbf{184}$$
$$R = Dis = (Me_3Si)_2CH, \; \mathbf{185}$$

(36)

Chalcogenation of a divalent germanium compound with styrene sulfide has been examined as an alternative route to the first free germanethione Tbt(Tip)Ge=S **165**[142] (Scheme 32) and later on allowed the synthesis of new base-stabilized germanethiones **187** and **188**[156] [Eq. (37)]. Phenyl isocyanate also may serve as a sulfur source leading to **165**, which was evidenced by electronic spectroscopy and underwent a subsequent [2 + 2] cycloaddition with phenyl isocyanate[157] (Scheme 36).

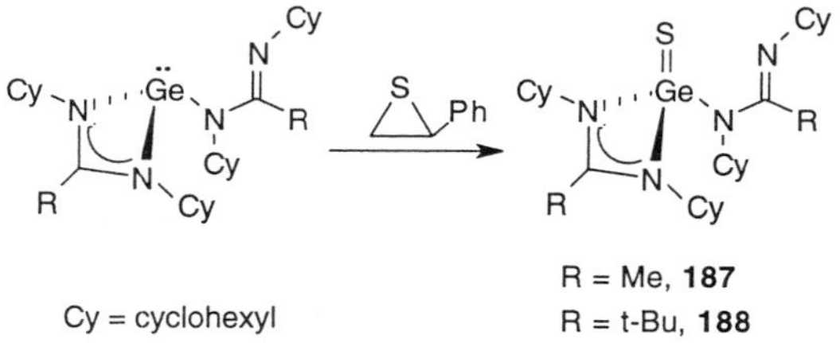

Cy = cyclohexyl

$$R = Me, \; \mathbf{187}$$
$$R = t\text{-}Bu, \; \mathbf{188}$$

(37)

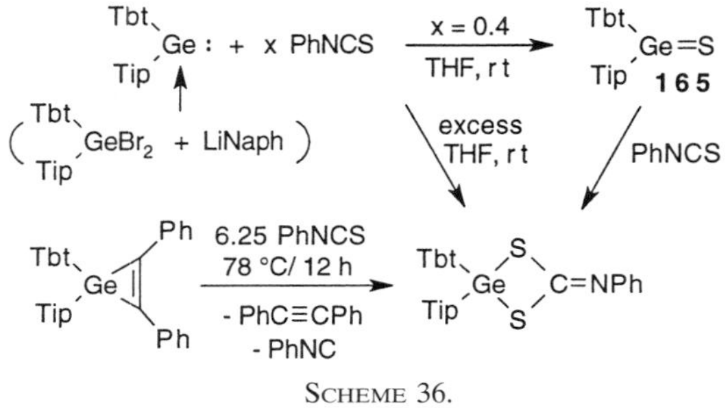

SCHEME 36.

3. *Chalcogen Exchange*

The presence of a fragment corresponding to the loss of selenium in mass spectroscopy under electronic impact of germaneselone **179** points to a thermal instability of the Ge=Se bond.[132] Heating a solution of **179** in benzene at 80°C in the presence of sulfur leads to a chalcogen exchange, likely via the divalent species $(ArO)_2Ge$ **141** [Eq. (38)].

$$(ArO)_2Ge{=}Se \xrightarrow[-Se]{} \left[(ArO)_2Ge:\right] \xrightarrow{1/8\ S_8} (ArO)_2Ge{=}S$$

$$\mathbf{179} \qquad\qquad\qquad\qquad\qquad \mathbf{178} \qquad\qquad (38)$$

Ar = 2,4,6-tris[(dimethylamino)methyl]phenyl

$\eta^4\text{-}Me_8taaGe{=}Se$ (or Te) **172** and **173** were converted into the terminal sulfido complex **171** when treated with ethylene sulfide. By analogy with the tin system,[158] a potential mechanism for this chalcogen exchange reaction likely involves the formation of a mixed 1,3-dichalcogenido-2-germacyclopentane intermediate followed by cycloreversion[158] [Eq. (39)].

$$(\eta^4\text{-}Me_8taa)Ge{=}E \xrightarrow[-CH_2{=}CH_2]{\overset{S}{\triangle}} (\eta^4\text{-}Me_8taa)Ge{=}S$$

E = Se **172**
Te **173**   **171**

$(\eta^4\text{-}Me_8taa)Ge$ =

$$(39)$$

B. *Physical Properties*

1. *X-Ray Studies*

X-ray structure analyses of germanetellones $Tbt(R)Ge{=}Te$ (R = Tip, Dis) are reminiscent of those of Tbt-substituted stable germanethione $Tbt(Tip)Ge{=}S$[142] and germaneselone $Tbt(Tip)Ge{=}Se$[143] and disclosed a trigonal planar geometry of the germatellurocarbonyl units of $Tbt(R)Ge{=}Te$, with the sum of bond angles around the germanium atom being 359°5 (R = Tip) and 360° (R = Dis) (Table II). The Ge–Te distances [d Ge=Te: 2.398(1) and 2.384(2) Å, respectively] are ca. 8% shortened in relation to those reported for typical Ge–Te single bonds (2.59–2.60 Å) and agree well with the sum of covalent radii for sp^2 hybridized germanium and tellurium atoms (2.39 Å).

A distorted tetrahedral geometry was revealed for Ge by X-ray investigation of terminal selenido complexes **181** and **182**.[156] The respective Ge=Se

TABLE II

SELECTED PHYSICAL DATA OF DOUBLY BONDED GERMANIUM COMPOUNDS

Compound	Ge=E (Å)	Bond angles (°)	NMR data	Ref.
$(Me_3Si)_2N^1$ Ge=N^4-SiMe$_3$ t-Bu-N^2 N^3 Me-Si t-Bu-N **142**	1.701(5)	N^{4}GeN^1 111.73(3); N^{4}GeN^2 125.0(2) N^{1}GeN^2 115.1(2); N^{2}GeN^3 78.5(2)		131
$(Me_3Si)_2N^1$ Ge=S t-Bu-N^2 N^3 Me-Si t-Bu-N **177**	2.090(2)	SGeN^1 113.7(1); SGeN^2 119.5(1) N^{1}GeN^2 115.6(2); N^{1}GeN^3 78.2(1)		131
Se / Cy ... Cy-N^2...Ge, N^1 t-Bu, t-Bu-N^3-Cy, Cy **181**	2.196(4)	SeGeN^1 120.0(5); SeGeN^2 122.3(4) SeGeN^3 121.4(4); N^{2}GeN^3 65.6(6)	δ ^{77}Se 1023.8	156
Se / Cy-N^2...Ge, N^1(SiMe$_3$)$_2$, Me-N^3-Cy **182**	2.2212(3)	SeGeN^1 120.52(5); SeGeN^2 115.18(5) SeGeN^3 121.03(5); N^{2}GeN^3 67.14(6)	δ ^{77}Se 1115.6	156
(pyridyl N^2) $(Me_3Si)_2C^2$ Ge=Se $(Me_3Si)_2C^1$ N^1 (pyridyl) **175**	2.2472(7)	SeGeC^1 119.43(11); SeGeC^2 119.30(10) C^{1}GeC^2 121.26(15); C^{1}GeN^1 67.16(3) N^{2}GeC^2 67.40(13)	δ ^{77}Se −97.6	154

bond lengths of 2.196(4) and 2.2212(3) Å correlated favorably with the comparable reports in the literature.[143,153]

Pentacoordinate Ge complexes **175** and **176**[154] were found to be isostructural in the solid state, with the only significant difference occurring in the Ge–element bond. For both structures, the observed Ge–E distances

TABLE II (continued)

Compound	Ge=E (Å)	Bond angles (°)	NMR data	Ref.
176	2.4795(5)	TeGeC^1 119.73(6); C^{1}GeC 120.55(12) C^{1}GeN^1 67.12(8)	δ ^{125}Te −460.9	154
Tbt(Tip)Ge=Te **184**	2.398(1)		δ ^{125}Te 1143	
Tbt(Dis)Ge=Te **185**	2.384(2)	TeGeC(Dis) 118.7(3); CGeC 118.0(4) TeGeC(Tbt) 123.3(3)	1009	144

[d Ge–Se = 2.247(1) Å and d Ge–Te = 2.480(1) Å] lie between the values calculated for respective Ge–E single [E = Se (2.39 Å); Te (2.59 Å)] and double bond lengths [E = Se (2.19 Å); Te (2.39 Å)] and match those reported for η^4-Me$_8$taaGe=E (E = Se, Te).[153] The Ge–N distances in **175**[154] [mean 2.169(3) Å] and in **176**[154] [2.171(2) Å], significantly larger than the sum of covalent radii of Ge and N (1.92 Å), correspond to a dative bonding situation. The intramolecular chelation of the nitrogen atoms leads to a pseudo–trigonal bipyramidal geometry with a trigonal planar arrangement of the chalcogen and the two carbon atoms around the germanium center (the sum of bond angles being 360°).

Physical properties of **177**[131] parallel those previously mentioned for the germanimine **142** (stemming from the same germylene **140**), particularly the formation of pairs of enantiomers in the solid state, an analogous marked shortening of the Ge=S distance [d Ge=S: 2.090(2) Å] in comparison to the common Ge–S single bond length (2.26 Å), and the temperature dependence of the ^{1}H NMR spectrum.

2. ^{77}Se and ^{125}Te NMR Studies

Terminal selenido complexes **181** and **182**[156] with a tetracoordinate germanium atom displayed downfield shifted signals in ^{77}Se NMR spectroscopy (δ ^{77}Se = 1023.8 and 1115.6 ppm vs Me$_2$Se, respectively) comparable to that of the structurally characterized Tbt(Tip)Ge=Se (δ ^{77}Se = 940.6 ppm). Similar deshielded singlets are observed in the ^{125}Te NMR spectra of the germanetellones Tbt(R)Ge=Te [δ ^{125}Te = 1143, (R = Tip); 1009 ppm (R = Dis)].[144] In contrast, pentacoordinate Ge chalcogenido complex **175**[154] exhibits a more shielded ^{77}Se NMR signal ($\delta\ ^{77}$Se $= -97.6$ ppm) but still at lower field than is typical for compounds containing a Ge–Se single bond [e.g., δ (H$_3$Ge)$_2$Se $= -612$ ppm] (Table II).

3. UV Studies

Interestingly, the green or blue-green germanetellones are not light-sensitive. In the electronic spectra, n–π^* transitions gave rise to characteristic absorption maxima at 636 nm for Tbt(Tip)Ge=Te[144] and 599 nm for Tbt(Dis)Ge=Te,[144] which are red shifted relative to those observed for germanethione Tbt(Tip)Ge=S (444 nm)[142] and germaneselone Tbt(Tip) Ge=Se (513 nm).[143]

C. Reactivity

The reactivity of stable Ge=E species has been previously reviewed.[4,7]

The regioselective addition of water and methanol across the Ge=E unsaturation of **178** and **179** produces the corresponding hydroxy- or methoxygermanethiol or -selenol[132] [Eq. (40)].

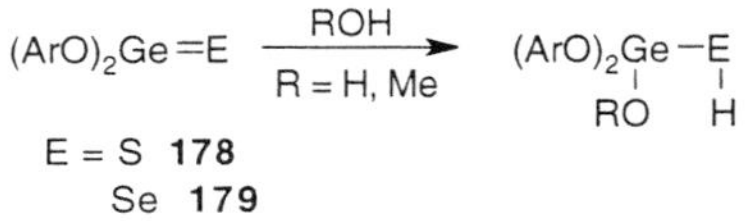

$$(ArO)_2Ge{=}E \xrightarrow[R = H, Me]{ROH} \underset{RO\quad H}{(ArO)_2Ge{-}E} \tag{40}$$

$$E = S\ \mathbf{178}$$
$$Se\ \mathbf{179}$$

An immediate reaction takes place between **178** or **179** and 3,5-di-*tert*-butyl-1,2-benzoquinone to lead to the dioxagermole **189**, the formal adduct of the germylene (ArO)$_2$Ge: with the quinone. A single electron mechanism, supported by an ESR study and depicted in Scheme 37, was postulated to account for the formation of **189**.[132]

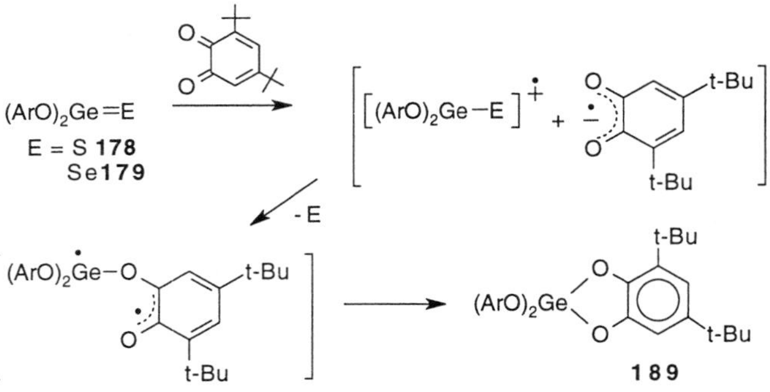

Ar = 2,4,6-tris[(dimethylamino)methyl]phenyl

SCHEME 37.

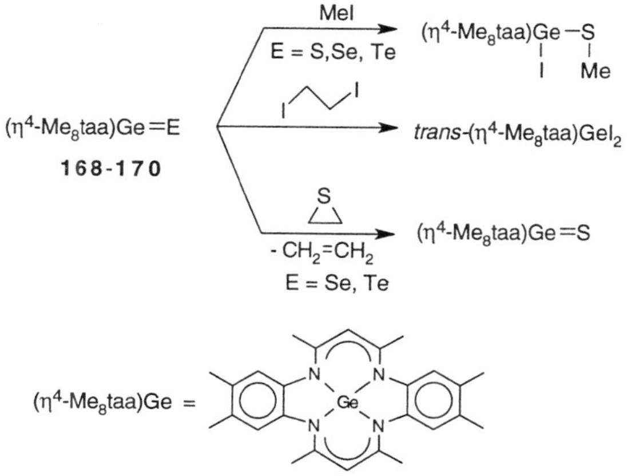

SCHEME 38.

Alkylation of the terminal chalcogenido complexes $(\eta^4\text{-Me}_8\text{taa})\text{Ge}=E$ (X = S, Se, Te) **168–170** with MeI afforded the corresponding methyl-chalcogenate derivatives in accordance with a previous regioselective addition across a base-stabilized $\text{Ge}=S$ double bond.[159] The diiodide complex $(\eta^4\text{-Me}_8\text{taa})\text{GeI}_2$ rather than a chalcogenate derivative is obtained in the reaction with 1,2-diiodoethane. As mentioned previously, a chalcogen exchange to give the terminal sulfido complex took place when $(\eta^4\text{-Me}_8\text{taa})\text{Ge}=E$ (E = Se, Te) were treated with ethylene sulfide[158] (Scheme 38).

Overheating the reaction mixture during the preparation of the base-stabilized germachalcogenones **174, 175** from the corresponding germylene led to the migration of a trimethylsilyl group to the chalcogen atom, yielding the compounds **190** and **191**[154] [Eq. (41)].

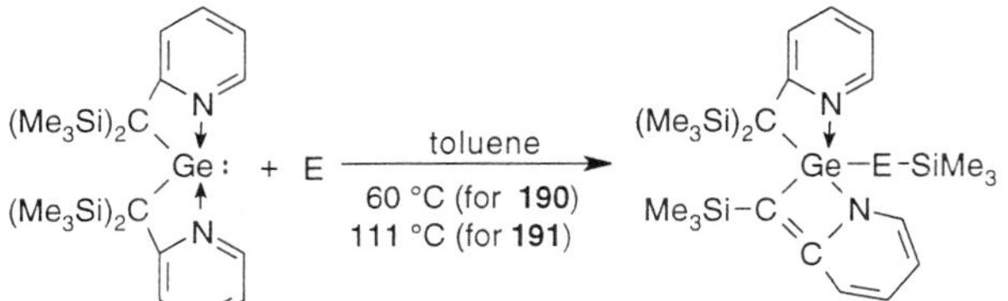

$$E = S \ \mathbf{190}$$
$$Se \ \mathbf{191}$$

(41)

Like their stable "heavy ketone" congeners, the germanetellones $\text{Tbt(R)Ge}=\text{Te}$ underwent [2 + 3] and [2 + 4] cycloadditions with mesitonitrile oxide and 2,3-dimethyl-1,3-butadiene, respectively[144] (Scheme 39).

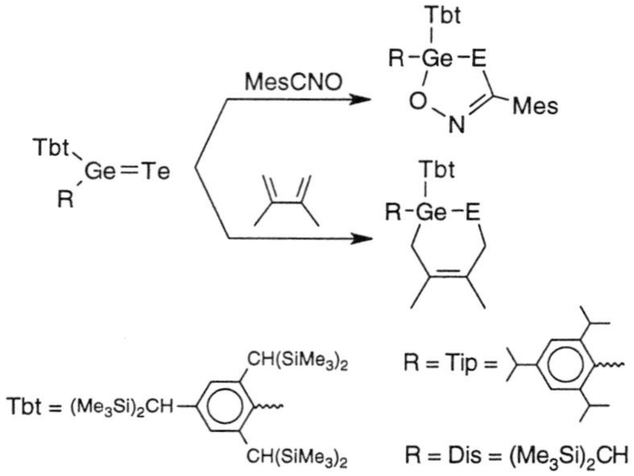

SCHEME 39.

X

GERMAALLENES (>Ge=C=X)

If doubly-bonded compounds $>M=X$ (M = Si, Ge, Sn) are now well known, this is not yet the case for the heavy allenic derivatives $>M=C=X$. Transient silaallenes $>Si=C=C$ [160–167] or silaazaallene $>Si=C=N-$ [168] were postulated some years ago, but the first stable derivative of this type, a formal 3-stanna-1-azaallene $>Sn=C=N-$, was isolated only in 1992 by Grützmacher et al. [169] The compound $Tbt(Mes)Si=C=NR$ [170] has also been obtained, but according to the authors it is closer to a silylene Lewis base complex than to a silacumulene. The first stable silaallene was only reported in 1993 by West et al. [171] and the first metastable $>Si=C=P-$ was characterized still more recently. [172]

In germanium chemistry, the first metastable germaphosphallene $>Ge=C=P-$ was described even more recently, in 1996, [173] and the first stable 1-germaallenes $>Ge=C=C<$ in 1998. [174,175]

Transient or stable 2-germaallenes $X=Ge=X$ are still unknown, except for compounds of the type $L_nM'=Ge=M'L_n$ (M' = transition metal) (Ref. 176 and references cited therein; see also Ref. 177 for a review), which are not the topics of this review.

A. 1-Germaallenes

1. Synthesis

192 was first postulated as intermediate in the reaction between an excess of germylene Tbt(Mes)Ge: and 9-(dichloromethylene)fluorene leading, in

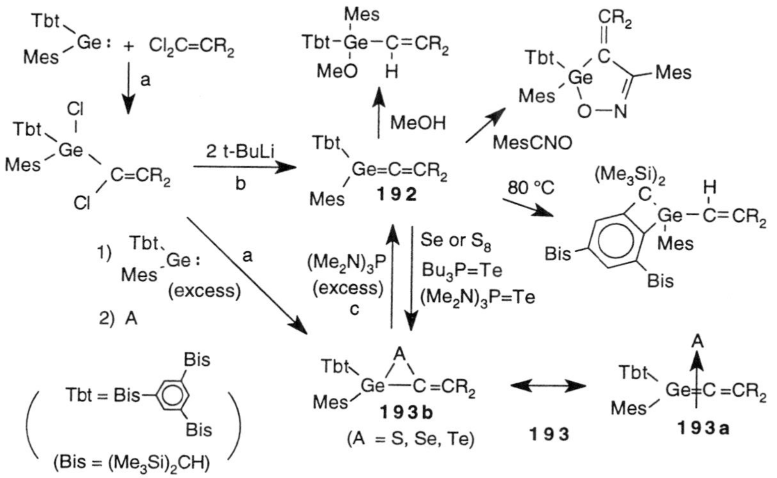

SCHEME 40.

the presence of tellurium, to **193**[178] (Scheme 40, route a). **192** has further been prepared as a colorless solid by the dechlorination of the corresponding dichloro compound by *t*-BuLi (route b).[175] This is the best route to **192**. When 50 equiv of $(Me_2N)_3P$ were added to **193**, the exclusive formation of **192** was also observed by ${}^1H$ and ${}^{13}C$ NMR (route c).[175] The X-ray of **193** (A = Te) shows that its geometry corresponds to the form **193a**, and thus a π-complex of tellurium with the Ge–C double bond of the corresponding 1-germaallene [for example, short Ge–C(sp) bond: 1.88 $\mathring{A}$[178]]. Note that a 3-alkylidene-1,2-thiagermirane analogous to **193** was previously prepared by Ando *et al.* by cycloaddition of a germylene with a thioketone.[179]

The germaallene **194** was prepared by reaction of fluoroalkynylgermane **195** with *tert*-butyllithium at −78°C in about 85% yield.[174] Elimination of LiF occurred at low temperature. **194** was the first allenic compound of germanium structurally characterized (Scheme 41).

2. Physical Properties

The ${}^{13}C$ NMR spectra of **192** and **194** show characteristic deshielded sp carbons at 243.5 ppm[175] and 235.1 ppm.[174] A similar value was found for the analogous silicon compound (233.6 ppm).[180] This chemical shift lies in the normal range and as expected, about 45 ppm upfield of that for germaphosphaallene $Mes_2Ge=C=PAr$ (280.9 ppm)[173] because of the presence of the more electronegative carbon than phosphorus (see Section X,B).

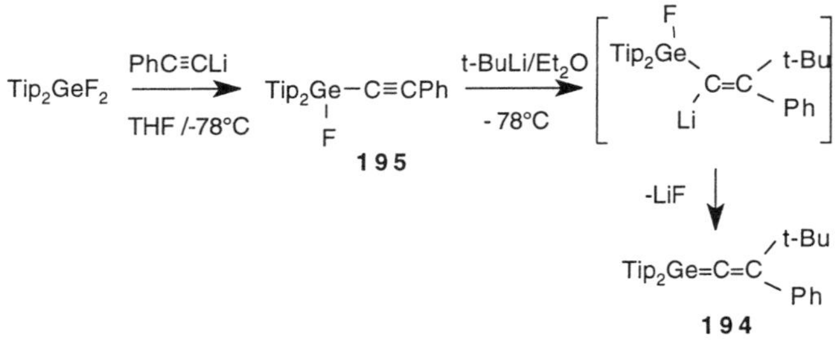

$$\text{Tip}_2\text{GeF}_2 \xrightarrow[\text{THF}/-78°C]{\text{PhC}\equiv\text{CLi}} \underset{\underset{\textbf{195}}{|}}{\text{Tip}_2\text{Ge}-\text{C}\equiv\text{CPh}} \xrightarrow[-78°C]{\text{t-BuLi/Et}_2\text{O}} \left[\begin{array}{c} \text{F} \\ | \\ \text{Tip}_2\text{Ge} \\ \diagdown \\ \text{Li} \end{array} \underset{\diagup}{\text{C}=\text{C}} \begin{array}{c} \diagup\text{t-Bu} \\ \diagdown\text{Ph} \end{array} \right]$$

$$\downarrow \text{-LiF}$$

$$\text{Tip}_2\text{Ge}=\text{C}=\text{C} \begin{array}{c} \diagup\text{t-Bu} \\ \diagdown\text{Ph} \end{array}$$

$$\textbf{194}$$

SCHEME 41.

The structure of **194** has been determined at $-140°C$ by X-ray crystallography.[174] The $Ge=C$ bond length [1.7783(2) Å] lies in the expected range. The $Ge=C=C$ skeleton is nonlinear: The bending at central carbon atom is 159.2°[174] greater than in its silicon analogue (172.0°).[180] The germanium atom is also more pyramidalized than the silicon in the corresponding silaallene (sum of angles: 348.4° at Ge, 357.2° at Si).

3. Reactivity

194 is stable in Et_2O or hexane solutions at $0°C$, but decomposed after 15 h at $25°C$. However, crystals of **194** decompose only at $90°C$.[174] A large difference of stability is observed with its silicon analogue, which is unchanged in boiling toluene.[180]

MeOH adds to the $Ge=C$ double bond of **192**.[175] [2 + 1] and [2 + 2] cycloadditions are also observed, respectively, with sulfur and mesitonitrile. An intramolecular cyclization occurred in benzene at $80°C$ (Scheme 40).[175]

B. 3-Germa-1-phosphaallene

1. Synthesis

As the simultaneous creation of $Ge=C$ and $P=C$ double bonds are unlikely, the $P=C$ double bond, much less reactive than the $Ge=C$ double bond, was formed first.[173] **196** was prepared by debromofluorination of **197** [obtained by reaction of $ArP=C(Br)Li$[181] with dimesityldifluorogermane] with n-butyllithium at low temperature (Scheme 42). The reaction, followed by ^{31}P NMR between $-90°C$ and room temperature, showed the immediate formation of the lithio compound **198**, which lost LiF at $-60°C$ to give the germaphosphaallene **196** in 65–70% yield. **196** was stable at $-50°C$ and dimerized slowly above this temperature. It was the first allenic compound of germanium to be characterized by physicochemical methods.

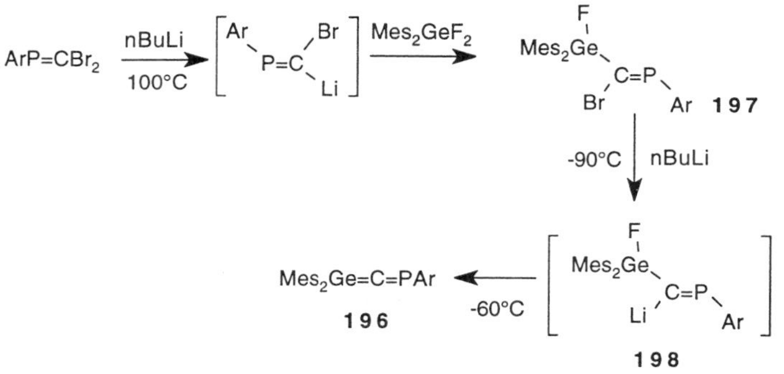

$$ArP{=}CBr_2 \xrightarrow[100°C]{nBuLi} \left[\begin{array}{c} Ar \quad\quad Br \\ \diagdown\!P{=}C\!\diagup \\ Li \end{array} \right] \xrightarrow{Mes_2GeF_2} \begin{array}{c} F \\ | \\ Mes_2Ge\diagdown \\ \quad\quad C{=}P\diagdown \\ Br \quad Ar \quad \mathbf{197} \end{array}$$

$$\mathbf{197} \xrightarrow[-90°C]{nBuLi} \left[\begin{array}{c} F \\ | \\ Mes_2Ge\diagdown \\ \quad\quad C{=}P\diagdown \\ Li \quad Ar \end{array} \right]$$

$$\mathbf{198} \xrightarrow{-60°C} Mes_2Ge{=}C{=}PAr \quad \mathbf{196}$$

SCHEME 42.

2. Physical Properties

The structure of **196** was proved by low-field shifts in ^{31}P NMR (δ: 240 ppm characteristic of a P_{II} derivative) (for reviews on ^{31}P NMR of P_{II} derivatives, see Ref. 182) and in ^{13}C NMR (δ: 280.9 ppm, ^1JCP: 54.3 Hz) for the allenic carbon.[173] Similar chemical shifts were observed for the allenic carbon of diphosphaallene $ArP{=}C{=}PAr$ (δ: 276.2 ppm, ^1JCP: 58 Hz),[183] arsaphosphaallene $ArP{=}C{=}AsAr$ (δ: 299 ppm),[184] phosphasilaallene $ArP{=}C{=}Si(Ph)Tip$ (Tip = 2,4,6-triisopropylphenyl, 269.1 ppm),[172] and the germaallenes **192**[175] and **194**.[174]

3. Reactivity

The structure of the germaphosphaallene **196** was proved by its reaction with MeOH and MeLi, which add exclusively to the $Ge{=}C$ double bond. A regioselective addition is observed according to the $Ge^{\delta+}{-}C^{\delta-}$ polarity.

In the absence of trapping reagent, **196** gives both the classical head-to-tail dimer **199a** via two $Ge{=}C$ bonds and the unexpected dimer **200** via a $Ge{=}C$ and a $P{=}C$ double bond in the ratio 12/88[173] (Scheme 43). A thermodynamic equilibrium was observed between **199a**, which was obtained immediately after reaction, and **199b**.

Preliminary restricted Hartree–Fock calculations have been performed on the dimers **201–204** of $H_2Ge{=}C{=}PH$.[173] The most stable dimers are the bicyclic compounds **201**, about 15 kcal/mol below **202**, which are in turn favored by about 18 kcal/mol with respect to **203**, and 43 kcal/mol with respect to **204** (Scheme 44). Very small differences are found between head-to-head or head-to-tail dimers for each series, and for the *cis* or

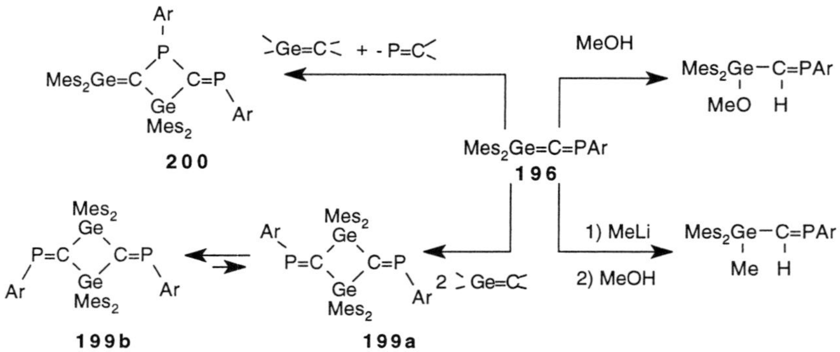

Ar
|
P
Mes$_2$Ge=C⟋ ⟍C=P
⟍Ge⟋ ⟍Ar
Mes$_2$

200

⟩Ge=C⟨ + - P=C⟨ ⟶ (MeOH) Mes$_2$Ge—C=PAr
| |
MeO H

Mes$_2$Ge=C=PAr

196

Mes$_2$
Ge
P=C⟋ ⟍C=P
Ar⟋ ⟍Ge⟋ ⟍Ar
Mes$_2$

199b

Ar Mes$_2$
⟍ Ge
P=C⟋ ⟍C=P
⟍Ge⟋ ⟍Ar
Mes$_2$

199a

2 ⟩Ge=C⟨

1) MeLi Mes$_2$Ge—C=PAr
2) MeOH | |
Me H

Scheme 43.

HP———GeH$_2$
| |
H$_2$Ge———PH

201

HP———GeH$_2$
| |
HP———GeH$_2$

HP⟍ ⟋GeH$_2$
⟍ /
H$_2$Ge⟍ ⟍PH

HP⟍ ⟋GeH$_2$
⟍ |
HP⟋ ⟍GeH$_2$

202

HP⟍ ⟋GeH$_2$
⟍ /
HP⟋ ⟍GeH$_2$

203

HP⟍ ⟋GeH$_2$
⟍ |
H$_2$Ge⟋ ⟍PH

HP———⟋GeH$_2$
⟍ /
H$_2$Ge⟍ ⟍PH

HP———⟋GeH$_2$
| |
HP⟋ ⟍GeH$_2$

204

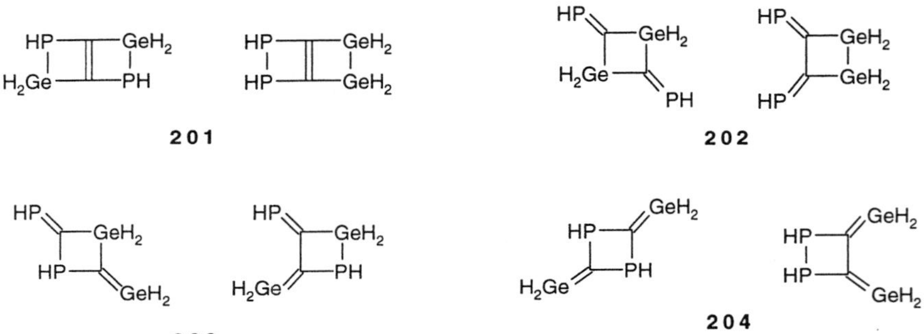

Scheme 44.

trans PH arrangement in agreement with the thermodynamic equilibrium observed for **199**. The use of very large groups in the experiments explains the discrepancies observed with calculations.[173]

XI

CONCLUSION

As shown in this review, the synthesis, physical properties, and reactivity of doubly bonded germanium species are now well known. In contrast, triply bonded derivatives of group 14 elements $-M\equiv X$ ($X = C, M, N, P$) are still unknown, even though it seems possible, using convenient groups, to stabilize such species. Note, however, although it is not the topic of this review, that a compound with a probable $Ge \equiv Mo$ triple bond ($ArGe \equiv Mo$-$(Co)_2Cp$, $Ar = 2,6\text{-}Mes_2C_6H_3$) has been isolated and structurally characterized [$Ge \equiv Mo$: 2.271(1) Å][185]; the shortening of about 0.35 Å in comparison with a Ge–Mo single bond is consistent with the presence of a triple bond.

Calculations performed on H_2CGe ($H_2C=Ge$:, $H_2Ge=C$:, and $HGe \equiv CH$)[186] predict that the *trans*-bent germyne isomer requires 7 kcal/mol to isomerize to the germavinylidene isomer. The linear isomer **205** has the shortest bond length (≈ 1.65 Å), consistent with a triple bond, and the length increases in the *trans*-bent germyne **206** (≈ 1.73 Å) to the germavinylidene **207** (≈ 1.80 Å). In this case the $Ge=C$ value is very close to that for germenes. Rather similar values have been obtained in the ground state potential energy surface of H_2SiC (Ref. 186 and references cited therein).

1.779 - 1.823 Å	1.711 - 1.753 Å	1.623 - 1.676 Å
205	**206**	**207**

In conclusion we can predict that the next review on low coordinate germanium will describe some triply bonded germanium derivatives.

REFERENCES

(1) Satgé, J. *Adv. Organomet. Chem.* **1982**, *21*, 241.

(2) Wiberg, N. *J. Organomet. Chem.* **1984**, *273*, 141.

(3) Barrau, J.; Escudié, J.; Satgé, J. *Chem. Rev.* **1990**, *90*, 283.

(4) Escudié, J.; Couret, C.; Ranaivonjatovo, H.; Satgé, J. *Coord. Chem. Rev.* **1994**, *130*, 427.

(5) (a) Chaubon, M. A.; Ranaivonjatovo, H.; Escudié, J.; Satgé, J. *Main Group Metal Chem.* **1996**, *19*, 145; (b) Kandri Rodi, A.; Ranaivonjatovo, H.; Escudié, J.; Kerbal, A. *Main Group Metal Chem.* **1996**, *19*, 199.

(6) Rivière, P.; Rivière-Baudet, M.; Satgé, J. In *Comprehensive Organometallic Chemistry*; Abel E. W.; Stone, F. G. A.; Wilkinson G., Eds; Pergamon Press: Oxford, 1995; Vol. 2, Chap. 5, pp. 137–216.

(7) Baines, K. M.; Stibbs, W. G. *Adv. Organomet. Chem.* **1996**, *39*, 275.

(8) Tsumuraya, T.; Batcheller, S. A.; Masamune, S. *Angew. Chem., Int. Ed. Engl.* **1991**, *30*, 902.

(9) Grev, R. S. *Adv. Organomet. Chem.* **1991**, *33*, 125.

(10) Driess, M.; Grützmacher, H. *Angew. Chem., Int. Ed. Engl.* **1996**, *35*, 828.

(11) (a) Meyer, H.; Baum, G.; Massa, W.; Berndt, A. *Angew. Chem., Int. Ed. Engl.* **1987**, *26*, 798; (b) Berndt, A.; Meyer, H.; Baum, G.; Massa, W.; Berger, S. *Pure Appl. Chem.* **1987**, *59*, 1011.

(12) Couret, C.; Escudié, J.; Satgé, J.; Lazraq, M. *J. Am. Chem. Soc.* **1987**, *109*, 4411.
(13) Lazraq, M.; Couret, C.; Escudié, J.; Satgé, J.; Soufiaoui, M. *Polyhedron* **1991**, *10*, 1153.
(14) Anselme, G.; Escudié, J.; Couret, C.; Satgé, J. *J. Organomet. Chem.* **1991**, *403*, 93.
(15) Couret, C.; Escudié, J.; Delpon-Lacaze, G.; Satgé, J. *Organometallics* **1992**, *11*, 3176.
(16) Tokitoh, N.; Kishikawa, K.; Okazaki, R. *J. Chem. Soc., Chem. Commun.* **1995**, 1425.
(17) Tokitoh, N.; Kishikawa, K.; Matsumoto, T.; Okazaki, R. *Chem. Lett.* **1995**, 827.
(18) Tokitoh, N.; Manmaru, T.; Okazaki, R. *Organometallics* **1994**, *13*, 167.
(19) Schumann, H.; Glanz, M.; Girgsdies, F.; Ekkehardt Hahn, F.; Tamm, M.; Grzegorzewski, A. *Angew. Chem., Int. Ed. Engl.* **1997**, *36*, 2232.
(20) Apeloig, Y.; Bendikov, M.; Yuzefovich, M.; Nakash, M.; Bravo-Zhivotovskii, D. *J. Am. Chem. Soc.* **1996**, *118*, 12228.
(21) Bravo-Zhivotovskii, D.; Zharov, I.; Kapon, M.; Apeloig, Y. *J. Chem. Soc., Chem. Commun.* **1995**, 1625.
(22) (a) Raabe, G.; Michl, J. *Chem. Rev.* **1985**, *85*, 419; (b) Raabe, G.; Michl, J. In *The Chemistry of Organosilicon Compounds*; Patai, S.; Rappoport, Z., Eds.; J. Wiley and Sons: New York, 1989; pp. 1015–1140.
(23) Brook, A. G.; Baines, K. M. *Adv. Organomet. Chem.* **1986**, *25*, 1.
(24) Brook, A. G.; Brook, M. A. *Adv. Organomet. Chem.* **1996**, *39*, 71.
(25) Toltl, N. P.; Leigh, W. J. *J. Am. Chem. Soc.* **1998**, *120*, 1172.
(26) Khabashesku, V. N.; Kudin, K. N.; Tamas, J.; Boganov, S. E.; Margrave, J. L.; Nefedov, O. M. *J. Am. Chem. Soc.* **1998**, *120*, 5005.
(27) Barrau, J.; Rima, G.; Satgé, J. *J. Organomet. Chem.* **1983**, *252*, C73.
(28) Barrau, J.; Rima, G.; El Amine, M.; Satgé, J. *J. Organomet. Chem.* **1988**, *345*, 39.
(29) Barrau, J.; Rima, G.; Satgé J. *Phosphorus, Sulfur, Silicon* **1995**, *107*, 99.
(30) Kabeta, K.; Powell, D. R.; Hanson, J.; West, R. *Organometallics* **1991**, *10*, 827.
(31) Wiberg, N.; Wagner S. *Z. Naturforsch.* **1996**, *51b*, 838.
(32) Chaubon, M. A.; Escudié, J.; Ranaivonjatovo, H.; Satgé, J. *J. Chem. Soc., Dalton Trans.* **1996**, 893.
(33) Wiberg, N.; Hwang-Park, H. S.; Mikulcik, P.; Müller, G. *J. Organomet. Chem.* **1996**, *511*, 239.
(34) Wiberg, N.; Hwang-Park, H. S. *J. Organomet. Chem.* **1996**, *519*, 107.
(35) Wiberg, N.; Wagner, G. *Chem. Ber.* **1986**, *119*, 1455 and 1467.
(36) Khabashesku, V. N.; Boganov, S. E.; Antic, D.; Nefedov, O. M.; Michl, J. *Organometallics* **1996**, *15*, 4714.
(37) Khabashesku, V. N.; Balaji, V.; Boganov, S. E.; Nefedov, O.M.; Michl, J. *J. Am. Chem. Soc.* **1994**, *116*, 320.
(38) Lazraq, M.; Escudié, J.; Couret, C.; Satgé, J.; Dräger, M.; Dammel, R. A. *Angew. Chem., Int. Ed. Engl.* **1988**, *27*, 828.
(39) Trinquier, G.; Barthelat, J. C.; Satgé, J. *J. Am. Chem. Soc.* **1982**, *104*, 5931.
(40) Grev, R. S.; Schaefer III, H. F. *Organometallics* **1992**, *11*, 3489.
(41) Windus, T. L.; Gordon, M. S. *J. Am. Chem. Soc.* **1992**, *114*, 9559.
(42) Kudin, K. N.; Margrave, J. L.; Khabashesku, V. N. *J. Phys. Chem. A* **1998**, *102*, 744.
(43) Dewar, M. J. S.; Grady, G. L.; Healy, E. F. *Organometallics* **1987**, *6*, 186.
(44) Allison, C. E.; McMahon, T. B. *J. Am. Chem. Soc.* **1990**, *112*, 1672.
(45) Jouany, C.; Trinquier, G. *Organometallics* **1997**, *16*, 3148.
(46) Maslowsky, E. In *Vibrational Spectra of Organometallic Compounds*; John Wiley and Sons: New York, 1977.
(47) Nefedov, O. M.; Maltsev, A. K.; Khabashesku, V. N.; Korolev, V. A. *J. Organomet. Chem.* **1980**, *201*, 123.
(48) Khabashesku, V. N.; Kudin, K. N.; Margrave, J. L. *J. Mol. Struct.* **1998**, *443*, 175.

(49) Dobbs, K. D.; Hehre, W. *Organometallics* **1986**, *5*, 2057.
(50) Jouany, C.; Mathieu, S.; Chaubon-Deredempt, M. A.; Trinquier, G. *J. Am. Chem. Soc.* **1994**, *116*, 3973.
(51) Wiberg, K. B.; Fenoglio, R. A. *J. Am. Chem. Soc.* **1968**, *90*, 3395.
(52) Lipnick, R. L.; Garbish, E. W. *J. Am. Chem. Soc.* **1973**, *95*, 6370.
(53) Furukawa, H.; Takeuchi, H.; Harada, I.; Tasumi, M. *Bull. Chem. Soc. Jpn.* **1983**, *56*, 392.
(54) Kishikawa, K.; Tokitoh, N.; Okazaki, R. *Chem. Lett.* **1996**, 695.
(55) Raasch, M. S. *J. Org. Chem.* **1970**, *35*, 3470.
(56) Bravo-Zhivotovskii, D.; Braude, V.; Stanger, A.; Kapon, M.; Apeloig, Y. *Organometallics* **1992**, *11*, 2326.
(57) Apeloig, Y.; Bravo-Zhivotovskii, D.; Zharov, I.; Panov, V.; Leigh, W. J.; Sluggett, G. W. *J. Am. Chem. Soc.* **1998**, *120*, 1398.
(58) Lazraq, M.; Couret, C.; Escudié, J.; Satgé, J.; Dräger, M. *Organometallics* **1991**, *10*, 1771.
(59) (a) Rivière-Baudet, M.; Satgé, J.; El Baz, F. *J. Chem. Soc., Chem. Commun.* **1995**, 1687; (b) Rivière-Baudet, M.; El Baz, F.; Satgé, J.; Khallaayoun, A.; Ahra, M. *Phosphorus, Sulfur, Silicon* **1996**, *112*, 203; (c) El Baz, F.; Rivière-Baudet, M.; Ahra, M. *J. Organomet. Chem.* **1997**, *548*, 123.
(60) Kishikawa, K.; Tokitoh, N.; Okazaki, R. *Chem. Lett.* **1998**, 239.
(61) Kira, M.; Iwamoto, T.; Maruyama, T.; Kabuto, C.; Sakurai, H. *Organometallics* **1996**, *15*, 3767.
(62) Schäfer, A.; Saak, W.; Weidenbruch, M.; Marsmann, H.; Henkel, G. *Chem. Ber. Recueil* **1997**, *130*, 1733.
(63) Weidenbruch, M.; Stürmann, M.; Kilian, H.; Pohl, S.; Saak, W. *Chem. Ber. Recueil* **1997**, *130*, 735.
(64) Sekiguchi, A.; Yamazaki, H.; Kabuto, C.; Sakurai, H. *J. Am. Chem. Soc.* **1995**, *117*, 8025.
(65) Nagase, S.; Kobayashi, K.; Nagashima, M. *J. Chem. Soc., Chem. Commun.* **1992**, 1302.
(66) Nagase, S. *Pure Appl. Chem.* **1993**, *65*, 675.
(67) Jutzi, P.; Leue, C. *Organometallics* **1994**, *13*, 2898.
(68) Lange, L.; Meyer, B.; Du Mont, W. W. *J. Organomet. Chem.* **1987**, *329*, C17.
(69) Heine, A.; Stalke, D. *Angew. Chem., Int. Ed. Engl.* **1994**, *33*, 113.
(70) Weidenbruch, M.; Grimm, F. T.; Herrndorf, M.; Schäfer, A.; Peters, K.; Von Schnering, H. G. *J. Organomet. Chem.* **1992**, *341*, 335.
(71) Weidenbruch, M.; Ritschl, A.; Peters, K.; Von Schnering, H. G. *J. Organomet. Chem.* **1992**, *438*, 39.
(72) Weidenbruch, M.; Ritschl, A.; Peters, K.; Von Schnering, H. G. *J. Organomet. Chem.* **1992**, *437*, C25.
(73) Weidenbruch, M.; Hagedorn, A.; Peters, K.; Von Schnering, H. G. *Angew. Chem., Int. Ed. Engl.* **1995**, *34*, 1085.
(74) Weidenbruch, M.; Hagedorn, A.; Peters, K.; Von Schnering, H. G. *Angew. Chem., Int. Ed. Engl.* **1996**, *129*, 401.
(75) Suzuki, H.; Okabe, K.; Uchida, S.; Watanabe, H.; Goto, M. *J. Organomet. Chem.* **1996**, *509*, 177.
(76) Baines, K. M.; Cooke, J. A. *Organometallics* **1991**, *10*, 3419.
(77) Baines, K. M.; Cooke, J. A. *Organometallics* **1992**, *10*, 3487.
(78) Ohtaki, T.; Ando, W. *Organometallics* **1996**, *15*, 3103.
(79) Ohgaki, H.; Fukaya, N.; Ando, W. *Organometallics* **1997**, *16*, 4956.
(80) Simons, R. S.; Pu, L.; Olmstead, M. M.; Power, P. P. *Organometallics* **1997**, *16*, 1920.
(81) Olmstead, M. M.; Pu, L.; Simons, R. S.; Power, P. P, *Chem. Commun.* **1997**, 1595.
(82) Baines, K. M.; Stibbs, W. G. *Coord. Chem. Rev.* **1995**, *145*, 157.
(83) Jacobsen, H.; Ziegler, T. *J. Am. Chem. Soc.* **1994**, *116*, 3667.

(84) (a) Kapp, J.; Remko, M.; Schleyer, P. v. R. *J. Am. Chem.* Soc. **1996**, *118*, 5745. (b) Kapp, J.; Remko, M.; Schleyer, P. v. R. *Inorg. Chem.* **1997**, 36, 4241.
(85) Liang, C.; Allen, L. C. *J. Am. Chem. Soc.* **1990**, *112*, 1039.
(86) (a) Sekiguchi, A.; Tsukamoto, M.; Ichinohe, M. *Science* **1997**, *275*, 60. (b) Sekiguchi, A.; Tsukamoto, M.; Ichinohe, M.; Fukaya, N. *Phosphorus, Sulfur, Silicon* **1997**, *124–125*, 323. (c) Ichinohe, M.; Fukaya, N; Sekiguchi, A. *Chem. Lett.* **1998**, 1045.
(87) Cotton, F. A.; Cowley, A. H.; Feng, X. *J. Am. Chem. Soc.* **1998**, *120*, 1795.
(88) Della Bona, M. A.; Cassani, M. C.; Keates, J. M.; Lawless, G. A.; Lappert, M. F.; Stürmann, M.; Weidenbruch, M. *J. Chem. Soc., Dalton Trans.* **1998**, 1187.
(89) Hudson, A.; Lappert, M. F.; Lednor, P. W. *J. Chem. Soc., Dalton Trans.* **1976**, 2369.
(90) Grev, R. S.; Schaefer III, H. F.; Baines, K. M. *J. Am. Chem. Soc.* **1990**, *112*, 9458.
(91) Tokitoh, N.; Suzuki, H.; Okazaki, R.; Ogana, K. *J. Am. Chem. Soc.* **1993**, *115*, 10428.
(92) Suzuki, H.; Tokitoh, N., Okazaki, R. *Bull. Chem. Soc. Jpn.* **1995**, *68*, 2471.
(93) Suzuki, H.; Tokitoh, N.; Okazaki, R.; Harada, J.; Ogawa, K.; Tomoda, S.; Goto, M. *Organometallics* **1995**, *14*, 1016.
(94) Ando, W.; Tsumuraya, T. *J. Chem. Soc., Chem. Commun.* **1989**, 770.
(95) Batcheller, S. A.; Masamune, S. *Tetrahedron Lett.* **1988**, *29*, 3383.
(96) Rivière-Baudet, M.; Khallaayoun, A.; Satgé, J. *Organometallics* **1993**, *12*, 1003.
(97) Rivière-Baudet, M.; Khallaayoun, A.; Satgé, J. *J. Organomet. Chem.* **1993**, *462*, 89.
(98) Escudié, J.; Couret, C.; Andrianarison, M.; Satgé, J. *J. Am. Chem. Soc.* **1987**, *109*, 386.
(99) Masamune, S.; Batcheller, S. A.; Park, W.; Davies, W. M.; Yamashita, O.; Ohta, Y.; Kabe, Y. *J. Am. Chem. Soc.* **1989**, *111*, 1888.
(100) McKillop, K. L.; Gillette, G. R.; Powell, D. R.; West, R. *J. Am. Chem. Soc.* **1992**, *114*, 5203.
(101) Toltl, N. P.; Leigh, W. J.; Kollegger, G. M.; Stibbs, W. G.; Baines, K. M. *Organometallics* **1996**, *15*, 3732.
(102) (a) Mochida, K.; Tokura, S. *Bull. Chem. Soc. Jpn.* **1992**, *65*, 1642. (b) Mochida, K.; Kanno, N.; Kato, R.; Kotani, M.; Yamauchi, S.; Wakasa, M.; Hayashi, H. *J. Organomet. Chem.* **1991**, *415*, 191.
(103) Okazaki, R.; West, R. *Adv. Organomet. Chem.* **1996**, *39*, 231.
(104) Dixon, C. E.; Netherton, M. R.; Baines, K. M. *J. Am. Chem. Soc.* **1998**, *120*, 10365.
(105) Kolleger, G. M.; Stibbs, W. G.; Vittal, J. J.; Baines, K. M. *Main Group Metal Chem.* **1996**, *19*, 317.
(106) Baines, K. M.; Cooke, J. A.; Dixon, C. E.; Liu, H. W.; Netherton, M. R. *Organometallics* **1994**, *13*, 631.
(107) Baines, K. M.; Cooke, J. A.; Vittal, J. J. *Heteroatom. Chem.* **1994**, *5*, 293.
(108) Dixon, C. E.; Liu, H. W.; Van der Kant, C. M.; Baines, K. M. *Organometallics* **1996**, *15*, 5701.
(109) Dixon, C. E.; Cooke, J. A.; Baines, K. M. *Organometallics* **1997**, *16*, 5437.
(110) Dixon, C. E.; Baines, K. M. *Phosphorus, Sulfur, Silicon* **1997**, *124–125*, 123.
(111) Fanta, A. D.; De Young, D. J.; Belzner, J.; West, R. *Organometallics* **1991**, *10*, 3466.
(112) Lazraq, M.; Escudié, J.; Couret, C.; Satgé, J. *Organometallics* **1992**, *11*, 555.
(113) Dixon, C. E.; Netherton, M. R.; Baines, K. M. *J. Am. Chem. Soc.* **1998**, *120*, 10365.
(114) Chaubon, M. A.; Escudié, J.; Ranaivonjatovo, H.; Satgé, J. *Chem. Commun.* **1996**, 2621.
(115) Masamune, S.; Sita, L. R. *J. Am. Chem. Soc.* **1985**, *107*, 6390.
(116) (a) Ranaivonjatovo, H.; Escudié, J.; Couret, C.; Satgé, J. *J. Chem. Soc., Chem. Commun.* **1992**, 1047; (b) Anselme, G.; Ranaivonjatovo, H.; Escudié, J.; Couret, C.; Satgé, J. *Organometallics* **1992**, *11*, 2748.
(117) Weidenbruch, M.; Schaefer, A.; Kilian, H.; Pohl, S.; Saak, W.; Marsmann, H. *Chem. Ber.* **1992**, *125*, 563.

(118) Drost, C.; Gehrhus, B.; Hitchcock, P. B.; Lappert, M. F. *Chem. Commun.* **1997**, 1845.
(119) Rivière-Baudet, M. *Main Group Metal Chem.* **1995**, *18*, 353.
(120) (a) Glidewell, C.; Lloyd, D; Lumbard, K. W. *J. Chem. Soc., Dalton Trans.* **1987**, 501; (b) Glidewell, C.; Lloyd, D.; Lumbard, K. W.; McKechnie, J. S. *Tetrahedron Lett.* **1987**, *28*, 343; (c) Glidewell, C.; Lloyd, D.; Lumbard, K. W.; McKechnie, J. S; Hursthouse, M. B.; Short, R. L. *J. Chem. Soc., Dalton Trans.* **1987**, 2981.
(121) Ossig, G.; Meller, A.; Freitag, S.; Müller, O.; Gornitzka, H.; Herbst-Irmer, R. *Organometallics* **1996**, *15*, 408.
(122) Ossig, G.; Meller, A.; Müller, O.; Herbst-Irmer, R. *Organometallics* **1996**, *15*, 5060.
(123) Rivière-Baudet, M.; Morère, A. *J. Organomet. Chem.* **1992**, *431*, 17.
(124) Dahrouch, M.; Rivière-Baudet, M.; Satgé, J.; Mauzac, M.; Cardin, C. J.; Thorpe, J. H. *Organometallics* **1998**, *17*, 623.
(125) Pfeiffer, J.; Maringgele, W.; Noltemeyer, M.; Meller, A. *Chem. Ber.* **1989**, *122*, 245.
(126) Rivière-Baudet, M.; Satgé, J.; Morère, A. *J. Organomet. Chem.* **1990**, *386*, C7.
(127) Veith, M.; Becker, S.; Huch, V. *Angew. Chem., Int. Ed. Engl.* **1990**, 29, 216.
(128) Meller, A.; Ossig, G.; Maringgele, W.; Stalke, D.; Herbst-Irmer, R.; Freitag, S.; Sheldrick, G. M. *J. Chem. Soc., Chem. Commun.* **1991**, 1123.
(129) Ando, W.; Ohtaki, T.; Kabe, Y. *Organometallics* **1994**, *13*, 434.
(130) Ohtaki, T.; Kabe, Y.; Ando, W. *Heteroatom. Chem.* **1994**, *5*, 313.
(131) Veith, M.; Rammo, A. *Z. Anorg. Allg. Chem.* **1997**, *623*, 861.
(132) Barrau, J.; Rima, G.; El Amraoui, T. *J. Organomet. Chem.* **1998**, *570*, 163.
(133) (a) Ang, H. G.; Lee, F. K. *J. Chem. Soc., Chem. Commun.* **1989**, 310; (b) Ang, H. G.; Lee, F. K. *J. Fluorine Chem.* **1989**, *43*, 435. (c) Ang, H. G.; Lee, F. K. *J. Fluorine Chem.* **1995**, *75*, 181.
(134) Escudié, J.; Couret, C.; Satgé, J.; Andrianarison, M.; Andriamizaka, J.-D. *J. Am. Chem. Soc.* **1985**, *107*, 3378.
(135) Ranaivonjatovo, H.; Escudié, J.; Couret, C.; Satgé, J. Dräger, M. *New J. Chem.* **1989**, *13*, 389.
(136) Ranaivonjatovo, H.; Escudié, J.; Couret, C.; Satgé, J. *J. Organomet. Chem.* **1991**, *415*, 327.
(137) Dräger, M.; Escudié, J.; Couret, C.; Ranaivonjatovo, H.; Satgé, J. *Organometallics* **1988**, 7, 1010.
(138) Kandri Rodi, A.; Declercq, J.-P.; Dubourg, A.; Ranaivonjatovo, H.; Escudié, J. *Organometallics* **1995**, *14*, 1954.
(139) Jutzi, P.; Schmidt, H.; Neumann, B.; Stammler, H.-G. *Organometallics* **1996**, *15*, 741.
(140) Lange, L.; Meyer, B.; DuMont, W. W. *J. Organomet. Chem.* **1987**, *329*, C17.
(141) Tokitoh, N.; Matsumoto, T.; Okazaki, R. *Chem. Lett.* **1995**, 1087.
(142) Tokitoh, N.; Matsumoto, T.; Manmaru, K.; Okazaki, R. *J. Am. Chem. Soc.* **1993**, *115*, 8855.
(143) Matsumoto, T.; Tokitoh, N.; Okazaki, R. *Angew. Chem., Int. Ed. Engl.* **1994**, *33*, 2316.
(144) Tokitoh, N.; Matsumoto, T.; Okazaki, R. *J. Am. Chem. Soc.* **1997**, *119*, 2337.
(145) Matsumoto, T.; Tokitoh, N.; Okazaki, R. *Chem. Commun.* **1997**, 1553.
(146) Nowek, A.; Sims, R.; Babinec, P.; Leszczynski, J. *J. Phys. Chem. A* **1998**, *102*, 2189.
(147) Antoniotti, P.; Grandinetti, F. *Gazz. Chim. Ital.* **1990**, *120*, 701.
(148) (a) So, S. P. *J. Phys. Chem. A* **1991**, *95*, 10658; (b) So, S. P. *J. Phys. Chem. A* **1993**, *97*, 4643.
(149) Trinquier, G.; Pelissier, M.; Saint-Roch, B.; Lavayssière, H. *J. Organomet. Chem.* **1981**, *214*, 169.
(150) Tokitoh, N.; Okazaki, R. *Main Group Chem. News* **1995**, *3*, 4.
(151) Tokitoh, N.; Matsumoto, T.; Okazaki, R. *Heterocycles* **1995**, *40*, 127.
(152) Veith, M.; Becker, S.; Huch, V. *Angew. Chem., Int. Ed. Engl.* **1989**, *28*, 1237.
(153) Kuchta, M. C.; Parkin, G. *J. Chem. Soc., Chem. Commun.* **1994**, 1351.

(154) Ossig, G.; Meller, A.; Brönneke, C.; Müller, O.; Schäfer, M.; Herbst-Irmer, R. *Organometallics* **1997**, *16*, 2116.
(155) Barrau, J.; Rima, G.; El Amraoui, T. *Inorg. Chim. Acta* **1996**, *241*, 9.
(156) Foley, S. R.; Bensimon, C.; Richeson, D. S. *J. Am. Chem. Soc.* **1997**, *199*, 10359.
(157) Tokitoh, N.; Kishikawa, K.; Manmaru, K.; Okazaki, R. *Heterocycles* **1997**, *44*, 149.
(158) Kuchta, M. C.; Parkin, G. *Chem. Commun.* **1996**, 1669.
(159) Veith, M.; Detemple, A.; Huch, V. *Chem. Ber.* **1991**, *124*, 1135.
(160) Ishikawa, M.; Sugisawa, H.; Fuchikami, T.; Kumada, M.; Yamabe, T.; Kawakami, H.; Fukui, K.; Ueki, Y.; Shizuka, H. *J. Am. Chem. Soc.* **1982**, *104*, 2872.
(161) Ishikawa, M.; Nishimura, K.; Ochiai, H.; Kumada, M. *J. Organomet. Chem.* **1982**, *236*, 7.
(162) Ishikawa, M.; Nomura, Y.; Tozaki, E.; Kunai, A.; Ohshita, J. *J. Organomet. Chem.* **1990**, *399*, 205.
(163) Ishikawa, M.; Yuzurika, Y.; Horio, T.; Kunai, A. *J. Organomet. Chem.* **1991**, *402*, C20.
(164) Ishikawa, M.; Horio, T.; Yuzuriha, Y.; Kunai, A.; Tsukihara, T.; Naitou, H. *Organometallics* **1992**, *11*, 597.
(165) Kerst, C.; Rogers, C. W.; Ruffolo, R.; Leigh, W. J. *J. Am. Chem. Soc.* **1982**, *104*, 2872.
(166) Kerst, C.; Ruffolo, R.; Leigh, W. J. *Organometallics* **1997**, *16*, 5804.
(167) Yin, I.; Klosin, J.; Ahboud, K. A.; Jones, W. M. *J. Am. Chem. Soc.* **1995**, *117*, 3298.
(168) Weidenbruch, M.; Brand-Roth, B.; Pohl, S.; Saak, W.; *Angew. Chem., Int. Ed. Engl.* **1989**, *29*, 90.
(169) Grützmacher, H.; Freitag, S.; Herbst-Irmer, R.; Sheldrick, G. S.; *Angew. Chem., Int. Ed. Engl.* **1992**, *31*, 437.
(170) Takeda, N.; Suzuki, H.; Tokitoh, N.; Okazaki, R.; Nagase, S. *J. Am. Chem. Soc.* **1997**, *119*, 1456.
(171) Miracle, G. E.; Ball, J. L.; Powell, D. R.; West. R. *J. Am. Chem. Soc.* **1993**, *115*, 11598.
(172) Rigon, L.; Ranaivonjatovo, H.; Escudié, J.; Dubourg, A.; Declercq, J.-P. *Chemistry, a European Journal* **1999**, *5*, 776.
(173) Ramdane, H.; Ranaivonjatovo, H.; Escudié, J.; Mathieu, S. *Organometallics* **1996**, *15*, 3070.
(174) Eichler, B. E.; Powell, D. R.; West, R. *Organometallics* **1998**, *17*, 2147.
(175) Tokitoh, N.; Kishikawa, K.; Okazaki, R. *Chem. Lett.* **1998**, 811.
(176) Kircher, P.; Huttner, G.; Heinze, K.; Schiemenz, B.; Zsolnai, L.; Büchner, M.; Driess, A. *Eur. J. Inorg. Chem.* **1998**, 703.
(177) Herrmann, W. A. *Angew. Chem. Int., Ed. Engl.* **1986**, *25*, 56.
(178) Kishikawa, K.; Tokitoh, N.; Okazaki, R. *Organometallics* **1997**, *16*, 5127.
(179) Ando, W.; Tsumuraya, T. *Organometallics* **1989**, *8*, 1467.
(180) Trommer, M.; Miracle, G. E.; Eichler, B. E.; Powell, D. R.; West, R. *Organometallics* **1997**, *16*, 5737.
(181) Goede, S. J.; Bickelhaupt, F. *Chem. Ber.* **1991**, *124*, 2677.
(182) Lochschmidt, S.; Schmidpeter, A. *Phosphorus, Sulfur, Silicon* **1986**, *29*, 73.
(183) Appel, R.; Fölling, P.; Josten, P.; Siray, M.; Winkhaus, V.; Knoch, F. *Angew. Chem., Int. Ed. Engl.* **1984**, *23*, 619.
(184) Ranaivonjatovo, H.; Ramdane, H.; Gornitzka, H.; Escudié, J. *Organometallics* **1998**, *17*, 1631.
(185) Simons, R. S.; Power, P. P. *J. Am. Chem. Soc.* **1996**, *118*, 11966.
(186) Stogner, S. M.; Grev, R. S. *J. Chem. Phys.* **1998**, *108*, 5458.

ADVANCES IN ORGANOMETALLIC CHEMISTRY, VOL. 44

The Chemistry of Phosphinocarbenes

DIDIER BOURISSOU and GUY BERTRAND

Laboratoire de Chimie de Coordination du CNRS
31077 Toulouse Cédex, France

I

INTRODUCTION

Carbenes are electron-deficient two-coordinate carbon compounds that have two nonbonding electrons at one carbon. In the ground state, the two unshared electrons may be either in the same orbital and have antiparallel spins (singlet state **S**), or in two different orbitals with parallel spins (triplet state **T**). They can be considered as typical representatives of reactive intermediates and have found a broad range of applications in synthetic chemistry.

As early as 1835, attempts to prepare the parent carbene (CH_2) by dehydration of methanol had been reported.[1] It is interesting to note that at that time the tetravalency of carbon was not established, and therefore the existence of stable carbenes was considered to be quite reasonable. At the very beginning of the twentieth century, Staudinger demonstrated that carbenes, generated from diazo compounds or ketenes, were highly reactive species.[2] It quickly became clear that their six-valence-electron shell, which defied the octet rule, was responsible for their fugacity. As a consequence,

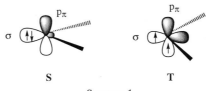

SCHEME 1.

the quest for stable carbenes became an unreasonable target, and indeed remained so for quite some time! In the 1950s, Breslow[3a] and Wanzlick[3b] realized that the stability of a carbene could be dramatically enhanced by the presence of amino substituents, but were not able to isolate a "monomeric" carbene. It was only in 1988 that our group reported the synthesis of a stable carbene, namely a (phosphino)(silyl)carbene.[4]

Since this discovery, a few types of other stable singlet carbenes have been described and reviewed[5]: imidazol-2-ylidenes,[6] 1,2,4-triazol-3-ylidenes,[7] imidazolidin-2-ylidenes,[8] acyclic diaminocarbenes,[9] thiazol-2-ylidenes,[10] and acyclic aminooxy- and aminothiocarbenes.[11]

With the exception of the title carbenes, which feature a π-donor and a

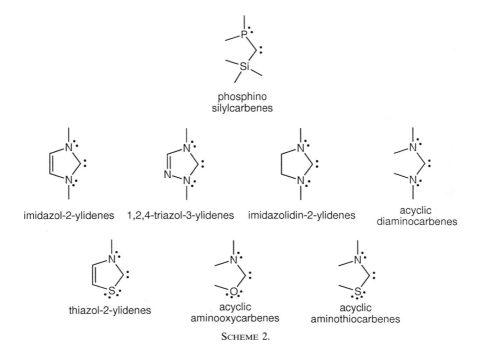

SCHEME 2.

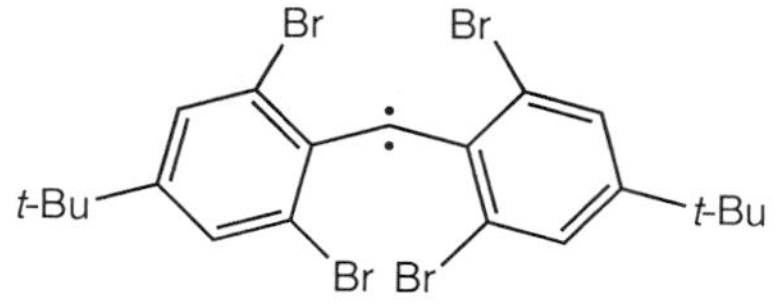

$t_{1/2}$ (20°C) = 16 seconds

SCHEME 3.

π-type-withdrawing substituent, all the stable singlet carbenes known to date bear two π-donor substituents, of which at least one is an amino group. In other words, the phosphinocarbenes are the only stable push–pull carbenes and therefore behave totally differently from the other isolated carbenes, but very much like most of the transient carbenes.

It should also be noted that persistent triplet carbenes have been discovered by Tomioka.[12]

In this review, we present successively the theoretical aspect, the synthesis, the structural features, the reactivity, and lastly, the ligand properties of phosphinocarbenes.

II

THEORETICAL STUDIES

Several theoretical investigations have been carried out for the phosphinocarbenes (Fig. 1).[13–16]

The first study by Hegarty and co-workers[13] concluded that the parent phosphinocarbene H_2PCH **Ia** has a singlet ground state, with a singlet–triplet gap of 3 kcal/mol, while later on, Hoffmann and Kuhler[14] predicted a somewhat larger gap (6.7 kcal/mol) using a more sophisticated level of theory. Together, the planar geometry of the singlet state along with the shortness of the PC bond (1.616 Å) indicate a strong interaction of the phosphorus lone pair with the vacant orbital of the carbenic center. In other words, the phosphorus–carbon bond has a multiple bond character, and the phosphorus vinyl ylide form **B** best describes this species. Interestingly, the structure **A** featuring a pyramidal phosphorus center is neither a minimum nor a saddle point on the potential energy surface. Moreover, the λ^5-phosphaacetylene structure **C** is the transition structure corresponding to the inversion at the carbenic center with an energy barrier of 10.3 kcal/mol. In contrast, calculations predict a pyramidal geometry at phosphorus and a long P–C bond (1.782 Å) for the triplet state. Moreover, the

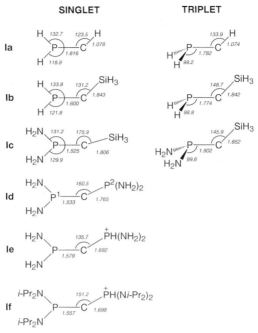

FIG. 1.   Calculated geometric parameters for singlet and triplet phosphinocarbenes (bond lengths in Å; bond angles in degrees).

PCH bond angle is broader (133.9°) than that of the singlet state (123.5°), as classically predicted for most carbenes.[17]

Dixon et al. theoretically investigated the role of silicon substitution at carbon (SiH$_3$) and of amino groups at phosphorus.[15] In the singlet state the silyl substituent (**Ib**) induces a broadening of the carbene angle (131.2°). This effect, which has been previously reported for other silylcarbenes,[18] corroborates the observation by Schoeller[19] and Pauling[20] that carbenes substituted by elements less electronegative than carbon preferentially adopt linear structures. Moreover, the shortness of the C–Si bond (1.843 Å), which is in the range typical for silicon–carbanion bond lengths, indicates a significant back donation of the carbene lone pair into the σ^* orbitals of

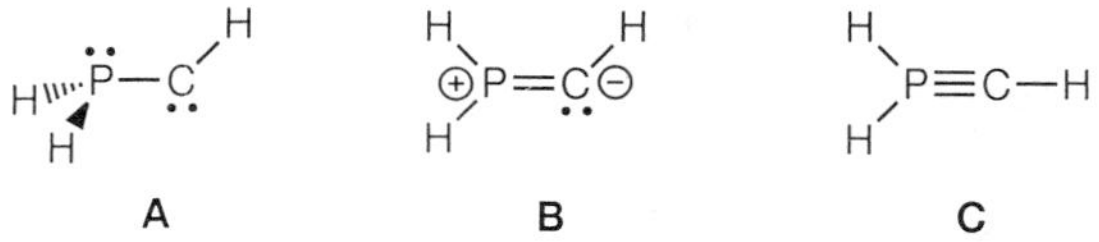

A   B   C

SCHEME 4.

the silyl group. Replacement of the hydrogen atoms by amino groups at phosphorus (**Ic**) leads to a shortening of the P–C and C–Si bonds, while the P–C–Si bond angle becomes almost linear. In this case, the carbene lone pair probably interacts (negative hyperconjugation) with the σ^* orbitals of both the silyl and the phosphino groups.[21] Therefore, the singlet state adopts a cumulenic structure. Note that the triplet state of both carbenes **Ib** and **Ic** is higher in energy, by 5.6 and 13.9 kcal/mol, respectively. Their geometries are quite comparable to that of the parent triplet phosphinocarbene **Ia**, indicating a weaker influence of the carbene substituents.

The most recent theoretical study, by Alhrichs and co-workers, deals with the di(phosphino)carbene **Id** and (phosphino)(phosphonio)carbenes **Ie,f**.[16] The optimized geometry of the di(phosphino)carbene **Id** is weakly bent (PCP angle: 160.5°) and highly unsymmetrical: Only one of the phosphorus centers (P^1) is in a planar environment, and it is much more closely bonded to the carbenic center than the other one (P^1C: 1.533 and P^2C: 1.765 Å). The atomic charges (P^1: +1.0, C: −0.8, P^2: +0.6) indicate that the short P^1C bond is a double bond reinforced by Coulombic attraction, while the nature of the molecular orbitals revealed a slight delocalization of the carbene lone pair into the low-lying σ^* (P–N) orbitals of the two phosphino substituents. The "distortion" from the symmetrical structure can be viewed as a second-order Jahn–Teller effect.

The σ^3P center of the (phosphino)(phosphonio)carbenes **Ie,f** is in a trigonal planar environment and the σ^3P–C bond length is very short. Interestingly, the σ^4P–C bond length is also short, in the range expected for phosphorus ylides, which indicates a degree of back donation of the carbene lone pair into the low-lying orbitals of the σ^4-phosphorus center. Therefore, (phosphino)(phosphonio)carbenes are strongly related to the isoelectronic (phosphino)(silyl)carbenes.

In summary, independently of the second carbene substituent, phosphinocarbenes have a singlet ground state, with a small singlet–triplet gap. They have a planar geometry at phosphorus and a short phosphorus–carbon bond, indicating an interaction between the phosphorus lone pair and the carbene vacant orbital. In the case of silyl- and phosphoniophosphinocarbenes, there is an additional interaction between the carbene lone pair and lowlying σ^* orbitals at the second substituent.

III

SYNTHESIS

Most of the transient and all the stable phosphinocarbenes **2** have been prepared from the corresponding diazo derivatives **1**. However, three other

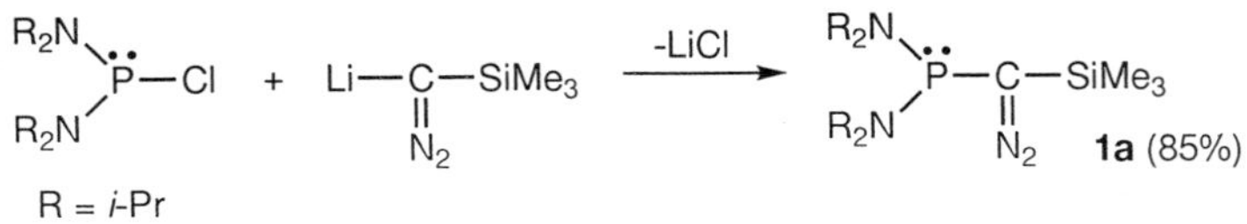

SCHEME 5.

types of compounds have been postulated as precursors, and in order to be comprehensive, they will also be described.

A. Nitrogen Elimination from α-Diazophosphines

In contrast to λ^5-phosphorus-substituted diazo derivatives, which have been known for a long time,[22] the synthesis of the first α-diazophosphine was reported as recently as 1985.[23] This compound, namely the [bis(diisopropylamino)phosphino](trimethylsilyl)diazomethane **1a**, was obtained by treatment of the lithium salt of trimethylsilyldiazomethane with 1 equiv of bis(diisopropylamino)chlorophosphine.

We first recognized that photolysis of **1a**, in the presence of a variety of trapping agents, led to products resulting from the transient formation of the phosphinocarbene **2a**.[23] It was only in 1988 that we discovered that flash thermolysis of **1a** at 250°C under vaccum afforded the desired species **2a** in 80% yield.[4] Compound **2a** is a red oily material, stable for several weeks at room temperature in benzene solution, which can be purified by quick distillation (bp 75–80°C under 10^{-2} mm Hg). Using this experimental procedure, the synthesis of **2a** is not straightforward. The experimental conditions used are drastic, and the effective temperature range for the thermolysis very narrow: Below 240°C, the starting diazo **1a** is not decomposed, and above 260°C, the carbene inserts into a CH bond of a diisopropylamino substituent. We recently found that irradiation at 300 nm of a degassed dry benzene solution of **1a** cleanly leads to **2a**.[24] Moreover, the [bis(dicyclohexylamino)phosphino](trimethylsilyl)carbene **2b** is cleanly accessible by irradiation at 300 nm, or even by heating a toluene solution of the diazo precursor **1b** for 16 h at 80°C.[25] Therefore, both carbenes **2a** and **2b** are now accessible in multigram quantities.

A number of phosphinocarbenes bearing various substituents at phospho-

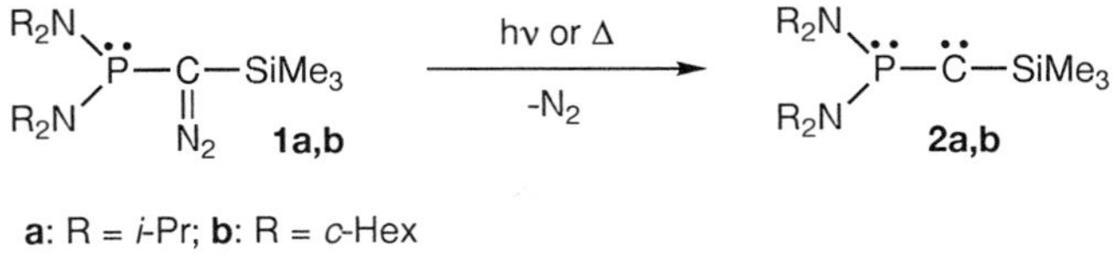

a: R = i-Pr; b: R = c-Hex

SCHEME 6.

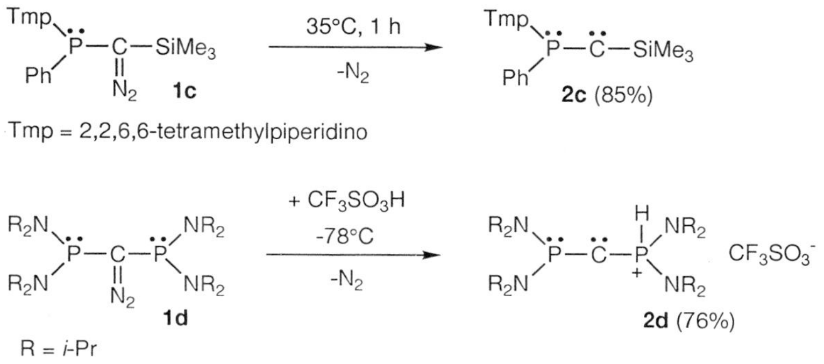

Tmp�
   P—C—SiMe₃ $\xrightarrow[-N_2]{35°C, 1 h}$ Tmp—P—C—SiMe₃
Ph N₂ **1c**   Ph **2c** (85%)

Tmp = 2,2,6,6-tetramethylpiperidino

R₂N—P—C—P—NR₂ $\xrightarrow[-N_2]{+\ CF_3SO_3H \\ -78°C}$ R₂N—P—C—P(H)—NR₂ CF₃SO₃⁻

R₂N N₂ NR₂ **1d**   R₂N NR₂ **2d** (76%)

R = i-Pr

SCHEME 7.

rus and carbon have been synthesized by thermolysis and photolysis of α-diazophosphines.[26] However, only a few of them are stable, and the influence of the substituents on their stability merits comment.

So far, all stable phosphinocarbenes feature two amino groups at phosphorus and a trimethylsilyl group at carbon with only two exceptions: (i) the [2, 2, 6, 6 - tetramethylpiperidino) (phenyl) phosphino] (trimethylsilyl) - carbene **2c**,[27] which is stable for 1 day in solution at room temperature, but for weeks at −20°C; and (ii) the [bis(diisopropylamino)phosphino]-[bis(diisopropylamino)phosphonio]carbene **2d**,[28] which is stable for years in the solid state (mp: 88°C).

The stabilizing effects of amino groups at phosphorus and silyl group at carbon can be well rationalized taking into account the vinyl ylide structure predicted by calculations (Section II). In other words, the positive charge at phosphorus and the negative charge at carbon are delocalized into the amino and silyl groups, respectively. Since silyl and phosphonio groups are isoelectronic and isovalent, the stability of **2d** is not surprising.

Of course, bulky substituents kinetically stabilize carbenes, but interestingly, during the course of our study, we realized that the stability of carbenes is often inversely proportional to the stability of the starting diazo compounds,[27] as illustrated in Table I.

B. α,β-Elimination from P-Halogenomethylene Phosphoranes

In 1981, Appel et al. postulated the transient formation of the (diphenylphosphino)(trimethylsilyl)carbene **2e** to explain the formation of the phosphaalkene **4e**, in the thermolysis of the P-chloromethylene phosphorane **3e**.[29] At that time, the authors did not recognize the carbene character of **2e** and simply named the intermediate a λ⁵-phosphaalkyne.

TABLE I

STABILITY OF THE CARBENES AND OF THE DIAZO PRECURSORS

R^1	R^2	R^3	Diazo stability	Carbene stability
i-Pr$_2$N	i-Pr$_2$N	SiMe$_3$	b.p. 85–90°C, 10^{-2} mm Hg	b.p. 75–80°C, 10^{-2} mm Hg, stable several weeks at 25°C
Tmp[a]	i-Pr$_2$N	SiMe$_3$	Few minutes at 25°C	Several weeks at 25°C
Tmp[a]	Me$_2$N	Si(i-Pr)$_3$	Several days at 25°C, 1 h at 35°C	Several weeks at 25°C
Tmp[a]	Me$_2$N	SiMe$_3$	Several days at 25°C, 1 h at 35°C	Several weeks at 25°C
Tmp[a]	Ph	SiMe$_3$	Few minutes at 25°C	Few hours at 25°C
c-Hex$_2$N	c-Hex$_2$N	SiMe$_3$	Stable 24 h at 70°C	Several weeks at 25°C
i-Pr$_2$N	i-Pr$_2$N	PR$_2$H^{+b}	Not observed at 25°C	m.p. 88°C Indefinitely at 25°C

[a] Tmp = 2,2,6,6-tetramethylpiperidino.
[b] R = i-Pr$_2$N.

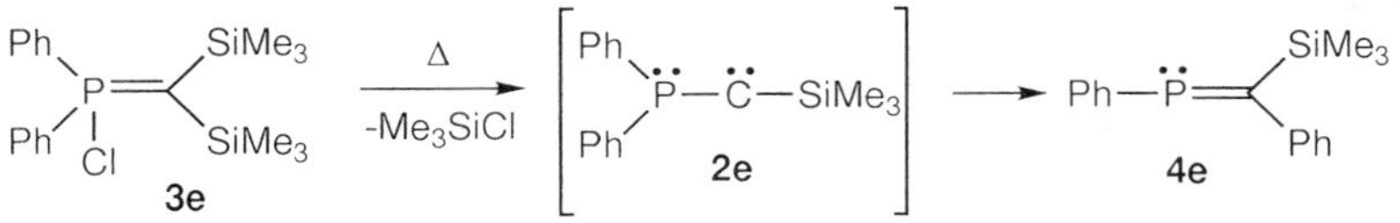

SCHEME 8.

In 1985, Fluck *et al.* reported that treatment of the *P*-fluoromethylene-phosphorane **3f** with 2 equiv of butyllithium at −95°C gave the 1,1,3,3-tetrakis(dimethylamino)-1λ^5,3λ^5-diphosphete **5f** in 34% yield.[30] The formation of this four-membered ring could result from a [2 + 2] head-to-tail dimerization of the transient phosphinocarbene **2f**. However, an alternative

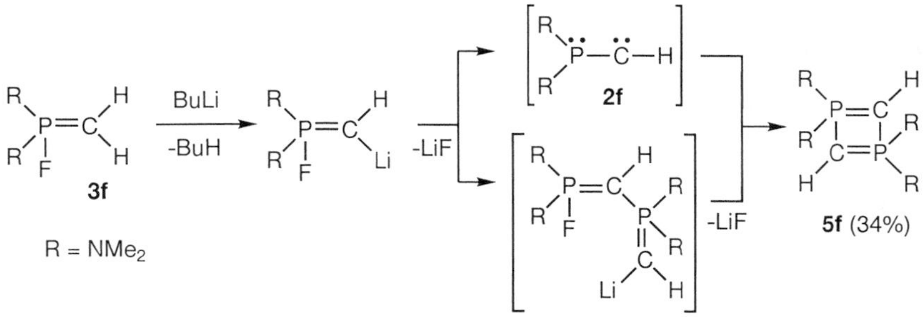

SCHEME 9.

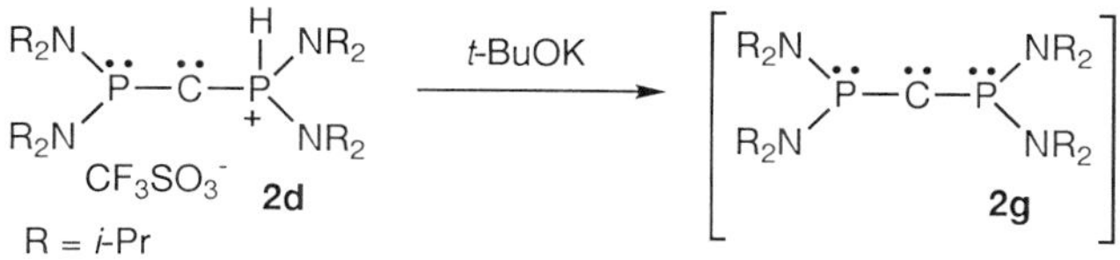

SCHEME 10.

mechanism, with intermolecular LiF elimination, which does not involve the transient formation of **2f**, is also possible.

C. *Miscellaneous Methods*

Two synthetic routes with limited applicability have also been reported:

Deprotonation of the stable (phosphino)(phosphonio)carbene **2d** gives rise to the transient bis(phosphino)carbene **2g**.[31] Obviously, this method is extremely restricted, but this is certainly the only example of the preparation of a carbene from a carbene!

The bis(diisopropylamino)carbene **7** has been obtained by treating the

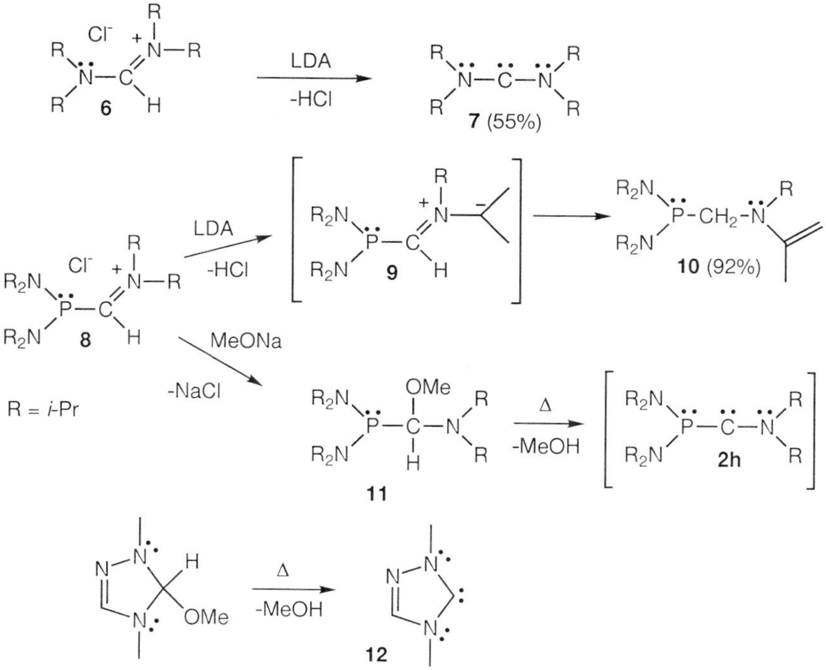

SCHEME 11.

corresponding formamidinium chloride **6** with LDA.[9] However, this synthetic approach appeared inappropriate to generate the (amino)(phosphino)carbene **2h**. Indeed, LDA reacts with **8**, but leads to the enamine **10** (92% yield), probably through the transient formation of the azomethine ylide **9**, followed by a H-rearrangement. In contrast, thermolysis at 160°C under vacuum of the hemi-aminal **11**, obtained in 90% yield by the reaction of sodium methoxide with **8**, does lead to the transient (amino)(phosphino)carbene **2h**.[32] This synthetic method has been developed by Enders *et al.* to generate carbenes of type **12**.[7]

IV

SPECTROSCOPIC DATA AND X-RAY ANALYSIS

In solution, multinuclear NMR spectroscopy is by far the most informative technique for analyzing the structure and bonding of phosphinocarbenes. In fact, prior to the synthesis and single crystal X-ray analysis of the (phosphino)(phosphonio)carbene **2d**,[28] the only spectroscopic evidence for the formation of carbenes came from NMR.

A. NMR Spectroscopy

Table II lists all pertinent chemical shifts and coupling constants for the known phosphinocarbenes and their respective diazo precursors. The (phosphino)(silyl)carbenes are all characterized by high field chemical shifts for phosphorus (-24 to -50 ppm) and silicon (-3 to -21 ppm), and low field chemical shifts for carbon (120 to 143 ppm) with large couplings to phosphorus (147 to 203 Hz).

Classical shielding arguments indicate an electron-rich phosphorus atom, or equally, an increase in coordination number. The silicon atom seems also to be electron-rich, while the carbon has a chemical shift in the range expected for a multiply bonded species. The coupling constant data are difficult to rationalize, as it is not possible to predict the influence of orbital, spin-dipolar, Fermi contact, or higher-order quantum mechanical contributions to the magnitude of the coupling constants. However, classical interpretation of the NMR data indicates that the (phosphino)(silyl)carbenes have a P–C multiple bond character.

The replacement of the trimethylsilyl group by an isoelectronic, isovalent phosphonio substituent produces the carbene **2d** with similar NMR spectroscopic characteristics (Table II). A quick comparison of the reduced $^2K_{pp}$ and $^2K_{PSi}$ coupling constants indicate that these values are also similar. Thus

TABLE II

PERTINENT CHEMICAL SHIFTS (IN ppm) AND COUPLING CONSTANTS (IN Hz) FOR THE
PHOSPHINOCARBENES AND THEIR DIAZO PRECURSORS

			Diazo 1	Carbene 2		
R^1	R^2	R^3	$\delta^{31}P$	$\delta^{31}P$	$\delta^{13}C(J_{PC})$	$\delta^{29}Si\ (J_{PSi})$
i-Pr$_2$N	i-Pr$_2$N	SiMe$_3$	+56.1	−40.0	142.7(159)	−19.7(59)
Tmp^a	i-Pr$_2$N	SiMe$_3$	+88.0	−49.7	145.5(203)	−21.3(70)
Tmp^a	Me$_2$N	SiMe$_3$	+83.8	−24.1	133.5(147)	−13.2(52)
Tmp^a	Me$_2$N	Si(i-Pr)$_3$	+88.7	−27.8	120.7(181)	−2.8(47)
Tmp^a	Ph	SiMe$_3$	+25.0	−38.1	136.9(147)	−17.0(27)
c-Hex$_2$N	c-Hex$_2$N	SiMe$_3$	+58.0	−31.4	139.3(160)	−19.7(59)
i-Pr$_2$N	i-Pr$_2$N	PR$_2$H^{+b}	not observed	+27.1	98.9(157)	c
i-Pr$_2$N	i-Pr$_2$N	PR$_2$Cl^{+b}	+73.9	+35.1	103.9(154)	d

^{a} Tmp = 2,2,6,6-tetramethylpiperidino.
^{b} R = i-Pr$_2$N.
^{c} J_{PP} = 121 Hz.
^{d} J_{PP} = 71 Hz.

the similarity of the silyl- and phosphoniophosphinocarbenes in solution is evident.

B. *Solid-State Structure*

The great advantage of the (phosphino)(phosphonio)carbene **2d** over its silyl analogues is that it can be crystallized. As we have seen, these two types of phosphinocarbenes are very similar; thus, conclusions from the X-ray analysis of **2d**[28] can probably be extended to the silylcarbenes. Ball and stick views of the molecule are shown in Fig. 2, and the pertinent geometric parameters are in the legend.

No interaction with the trifluoromethanesulfonate anion is observed, confirming the ionic character of **2d**. The P2–C2 bond length [1.548(4) Å] is in the range expected for a phosphorus–carbon triple bond,[33] the P2 atom is planar, and the value of the P1–C2–P2 angle is rather large [165.1(4)–164.1(4)°]. Because of a disorder associated with the phosphonio part [s.o.f. = 0.62(1)], the values of the P1–C2 and P1a–C2 bond lengths [1.605(5) and 1.615(5) Å] could well not be accurate. A riding model[34] allowed us to estimate the lower and upper limits of these bond lengths, 1.607 and 1.709 Å, respectively. The computed bond distances[16] (1.698 Å) strongly favor the upper limit of the experimental result. Either way, the P1–C2 bond length is far too short for a phosphorus–carbon single bond and is more in the range observed for phosphorus ylides.[35]

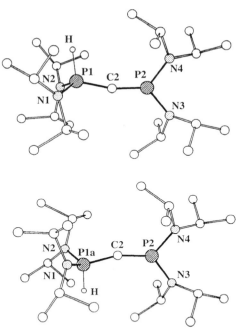

FIG. 2.   Ball-and-stick view of the two units of **2d**. Selected bond distances (Å) and bond angles (deg) are as follows: P1–N1 1.635(4), P1–N2 1.641(4), P1a–N1 1.622(5), P1a–N2 1.638(5), P1–C2 1.605(5), P1a–C2 1.616(5), P2–C2 1.548(4), P2–N3 1.632(3), P2–N4 1.635(3); P1–C2–P2 165.1(4), P1a–C2–P2 164.1(4), C2–P2–N3 126.3(2), C2–P2–N4 126.7(2), N3–P2–N4 107.0(2).

Although it is not known whether this is a dynamic or a static disorder, it is clear from Fig. 2 that the two units result from an inversion at the central carbon, followed by a 180° rotation around the P1–C2 bond. The observation that this carbene has difficulty in maintaining one discrete form in the solid state could well be consistent with the low value for the inversion barrier at carbon, calculated by Hegarty *et al.*[13] (see Section II).

V

REACTIVITY OF PHOSPHINOCARBENES R₂P–C–X

It appears that the reactivity (as well as the stability) of phosphinocarbenes (R$_2$P–C–X) is strongly dependent on the nature of the other carbene substituent (X), and therefore this section will be subdivided with respect

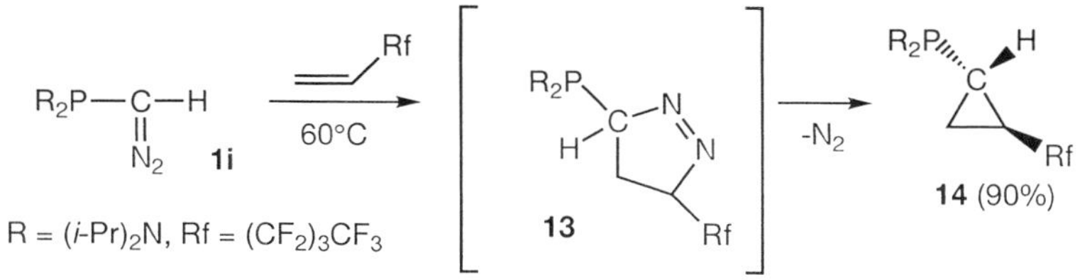

SCHEME 12.

to the nature of X (going from the left side to the right side of the periodic table).

A. X = Group 1 Element = H

Not much is known about the reactivity of the phosphinocarbene **2i**. Problems arise, at least in part, from the high 1,3-dipolar reactivity of the diazo precursor **1i**, which hides any carbene reactivity. Indeed, although **1i** is stable in a toluene solution at 60°C for hours, the addition of an electron-poor olefin, such as a perfluoroalkyl-monosubstituted alkene, induces the exclusive formation of the thermodynamically more stable *anti*-isomer of the cyclopropane **14** (see Section V,B,3,a).[36] This clearly demonstrates that the cyclopropanation reaction does not involve the carbene **2i**, but that an initial [2 + 3]-cycloaddition occurs leading to the pyrazoline **13**, which subsequently undergoes a classical N_2 elimination.[37]

In contrast, the formation of the phosphinonitrile **16** (30% yield) during the attempted distillation (90°C/10^{-2} mm Hg) of the bis(diisopropylamino)-

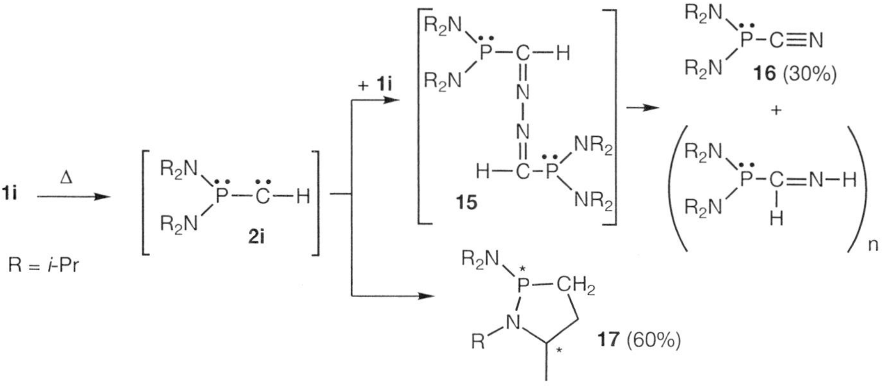

SCHEME 13.

phosphinodiazomethane **1i** probably involves the carbene **2i**.[4] Indeed, the reaction of **2i** with the starting diazo derivative **1i** could give the acetaldazine **15**, which would rearrange into the observed nitrile **16** and the corresponding imine. Interestingly, pyrolysis of **1i** at 250°C under vacuum led to the five-membered heterocycles **17** as a 70/30 mixture of diastereomers (60% yield) along with **16** (20% yield).[4] Clearly, these heterocycles **17** result from an intramolecular carbene insertion into a C–H bond of a phosphorus substituent.

B. X = Group 14 Element

1. (Phosphino)(alkyl)carbenes (X = R)

Only one report has dealt with the chemistry of (phosphino)(alkyl)carbenes. Regitz and Binger[38] found that the flash vacuum thermolysis (300°C/ 10^{-4} mm Hg) of the (di-*tert*-butylphosphino)(*tert*-butyl)diazomethane **1k** led to the formation of the *tert*-butylphosphaalkyne **18k**. They postulated that the initially formed phosphinocarbene **2k** undergoes a fragmentation reaction; the two *tert*-butyl groups bonded to phosphorus would be eliminated in the form of 2-methylpropene and hydrogen. Although a similar fragmentation reaction[38] leading to the trimethylsilylphosphaalkyne **18l** has been observed, starting from the (di-*tert*-butylphosphino)(trimethylsilyl)diazomethane **1l**, this reaction cannot be considered as typical behavior of phosphinocarbenes. It is more likely due to the particular properties of the *tert*-butyl groups at phosphorus and probably also to the rather drastic experimental conditions used.

2. (Phosphino)(acyl)carbenes (X = C(O)R)

The formation of the phosphoranyl alkyne **19**, in the spontaneous room temperature decomposition of the phosphinodiazoketone **1m**, has been explained in terms of an intramolecular Wittig-type reaction involving the phosphorus-vinyl-ylide form of the phosphinocarbene **2m**.[39]

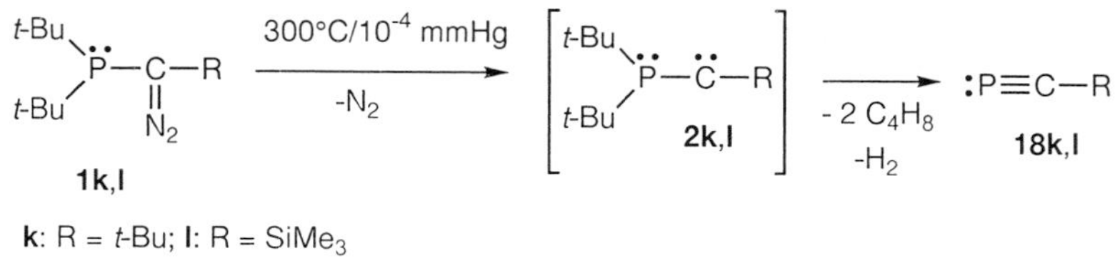

k: R = *t*-Bu; l: R = SiMe₃

SCHEME 14.

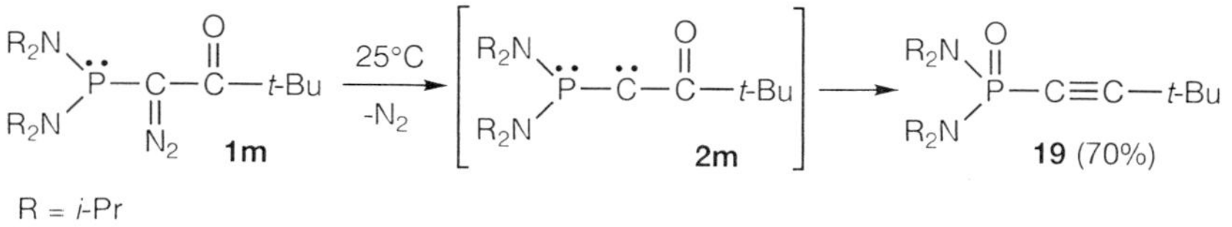

SCHEME 15.

3. (Phosphino)(silyl)carbenes

These are by far the most studied phosphinocarbenes, mainly because of their stability. As often as possible, their reactivity will be compared with that of "standard" transient carbenes.

a. *[1+2] Cycloadditions to multiple bonds [1 + 2] cycloadditions to olefins.*   Both singlet and triplet transient carbenes react with olefins to give cyclopropanes, although with totally different mechanisms, which is apparent from the stereochemistry of the reaction. Moreover, it has long been known that nucleophilic carbenes, in which the singlet state is stabilized by interaction of the vacant *p* orbital with the lone pair of a heteroatom substituent, do not react with electron-rich alkenes, but with electrophilic olefins.

The nucleophilic character of stable (phosphino)(silyl)carbenes **2a,b** was recognized early on, since they do not react with simple alkenes (e.g., cyclohexene, pentenes), but do react with dimethyl fumarate and methyl acrylate[40]; this has been confirmed by competitive cyclopropanation reactions with a variety of *p*-substituted styrenes.[36] All these reactions occur at room temperature, with the corresponding cyclopropenes **20–24** being obtained in high yields. Interestingly, in each case, only one diastereomer is obtained. The NMR data for **21–24** are consistent with a "*syn*-attack" of the phosphinocarbenes **2a,b**, and this has been confirmed by a single-crystal X-ray diffraction study of **23b**.[36] So far, this stereoselectivity is not well understood, especially since steric effects should favor "*trans*-attack" [the bis(amino)phosphino group being more sterically demanding than the trimethylsilyl group]. The reaction of **2a,b** with dimethyl fumarate occurs with retention of the stereochemistry about the double bond. However, since dimethyl maleate does not react with **2a,b** [the difference in reactivity between *cis* and *trans* olefins toward carbenes has already been noted by Moss *et al.* with the methoxy(phenyl)carbene],[41] the chemical method developed by Skell[42] for distinguishing between singlet and triplet carbenes could not be used. Of course, the singlet nature of **2a,b** was not debatable from the calculations and the spectroscopic data, but it was of importance

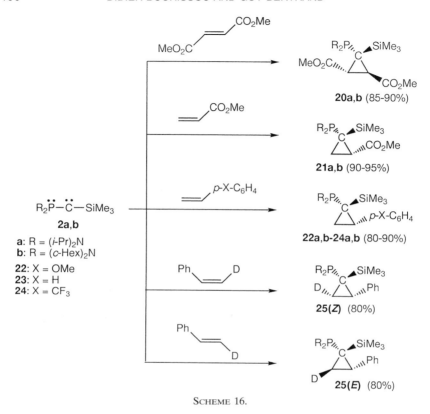

SCHEME 16.

to bring evidence that a concerted mechanism was involved in the cyclopropanation reactions, and hence the genuine carbene nature of **2a,b**. Using pure Z and E monodeuterated styrene isomers, we have observed the stereospecific formation of the corresponding cyclopropanes **25(Z)** and **25(E)**.[36] The concerted nature of the cyclopropanation reactions has been corroborated by the results observed by reacting **2a,b** with a variety of carbon–heteroatom double and triple bonds (vide infra).

[1 + 2] Cycloadditions to Carbonyl Derivatives. Electrophilic transient carbenes are known to react with carbonyl derivatives through the oxygen lone pair to give carbonyl ylides **26**.[43] These 1,3-dipolar species are usually characterized by [3 + 2]-cycloaddition reactions or can even be isolated[44]; a small amount of the corresponding oxiranes is sometimes obtained.[43a,45] To date, no reaction of transient nucleophilic carbenes with carbonyl derivatives has been reported.

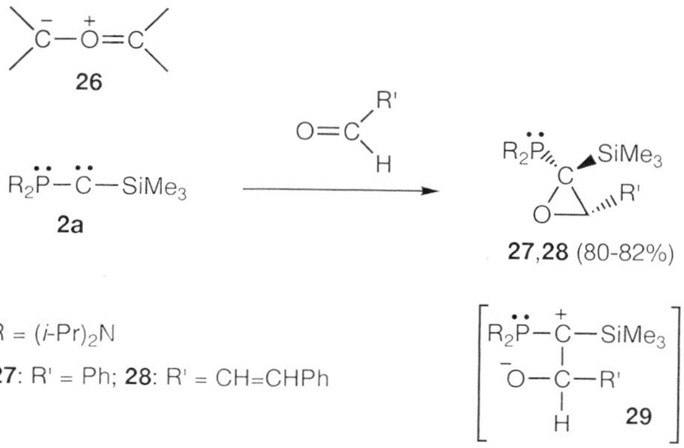

$$\sum C - \overset{+}{O} = C \lesssim$$
26

$$O = C \overset{R'}{\underset{H}{\diagdown}}$$

$$R_2\overset{..}{P} - \overset{..}{C} - SiMe_3$$
2a

$$R_2\overset{..}{P}_{\prime\prime\prime} \overset{SiMe_3}{\underset{O}{\diagup}} \underset{R'}{\diagdown}$$
27,28 (80-82%)

R = (*i*-Pr)₂N
27: R' = Ph; **28**: R' = CH=CHPh

$$\left[\begin{array}{c} R_2\overset{..}{P} - \overset{+}{C} - SiMe_3 \\ | \\ \overset{-}{O} - \overset{|}{C} - R' \\ | \\ H \quad \mathbf{29} \end{array} \right]$$

SCHEME 17.

The (phosphino)(silyl)carbene **2a** readily and cleanly adds to benzalde-hyde and cinnamaldehyde, affording the oxiranes **27** and **28**, as single diaste-reomers.[40] These results strongly suggest a concerted mechanism, since the formation of a zwitterionic intermediate, such as **29**, would result in the formation of a phosphoryl alkene via oxygen atom attack at the phosphorus center. Note that **2a** does not react with ketones, which is in line with its nucleophilic character.

[1 + 2] Cyclodditions to Carbon–Heteroatom Triple Bonds. Transient electrophilic carbenes are known to react with nitriles to give transient[46] or even stable nitrile ylides **30**.[47] No reaction of transient nucleophilic carbenes with nitriles has been reported.

The first example of azirine formation from a carbene and a nitrile has

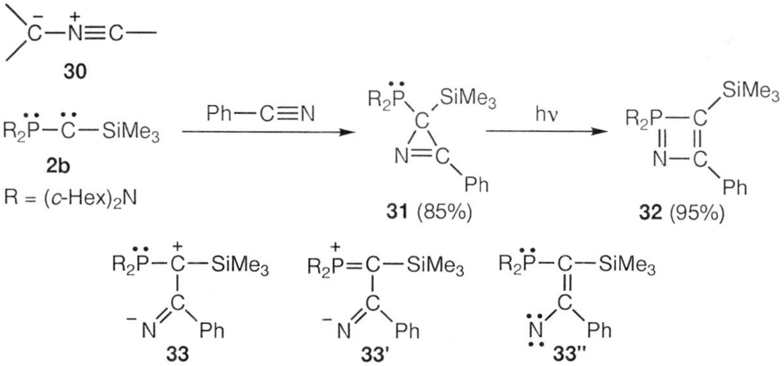

$$\sum C - \overset{+}{N} \equiv C -$$
30

$$R_2\overset{..}{P} - \overset{..}{C} - SiMe_3$$
2b
R = (*c*-Hex)₂N

$$\xrightarrow{Ph-C\equiv N}$$

$$R_2\overset{..}{P} \overset{SiMe_3}{\underset{N=C}{\diagup}} \overset{}{\underset{Ph}{\diagdown}}$$
31 (85%)

$$\xrightarrow{h\nu}$$

$$R_2P - \overset{SiMe_3}{\underset{N-C}{\overset{||}{C}}} \overset{}{\underset{Ph}{\diagdown}}$$
32 (95%)

$$R_2\overset{..}{P} - \overset{+}{C} - SiMe_3 \\ | \\ \overset{-}{N} \overset{C}{\diagdown} Ph$$
33

$$R_2\overset{..}{P} = \overset{+}{C} - SiMe_3 \\ | \\ \overset{-}{N} \overset{C}{\diagdown} Ph$$
33'

$$R_2\overset{..}{P} - \overset{}{C} - SiMe_3 \\ || \\ \overset{..}{N} \overset{C}{\diagdown} Ph$$
33''

SCHEME 18.

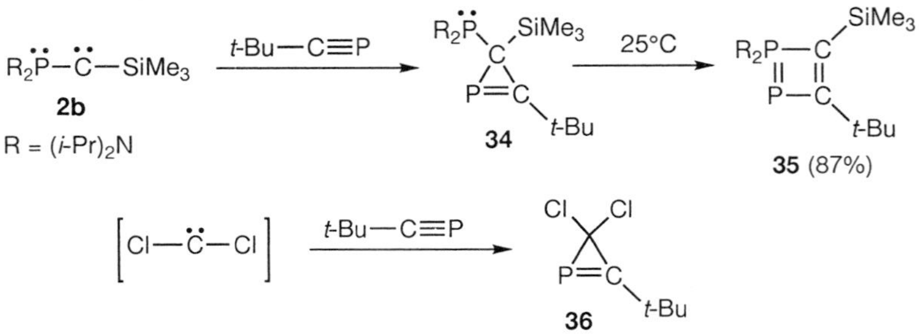

SCHEME 19.

been observed by addition of the phosphinocarbene **2b** to benzonitrile.[48,49] It is quite likely that the formation of the azirine **31** results from a concerted [1 + 2]-cycloaddition. A stepwise mechanism, involving the initial nucleophilic attack of the carbene at the carbon atom of the nitrile, would have led to the 1,3-dipole **33**, which can also be regarded as the azabetaine **33′**, or the vinyl nitrene **33″**. The ring closure of vinyl nitrenes to produce azirines is known,[50] but it has been shown that these unsaturated species are efficiently trapped by phosphines to give the phosphazene adducts[51]; therefore, in the case of the vinyl nitrene **33″** an intramolecular reaction of this type should have led to the azaphosphete **32**.[52] Interestingly, irradiation of the phosphinoazirine **31** at 254 nm led to the azaphosphete **32** in 98% yield.[48,49]

Similarly, the (phosphino)(silyl)carbene **2a** reacts at −30°C with a slight excess of the *tert*-butylphosphaalkyne cleanly affording the 2-phosphino-2*H*-phosphirene **34**.[53] The reaction leading to **34** is strictly analogous to that observed on reacting the transient dichlorocarbene with the *tert*-butylphosphaalkyne, in which the 2*H*-phosphirene **36** was obtained.[54] The three-membered heterocycle **34** appeared to be rather unstable and rearranged, quantitatively, to afford the $1\lambda^5,2\lambda^3$-diphosphete **35** after 3 h at room temperature.[55] Once again, these results as a whole indicate that a concerted [1 + 2]-cycloaddition process is involved in the formation of the 2*H*-phosphirene **34**.

b. *Insertion into C–H Bonds.* Insertion into C–H bonds is a characteristic and well-documented feature of singlet and triplet carbenes; however, here also, the mechanism of the reaction is totally different depending on the spin multiplicity. Triplet carbenes first abstract hydrogen to give radicals, while consideration of the rule of spin conservation leads to the formulation of the insertion of a singlet carbene as a one-step process involving a three-center cyclic transition state.[56]

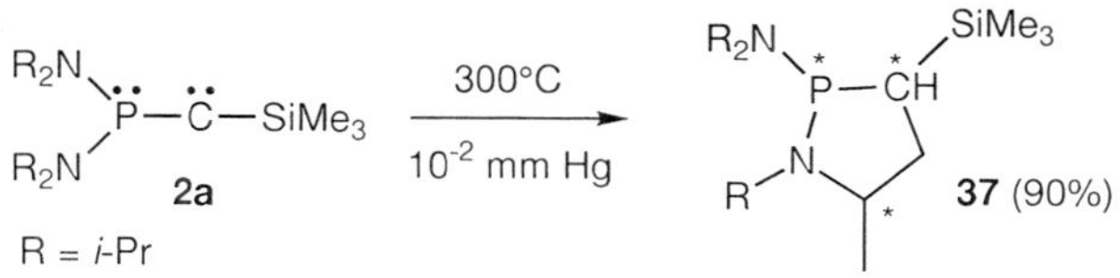

SCHEME 20.

The (phosphino)(silyl)carbene **2a** has been shown to exhibit this diagnostic reactivity. Under rather drastic conditions (300°C, 10^{-2} mm Hg), **2a** has been converted into the azaphospholidine **37**, in high yield, as a mixture of four diastereomers, which result from the presence of three asymmetric centers.[4] The high regioselectivity of the carbene insertion (no formation of a four-membered ring was detected) is intriguing, especially since the same regioselectivity is observed upon thermolysis of bis(diisopropylamino)phosphinodiazomethane **1i**.[4] Yet, in contrast, four-membered heterocycles were obtained with other phosphinocarbenes (Sections V,C,1 and 3).

c. *Insertion into A–X Single Bonds.*   The insertion of (phosphino)(silyl) carbenes **2a** into polarized A–X single bonds is more controversial. Indeed, chlorotrimethylsilane,[4] *P*-chlorophosphaalkene,[57] and dimethylamine[4] cleanly react with the [bis(diisopropylamino)phosphino](trimethylsilyl)carbene **2a**, giving rise to the λ^5-phosphorus derivatives **39**, and not the expected insertion products **38**. At first glance, derivatives **39** result from a 1,2-addition of A–X reagents across the polarized PC-multiple bond of **2a**. However, it is more likely that a carbene insertion into the A–X bonds takes place, giving **38**, which subsequently undergoes a rearrangement affording **39**; such 1,2-shifts have already been exemplified.[58]

d. *Reactions with Group 14 Lewis Acids.*   Addition of trimethylsilyl trifluoromethanesulfonate to the phosphinocarbene **2a** proceeds cleanly to

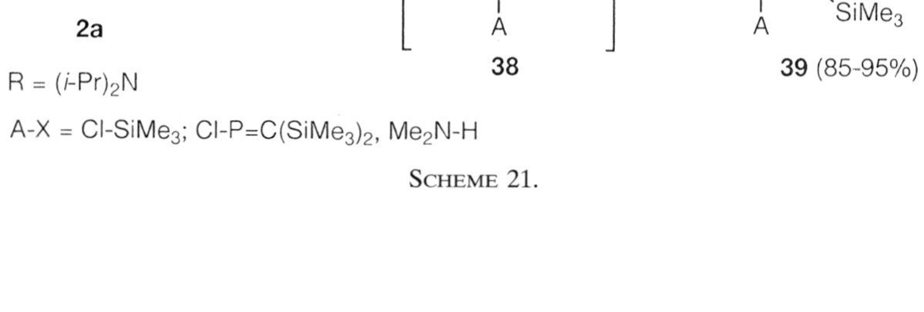

SCHEME 21.

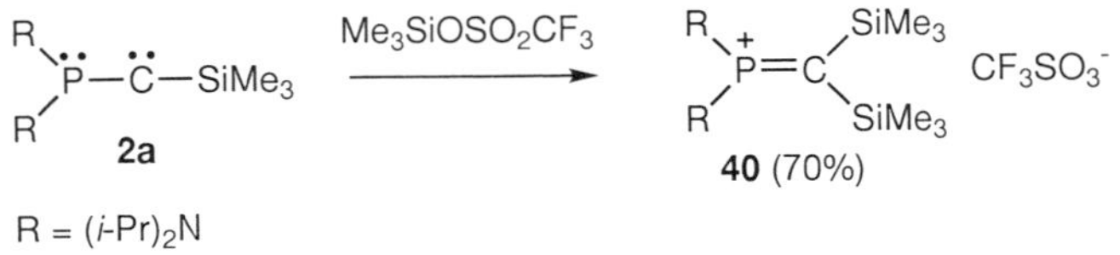

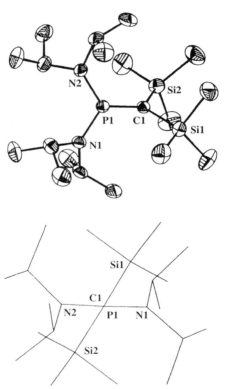

$$R = (i\text{-Pr})_2N$$

SCHEME 22.

FIG. 3. Thermal ellipsoid diagram (30% probability) of the methylenephosphonium salt **40** showing the atom numbering scheme, and simplified view of the molecule showing the twist angle. Pertinent bond lengths (Å) and bond angles (deg) are as follows: P1–C1 1.620(3), P1–N1 1.615(3), P1–N2 1.610(3), C1–Si1 1.875(3), C1–Si2 1.913(3); N1–P1–C1 123.4(1), N2–P1–C1 124.4(1), N1–P1–N2 112.2(1), Si1–C1–P1 121.7(1), Si2–C1–P1 119.3(1), Si1–C1–Si2 119.1(1).

give a new type of phosphorus cation, namely, the methylene phosphonium salt **40**.[59,60]

This extremely air-sensitive compound, which is valence isoelectronic to an olefin, has been structurally characterized by X-ray diffraction. It has a short carbon–phosphorus double bond (1.62 Å); the phosphorus and carbon atoms adopt a trigonal planar geometry with a dihedral angle of 60° (Fig. 3). This value is significantly larger than that reported for the most crowded olefin.[61] Formally, this compound can be viewed as the product of a carbene–carbenoid coupling between bis(trimethylsilyl)carbene and bis(diisopropylamino)phosphenium triflate. Note that another route to methylenephosphonium salt has been reported by Grützmacher et al.[62]

e. *Reactions with Group 13 Lewis Acids.*   Due to their nucleophilic character, (phosphino)(silyl)carbenes react with a variety of Lewis acids. The fate of these reactions appears to be strongly dependent on the nature of the acids, but it seems quite likely that in all cases the first step is the formation of a carbene–Lewis acid adduct. In fact, compounds of this type **41–43** have been isolated using aluminum, gallium, and indium trichlorides.[63]

The deshielded ^{31}P and ^{13}C NMR chemical shifts observed for **41–43** ($\delta^{31}P \approx +130$, $\delta^{13}C \approx +76$, $J_{PC} \approx 85$ Hz) are consistent with the presence of a P=C double bond and positive charge development at phosphorus; these spectroscopic data are in fact very similar to those observed for the methylenephosphonium salt **40**.[59,60] This similarity is reinforced by the X-ray analysis of the gallium adduct **42** (Fig. 4): (i) The phosphorus and carbon atoms adopt a trigonal planar geometry, (ii) there is a twist angle between the two planes of 34.1°, and (iii) the phosphorus–carbon bond distance is rather short (1.61 Å).

When triethylborane is used, the borane–carbene adduct **44** is stable in solution for several weeks at −20°C, but only for a day at room temperature.[25] Elimination of diethyl(dicyclohexylamino)borane occurs leading to

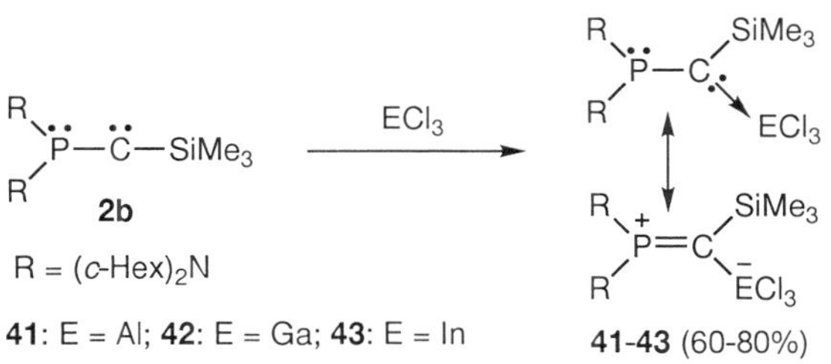

R = (*c*-Hex)$_2$N

41: E = Al; **42**: E = Ga; **43**: E = In

41-43 (60-80%)

SCHEME 23.

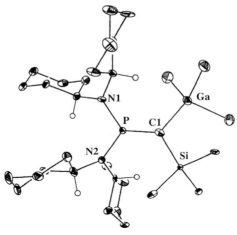

FIG. 4. Ortep view of the gallium adduct **42** showing the atom numbering scheme. Pertinent bond lengths (Å) and bond angles (deg) are as follows: P–C1 1.601(9), P–N1 1.622(7), P–N2 1.619(8), C1–Si 1.886(8), C1–Ga 1.978(10); N1–P–C1 125.1(5), N2–P–C1 121.4(4), N1–P–N2 113.4(4), Si–C1–P 121.8(6), Ga–C1–P 122.6(5), Si–C1–Ga 115.7(5).

the phosphaalkene **46**, which was isolated in near quantitative yield. This fragmentation can be rationalized by a classical migration of an ethyl group from the four-coordinate boron atom to the electron deficient α-carbon[64] forming **45**, followed by a 1,2-elimination of diethyl(dicyclohexylamino)-borane.

Lastly, when dimesitylfluoroborane,[65] trimethoxyborane,[65] and trimethyl-aluminum, -gallium, and -indium[63] were used, phosphorus ylides **47–51** were obtained in good yields; no intermediates were spectroscopically detected. Once again, it is reasonable to postulate the formation of carbene–Lewis acid adducts, followed by a 1,2-migration; finally, instead of undergoing a 1,2-elimination, a classical methylenephosphine–phosphorus ylide conversion[58] would lead to **47–51**.

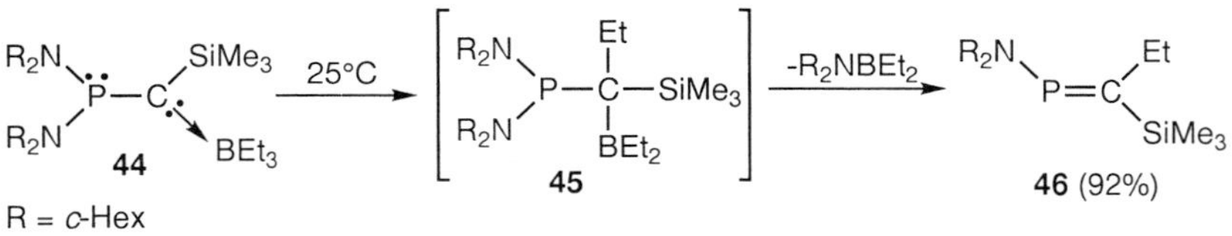

SCHEME 24.

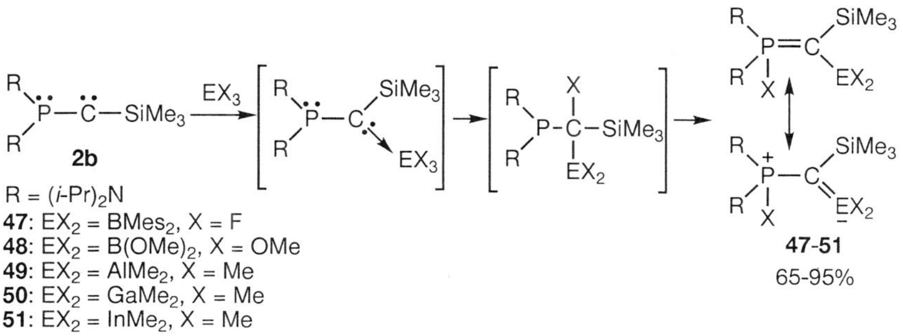

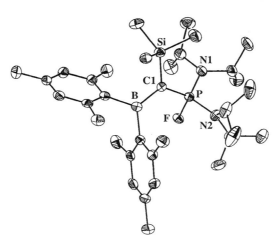

SCHEME 25.

R = (*i*-Pr)$_2$N
47: EX$_2$ = BMes$_2$, X = F
48: EX$_2$ = B(OMe)$_2$, X = OMe
49: EX$_2$ = AlMe$_2$, X = Me
50: EX$_2$ = GaMe$_2$, X = Me
51: EX$_2$ = InMe$_2$, X = Me

47-51
65-95%

Although numerous so-called stabilized phosphorus ylides, in which the negative charge is delocalized into an organic or organometallic framework have been studied, compounds **47**[65] (Fig. 5) and **50**[63] were the first examples of phosphorus ylides *C*-substituted by a group 13 element to be studied by X-ray diffraction. These compounds are of interest since they can also be considered as boron- and gallium–carbon double bonded compounds, *C*-substituted by a phosphonio group. Indeed, the boron–carbon bond length in **47** (1.52 Å) is shorter than a usual boron–carbon single bond (1.58–1.62 Å), but a little longer than a boron–carbon double bond, e.g.,

FIG. 5. Molecular structure of the boron-substituted phosphorus ylide **47** showing the atom numbering scheme. Pertinent bond lengths (Å) and bond angles (deg) are as follows: B–C1 1.525(4), P–C1 1.696(2), C1–Si 1.872(3), P–F 1.585(1), P–N1 1.663(2), P–N2 1.655(2); P–C1–Si 123.3(1), P–C1–B 117.9(2), Si–C1–B 118.8(2).

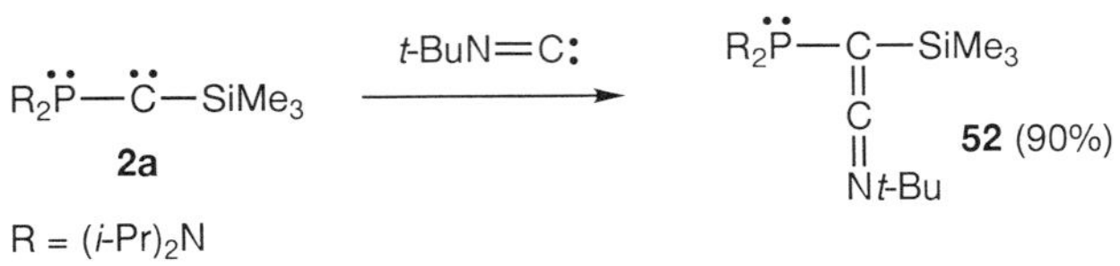
t-BuN=C:
R2P—C—SiMe3
2a
R = (i-Pr)2N
R2P—C—SiMe3
C
Nt-Bu
52 (90%)

SCHEME 26.

$Mes_2{=}CH_2^-$ (1.44 Å)[66]; this boron–carbon π-type interaction was confirmed by NMR in solution. In the same way, the gallium–carbon bond length in **50** (1.93 Å) is the shortest Ga–C distance so far reported.

f. *Reactions with Carbenoid Species.* To date, no dimerization or genuine carbene–carbene coupling reactions to give the corresponding alkenes have been reported for stable phosphinocarbenes. However, it appears that *tert*-butyl isocyanide is one of the very rare reagents that react with almost all of the stable phosphinocarbenes reported so far.[27] For example, it reacts with **2a**, even at −78°C, affording the ketenimine **52**, which was isolated after treatment with elemental sulfur as its thioxophosphoranyl analogue in 90% yield.[4] The great reactivity of isonitriles toward phosphinocarbenes can easily be explained in terms of steric factors: The reactive site of RN=C: is comparatively unhindered.

Reactions involving the heavier congeners of carbenes have been observed. (Phosphino)(silyl)carbene **2n** reacts with germanium(II) and tin(II) compounds **53**, affording *C*-germyl and *C*-stannylphosphaalkenes **55**, respectively, in 42 to 78% yields.[67] By analogy with the reaction of carbenes **2** with isonitriles, it is reasonable to postulate the primary formation of germa- and stannaethenes **54**; a subsequent 1,3-shift of the dimethylamino group from phosphorus to germanium or tin produces derivatives **55**. These

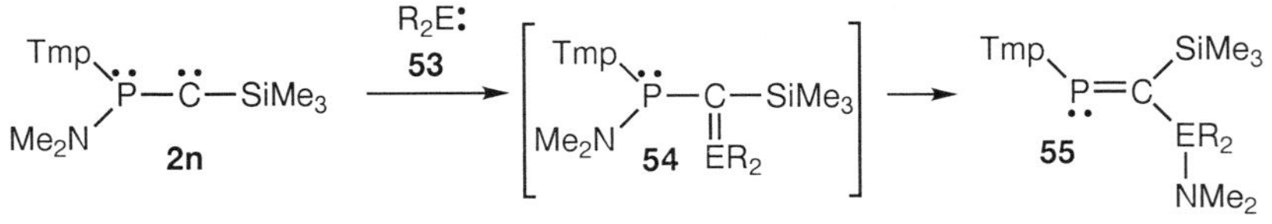
Tmp
Me2N
P—C—SiMe3
2n
R2E:
53
Tmp
Me2N
P—C—SiMe3
54 ER2
Tmp
P=C
SiMe3
55
ER2
NMe2

R	E	yield
$(Me_3Si)_2N$	Ge	78%
Mes*NH	Ge	46%
$(Me_3Si)_2N$	Sn	42%
$(Me_3Si)_2CH$	Sn	48%

Tmp: 2,2,6,6-Tetramethylpiperidino
Mes*: 2,4,6-tri-*tert*-butylphenyl

SCHEME 27.

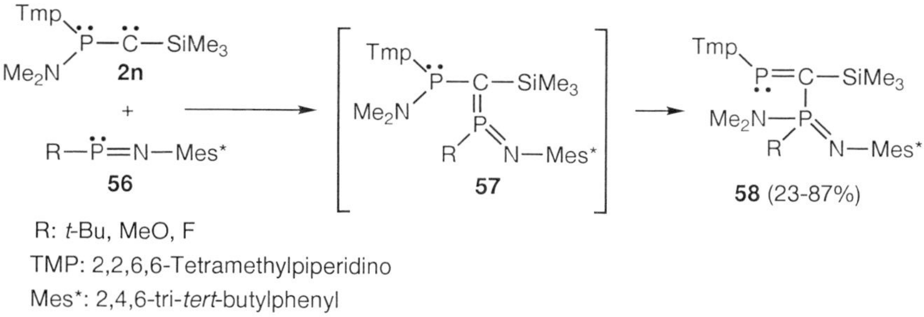

R: t-Bu, MeO, F
TMP: 2,2,6,6-Tetramethylpiperidino
Mes*: 2,4,6-tri-tert-butylphenyl

SCHEME 28.

two reactions are highly chemoselective, since we only observed the migration of the smallest phosphorus substituent. The driving force of this rearrangement is probably the reluctance of germanium and tin to form double bonds.[68]

Because of the existence of two energetically close HOMOs π and n, the orbital sequence HOMO/LUMO of iminophosphines can be π/π^*, inducing an alkene-like behavior, but also n/π^*, leading to behavior analogous to that of a carbene. Indeed, phosphinocarbene 2n reacts with iminophosphines 56 leading to the phosphaalkenes 58 (23–87% yield).[69] These results are strictly analogous to those observed in the reaction of the carbene 2n with germylenes and stannylenes, and therefore it is quite likely that a carbene–carbenoid coupling-type reaction occurs, leading to the transient (methylene)(imino)phosphoranes 57, which would undergo a subsequent migration of a dimethylamino substituent from the λ^3-phosphorus to the positively charged λ^5-phosphorus atom. Once again, these reactions are highly chemoselective, since we only observed the migration of the smallest phosphorus substituent.

g. *Reactions with Lewis Bases.* Carbenoids such as isocyanides or even germylenes or stannylenes can react either as Lewis bases or acids because of the presence of a lone pair and a vacant orbital, respectively. In order to show the presence of an available vacant orbital in phosphinocarbenes,

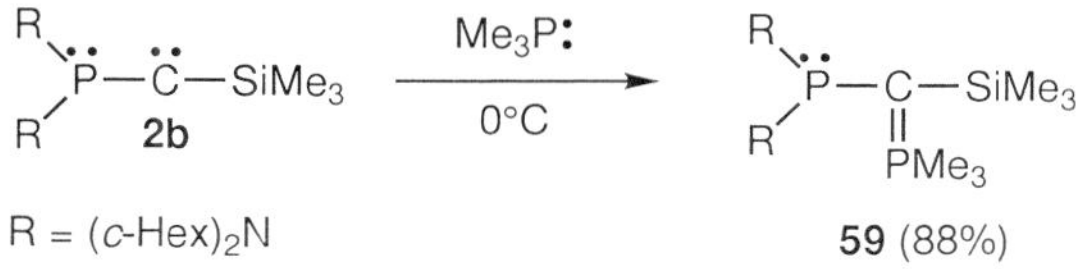

R = (c-Hex)$_2$N

SCHEME 29.

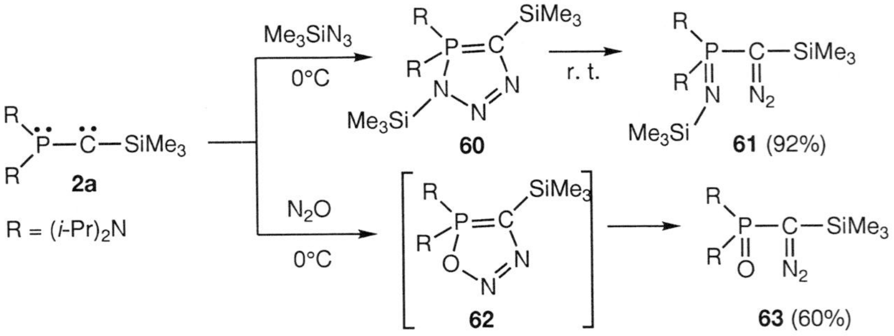

SCHEME 30.

a crucial point in proving their genuine carbene nature, it was of primary importance to study their reaction with Lewis bases. Note that transient carbenes, including nucleophilic carbenes such as oxacarbenes,[70] react with Lewis bases to afford ylides.[71] This reaction has been widely used to spectroscopically characterize these electron-deficient species and is a preparative method for *C*-dihalogeno phosphorus ylides.[35, 72] Indeed, the stable [bis(dicyclohexylamino)phosphino](trimethylsilyl)carbene **2b** cleanly reacts at 0°C with trimethylphosphine, affording the phosphorus ylide **59**.[36]

h. *Reactions with 1,3-Dipoles.* Because of the possible hypervalency of phosphorus and the phosphorus–carbon multiple bond character of phosphinocarbenes, trimethylsilylazide and nitrogen oxide react with **2a** affording the diazo derivatives **61** and **63** in 92 and 86% yield, respectively. These results have been explained by [2 + 3] cycloaddition reactions followed by ring-opening of the resulting five-membered rings **60** and **62**.

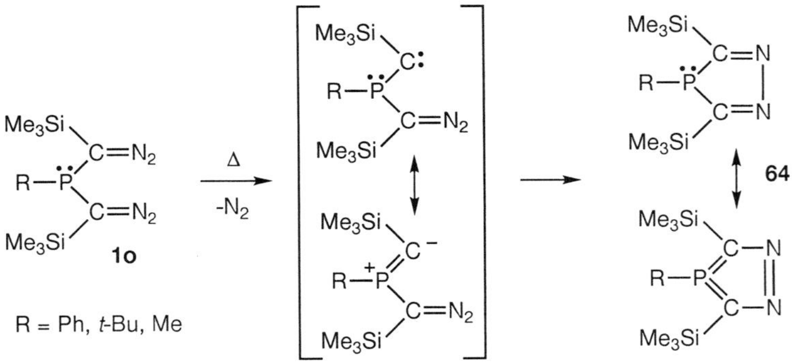

SCHEME 31.

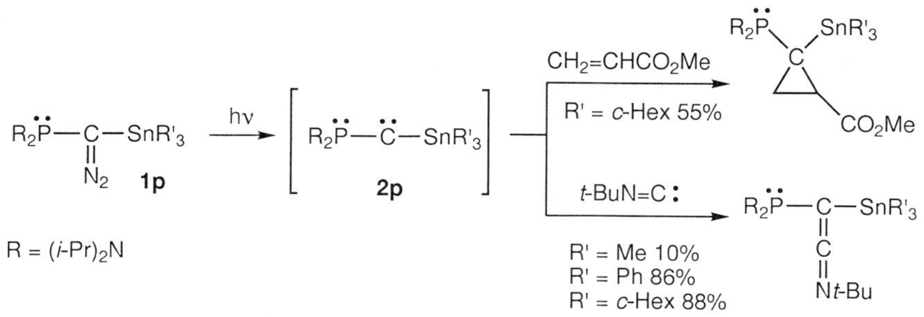

SCHEME 32.

Indeed, the initial $1,2,3,4\lambda^5$-triazaphosphole **60** was characterized in solution at 4°C.[4]

It has been reported that thermolysis of the bis(diazomethyl)phosphines **1o** leads to the formation of the diazaphospholes **64**. This can be explained by either an intramolecular [2 + 3] cycloaddition process involving the diazo group, or a diazo–carbene coupling reaction.

4. (Phosphino)(stannyl)carbenes (X = SnR₃)

In contrast to silylcarbenes, the analogous stannylcarbenes **2p** are not stable, which explains why they have attracted little interest. Their instability is probably due to the long carbon–tin bond, which does not allow sufficient steric protection of the carbene center. Their reactivity seems to be quite similar to that of stable (phosphino)(silyl)carbenes: Cyclopropanation reactions have been reported with methyl acrylate as well as coupling reactions with tert-butyl isonitrile.[73]

C. X = Group 15 Element

1. (Phosphino)(amino)carbenes (X = NR₂)

Vacuum thermolysis (160°C) of the hemiaminal **11** generates the azaphosphetane **65** in 85% yield.[32] This product clearly results from the intramolecular insertion of the transient (amino)(phosphino)carbene **2h** into the C–H bond of a diisopropylamino group bonded to phosphorus. Note that the four-membered heterocycle **65** is formed exclusively in spite of the ratio of six methyl-CH bonds to one methine–CH bond, and that only one of the two possible diastereomers is detected. The same regio- and diastereoselectivity have already been observed with the di(phosphino)carbene **2g**,[74] but is in marked contrast to the exclusive formation of five-membered rings

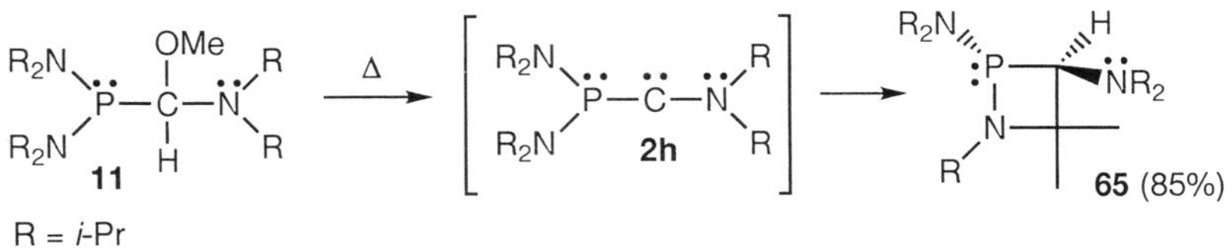

SCHEME 33.

17 (as two diastereomers)[4] and **37** (as four diastereomers)[4] in the thermolysis of the diazophosphine **1i** and (phosphino)(silyl)carbene **2a**, respectively (see Sections V,A and V,B,3,b). The regio- and diastereoselectivity are therefore strongly dependent on the nature of the second substituent of the phosphinocarbenes; so far this has not been rationalized. Note that CH-insertions are typical reactions for transient singlet carbenes, but have never been observed with di(amino)carbenes.

2. (Phosphino)(nitroso)carbenes (X = NO)

An exchange reaction occurs when nitrosylchloride is reacted with (phosphino)(trimethylsilyl)diazomethane **1a** to yield an unstable α-nitroso α-phosphinodiazo derivative **1q** and chlorotrimethylsilane.[75] Spontaneous loss of N_2 leads to the transient (nitroso)(phosphino)carbene **2q**, which can also be regarded as the phosphorus vinyl ylide **2q'** or even better as the λ^3-

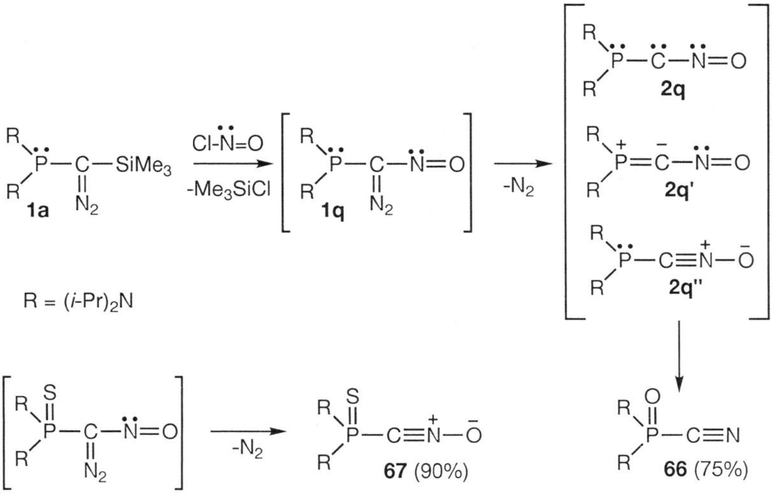

SCHEME 34.

phosphinonitrile oxide **2q″** (because of the superior π-donor ability of nitrogen compared to phosphorus).[32,76] Subsequently, a rearrangement, via a formal 1,3-oxygen shift, affords the corresponding phosphoranyl nitrile **66** in 75% isolated yield. Interestingly, note that the (thioxophosphoranyl) (nitroso)diazomethane analogue spontaneously decomposes to afford the nitrile oxide **67**.[75]

3. Di(phosphino)carbenes (X = PR₂)

a. *Intramolecular reactivity.* The fate of the transient di(phosphino)carbene **2g** appeared to be strongly dependent on the experimental conditions used for its generation. Attempted distillation or irradiation at 300 nm of a benzene solution of the diazo precursor **1g** afforded the corresponding phosphaalkene **68** in high yield.[39] Note that the intermediacy of a carbene in this reaction has been proved by generating **2g** via deprotonation of the stable (phosphino)(phosphonio)carbene **2d**.[31] Although this type of 1,2-shift is very typical for "normal" carbenes, to date, this is one of only two[29] reported examples involving a phosphinocarbene. Note that ab initio calculations predict that the phosphaalkene **68′** (R = NH₂ instead of Ni-Pr₂) is 53 kcal/mol more stable than the carbene **2g′**.[16]

In marked contrast, heating the bis[bis(diisopropylamino)phosphino]diazomethane **1g** in refluxing toluene affords the four-membered ring **69** as a single diastereomer in 92% yield. The structure of **69** has been confirmed by an X-ray analysis of the product **70**, obtained after reaction with elemental sulfur.[74] The formation of the 1,2λ⁵-azaphosphetane **69** can be explained

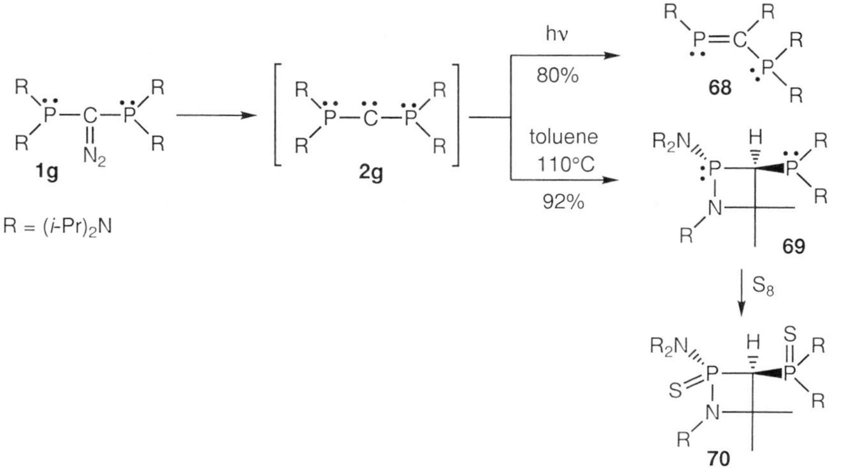

SCHEME 35.

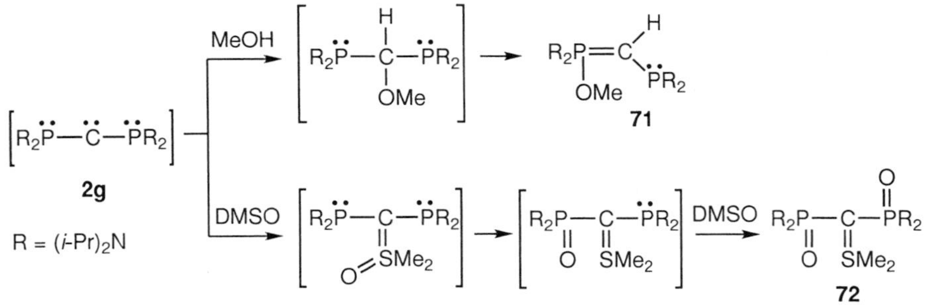

MeOH
DMSO
DMSO
2g
R = (i-Pr)2N
71
72

SCHEME 36.

in terms of insertion of the singlet carbene center of **2g** into a C–H bond of an isopropyl substituent. The selectivity of this reaction has already been discussed.

b. *Intermolecular reactivity.* The di(phosphino)carbene **2g** can be efficiently trapped by methanol or dimethylsulfoxide.[39] **71** probably results from the insertion of the carbene **2g** into the O–H bond of the alcohol followed by a 1,2-OMe shift. The [1 + 1] addition of **2g** to the sulfur atom, subsequent 1,3-oxygen shift, and oxidation of the second phosphorus atom by excess sulfoxide explain the formation of **72**.

4. *(Phosphino)(phosphoryl)carbenes [X = P(O)R$_2$]*

Among the very rare phosphinocarbenes that undergo dimerization are the (phosphino)(phosphoryl)carbenes **2r**. Two examples of "direct" head-to-tail dimerization leading to the corresponding 1λ^5,3λ^5-disphosphetes **73$_{1,2}$** have been reported and occur as a result of the multiple bond character of transient carbenes **2$_{1,2}$**.[77]

1λ^5,3λ^5-Diphosphetes **73$_{5,6}$** have also been obtained during the thermoly-

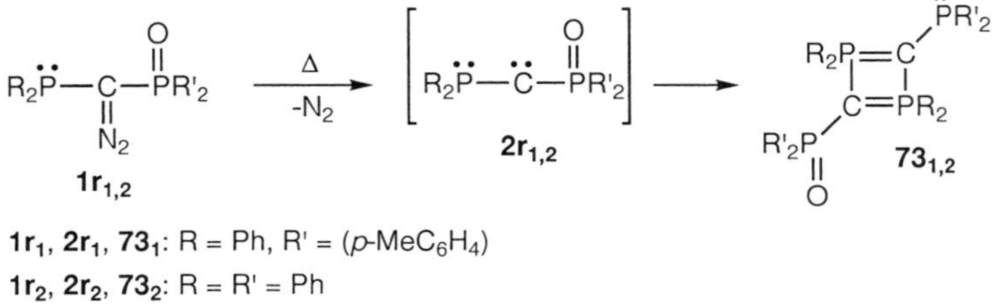

Δ
-N2
1r1,2
2r1,2
731,2
1r1, 2r1, 731: R = Ph, R' = (p-MeC6H4)
1r2, 2r2, 732: R = R' = Ph

SCHEME 37.

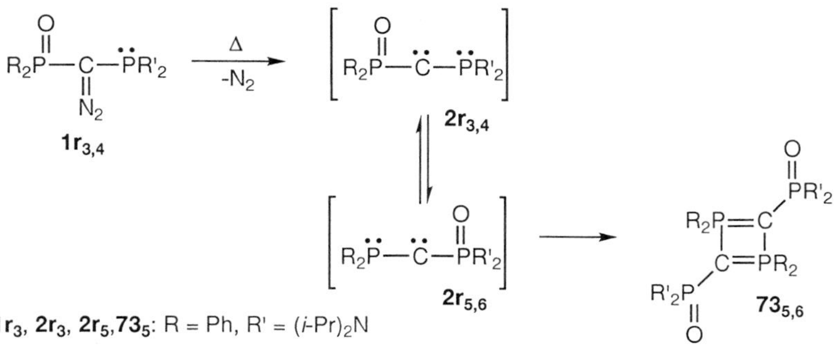

$1r_3$, $2r_3$, $2r_5$,73_5: R = Ph, R' = (*i*-Pr)$_2$N

$1r_4$, $2r_4$, $2r_6$,73_6: R = Ph, R' = Me$_2$N

SCHEME 38.

sis of (phosphino)(phosphoranyl)diazomethanes $\mathbf{1r_{3,4}}$.[77] However, $\mathbf{73_{5,6}}$ does not come from the dimerization of the initially generated phosphinocarbenes $\mathbf{2r_{3,4}}$ but from the isomeric $\mathbf{2r_{5,6}}$, which result from a 1,3-oxygen shift from phosphorus to phosphorus. These results suggest a possible equilibrium between the isomeric (phosphino)(phosphoranyl)carbenes $\mathbf{2r_{3,4}}$ and $\mathbf{2r_{5,6}}$; dimerization occurs for the less sterically hindered species.

5. *(Phosphino)(phosphonio)carbenes [X = PR_3^+]*

This is the second type of stable phosphinocarbene to have been reported. In contrast to (phosphino)(silyl)carbenes, very little is known concerning their reactivity. No simple reactions typical of a carbene-like behavior have been reported. However, the reactivity at the periphery of the carbene center makes these derivatives powerful building blocks for novel compounds.

First, it should be noted that the (phosphino)(phosphonio)carbene **2d** is stable with respect to the possible 1,2-migration of the hydrogen atom from phosphorus to the carbene centre, although calculations on the amino analogue (H_2N instead of *i*Pr_2N) predict **74** to be more stable than **2d** by 17 kcal/mol at the MP2 level[16]; the energy barrier for the 1,2-shift is probably too high.

Interestingly, the formation of both (phosphino)(phosphonio)carbenes

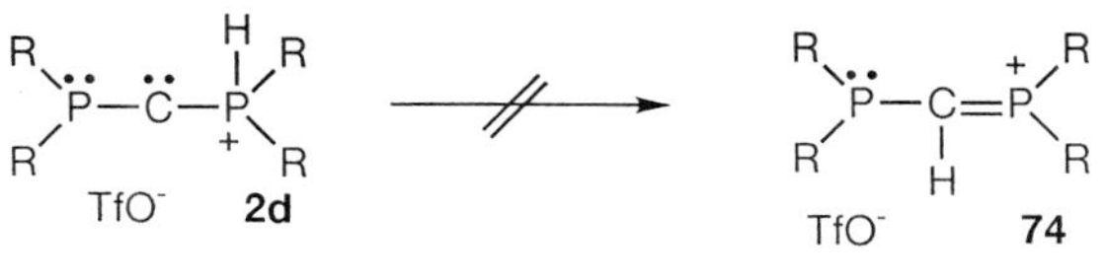

R = (*i*-Pr)$_2$N; Tf = CF_3SO_2

SCHEME 39.

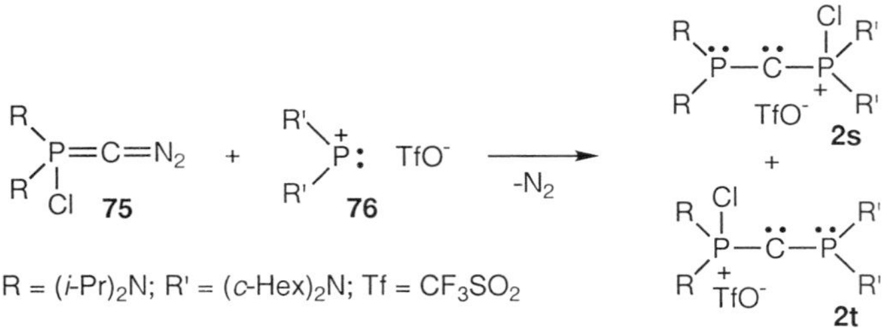

$$R = (i\text{-Pr})_2N; \ R' = (c\text{-Hex})_2N; \ Tf = CF_3SO_2$$

SCHEME 40.

2s and **2t** in the reaction of the diazocumulene **75** with the bis(dicyclohexyl-amino)phosphenium salt **76** strongly suggests the facile 1,3-chlorine shift between the two phosphorus centers.[78]

Addition of sodium tetrafluoroborate to **2d** led to the carbodiphosphor-ane **77** possessing a P–H bond, which was isolated in 70% yield.[31] Although **77** is thermally quite stable as a solid (mp 116°C), it slowly rearranges in solution at room temperature (1 week) into its isomeric phosphorus ylide **78**. This last result was not surprising, since calculations predicted **78** to be 19 kcal/mol more stable than **77**.[16]

Carbodiphosphoranes ($R_3P{=}C{=}PR_3$) are known,[79] but ylides with a P–H bond are rare.[80] Therefore, the spectroscopic characterization of **77** was unexpected. Even more surprising was the characterization of the carbodiphosphorane **79** featuring two P–H bonds.[31] This compound, pre-pared by treatment of **2d** with *tert*-butyllithium, rearranged in solution at room temperature over a period of 16 h to afford the phosphorus ylide **80** with one remaining P–H bond. This compound was also unstable and transformed completely into the diphosphinomethane **81** overnight. Note that calculations for the model compounds where $R = NH_2$ predicted **79** to be 28 kcal/mol less stable than **80**, which is also 34 kcal/mol above **81**.[16] The surprising stability of **79** and **80** is probably due to the presence of bulky substituents, since tetracoordinate phosphorus atoms can more readily accommodate the increased steric constraints than can their tricoor-dinate counterparts.

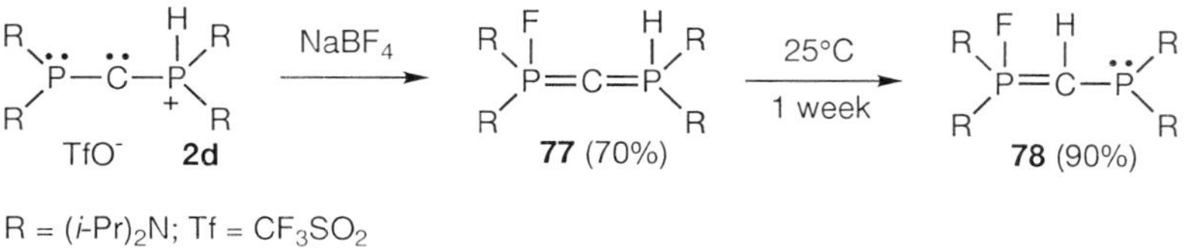

$$R = (i\text{-Pr})_2N; \ Tf = CF_3SO_2$$

SCHEME 41.

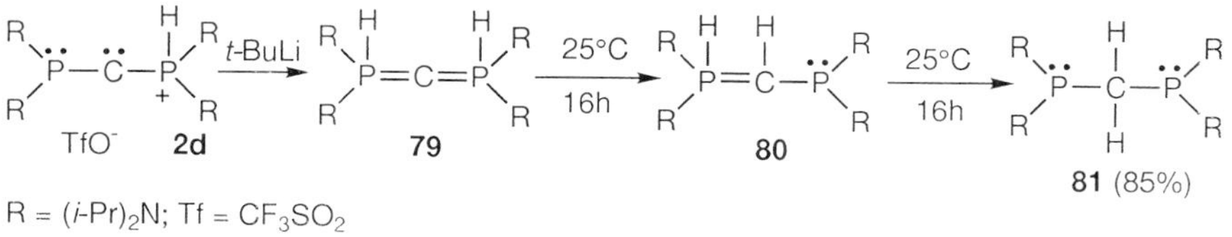

SCHEME 42.

The addition of fluoride ions to **2d**, giving **77**, demonstrates the strong Lewis acid character of the three-coordinate phosphorus center. Therefore, by analogy with the reaction of DBU or DBN with phosphenium salts,[81] we attempted to prepare an adduct of type **82**, using the *P*-chloro(phosphonio)carbene **2u**. In fact, the cationic unsaturated tricyclic adduct **85** was obtained in 50% yield.[82] This reaction is believed to proceed as expected via initial nucleophilic attack of DBN at the σ^3-phosphorus center of **2u**, resulting in the formation of **82**. The basicity of the carbodiphosphorane carbon is then sufficient to abstract the relatively acidic proton in the α-position to the iminium function, giving **83**. Cyclization then occurs by attack of the enamine on the phosphonio center affording **84**; lastly, excess DBN easily removes HCl, yielding product **85**. Overall the reaction can be considered as the 1,3-dinucleophilic attack of DBN on the (phosphino)-(phosphonio)carbene **2u**. Note that deprotonation of **85** leads to **86**, one of the very rare examples of cyclic carbodiphosphoranes.[83]

As already indicated (Section III,C), it is also possible to abstract the

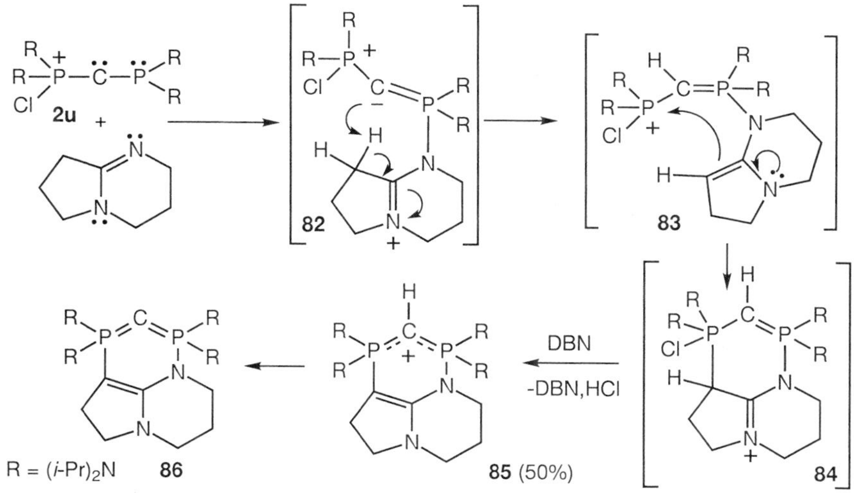

SCHEME 43.

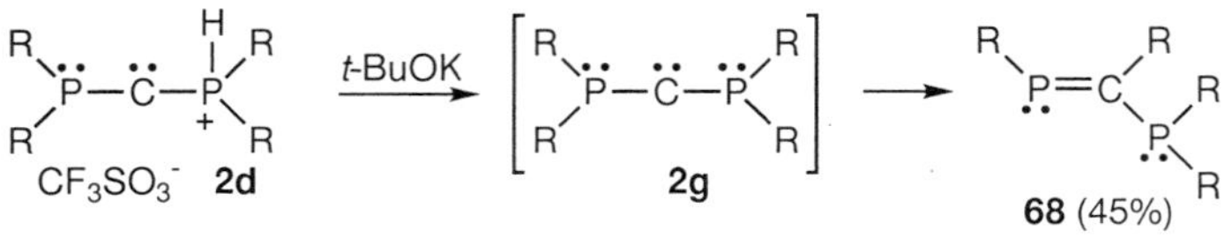

SCHEME 44.

proton of **2d** using a strong base such as potassium *tert*-butoxide to generate the transient di(phosphino)carbene **2g**, which rearranges into the phosphaalkene **68**.[31]

The only carbene-like reaction reported so far is the low-temperature addition of *tert*-butyl isocyanide to carbenes **2d** and **2u**.[78] From the *P*-hydrogenophosphonio carbene **2d**, the heterocycle **89** was isolated in high yield. It is believed that the initial coupling product **87d** rapidly inserts a further equivalent of isocyanide into the P–H bond, leading to the intermediate **88**, which then undergoes rapid elimination of diisopropylamine. When the same reaction was performed with the *P*-chloro(phosphonio)carbene **2u**, a 1/1 mixture of keteneimine **90** and phosphinonitrile was obtained. This result can be explained by the cleavage of the carbene–isocyanide coupling product **87u** by residual HCN, inherently present in the *t*-BuNC.

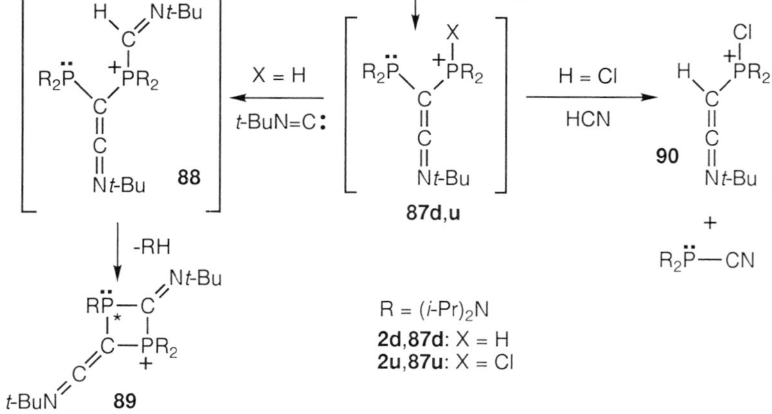

SCHEME 45.

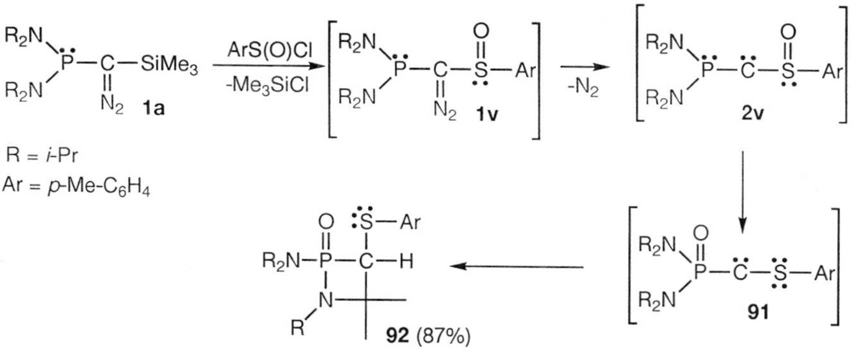

SCHEME 46.

D. *X = Group 16 Element = S(O)R*

Elimination of trimethylchlorosilane and nitrogen occurs when the (phosphino)(silyl)diazomethane **1a** is reacted with *para*-toluenesulfinyl chloride at low temperature. The formation of the four-membered heterocycle **92,** obtained in 87% yield, can be rationalized by a multiple-step mechanism involving the formation of the (phosphino)(sulfinyl)carbene **2v**. The insertion of the (phosphoryl)(sulfenyl)carbene **91**, resulting from a 1,3-oxygen shift from sulfur to phosphorus in **2v**, into a carbon–hydrogen bond of a diisopropylamino group readily accounts for the formation of **92**.[84]

VI

LIGAND PROPERTIES OF PHOSPHINOCARBENES

Direct complexation of a free phosphinocarbene has not yet been reported. However, a few complexes featuring a phosphinocarbene ligand in a number of different coordination modes (A, B, and C) have been described.

A. *Phosphinocarbenes as 2-Electron Ligands (Coordination Mode A)*

In the reaction of the cationic carbyne complex $[(OC)_5WCNEt_2] \cdot BF_4$ with potassium methylphenylphosphide, the two phosphinocarbene complexes **93** and **94** have been isolated as by-products in 4 and 3.5% yield, respectively.[85a] Here, the phosphinocarbenes simply act as 2-electron donors via the carbenic center. It is also interesting to note that this strategy can be used to prepare the analogous arsinocarbene complexes.[85b]

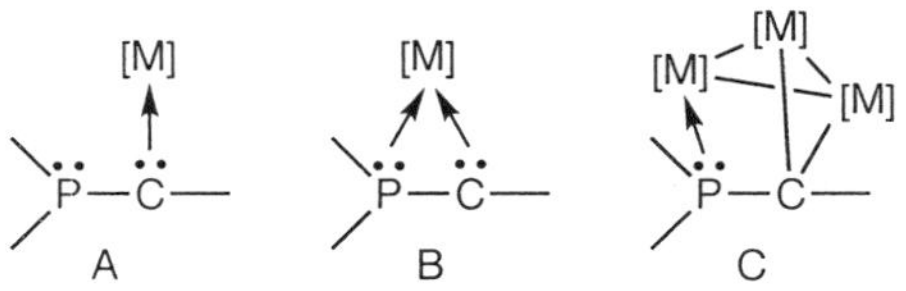

SCHEME 47.

B. *Phosphinocarbenes as 4-Electron Ligands (Coordination Mode B)*

Interestingly, on warming a solution of **94** the phosphine ligand dissociates to afford the new complex **95**. An X-ray diffraction study of the latter complex indicates that the phosphinocarbene behaves as a 4-electron ligand (coordination mode B).[86]

Kreissl *et al.* have developed two different strategies for the conversion of neutral carbyne complexes into cationic phosphinocarbene complexes **97** of type B. The first approach relied on the decarbonylation of tungstaphosphabicyclo[1.1.0]butanone complexes **96**.[87]

The second synthetic method is much shorter. The carbene complexes **97** are prepared in one step by treatment of the corresponding carbyne derivatives with a chlorophosphine in the presence of a Lewis acid.[88] This strategy can also be used to prepare the analogous arsinocarbene complexes.[89]

Note also that a related complex **98** has been synthesized by reacting a carbyne complex with a transient terminal phosphinidene complex.[90]

In 1982, Cotton *et al.* prepared the first phosphinocarbene tantalum complex **99** from the reaction of $Ta_2Cl_6(SMe_2)_3$ with bis(diphenylphosphino)acetylene. The most remarkable feature of this reaction is the formal dimerization of the alkyne moiety.[91]

Some years later, Gibson and Green reported that Cp*TaCl₄[92] and even

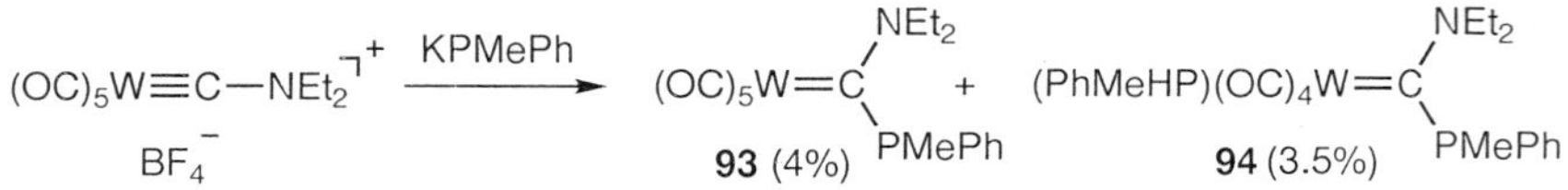

SCHEME 48.

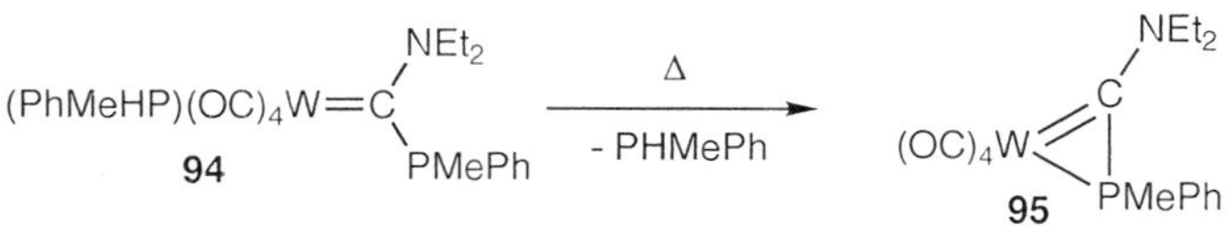

SCHEME 49.

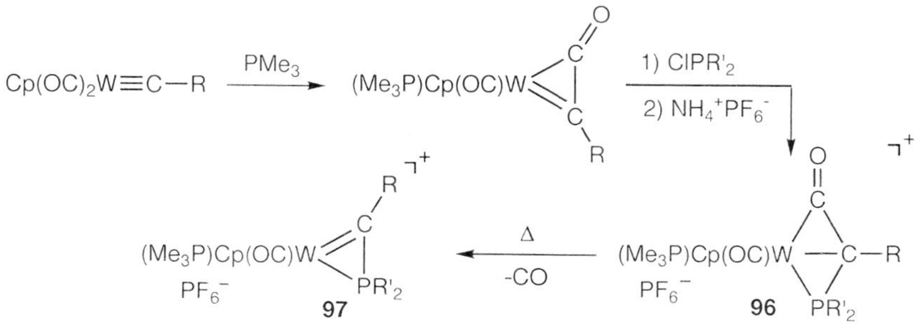

SCHEME 50.

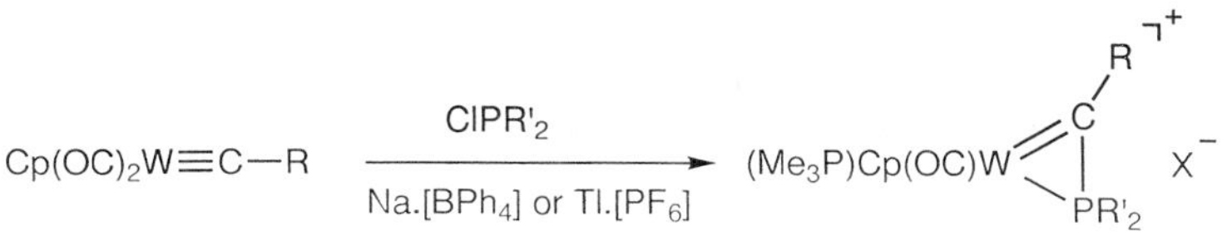

97a (85%) R = p-MeC$_6$H$_4$, R' = Me
97b (80%) R = R' = Me

SCHEME 51.

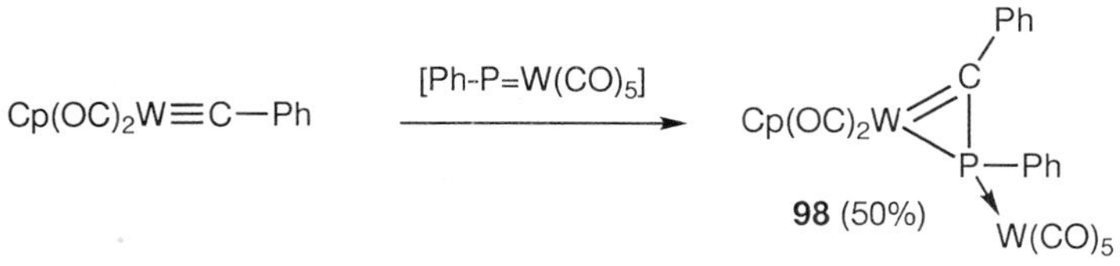

98 (50%)

SCHEME 52.

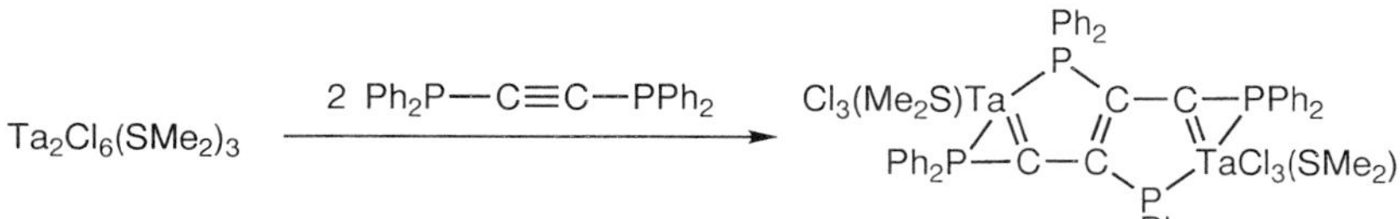

99 (80%)

SCHEME 53.

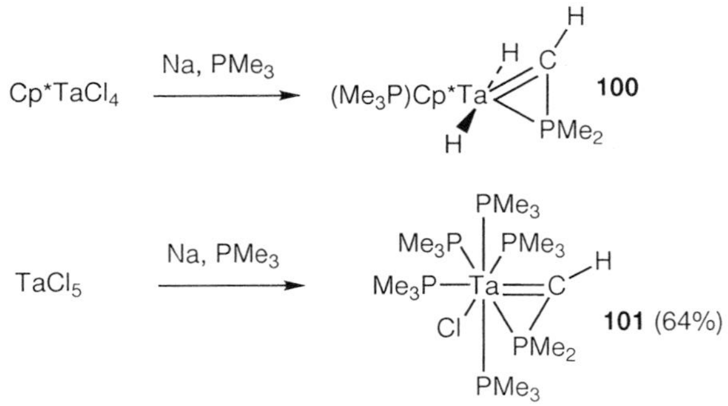

SCHEME 54.

$TaCl_5{}^{93}$ reacted with metallic sodium in neat trimethylphosphine to give the phosphinocarbene tantalum complexes **100** and **101**, respectively. These reactions are the first examples of double activation of coordinated trimethylphosphine via oxidative cleavage of a substituent methyl C–H bond. A similar process was also observed in the reduction of tantalum pentabromide with magnesium turnings in the presence of dimethylphenylphosphine.[94]

X-ray diffraction studies of the η^2-(R_2PCR') complexes reported so far have revealed a metal–carbon distance within the range for a metal–carbon double bond (resonance form B1). However, the phosphorus–carbon bond length is considerably shortened from that expected for a P–C single bond, which suggests a significant contribution from the resonance form B2 in which the metallacycle may really be regarded as a coordinated λ^5-phosphaalkyne. Calculations performed by Gibson *et al.* have concluded that resonance form B best describes this type of complex.[95] Typical for these complexes are the low field methylidyne carbon ^{13}C resonance (170 to 244 ppm) and the high field ^{31}P signal (−155 to −110 ppm) (Table III).

The η^2-phosphinocarbene complexes of tungsten show ambiphilic behavior. With Lewis acids such as MeS^+, electrophilic attack occurs at the metal–carbon double bond affording the dicationic tunstaphosphathiabicyclo[1.1.0]butane complexes **102**.[96,97] On the other hand, nucleophiles such as trialkyl phosphines or the cyclopentadienyl anion $C_5H_5^-$ add at the carbenic center, affording phosphoranylidene complexes **103**[97] or tungstaphos-

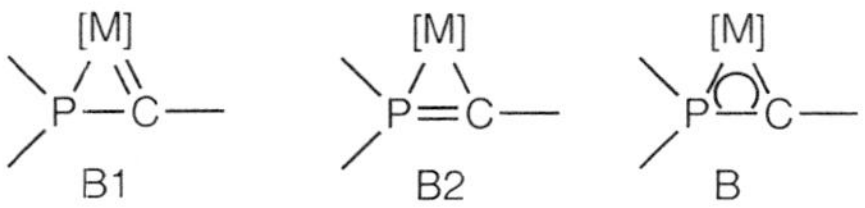

SCHEME 55.

TABLE III

SELECTED ^{13}C NMR, ^{31}P NMR, AND X-RAY DATA FOR η^2-(R_2PCR') COMPLEXES

| Complex | Bond lengths (Å) | | | ^{13}C NMR | | ^{31}P NMR | References |
	MC	MP	PC	δ (ppm)^a	$J_{P,C}$ (Hz)	δ (ppm)^b	
95	2.108	2.465	1.807				86
96a				223.1	45.0	−152.6	87b
96b				215.3	34.2	−110.7	87b
96c				217.1	41.8	−125.0	87
97a	2.032	2.377	1.737	243.9	39.5	−145.1	88b
99	2.018	2.471	1.683				91
100	2.005	2.480	1.714	191.2		−114.3	92
101				187.8	73	−136.9	93c
109	2.015	2.516	1.716	193.5	78	−135.2	93a,b
110	2.026	2.495	1.704	173.1	50	−118.2	93c
111				170.0		−124.0	93c

^{a} Chemical shift for the methylidyne carbon in ppm relative to TMS.

^{b} Chemical shift for the phosphorus atom of the metallacycle in ppm relative to H_3PO_4.

phacyclopropane complexes **104**,[98] respectively. Lastly, a third reaction pathway is observed on reaction with the thiocyanate anion or isonitriles: cationic tunstaphosphathiabicyclo[1.1.0]butane complexes **105** are formed via carbonyl substitution.[99]

In the case of the tantalum complexes **100**, reversible hydrogen migration may occur at room temperature in the presence of carbon monoxide, or at 70°C with dihydrogen or dimethylphosphinoethane to afford complexes **106**, **107**, and **108**, respectively.[92,95] In contrast, the phosphametallacycle remains intact when **100** is treated with halogenated reagents such as CH_3X (X = Cl, Br, I).[92,95]

Green *et al.* observed that the tantalum complex **101** is converted into the bicyclic complex **109** after prolonged exposure to sodium and trimethylphosphine.[93c] Two phosphine ligands of **101** are easily displaced by butadiene, affording the complex **110**, while treatment of **109** with dihydrogen leads to the reduction of the metallic center. In both cases, the metallacycle remains intact.[93c]

C. *Phosphinocarbenes as 4-Electron Ligands (Coordination Mode C)*

Controlled pyrolysis of $Os_3(CO)_{11}(PMe_3)$ in high-boiling hydrocarbon solvents was reported by Deeming *et al.* to give the dihydrido-complex **112** (60% yield), which contains the Me_2PCH fragment bridging the three metal centers (complex of type C).[100] The analogous complexes $H_2Os_3(Me_2PCH)$-

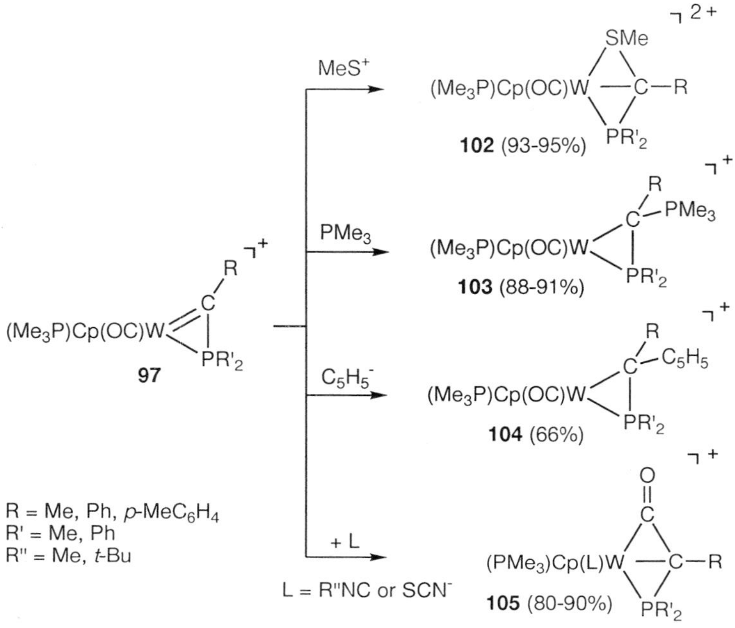

SCHEME 56.

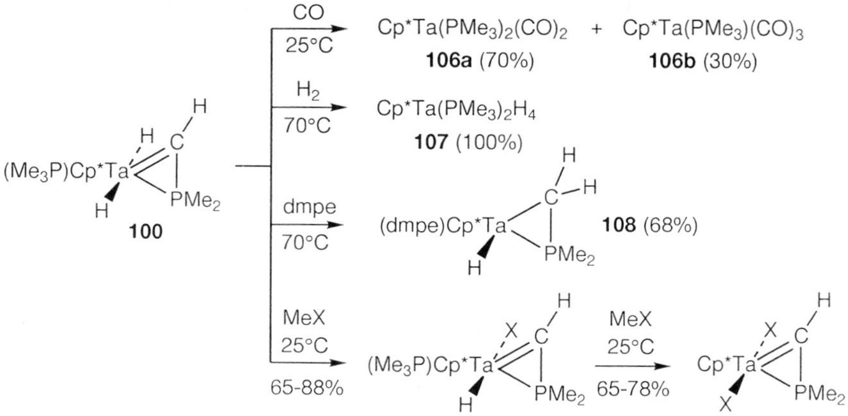

SCHEME 57.

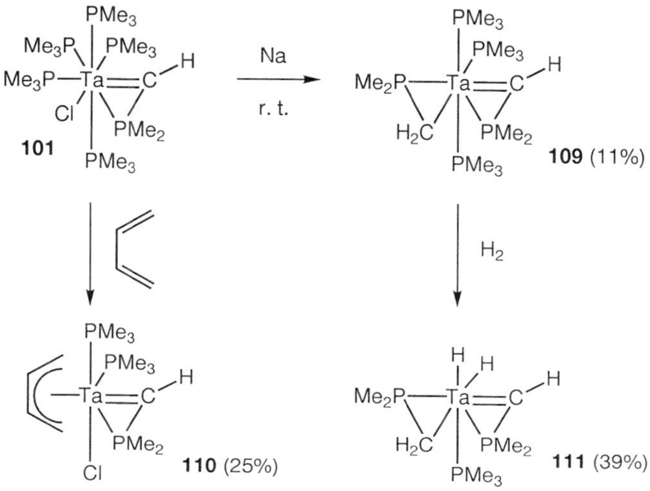

SCHEME 58.

$(CO)_8(PMe_3)$, $H_2Os_3(Et_2PCH)(CO)_9$, and $H_2Os_3(Et_2PCMe)(CO)_8(PEt_3)$ have also been obtained using the same method.[100]

VII

CONCLUSION

To date, the (phosphino)(silyl)- and (phosphino)(phosphonio)carbenes are the only stable carbenes that feature a π-donor and a π-type-withdrawing substituent. They have a singlet ground state with a planar environment at phosphorus and a short phosphorus–carbon bond distance. However, because of the reluctance of phosphorus to keep this planar geometry, these "push–pull" carbenes behave in a manner very close to that of most of the transient carbenes.

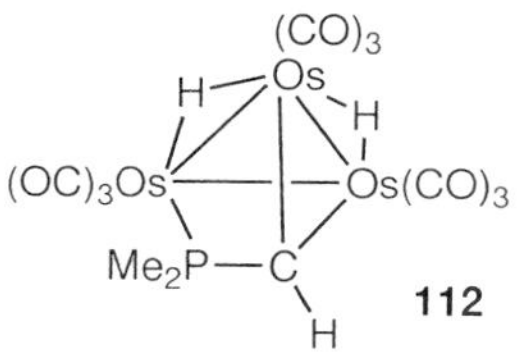

SCHEME 59.

Having in hand reasonable quantities of these unique building blocks has allowed for the discovery of new reactions and therefore the synthesis of original compounds. Their lack of reactivity toward transition metal centers is probably due to the excessive steric hindrance about the carbene center.

In the past few years, a few other types of stable carbene have been discovered. All of them bear two π-donor substituents, of which at least one is an amino group. In contrast to the phosphinocarbenes, they do not feature much of the reactivity of their transient cousins. However, they are excellent ligands for metal centers, and some of the complexes prepared show promising catalytic activity.[5a]

It is certain that in the near future other types of stable carbene, combining the reactivity of "usual carbenes" and good ligand properties, will be prepared.

ACKNOWLEDGMENT

G.B. is grateful to Dr. Antoine Baceiredo, who has been involved in the phosphinocarbene story since the very beginning, and to all his highly motivated co-workers (their names appear in the references) for their intensive and exciting cooperation, which transformed his dream of a stable carbene into a reality. Thanks are also due to the CNRS for financial support.

REFERENCES

(1) Dumas, J. B.; Péligot, E. *Ann. Chim. Phys.* **1835**, *58*, 5.
(2) Staudinger, H.; Kupfer, O. *Ber. Dtsch. Chem. Ges.* **1912**, *45*, 501.
(3) (a) Breslow, R. *J. Am. Chem. Soc.* **1958**, *80*, 3719; (b) Wanzlick, H.-W. *Angew. Chem.* **1962**, *74*, 129.
(4) Igau, A.; Grutzmacher, H.; Baceiredo, A.; Bertrand, G. *J. Am. Chem. Soc.* **1988**, *110*, 6463.
(5) (a) Herrmann, W. A.; Köcher, C. *Angew. Chem., Int. Ed. Engl.* **1997**, *36*, 2162; (b) Arduengo, A. J., III; Krafczyk, R. *Chem. Zeit.* **1998**, *32*, 6.
(6) Arduengo, A. J., III; Harlow, R. L.; Kline, M. *J. Am. Chem. Soc.* **1991**, *113*, 361.
(7) Enders, D.; Breuer, K.; Raabe, G.; Runsink, J.; Teles, J. H.; Melder, J. P.; Ebel, K.; Brode, S. *Angew. Chem., Int. Ed. Engl.* **1995**, *34*, 1021.
(8) Arduengo, A. J., III; Goerlich, J.; Marshall, W. *J. Am. Chem. Soc.* **1995**, *117*, 11027.
(9) Alder, R. W.; Allen, P. R.; Murray, M.; Orpen, G. *Angew. Chem., Int. Ed. Engl.* **1996**, *35*, 1121.
(10) Arduengo, A. J., III; Goerlich, J. R.; Marshall, W. J. *Liebigs Ann.* **1997**, 365.
(11) Alder, R. W.; Buts, C. P.; Orpen, A. G. *J. Am. Chem. Soc.* **1998**, *120*, 11526.
(12) Tomioka, H. *Acc. Chem. Res.* **1997**, *30*, 315.
(13) Nguyen, M. T.; McGinn, M. A.; Hegarty, A. F. *Inorg. Chem.* **1986**, *25*, 2185.
(14) Hoffmann, M. R.; Kuhler, K. *J. Chem. Phys.* **1991**, *94*, 8029.
(15) Dixon, D. A.; Dobbs, K. B.; Arduengo, A. J., III; Bertrand, G. *J. Am. Chem. Soc.* **1991**, *113*, 8782.
(16) Treutler, O.; Ahlrichs, R.; Soleilhavoup, M. *J. Am. Chem. Soc.* **1993**, *115*, 8788.

(17) See, for example, Richards, C. A., Jr; Kim, S. J.; Yamaguchi, Y.; Schaefer, H. F., III *J. Am. Chem. Soc.* **1995**, *117*, 10104.

(18) See, for example, Chedekel, M. R.; Skoglund, M.; Kreeger, R. L.; Shechter, H. *J. Am. Chem. Soc.* **1976**, *98*, 7846.

(19) Schoeller, W. W. *J. Chem. Soc., Chem. Commun.* **1980**, 124.

(20) Pauling, L. *J. Chem. Soc., Chem. Commun.* **1980**, 688.

(21) Römer, B.; Gatev, G. G.; Zhong, M.; Brauman, J. I. *J. Am. Chem. Soc.* **1998**, *120*, 2919.

(22) Regitz, M. "Carbene (carbenoide)," in *Methoden der Organischen Chemie (Houben-Weyl)*; Regitz, M., Ed.; Georg Thieme Verlag: Stuttgart, **1989**, 2.

(23) Baceiredo, A.; Bertrand, G.; Sicard, G. *J. Am. Chem. Soc.* **1985**, *107*, 4781.

(24) Baceiredo, A.; Bertrand, G. unpublished results.

(25) Alcaraz, G.; Reed, R.; Baceiredo, A.; Bertrand, G. *J. Chem. Soc., Chem. Commun.* **1993**, 1354.

(26) (a) Bertrand, G.; Reed, R. *Coord. Chem. Rev.* **1994**, *137*, 323; (b) Réau, R.; Bertrand, G. in *Methoden der Organischen Chemie (Houben-Weyl)*; Regitz, M., Ed.; Georg Thieme Verlag: Stuttgart, **1996**, *E17a*, 794.

(27) (a) Gilette, G.; Baceiredo, A.; Bertrand, G. *Angew. Chem., Int. Ed. Engl.* **1990**, *29*, 1429; (b) Gilette, G.; Igau, A.; Baceiredo, A.; Bertrand, G. *New. J. Chem.* **1991**, *15*, 393; (c) Bertrand, G. *Heteroatom.* **1991**, *2*, 29.

(28) (a) Soleilhavoup, M.; Baceiredo, A.; Treutler, O.; Ahlrichs, R.; Nieger, M.; Bertrand, G. *J. Am. Chem. Soc.* **1992**, *114*, 10959; (b) Soleilhavoup, M.; Alcaraz, G.; Réau, R.; Baceiredo, A.; Bertrand, G. *Phosphorus Sulfur* **1993**, *76*, 49.

(29) Appel, R.; Peters, J.; Schmitz, R. *Z. Anorg. Allg. Chem.* **1981**, *475*, 18 and 6204.

(30) Svara, J.; Fluck, E.; Riffel, H. *Z. Naturforsch.,* **1985**, *40b*, 1258; (b) Neumüller, B.; Fluck, E. *Phosphorus Sulfur* **1986**, *29*, 23.

(31) Soleilhavoup, M.; Baceiredo, A.; Bertrand, G. *Angew. Chem., Int. Ed. Engl.* **1993**, *32*, 1167.

(32) Goumri, S.; Leriche, Y.; Gornitzka, H.; Baceiredo, A.; Bertrand, G. *Eur. J. Inorg. Chem.* **1998**, 1539.

(33) *Multiple Bonds and Low Coordination in Phosphorus Chemistry*; Regitz, M.; Scherer, O. J., Eds.; Georg Thieme Verlag: Stuttgart, 1990.

(34) Dunitz, J. D.; Maverick, E. F.; Trueblood, K. N. *Angew. Chem., Int. Ed. Engl.* **1988**, *27*, 880.

(35) *Ylides and Imines of Phosphorus*; Johnson, A. W.; Kaska, W. C.; Starzewski, K. A. O.; Dixon, D. A., Eds.; Wiley: New York, 1993.

(36) Goumri, S.; Gornitzka, H.; Baceiredo, A.; Bertrand, G. unpublished results.

(37) Regitz, M. In *1,3-Dipolar Cycloaddition Chemistry*; Padwa, A., Ed.; Wiley Interscience: New York, 1984, Vol. 2, pp. 393–461.

(38) Regitz, M.; Binger, P. *Angew. Chem., Int. Ed. Engl.* **1988**, *27*, 1484.

(39) Baceiredo, A.; Igau, A.; Bertrand, G.; Menu, M. J.; Dartiguenave, Y.; Bonnet, J. J. *J. Am. Chem. Soc.* **1986**, *108*, 7868.

(40) Igau, A.; Baceiredo, A.; Trinquier, G.; Bertrand, G. *Angew. Chem., Int. Ed. Engl.* **1989**, *28*, 621.

(41) Moss, R. A.; Shen, S.; Hadel, L. M.; Kmiecik-Lawrynowicz, G.; Wlostowskaj, J.; Krogh-Jespersen, K. *J. Am. Chem. Soc.* **1987**, *109*, 4341.

(42) (a) Skell, P. S.; Woodworth, R. C. *J. Am. Chem. Soc.* **1956**, *78*, 4496; (b) Skell, P. S. *Tetrahedron* **1985**, *41*, 1427.

(43) (a) De March, P.; Huisgen, R. *J. Am. Chem. Soc.* **1982**, *104*, 4952; (b) De March, P.; Huisgen, R. *J. Am. Chem. Soc.* **1982**, *104*, 4953.

(44) Wulfman, D. S.; Pauling, B. In *Reactive Intermediates*; Abramovitch, R. A., Ed.; Plenum: New York, 1980, Vol. 1, p. 321.
(45) Regitz, M. *Synthesis* **1972**, 351.
(46) Padwa, A.; Hornbuckle, S. F. *Chem. Rev.* **1991**, *91*, 263.
(47) Janulis, E. P.; Wilson, S. R.; Arduengo, A. J., III. *Tetrahedron Lett.* **1984**, 405.
(48) Alcaraz, G.; Wecker, U.; Baceiredo, A.; Dahan, F.; Bertrand, G. *Angew. Chem., Int. Ed. Engl.* **1995**, *34*, 1246.
(49) Piquet, V.; Baceiredo, A.; Gornitzka, H.; Dahan, F.; Bertrand, G. *Chem. Eur. J.* **1997**, *11*, 1757.
(50) (a) Hassner, A. In *Azides and Nitrenes: Reactivity and Utility*; Scriven, E. F. V., Ed.; Academic Press: Orlando, 1984, Vol. 2; (b) Hassner, A.; Wiegand, N. H.; Gottlieb, H. E. *J. Org. Chem.* **1986**, *51*, 3176.
(51) (a) Padwa, A.; Smolanoff, J.; Tremper, A. *J. Org. Chem.* **1976**, *41*, 543; (b) Nishiwaki, T.; Nakano, A. *J. Chem. Soc. C* **1970**, 1825; (c) Nishiwaki, T.; Fugiyama F. *J. Chem. Soc., Perkin Trans. 1* **1974**, 1456.
(52) For a review on nonantiaromatic four-π-electron four-membered heterocycles, see: Bertrand, G. *Angew. Chem., Int. Ed. Engl.* **1998**, *37*, 270.
(53) Sanchez, M.; Réau, R.; Marsden, C. J.; Regitz, M.; Bertrand, G. *Chem. Eur. J.* **1999**, *5*, 274.
(54) (a) Wagner, O.; Ehle, M.; Regitz, M. *Angew. Chem., Int. Ed. Engl.* **1989**, *28*, 225; (b) Memmesheimer, H.; Regitz, M. *Rev. Heteroatom. Chem.* **1994**, *10*, 61.
(55) Armbrust, R.; Sanchez, M.; Réau, R.; Bergsträsser, U.; Regitz, M.; Bertrand, G. *J. Am. Chem. Soc.* **1995**, *117*, 10785.
(56) See, for example, Bethell, D. *Adv. Phys. Org. Chem.* **1969**, *7*, 153.
(57) Igau, A.; Baceiredo, A.; Bertrand, G.; Kuhnel Lysek, K.; Niecke, E. *New. J. Chem.* **1989**, *13*, 359.
(58) Appel, R.; Huppertz, M.; Westerhaus, A. *Chem. Ber.* **1983**, *116*, 114.
(59) Igau, A.; Baceiredo, A.; Grützmacher, H.; Pritzkow, H.; Bertrand, G. *J. Am. Chem. Soc.* **1989**, *111*, 6853.
(60) Guerret, O.; Bertrand, G. *Acc. Chem. Res.* **1997**, *30*, 486.
(61) Sakurai, H.; Ebata, K.; Kabuto, C.; Nakadaira, Y. *Chem. Lett.* **1987**, 301.
(62) (a) Grützmacher, H.; Pritzkow, H. *Angew. Chem., Int. Ed. Engl.* **1991**, *30*, 709; (b) Grützmacher, H.; Marchand, C. M. *Coord. Chem. Rev.* **1997**, *163*, 287.
(63) Cowley, A.; Gabbaï, F.; Carrano, C.; Mokry, L.; Bond, M.; Bertrand, G. *Angew. Chem., Int. Ed. Engl.* **1994**, *33*, 578.
(64) Brown, H. C.; Kramer, G. W.; Levy, A. B.; Midland, M. M. *Organic Syntheses via Boranes*; Wiley Interscience: New York, 1975.
(65) Locquenghien, K. H. v.; Baceiredo, A.; Boese, R.; Bertrand, G. *J. Am. Chem. Soc.* **1991**, *113*, 5062.
(66) Olmstead, M. M.; Power, P. P.; Weese, K. J.; Doedens, R. J. *J. Am. Chem. Soc.* **1987**, *109*, 2541.
(67) Romanenko, V. D.; Gudima, A. O.; Chernega, A. N.; Bertrand, G. *Inorg. Chem.* **1992**, *31*, 3493.
(68) Barrau, J.; Escudié, J.; Satgé, J. *Chem. Rev.* **1990**, *90*, 283.
(69) Romanenko, V. D.; Gudima, A. O.; Chernega, A. N.; Sotiropoulos, J. M.; Alcaraz, G.; Bertrand, G. *Bull. Soc. Chim. Fr.* **1994**, *131*, 748.
(70) Moss, R. A.; Xue, S.; Liu, W. *J. Am. Chem. Soc.* **1994**, *116*, 1583.
(71) (a) Padwa, A.; Hornbuckle, S. F. *Chem. Rev.* **1991**, *91*, 263; (b) Trinquier, G.; Malrieu, J. P. *J. Am. Chem. Soc.* **1987**, *109*, 5303; (c) Jackson, J. E.; Platz, M. S. In *Advances in Carbene Chemistry*; Brinker, U. H., Ed.; JAI Press: Greenwich, 1994, Vol. 1, p. 89.

(72) Bestmann, H. J.; Zimmermann, R. In *Methoden der Organischen Chemie* (*Houben-Weyl*); Regitz, M., Ed.; Georg Thieme Verlag: Stuttgart, 1982, Vol. E1, pp. 616–618.

(73) Emig, N.; Tejeda, J.; Réau, R.; Bertrand, G. *Tetrahedron Lett.* **1995**, *36*, 4231.

(74) Menu, M. J.; Dartiguenave, Y.; Dartiguenave, M.; Baceiredo, A.; Bertrand, G. *Phosphorus and Sulfur* **1990**, *47*, 327.

(75) Sicard, G.; Baceiredo, A.; Crocco, G.; Bertrand, G. *Angew. Chem., Int. Ed. Engl.* **1988**, *27*, 301.

(76) Kapp, J.; Schade, C.; El-Nahasa, A. M.; Schleyer, P. v. R. *Angew. Chem., Int. Ed. Engl.* **1996**, *35*, 2236, and references therein.

(77) Keller, H.; Maas, G.; Regitz, M. *Tetrahedron Lett.* **1986**, *27*, 1903.

(78) Dyer, P.; Baceiredo, A.; Bertrand, G. *Inorg. Chem.* **1996**, *35*, 46.

(79) Bestmann, H. J.; Zimmermann, R. In *Comprehensive Organic Synthesis*; Trost, B. M.; Fleming, I.; Winterfeld, T., Eds.; Pergamon: Oxford, 1991, Vol. 6, p. 171.

(80) Kolodyazhnyi, O. I.; Kukahar, V. P. *Russ. Chem. Rev.* **1983**, *52*, 1096.

(81) (a) Reed, R. W.; Réau, R.; Dahan, F.; Bertrand, G. *Angew. Chem., Int. Ed. Engl.* **1993**, *32*, 3099; (b) Bouhadir, G.; Reed, R. W.; Réau, R.; Bertrand, G. *Heteroatom.* **1995**, *6*, 371.

(82) Dyer, P.; Guerret, O.; Dahan, F.; Baceiredo, A.; Bertrand, G. *J. Chem. Soc., Chem. Commun.* **1995**, 2339.

(83) Schmidbaur, H.; Costa, T.; Milewski-Mahrla, B.; Schubert, B. *Angew. Chem., Int. Ed. Engl.* **1980**, *19*, 555.

(84) Sicard, G.; Grützmacher, H.; Baceiredo, A.; Fischer, J.; Bertrand, G. *J. Org. Chem.* **1989**, *54*, 4426.

(85) (a) Fischer, E. O.; Restmeier, R. Z. *Naturforsch.* **1983**, *38b*, 582; (b) Fischer, E. O.; Wittmann, D.; Himmelreich, D.; Schubert, U.; Ackermann, K. *Chem. Ber.* **1982**, *115*, 3141.

(86) Fischer, O. J.; Restmeier, R.; Ackermann, K. *Angew. Chem., Int. Ed. Engl.* **1983**, *22*, 411.

(87) (a) Kreissl, F. R.; Wolfgruber, M.; Sieber, W. J. *J. Organometal. Chem.* **1984**, *270*, C4; (b) Kreissl, F. R.; Wolfgruber, M. Z. *Naturforsch.* **1988**, *43b*, 1307.

(88) (a) Kreissl, F. R.; Ostermeier, J.; Ogric, C. *Chem. Ber.* **1995**, *128*, 289; (b) Lehotkay, Th.; Wurst, K.; Jaitner, P.; Kreissl, F. R. *J. Organometal. Chem.* **1996**, *523*, 105; (c) Kreissl, F. R.; Ostermeier, J.; Ogric, C. *Chem. Ber.* **1995**, *128*, 289.

(89) Kreissl, F. R.; Lehotkay, T.; Ogric, C.; Herdtweck, E. *Organometallics* **1997**, *16*, 1875.

(90) Tran Huy, N. H.; Fischer, J.; Mathey, F. *Organometallics* **1988**, *7*, 240.

(91) Cotton, F. A.; Falvello, L. R.; Najjar, R. C. *Organometallics* **1982**, *1*, 1640.

(92) Kee, T. P.; Gibson, V. C.; Clegg, W. *J. Organometal. Chem.* **1987**, *325*, C14.

(93) (a) Gibson, V. C.; Grebenik, P. D.; Green, M. L. H. *J. Chem. Soc., Chem. Commun.* **1983**, 1101; (b) Grebenik, P. D.; Prout, K. *J. Chem. Soc. Dalton Trans.* **1985**, 2025; (c) Green, M. L. H.; Hare, P. M.; Bandy, J. A. *J. Organometal. Chem.* **1987**, *330*, 61.

(94) Hovnanian, N.; Hubert-Pfalzgraf, L. G.; Le Borgne, G. *Inorg. Chem.* **1985**, *24*, 4647.

(95) Anstice, H. M.; Fielding, H. H.; Gibson, V. C.; Housecroft, C. E.; Kee, T. P. *Organometallics* **1991**, *10*, 2183.

(96) Ostermeier, J.; Schütt, W.; Kreissl, F. R. *J. Organometal. Chem.* **1992**, *436*, C17.

(97) Ostermeier, J.; Hiller, W.; Kreissl, F. R. *J. Organometal. Chem.* **1995**, *491*, 283.

(98) Lehotkay, Th.; Ostermeier, J.; Ogric, C.; Kreissl, F. R. *J. Organometal. Chem.* **1996**, *520*, 59.

(99) Ogric, C.; Ostermeier, J.; Heckel, M.; Hiller, W.; Kreissl, F. R. *Inorg. Chim. Acta* **1994**, *222*, 77; (b) Lehotkay, Th.; Wurst, K.; Jaitner, P.; Kreissl, F. R. *J. Organometal. Chem.* **1998**, *553*, 103.

(100) (a) Deeming, A. J.; Underhill, M. *J. Chem. Soc. Dalton Trans.* **1973**, 2727; (b) Deeming, A. J. *J. Organometal. Chem.* **1977**, *128*, 63.

ADVANCES IN ORGANOMETALLIC CHEMISTRY, VOL. 44

Zwitterionic Pentacoordinate Silicon Compounds

REINHOLD TACKE, MELANIE PÜLM, and BRIGITTE WAGNER

Institut für Anorganische Chemie
Universität Würzburg
am Hubland
D-97074 Würzburg, Germany

I

INTRODUCTION

The study of compounds containing pentacoordinate silicon atoms currently represents one of the main areas of research in silicon chemistry. This is evident from the numerous reviews and proceedings published on this topic in recent years.[1–12] Most of the pentacoordinate silicon compounds described in the literature are either salts with $\lambda^5 Si$-silicate anions or neutral silicon complexes with a 4+1 coordination to silicon. This review deals with a completely different class of pentacoordinate silicon compounds: zwitterionic $\lambda^5 Si$-silicates. These molecular compounds contain a pentacoordinate (formally negatively charged) silicon atom and a tetracoordinate (formally positively charged) nitrogen atom.

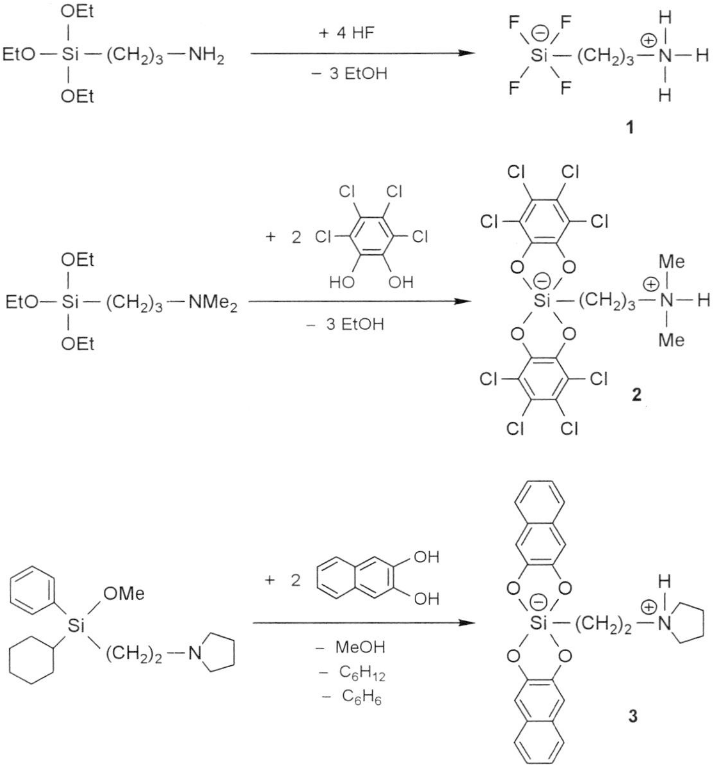

SCHEME 1.

To the best of our knowledge, compound **1** (Scheme 1) represents the first zwitterionic $\lambda^5 Si$-silicate mentioned in the literature[13–15]; however, it has only been characterized by elemental analyses. The spirocyclic compound **2** (Scheme 1) was the first zwitterionic $\lambda^5 Si$-silicate to be structurally characterized in the solid state (crystal structure analysis of $2 \cdot CH_3CN$).[16] Purely accidentally, we observed that treatment of cyclohexyl(methoxy)phenyl(2-pyrrolidinoethyl)silane with 2,3-dihydroxynaphthalene in acetonitrile at room temperature yields the zwitterionic $\lambda^5 Si$-silicate **3** (Scheme 1), which is formed by two remarkable Si–C cleavage reactions (formation of benzene and cyclohexane) and one Si–O cleavage reaction (formation of methanol).[17] Compound **3** has been demonstrated to exist not only in the solid state (crystal structure analysis of $3 \cdot CH_3CN$) but also in solution (^1H, ^{13}C, and ^{29}Si NMR).[17] Stimulated by these surprising results, we have been interested for about 10 years in the systematic development of the chemistry of zwitterionic λ^5Si-silicates. In this review, we report on the syntheses, structures, and properties of compounds of this particular type. Almost all results described in this article (which does not lay claim to completeness) come from our own laboratory.

II

ACYCLIC λ^5Si-SILICATES WITH AN SiF_4C OR SiF_3C_2 SKELETON

The acyclic (ammonioorganyl)tetrafluorosilicates **4**,[18] **5**,[19] **6–8**,[20] and **9–16**[21] represent zwitterionic λ^5Si-silicates with an SiF_4C skeleton. Compounds of this particular formula type can be synthesized as demonstrated in Scheme 2 for compounds **6** (with NH group) and **7** (without NH group). Typically, the syntheses were carried out at 0°C using mixtures of hydrofluoric acid and ethanol as the HF source.

| 4: n = 1 | 6: R = H | 8 |
| 5: n = 2 | 7: R = Me | |

	9	**10**	**11**	**12**	**13**	**14**	**15**
n	5	6	7	8	9	10	11

16

The acyclic (ammonioorganyl)trifluoro(organyl)silicates **17**,[22] **18**,[22] **19**,[20] **20**,[20] **21**,[23] and **22**[20] are zwitterionic λ^5Si-silicates with an SiF_3C_2 skeleton. As illustrated for **19** and **20** in Scheme 3, compounds **17–22** were prepared analogously to the syntheses of **4–16** (compare Scheme 2).

| 17: R = Me | 19: R = H | 21: R = Me |
| 18: R = Ph | 20: R = Me | 22: R = ^tBu |

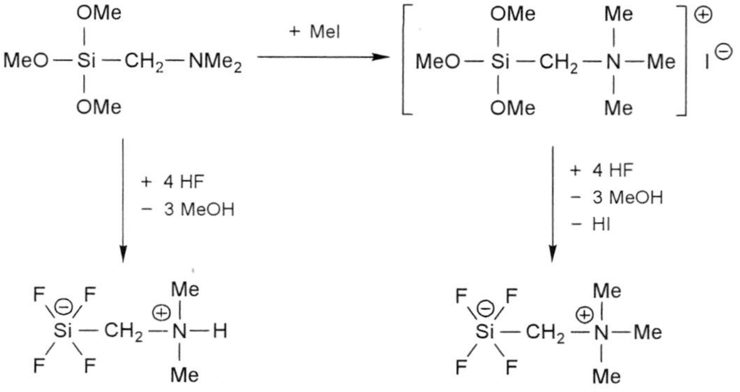

SCHEME 2.

The zwitterionic $\lambda^5 Si$-fluorosilicates **4–22** were isolated as crystalline solids. Compounds **4–8**, **13**, and **17–22** were structurally characterized in the solid state by single-crystal X-ray diffraction. In contrast to the achiral zwitterions **4–16**, the zwitterions **17–22** are chiral, the respective crystals consisting of pairs of enantiomers [(A)- and (C)-enantiomers]. In all cases, the Si-coordination polyhedron was found to be a somewhat distorted trigonal bipyramid, with fluorine atoms in the two axial sites. This is illustrated for **6** and **19** in Fig. 1. Selected geometric parameters for compounds **4–8**, **13**, and **17–22** are listed in Table I. As can be seen from these data, the axial Si–F distances [1.647(2)–1.743(1) Å] are significantly longer than the equatorial ones [1.589(2)–1.638(1) Å]. The Si–C1 distances amount

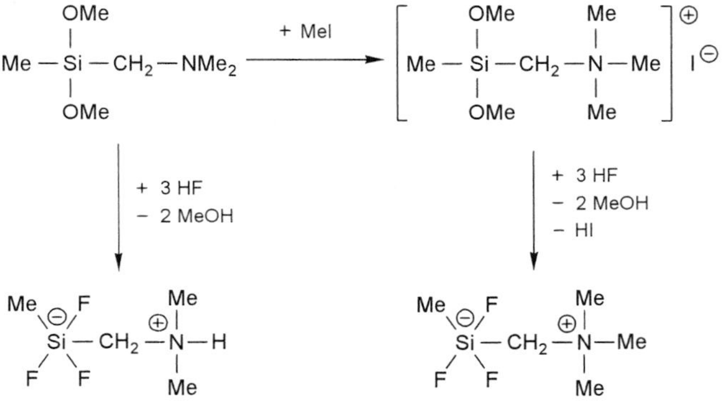

SCHEME 3.

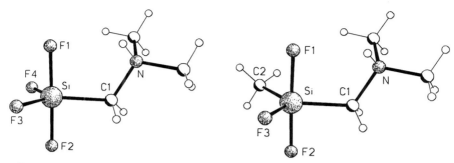

FIG. 1.   Molecular structures of **6** (left) and **19** (right) in the crystal. For selected bond distances and angles, see Table I.

to 1.872(4)–1.918(3) Å. For compounds **17–22**, the Si–C1 distances are generally longer than the corresponding Si–C2 distances [1.851(3)–1.914(4) Å]. It is likely that the peculiarly long Si–C distances observed for **22** [1.914(4) Å, 1.918(3) Å] are the result of steric bulkiness of the *t*-butyl and (tetramethylpiperidinio)methyl group.

TABLE I

SELECTED BOND DISTANCES (Å) AND ANGLES (°) FOR COMPOUNDS **4–8**, **13**, **17**, **18**, **19-I**, **19-II**, AND **20–22** AS DETERMINED BY SINGLE-CRYSTAL X-RAY DIFFRACTION[a,b]

Compound	Si–F1	Si–F2	Si–F3	Si–F4	Si–C1	Si–C2	F1–Si–F2
4 (M I)	1.692(2)	1.647(2)	1.589(2)	1.605(2)	1.882(3)	—	178.8(1)
4 (M II)	1.697(2)	1.648(2)	1.598(2)	1.595(2)	1.879(3)	—	179.8(1)
5	1.708(2)	1.667(2)	1.627(2)	1.607(2)	1.872(4)	—	174.8(1)
6 (M I)	1.690(1)	1.671(1)	1.623(2)	1.606(2)	1.893(3)	—	178.34(9)
6 (M II)	1.708(2)	1.664(2)	1.619(2)	1.606(1)	1.888(2)	—	178.45(9)
7	1.657(4)	1.678(4)	1.607(3)	1.607(3)	1.894(6)	—	178.3(2)
8	1.678(2)	1.649(2)	1.601(3)	1.606(2)	1.888(4)	—	179.7(1)
13	1.691(2)	1.680(2)	1.612(2)	1.600(2)	1.877(3)	—	177.28(10)
17	1.738(2)	1.689(2)	1.630(1)	—	1.899(3)	1.854(3)	173.1(1)
18	1.715(2)	1.681(2)	1.633(2)	—	1.896(3)	1.870(4)	174.5(1)
19-I[c]	1.743(1)	1.689(1)	1.638(1)	—	1.903(2)	1.860(2)	174.79(6)
19-II[d]	1.739(1)	1.689(2)	1.629(2)	—	1.897(2)	1.851(3)	174.31(9)
20	1.694(2)	1.696(2)	1.629(2)	—	1.917(3)	1.856(3)	173.6(1)
21	1.7257(7)	1.6999(8)	1.6212(8)	—	1.9098(12)	1.8639(13)	173.51(4)
22	1.709(2)	1.707(2)	1.624(2)	—	1.918(3)	1.914(4)	172.4(1)

[a] For the atomic numbering scheme, see Fig. 1. In all structures, the non-hydrogen atom bound to C1 (**4**, **6–8**, **13**, **17–22**: N atom; **5**: C atom) is pointing to F1.

[b] For compounds **4** and **6**, two crystallographically independent molecules (M I, M II) were observed.

[c] Space group $P2_1/n$.

[d] Space group $P2_1/c$.

TABLE II

^{29}Si Chemical Shifts for Compounds **4**, **6–8**, **21**, and **22** in Solution (CD$_3$CN) and/or
in the Crystal^a

Compound	δ_{solution}	$\delta_{\text{crystal}}{}^b$	Compound	δ_{solution}	$\delta_{\text{crystal}}{}^b$
4	−122.9	−121.1	**8**	−123.5	d
6	−123.8	−120.6	**21**	−87.5	−88.1
7	c	−128.8	**22**	−93.6	−97.2

^{a} Chemical shifts in ppm; spectra recorded at room temperature.
^{b} Isotropic chemical shifts obtained by ^{29}Si CP/MAS NMR experiments.
^{c} Experiment failed.
^{d} No experiment performed.

As can be seen from Table I, in most cases the axial Si–F1 distances are significantly longer than the respective axial Si–F2 distances. This weakening of the Si–F1 bond can be interpreted in terms of dipolar interactions between the fluorine atom F1 and the ammonium moiety. This suggestion is supported by the results of ab initio studies of **4**, **6**, **7**, **17**, **19**, and some related model species.[20] In general, intra- and/or intermolecular N–H···F hydrogen bonds play an important role in the structural chemistry of **4–6**, **8**, **13**, **17–19**, **21**, and **22**.

With only one remarkable exception, the results of SCF/SVP geometry optimizations of **4**, **6**, **7**, **17**, and **19** are in good agreement with the experimentally established crystal structures of these compounds.[20] Generally, the calculated Si–C1 distances are significantly longer than those determined by single-crystal X-ray diffraction (difference 0.03–0.055 Å). However, as shown by ab initio calculations with related model species, these apparent discrepancies can easily be explained by intermolecular interactions of the zwitterions with neighboring molecules in the crystal.

Compounds **4–22** were studied by solution-state ^{1}H, ^{13}C, ^{19}F, and ^{29}Si NMR experiments at room temperature (**4–8**, **16–22**, CD$_3$CN; **9–15**, CD$_3$OD). The ^{19}F NMR spectra are characterized by a more or less sharp resonance signal, indicating a rapid fluorine exchange. ^{29}Si NMR spectra could only be obtained for compounds **4**, **6**, **8**, **21**, and **22**. Except for **8** [quint, $^1J(\text{SiF}) = 204.4$ Hz], **21** [q, $^1J(\text{SiF}) = 242.3$ Hz], and **22** [q, $^1J(\text{SiF}) = 261.1$ Hz], SiF couplings were not resolved, indicating again a rapid fluorine exchange. As can be seen from Table II, the ^{29}Si chemical shifts of **4**, **6**, **8**, **21**, and **22** in solution are in good agreement with the isotropic ^{29}Si chemical shifts obtained for **4**, **6**, **7**, **21**, and **22** by solid-state ^{29}Si CP/MAS NMR experiments.

VT ^{19}F NMR studies of **22** in CD_2Cl_2 have shown that the just-mentioned ligand exchange can be reduced significantly by cooling. The low-temperature ($-90°C$) H and ^{19}F NMR spectra of **22** confirm the structure determined by single-crystal X-ray diffraction. Two different dynamic processes were studied by VT ^{19}F NMR experiments: the rotation of the ammoniomethyl moiety around the Si–C bond ($\Delta H^{\ddagger} = 40.9 \pm 1.2$ kJ mol^{-1}; $\Delta S^{\ddagger} = -15.4 \pm 0.6$ J mol^{-1} K^{-1}) and a pseudorotation process at the pentacoordinate silicon atom ($\Delta H^{\ddagger} = 44.4 \pm 1.6$ kJ mol^{-1}; $\Delta S^{\ddagger} = -16.5 \pm 0.7$ J mol^{-1} K^{-1}).[20]

In order to get more information about the dynamic behavior of the zwitterionic λ^5Si-silicates described in this chapter, ab initio studies with the model molecules **23** and **24** were performed.[20] These investigations are based on the search for local minima of these species and transition states for their interconversion. The calculated pathways for the intramolecular ligand exchange of **23** and **24** demonstrate that not only the Berry pseudorotation at the pentacoordinate silicon atom has to be considered for the intramolecular dynamic processes of **4–22**: The torsion around the Si–C bond of the ammoniomethyl moiety is also important.

Some of the compounds described in this chapter were studied for specific physical properties. Surface tension measurements with solutions of **9–16** in 0.01 M hydrochloric acid demonstrated that these zwitterionic λ^5Si-silicates are highly efficient surfactants.[21] These compounds contain a polar (zwitterionic) hydrophilic moiety and a long lipophilic *n*-alkyl group. Increase of the *n*-alkyl chain length (**9–15**) was found to result in an increase of surface activity. The equilibrium surface tension vs concentration isotherms for **9** and **16** were analyzed quantitatively and the surface thermodynamics of these surfactants interpreted on the molecular level. Furthermore, preliminary studies demonstrated that aqueous solutions of **9–16** lead to a hydrophobizing of glass surfaces.[21]

Compounds **9–16** were found to undergo an acidic reaction in aqueous solution.[21] Possibly, these zwitterionic λ^5Si-silicates behave as Lewis acids reacting with the Lewis base OH$^-$ to generate protons (2 H$_2$O $\rightleftarrows$ OH$^-$ + H$_3$O$^+$). This interpretation is supported by the observation that the related compounds **6** and **7** react with [HNMe$_3$]F in aqueous solution to yield the ionic λ^6Si-silicates **25** and **26**, respectively (Scheme 4; the Lewis bases F$^-$ and OH$^-$ are isoelectronic).[24] The identity of these hexacoordinated silicon

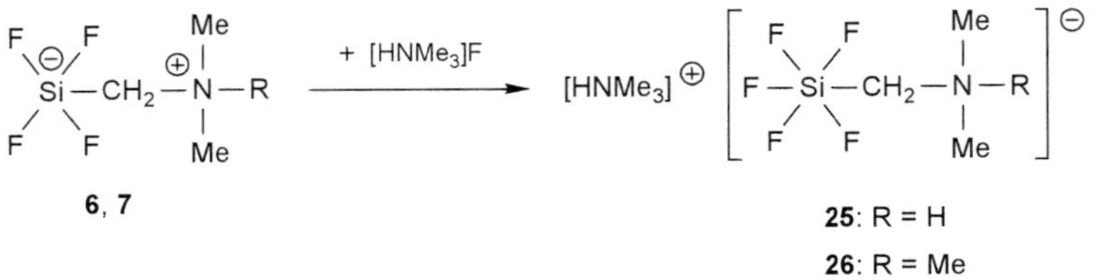

SCHEME 4.

compounds (isolated as the monohydrates **25** · H_2O and **26** · H_2O) was established by single-crystal X-ray diffraction.

As will be explained in Sections III,D, III,E, V, and VII, the acyclic zwitterionic $\lambda^5 Si$-silicates **6**, **19**, **21**, and **22** were used as starting materials for the synthesis of related zwitterionic spirocyclic and monocyclic $\lambda^5 Si$-silicates. These compounds were obtained, along with Me_3SiF, by reaction of **6**, **19**, **21**, and **22** with bis(trimethylsilyl) derivatives of various diols and some of their monothio analogues.

III

SPIROCYCLIC $^5 Si$-SILICATES WITH AN SiO_4C SKELETON

A. *Compounds Containing Two Bidentate Ligands of the Benzene-1,2-diolato(2−) or Naphthalene-2,3-diolato(2−) Type*

The pentacoordinate silicon compounds **3**,[17] **27**,[25] **28**,[26] **29**,[27,28] **30**,[29] **31**,[30] **32**,[31] **33**,[25,32] **34**,[33] **35**,[33,34] **36**,[27,32] **37**,[27] and **38**[35] are spirocyclic zwitterionic $\lambda^5 Si$-silicates with an SiO_4C skeleton. The chiral zwitterions contain two bidentate ligands of the benzene-1,2-diolato(2−) or naphthalene-2,3-diolato(2−) type.

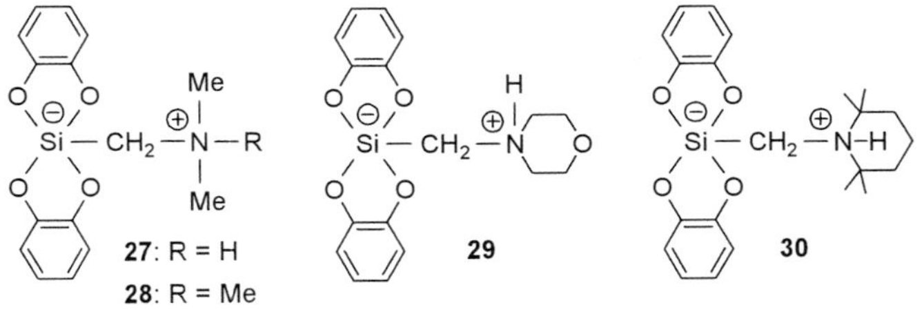

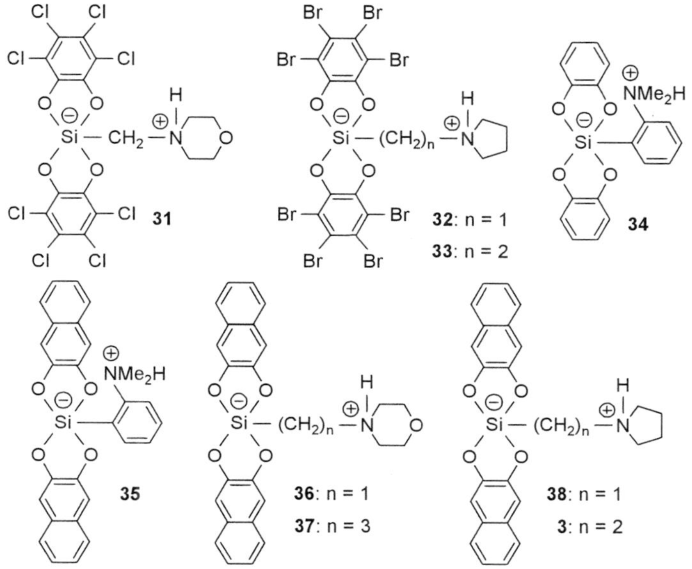

Structures **31**, **32**: n = 1, **33**: n = 2, **34**, **35**, **36**: n = 1, **37**: n = 3, **38**: n = 1, **3**: n = 2

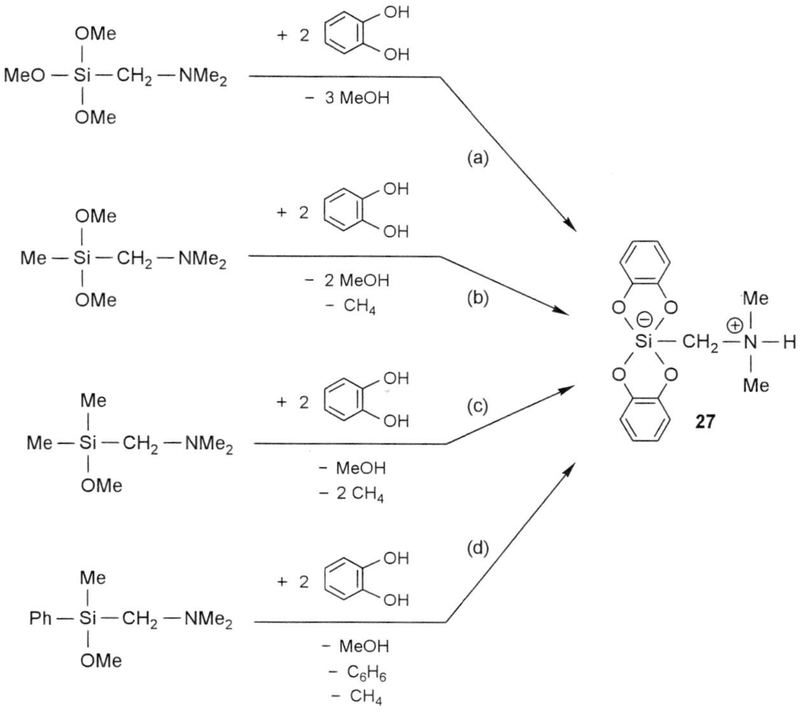

SCHEME 5.

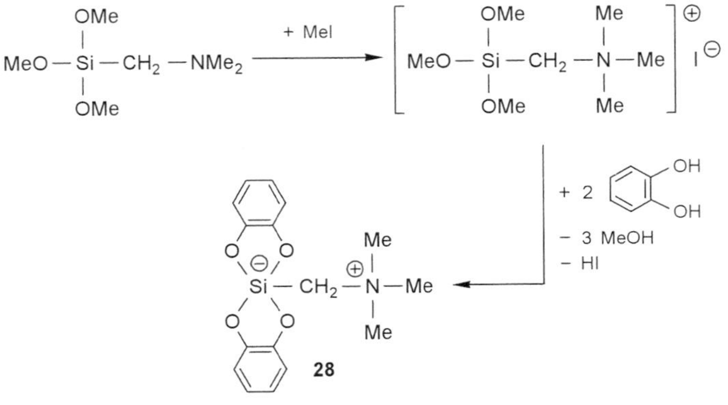

SCHEME 6.

Compounds **3** and **29–38** were prepared analogously to the synthesis of **27** according to Scheme 5, method a (cleavage of three Si–O bonds). In most cases, the zwitterionic products were formed spontaneously in precipitation reactions. Generally, the yields were very high. The synthesis of **28** (an *N*-methyl derivative of **27**) involves also cleavage of three Si–O bonds (Scheme 6). As demonstrated for **27** in Scheme 5 (methods b–d), Si–C cleavage reactions (cleavage of one Si–C bond and two Si–O bonds or cleavage of two Si–C bonds and one Si–O bond) were also used for the preparation of some of these compounds (**3**, **27**, **29**, **31**, **34–38**). As these syntheses involve relatively slow Si–C cleavage reactions as rate-determining steps, product formation was significantly slower than that observed for the alternative syntheses that only involve Si–O cleavage reactions. In many cases, Si–C cleavage (e.g., synthesis of **27** according to methods b–d) yielded directly well-crystallized products (crystal quality suitable for single-crystal X-ray diffraction) that did not need further purification by recrystallization. However, these syntheses (especially those involving cleavage of two Si–C bonds) gave lower yields. All syntheses were carried out in acetonitrile at room temperature and the products **3** · CH_3CN, **27–30**, **31** · CH_3CN, **32** · CH_3CN, **33** (two different phases; **33-I, 33-II**), **33** · H_2O, **34**, **35** · $\frac{1}{2}CH_3CN$, **36** · CH_3CN, **37** · $\frac{1}{2}CH_3CN$, and **38** · $\frac{1}{4}CH_3CN$ were isolated as crystalline solids. The synthesis of **36** was also carried out in acetone and nitromethane to yield the crystalline solvates **36** · Me_2CO and **36** · $MeNO_2$, respectively.

Compounds **3** · CH_3CN, **27–30**, **31** · CH_3CN, **32** · CH_3CN, **33-I**, **33–II**, **33** · H_2O, **34**, **35** · $\frac{1}{2}CH_3CN$, **36** · CH_3CN, **36** · Me_2CO, **36** · $MeNO_2$, **37** · $\frac{1}{2}CH_3CN$, and **38** · $\frac{1}{4}CH_3CN$ were structurally characterized by single-

TABLE III

Selected Bond Distances (Å) and Angles (°) for Compounds $3 \cdot CH_3CN$, 27–30, $31 \cdot CH_3CN$, $32 \cdot CH_3CN$, 33-I, 33-II, $33 \cdot H_2O$, 34, $35 \cdot \frac{1}{2}CH_3CN$, $36 \cdot CH_3CN$, $36 \cdot Me_2CO$, $36 \cdot MeNO_2$, $37 \cdot \frac{1}{2}CH_3CN$, and $38 \cdot \frac{1}{4}CH_3CN$ as Determined by Single-Crystal X-Ray Diffraction[a,b]

Compound	Si–O1	Si–O2	Si–O3	Si–O4	Si–Cl	O1–Si–O3	O2–Si–O4
$3 \cdot CH_3CN$	1.758(2)	1.741(2)	1.728(2)	1.752(2)	1.879(3)	153.6(1)	150.1(1)
27 (M I)	1.791(2)	1.695(2)	1.765(2)	1.714(2)	1.890(2)	176.75(6)	118.86(6)
27 (M II)	1.794(2)	1.701(2)	1.765(2)	1.713(2)	1.892(2)	176.45(6)	119.37(7)
28 (M I)	1.771(3)	1.707(3)	1.781(3)	1.700(3)	1.908(5)	174.5(2)	123.8(2)
28 (M II)	1.766(3)	1.712(3)	1.793(3)	1.711(3)	1.898(4)	173.8(2)	124.6(2)
29 (M I)	1.797(2)	1.692(2)	1.761(2)	1.708(2)	1.896(3)	178.08(10)	120.68(10)
29 (M II)	1.808(2)	1.691(2)	1.766(2)	1.700(2)	1.893(3)	178.76(9)	117.97(10)
30 (M I)	1.7978(12)	1.6971(12)	1.7753(12)	1.6965(13)	1.902(2)	178.46(6)	119.23(6)
30 (M II)	1.7978(13)	1.6963(13)	1.7739(13)	1.6956(12)	1.899(2)	177.19(6)	120.81(6)
$31 \cdot CH_3CN$	1.7513(10)	1.7309(10)	1.7455(10)	1.7421(10)	1.898(2)	157.76(5)	152.54(5)
$32 \cdot CH_3CN$	1.788(3)	1.708(3)	1.770(3)	1.702(4)	1.888(6)	176.0(2)	121.3(2)
33-I[c]	1.783(9)	1.706(9)	1.754(9)	1.712(9)	1.88(1)	171.4(5)	122.2(5)
33-II[d]	1.753(17)	1.727(15)	1.741(16)	1.699(18)	1.885(32)	157.9(8)	141.6(8)
$33 \cdot H_2O$ (M I)	1.78(2)	1.78(2)	1.75(2)	1.75(2)	1.87(3)	152(1)	151(1)
$33 \cdot H_2O$ (M II)	1.74(2)	1.74(2)	1.75(2)	1.75(2)	1.87(3)	156(1)	149(1)
34	1.769(2)	1.707(2)	1.745(2)	1.738(2)	1.878(2)	160.65(8)	144.68(8)
$35 \cdot \frac{1}{2}CH_3CN$	1.768(2)	1.720(2)	1.738(2)	1.752(2)	1.886(2)	155.9(1)	149.3(1)
$36 \cdot CH_3CN$ (M I)	1.780(2)	1.702(2)	1.770(2)	1.699(2)	1.890(2)	177.21(9)	118.06(9)
$36 \cdot CH_3CN$ (M II)	1.783(2)	1.702(2)	1.774(2)	1.699(2)	1.893(2)	177.05(9)	118.35(9)
$36 \cdot Me_2CO$	1.790(3)	1.697(3)	1.770(3)	1.702(3)	1.888(4)	176.9(1)	119.2(1)
$36 \cdot MeNO_2$	1.781(6)	1.689(6)	1.754(6)	1.709(6)	1.875(8)	173.2(3)	129.6(3)
$37 \cdot \frac{1}{2}CH_3CN$	1.759(3)	1.735(3)	1.788(3)	1.719(3)	1.870(3)	158.9(1)	140.6(1)
$38 \cdot \frac{1}{4}CH_3CN$	1.802(3)	1.693(3)	1.754(3)	1.709(3)	1.906(5)	172.8(2)	128.4(2)

[a] For the atomic numbering scheme, see Fig. 2. In the structures of 3, 27–33, and 36–38, the non-hydrogen atom bound to C1 (27–32, 36, 38: N atom; 3, 33, 37: C atom) is pointing to O1; in the structures of 34 and 35, the N-linked C atom of the phenyl group is pointing to O1.

[b] For compounds 27–30, $33 \cdot H_2O$, and $36 \cdot CH_3CN$, two crystallographically independent molecules (M I, M II) were observed.

[c] Space group $P2_1/n$.

[d] Space group $Pna2_1$.

crystal X-ray diffraction. In all cases, the crystals were found to be built up by pairs of enantiomers [(Λ)- and (Δ)-enantiomers]. Selected geometric parameters for the Si-coordination polyhedra of these compounds are listed in Table III. As can be seen from these data, the Si-coordination polyhedra can adopt quite different geometries. This is demonstrated for 27, $31 \cdot CH_3CN$, and 34 in Fig. 2. The Si-coordination polyhedra of compounds 27–30, $32 \cdot CH_3CN$, 33-I, $36 \cdot CH_3CN$, $36 \cdot Me_2CO$, $36 \cdot MeNO_2$, and $38 \cdot \frac{1}{4}CH_3CN$ are best described as almost ideal or less distorted trigonal bipyramids, the axial sites being occupied by the oxygen atoms O1 and O3. In contrast, the Si-coordination polyhedra of compounds $3 \cdot CH_3CN$,

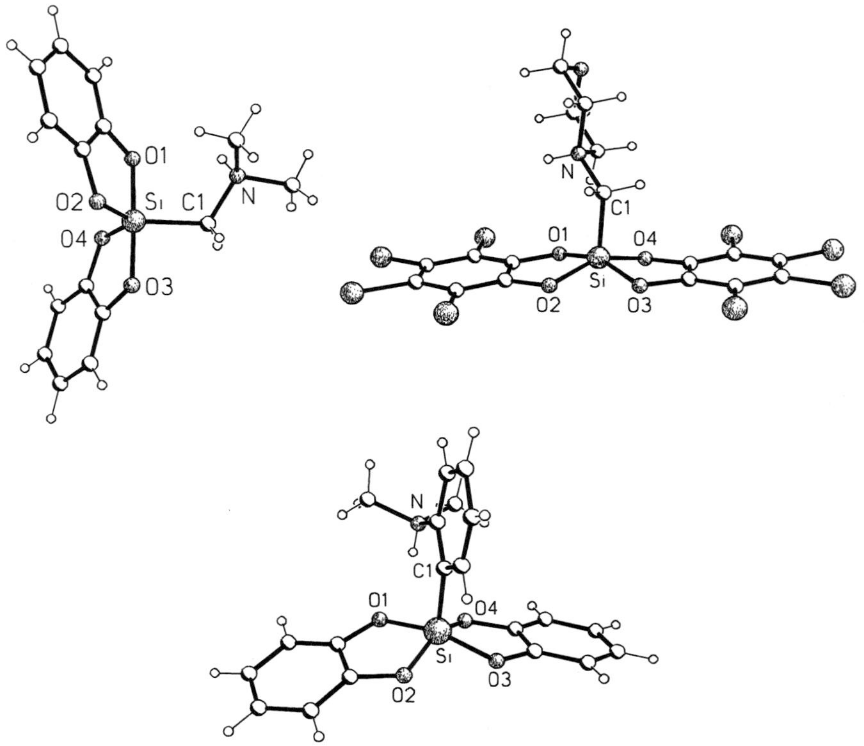

FIG. 2. Molecular structures of **27** (top left), **31** (solvate **31** · CH₃CN, top right), and **34** (bottom) in the crystal. For selected bond distances and angles, see Table III.

31 · CH_3CN, **33-II**, **33** · H_2O, **34**, **35** · $\frac{1}{2}CH_3CN$, and **37** · $\frac{1}{2}CH_3CN$ represent nearly ideal or (more or less) distorted square pyramids, with the carbon atom C1 in the apical position. In most cases, the geometries of the *Si*-coordination polyhedra in the crystal are located on the Berry-pseudorotation coordinate (pivot atom C1).

As can be seen from the geometric data for **33-I**, **33-II**, and **33** · H_2O, a particular zwitterion can even adopt quite different structures in the crystal. The respective Berry distortions (distortion of the ideal trigonal bipyramid toward the ideal square pyramid along the pathway of the Berry pseudorotation) amount to 34.9% (**33-I**), 70% (**33-II**), 86.2% (**33** · H_2O, molecule II), and 96.3% (**33** · H_2O, molecule I). These findings are in good agreement with results of ab initio studies with the anionic model species **39**. As shown by SCF/SVP geometry optimizations, the energetically preferred trigonal-bipyramidal structure of **39** (C_2 symmetry, local minimum) is only 5.8 kJ mol^{-1} more stable than the alternative square-pyramidal structure (C_s symmetry, transition state).[36] Obviously, the geometries of the *Si*-coordination polyhedra of the zwitterions **3** and **27–38** in the crystal are mainly governed by the crystal packing (including hydrogen bonds) rather than by the nature of the ligands bound to the silicon atom.

39

The Si–O distances determined for the zwitterions **3** and **27–38** amount to 1.689(6)–1.808(2) Å. In the case of the (more or less distorted) trigonal-bipyramidal structures, the axial Si–O distances [Si–O1, Si–O3; 1.754(9)–1.808(2) Å] are significantly longer than the equatorial ones [Si–O2, Si–O4; 1.689(6)–1.712(3) Å]. The Si–C distances of **3** and **27–38** amount to 1.870(3)–1.908(5) Å.

NMR studies (^{1}H, ^{13}C, ^{29}Si) have shown that the zwitterions **3** and **27–38** also exist in solution (**3**, **27–29**, **31–38**, [D$_6$]DMSO; **30**, CDCl$_3$). The NMR data, in context with the results of ab initio studies of **39**, can be interpreted in terms of a rapid Berry-type pseudorotation at room temperature, leading to an interconversion of the (Λ)- and (Δ)-enantiomers. The ^{29}Si chemical shifts of the zwitterions **3** and **27–38** are listed in Table IV. In most cases, these values are very similar to the isotropic ^{29}Si chemical shifts obtained for the respective crystalline compounds by solid-state ^{29}Si CP/MAS experiments. However, the ^{29}Si chemical shifts observed for **31** (δ −131.9) and **32** (δ −123.0) in solution differ substantially from those determined for **31** · CH$_3$CN (δ −85.1) and **32** · CH$_3$CN (δ −84.8) by the solid-state NMR experiments. Furthermore, the solution-state NMR data of **31** and **32** differ significantly from all the other ^{29}Si chemical shifts listed in Table IV. These results are indicative of special structural features of the zwitterions **31** and **32** in solution and might be explained in terms of hexacoordination of the silicon atoms of these molecules. However, further studies have to be performed to check this (rather speculative) hypothesis.

In the zwitterions **3** and **27–38**, two bidentate symmetric diolato(2−) ligands are bound to the silicon atom, whereas the derivatives **40–43** contain two unsymmetrically substituted diolato(2−) ligands. Because of this unsymmetric substitution pattern, two sets of resonance signals were observed

TABLE IV

^{29}Si Chemical Shifts for Compounds $3 \cdot CH_3CN$, 27–30, $31 \cdot CH_3CN$, $32 \cdot CH_3CN$, 33-I, 33-II, $33 \cdot H_2O$, 34, $35 \cdot \frac{1}{2}CH_3CN$, $36 \cdot CH_3CN$, $36 \cdot Me_2CO$, $36 \cdot MeNO_2$, $37 \cdot \frac{1}{2}CH_3CN$, and $38 \cdot \frac{1}{4}CH_3CN$ in Solution ($[D_6]DMSO$) and/or in the Crystal[a]

Compound	$\delta_{solution}$	$\delta_{crystal}$[b]	Compound	$\delta_{solution}$	$\delta_{crystal}$[b]
$3 \cdot CH_3CN$	-76.6	[c]	$33 \cdot H_2O$	-79.1	-80.8
27	-85.9	-84.8	34	-88.6	[c]
28	-87.3	[c]	$35 \cdot \frac{1}{2}CH_3CN$	-90.2	-88.6
29	-85.6	-87.3	$36 \cdot CH_3CN$	-85.3	-85.5
30	-87.9[d]	-86.7	$36 \cdot Me_2CO$	-85.3	-85.3
$31 \cdot CH_3CN$	-131.9	-85.1	$36 \cdot MeNO_2$	-85.3	-83.6
$32 \cdot CH_3CN$	-123.0	-84.8	$37 \cdot \frac{1}{2}CH_3CN$	-76.4	-75.5
33-I[e]	-79.1	-73.5	$38 \cdot \frac{1}{4}CH_3CN$	-87.0	[c]
33-II[f]	-79.1	-79.7			

[a] Chemical shifts in ppm; spectra recorded at room temperature.
[b] Isotropic chemical shifts obtained by ^{29}Si CP/MAS NMR experiments.
[c] No experiment performed.
[d] Solvent $CDCl_3$.
[e] Space group $P2_1/n$.
[f] Space group $Pna2_1$.

in the ^1H, ^{13}C, and ^{29}Si NMR spectra of 40,[37,38] 41,[39] 42,[37,38] and 43[39] at room temperature (solvent $[D_6]DMSO$). The intensity ratios of the respective resonance signals were $1:1.1$ (40, 42, 43) and $1:1.2$ (41). These results indicate the presence of two NMR-spectroscopically distinguishable species (isomers) in solution. In contrast, in the NMR spectra of 3 and 27–38 only one set of resonance signals was observed.

As shown by VT ^1H NMR experiments, the two distinguishable species present in solutions of 40–43 were found to interconvert (coalescence phenomena) at higher temperatures. Provided that this isomerization is based

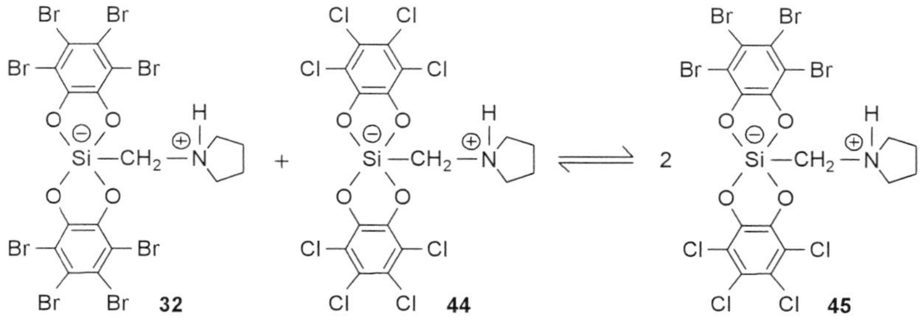

Scheme 7.

on an intramolecular ligand exchange, the activation free enthalpies for this process amount to ca. 70–75 kJ mol^{-1}.[37–39] This is in accordance with the results of ab initio studies with the model species **39**. This anion was shown to undergo an intramolecular ligand exchange via a trigonal-bipyramidal transition state, with the hydrogen atom in an axial position (energy barrier 66.5 kJ mol^{-1}).[36] The coalescence phenomena observed in the ^{1}H NMR studies of **40–43** might be explained in terms of an analogous ligand exchange process via a trigonal-bipyramidal transition state, with the carbon atom in an axial position.

Some of the zwitterionic λ^5Si-silicates described in this chapter were studied for their chemical properties. As shown for **32, 44**, and **45** in Scheme 7, compounds of this particular formula type undergo an intermolecular exchange of their benzene-1,2-diolato(2−) ligands in solution at room temperature.[38] Solution-state NMR studies ([D$_6$]DMSO; ^{1}H, ^{13}C, ^{29}Si) and FD MS experiments provided evidence for the equilibrium **32** + **44** ⇌ 2 **45**.

Generally, the pentacoordinate silicon compounds described in this chapter are sensitive to water and very easily undergo hydrolytic Si–O cleavage reactions in solution. This has been used for the synthesis of the octa(silasesquioxane) **46**, which was obtained in 90% yield by treatment of compound **35** with water in boiling acetonitrile (Scheme 8).[34]

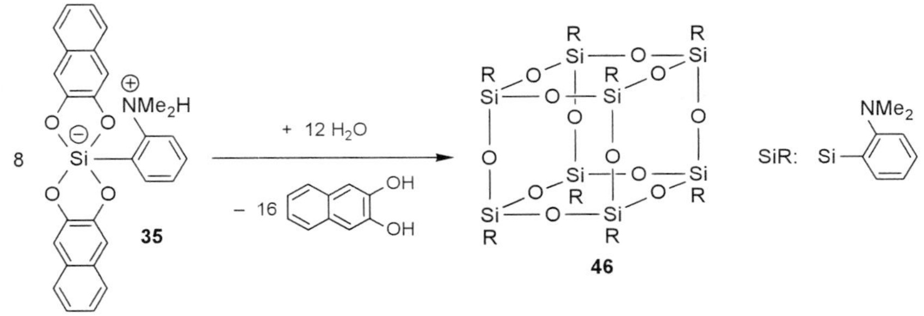

SCHEME 8.

B. *Compounds Containing Two Bidentate Ligands of the Ethene-1,2-diolato(2−) Type*

The spirocyclic zwitterionic λ^5Si-silicates **47**[40,41] and **48**[42] contain two *cis*-1,2-diphenylethene-1,2-diolato(2−) ligands. This particular type of ligand formally derives from the diol *cis*-HO(Ph)C=C(Ph)OH, a tautomer of benzoin.

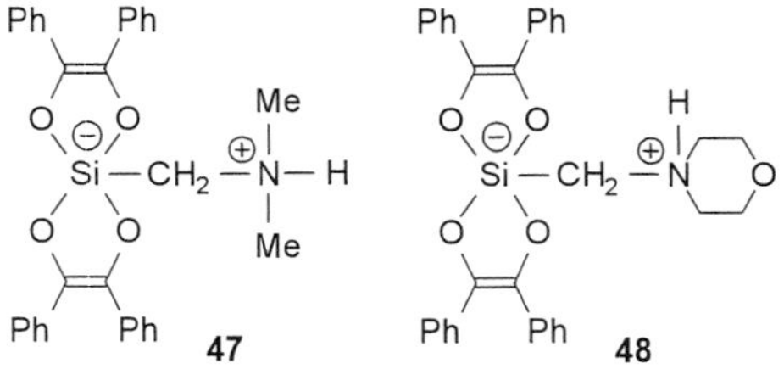

As shown for **47** in Scheme 9, different methods were applied for the syntheses of these compounds (**48** synthesized according to methods a–c). All reactions were carried out in acetonitrile at room temperature and compounds **47**·½CH₃CN and **48** were isolated as crystalline solids. The benzoin-mediated Si–C cleavage reactions (methods b–d in Scheme 9) are especially remarkable.

Single-crystal X-ray diffraction studies have shown that the *Si*-coordination polyhedra of **47**·½CH₃CN and **48** are slightly distorted trigonal bipyramids, with oxygen atoms in the axial positions. This is illustrated for the zwitterion **47** in Fig. 3. The zwitterions are chiral, the crystals of **47**·½CH₃CN and **48** consisting of pairs of (Λ)- and (Δ)-enantiomers. Selected geometric parameters for these compounds are listed in Table V. As can be seen from these data, the axial Si–O distances [1.763(2)–1.801(2) Å] are significantly longer than the equatorial ones [1.680(2)–1.704(2) Å]. The Si–C distances

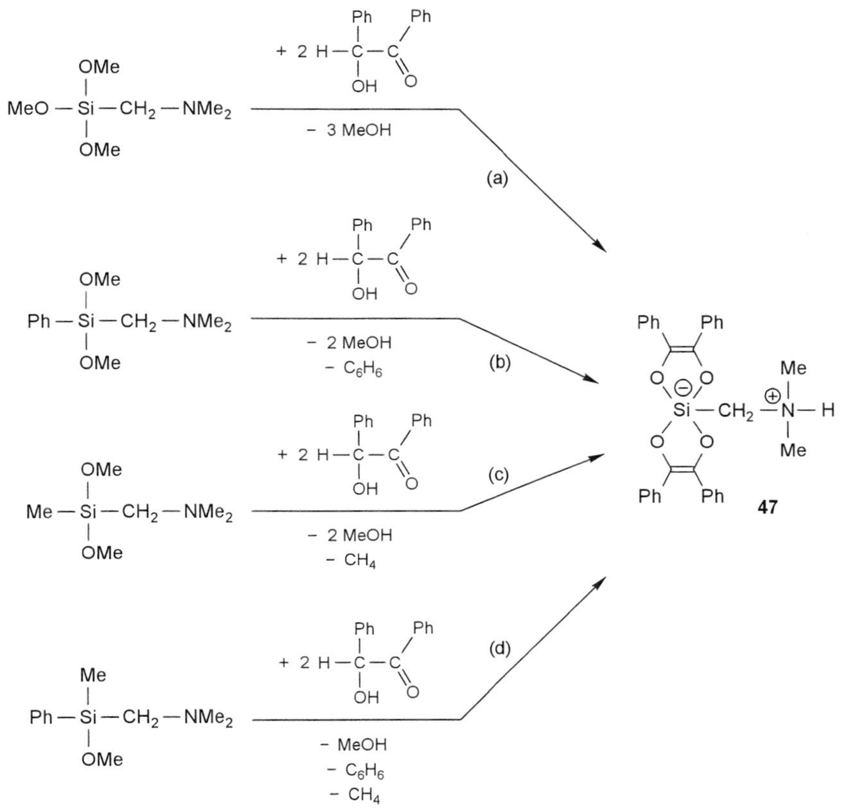

SCHEME 9.

amount to 1.897(3) and 1.902(3) Å, respectively. The Si–O and Si–C distances of $47 \cdot \frac{1}{2}CH_3CN$ and **48** are quite similar to those observed for the zwitterionic $\lambda^5 Si$-silicates **27–30, 32** $\cdot$ CH$_3$CN, **33-I, 36** $\cdot$ CH$_3$CN, **36** $\cdot$ Me$_2$CO, **36** $\cdot$ MeNO$_2$, and **38** $\cdot \frac{1}{4}$CH$_3$CN (see Section III,A); the latter compounds contain two bidentate ligands of the benzene-1,2-diolato(2−) or naphthalene-2,3-diolato(2−) type and their Si-coordination polyhedron is also a trigonal bipyramid.

As shown by ab initio studies with the anionic model species **49**, the energy difference between the trigonal-bipyramidal (C_2 symmetry) and the alternative square-pyramidal structure (C_{2v} symmetry) is rather small (6.9 kJ mol^{-1}).[40] From this result one may conclude that related zwitterionic $\lambda^5 Si$-silicates, such as **47** and **48**, should also have a chance to exist with a square-pyramidal Si-coordination polyhedron in the solid state. The square-

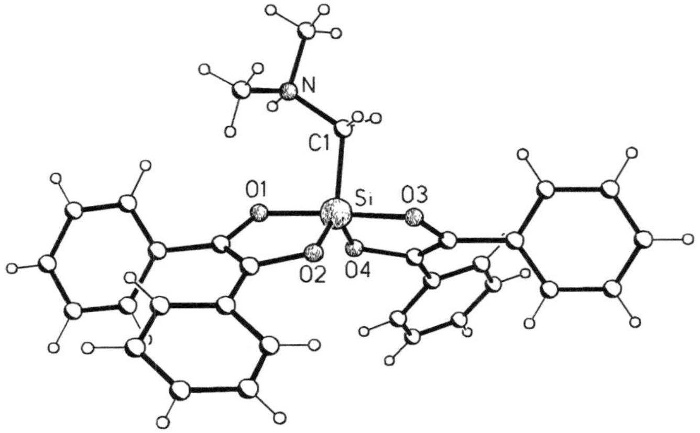

FIG. 3.   Molecular structure of **47** in the crystal of **47** · ½CH₃CN. For selected bond distances and angle, see Table V.

pyramidal geometry of **49** represents a transition state for a Berry-type pseudorotation process that allows a conversion of the trigonal-bipyramidal (Λ)- and (Δ)-enantiomers of **49** (local minima) into each other. Quite similar results were obtained for the related anionic model species **39** (see Section III,A).

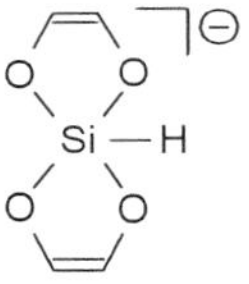

49

NMR studies ([D₆]DMSO; ¹H, ¹³C, ²⁹Si) have shown that the zwitterions **47** and **48** also exist in solution. The NMR data, in context with the results

TABLE V

SELECTED BOND DISTANCES (Å) AND ANGLES (°) FOR COMPOUNDS **47** · ½CH₃CN AND **48** AS DETERMINED BY SINGLE-CRYSTAL X-RAY DIFFRACTION[a]

Compound	Si–O1	Si–O2	Si–O3	Si–O4	Si–C1	O1–Si–O3
47 · ½CH₃CN	1.788(2)	1.680(2)	1.763(2)	1.704(2)	1.897(3)	177.5(1)
48	1.801(2)	1.691(3)	1.771(2)	1.695(3)	1.902(3)	176.93(8)

[a] For the atomic numbering scheme, see Fig. 3. In both structures, the nitrogen atom bound to C1 is pointing to O1.

of ab initio calculations with **49**, can be interpreted in terms of a dynamic behavior of **47** and **48** in solution (Berry-type pseudorotation).

The ^{29}Si chemical shifts of **47** and **48** are as follows: **47**, δ −87.6 (solid state); **48**, δ −89.6 ([D_6]DMSO), δ −85.6 (solid state). These data are quite similar to those obtained for the related compounds described in Section III,A (see Table IV).

C. Compounds Containing Two Bidentate Ligands of the Ethane-1,2-diolato(2−) or meso-Oxolane-3,4-diolato(2−) Type

The pentacoordinate silicon compounds **50**,[43] **51**,[43] and **52**[9] are spirocyclic zwitterionic λ^5Si-silicates with an SiO_4C skeleton. They contain two ethane-1,2-diolato(2−) ligands or two *meso*-oxolane-3,4-diolato(2−) ligands.

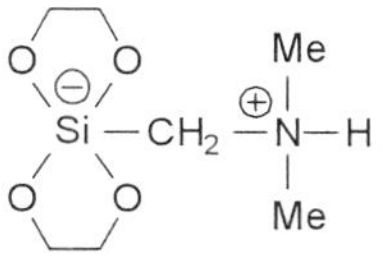

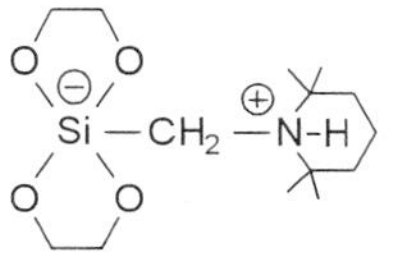

The method used for the syntheses of **50–52** is illustrated for compound **50** in Scheme 10. The syntheses of **50** and **51** were carried out without solvent at higher temperatures (removal of the methanol by distillation), whereas compound **52** was prepared in acetonitrile at room temperature. The products were isolated as crystalline solids, **50** as the solvate **50** $\cdot\frac{1}{2}$H OCH$_2$CH$_2$OH.

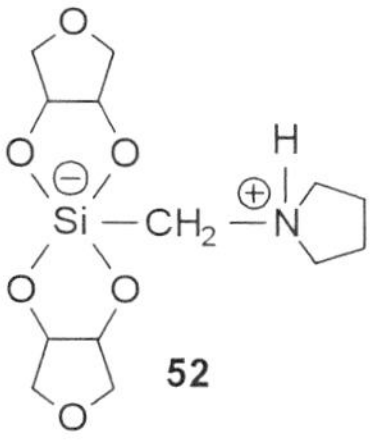

As shown by single-crystal X-ray diffraction, the Si-coordination polyhedra of **50** $\cdot\frac{1}{2}$HOCH$_2$CH$_2$OH, **51**, and **52** are distorted trigonal bipyramids, with oxygen atoms in both axial sites. This is shown for **50** $\cdot\frac{1}{2}$HOCH$_2$CH$_2$OH in Fig. 4. The zwitterions are chiral, the crystals of **50** $\cdot\frac{1}{2}$HOCH$_2$CH$_2$OH, **51**, and **52** consisting of pairs of (Λ)- and (Δ)-enantiomers. Selected geometric parameters for these compounds are listed in Table VI. As can be seen from these data, the axial Si–O distances [1.729(2)–1.7725(12) Å] are significantly

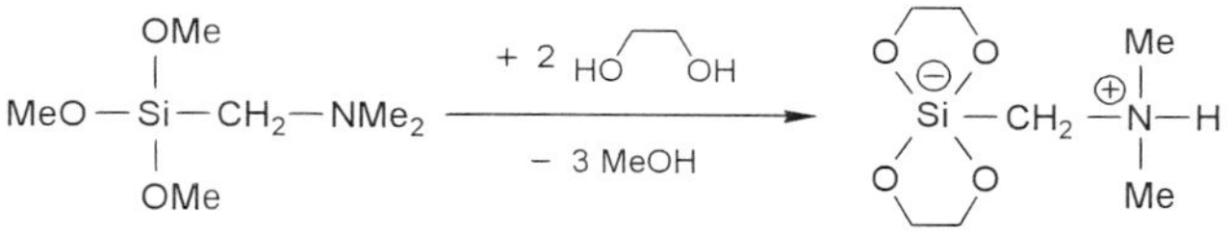

SCHEME 10.

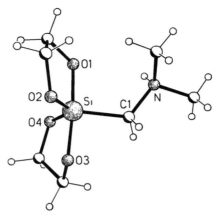

Fig. 4.   Molecular structure of **50** in the crystal of **50**·½HOCH₂CH₂OH. For selected bond distances and angle, see Table VI.

longer than the equatorial ones [1.6782(13)–1.706(2) Å]. The Si–C distances amount to 1.907(3)–1.922(2) Å.

NMR studies ([D₆]DMSO; ¹H, ¹³C, ²⁹Si) demonstrated that the zwitter-ions **50–52** also exist in solution. The ²⁹Si chemical shifts of these compounds are as follows: **50**·½HOCH₂CH₂OH, δ −91.8 ([D₆]DMSO), δ −91.3 (solid state); **51**, δ −93.0 (CDCl₃), δ −92.7 (solid state); **52**, δ −85.7 ([D₆]DMSO).

D. *Compounds Containing Two Diolato(2−) Ligands Derived from α-Hydroxycarboxylic Acids*

The spirocyclic pentacoordinate silicon compounds **53–56**,[44] **57–59**,[45] **60**,[46] **61**,[9] **62**,[8] **63**,[46] **64**,[46] **65–67**,[47] **68**,[48] **69**,[49] **70**,[9] **71**,[50] **72**,[51] **73**,[8] and **74**[43] represent zwitterionic $\lambda^5 Si$-silicates (**53–70**) or biszwitterionic $\lambda^5 Si,\lambda^5 Si'$-disilicates (**71–74**) with SiO_4C skeletons. Compounds **53–71** contain bidentate diol-

TABLE VI

Selected Bond Distances (Å) and Angles (°) for Compounds **50**·½HOCH₂CH₂OH, **51**, and **52** as Determined by Single-Crystal X-Ray Diffraction[a]

Compound	Si–O1	Si–O2	Si–O3	Si–O4	Si–C1	O1–Si–O3
50·½HOCH₂CH₂OH	1.7585(14)	1.6782(13)	1.7697(14)	1.6939(13)	1.920(2)	175.36(6)
51	1.7725(12)	1.6861(14)	1.7466(12)	1.6869(13)	1.922(2)	177.98(6)
52	1.730(2)	1.706(2)	1.764(2)	1.689(2)	1.909(3)	171.03(9)

[a] For the atomic numbering scheme, see Fig. 4. In all structures, the nitrogen atom bound to C1 is pointing to O1.

ato(2−) ligands that derive from α-hydroxycarboxylic acids [glycolic acid, 2-methyllactic acid, benzilic acid, citric acid, (R)-mandelic acid]; compounds **72–74** contain two tetradentate tetraolato(4−) ligands that derive from L- or D-tartaric acid.

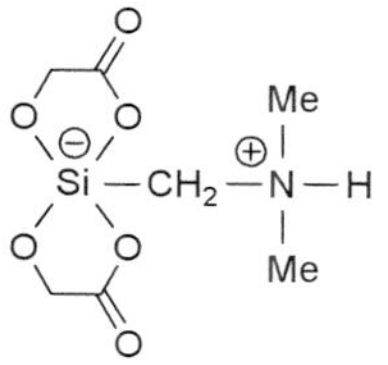

53

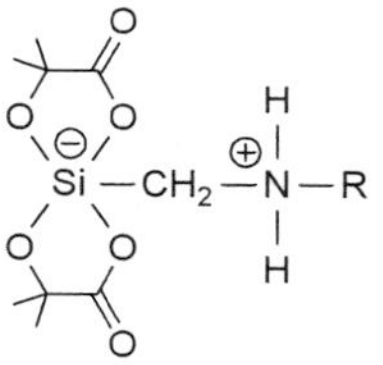

54: R = H
55: R = Me

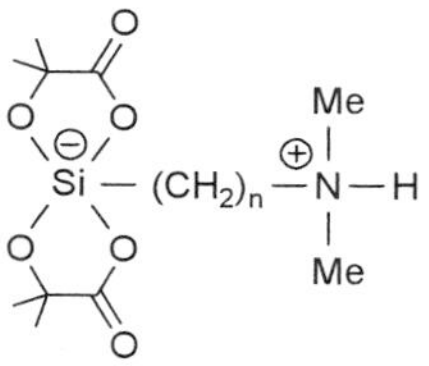

56: n = 1   **57**: n = 2
58: n = 3   **59**: n = 4

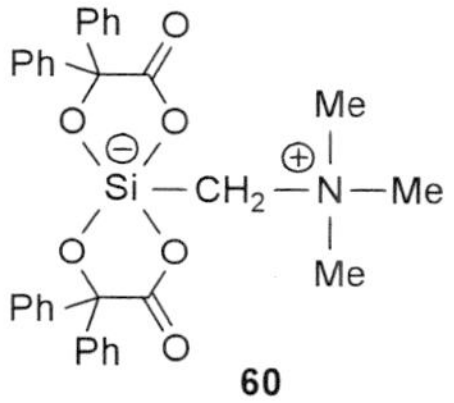

60

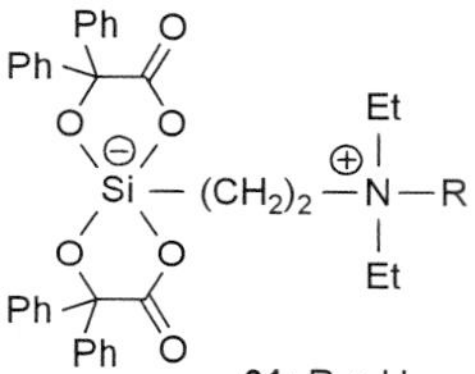

61: R = H
62: R = Et

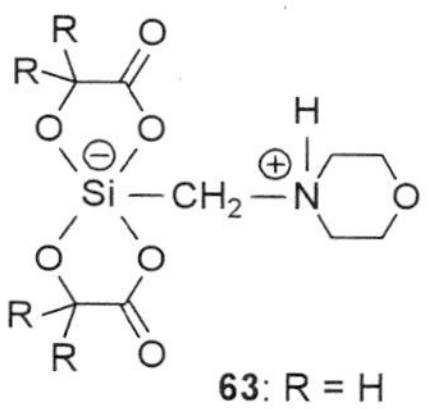

63: R = H
64: R = Ph

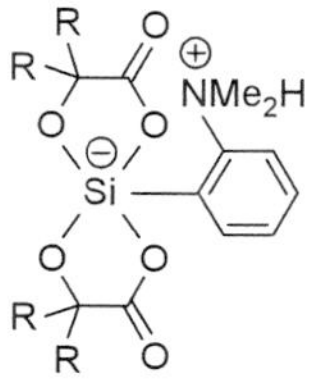

65: R = H
66: R = Me
67: R = Ph

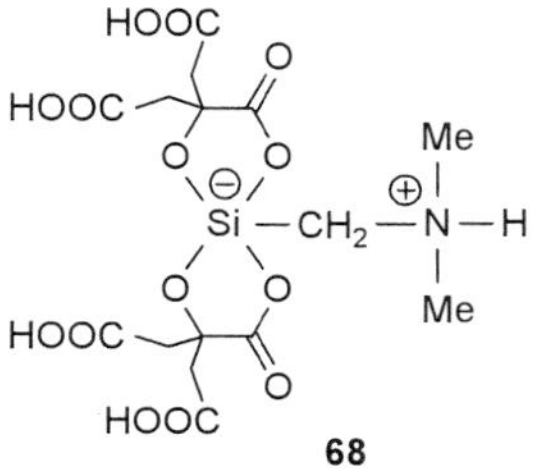

68

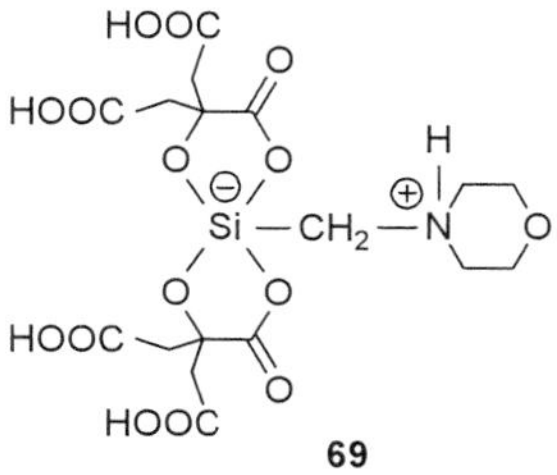

69

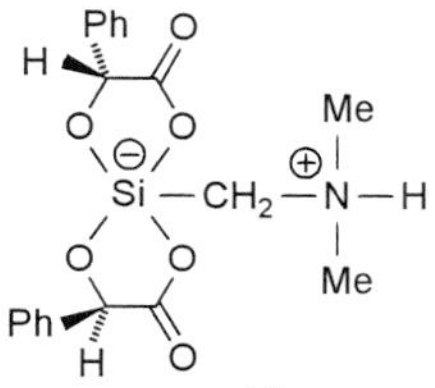

70

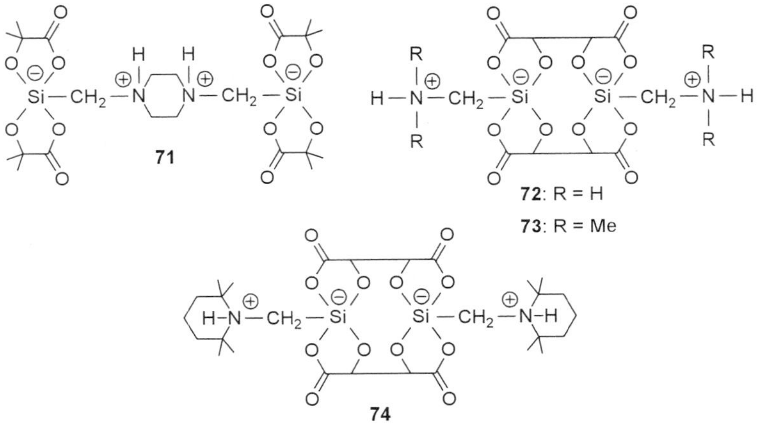

Compounds **53–55**, **58**, **59**, **61**, and **63–70** were prepared analogously to the syntheses of **56** and **57** according to Schemes 11 and 12, methods a (cleavage of three Si–O bonds). All reactions were carried out in acetonitrile at room temperature and the products isolated in high yields as crystalline solids. Compound **56** was obtained as the crystalline acetonitrile solvate **56** · 2CH$_3$CN. Compounds **54**, **56**, **63**, **68**, and **69** were isolated, after crystalli-

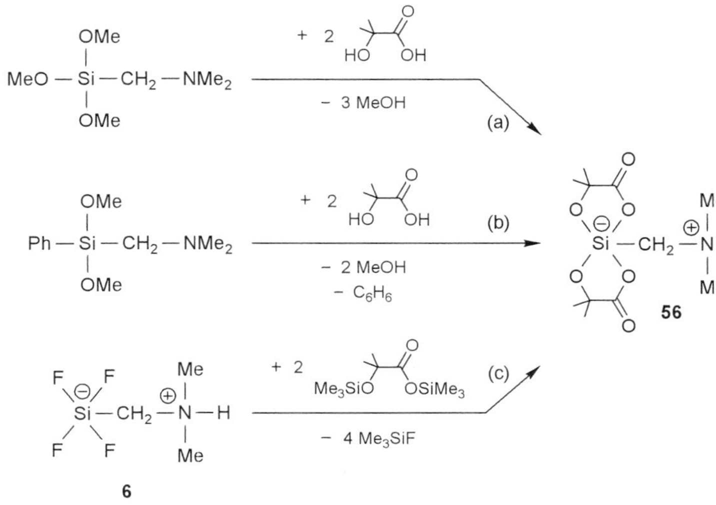

SCHEME 11.

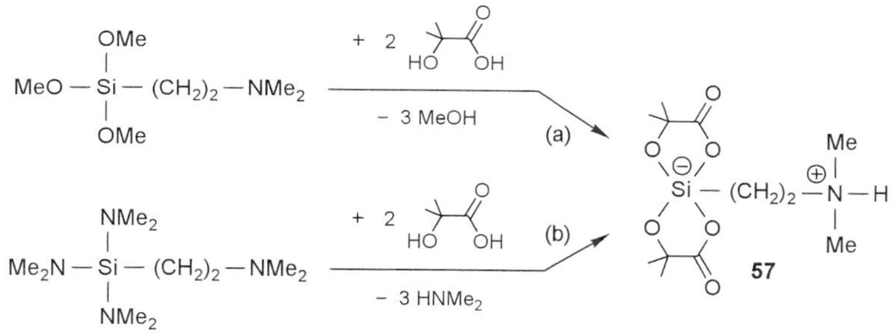

Scheme 12.

zation from water (slow evaporation of the water at room temperature over a period of several days) as crystalline monohydrates. The zwitterions 53–59, 61, and 63–70 are chiral, and compounds 53, 54 · H$_2$O, 55, 56 · H$_2$O, 56 · 2CH$_3$CN, 57–59, 61, 63 · H$_2$O, 64–67, 68 · H$_2$O, and 69 · H$_2$O were isolated as racemic mixtures, whereas compound 70 was obtained as the diastereomerically and enantiomerically pure optically active product (Λ,R)-70.

As shown for the synthesis of 60 in Scheme 13, the preparation of the quaternary ammonium compounds 60 and 62 also involves cleavage of three Si–O bonds. The syntheses were again performed in acetonitrile at room temperature and compounds 60 and 62 · CH$_3$CN isolated in high yields as crystalline products.

As demonstrated for compounds 56 and 57 in Scheme 11 (method b: cleavage of two Si–O bonds and one Si–C bond; method c: cleavage of four Si–F bonds) and Scheme 12 (method b: cleavage of three Si–N bonds),

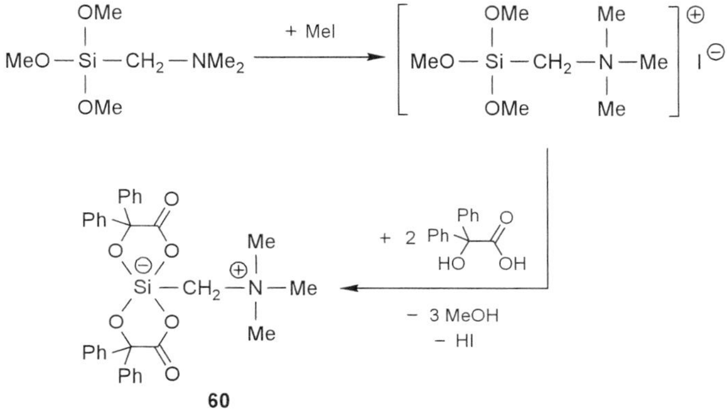

Scheme 13.

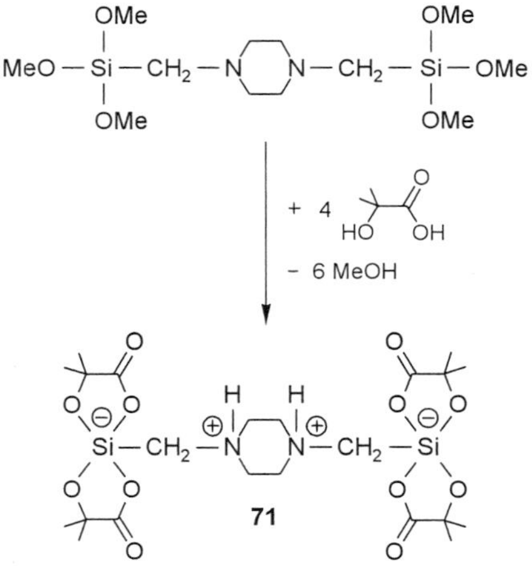

SCHEME 14.

alternative methods were also used for the synthesis of some of the compounds described in this section. All these reactions were again performed in acetonitrile at room temperature and the products isolated in good yields as crystalline solids.

The biszwitterionic λ^5Si,λ^5Si'-disilicate **71** was prepared according to Scheme 14 (solvent acetone/water, room temperature) and isolated, after crystallization from water, as the octahydrate *meso*-**71** · 8H$_2$O.

The biszwitterionic λ^5Si,λ^5Si'-disilicates **72–74** were synthesized according to Scheme 15 in water (**72**, **73**) or methanol (**74**) at room temperature. For these syntheses, the following tartaric acids were used: **72**, (R,R)-

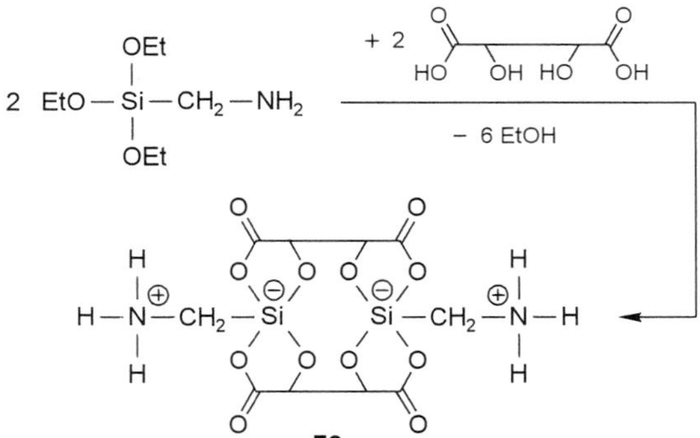

SCHEME 15.

tartaric acid; **73**, *rac*-tartaric acid; **74**, (*S,S*)-tartaric acid. After crystallization from water, the hydrates (Δ,Δ,*R,R,R,R*)-**72** · 3H$_2$O, *rac*-(Δ,Δ,*R,R,R,R/* Λ,Λ,*S,S,S,S*)-**73** · H$_2$O, and (Λ,Λ,*S,S,S,S*)-**74** · 2H$_2$O were isolated in good yields as crystalline solids. The formation of the optically active compounds (Δ,Δ,*R,R,R,R*)-**72** · 3H$_2$O and (Λ,Λ,*S,S,S,S*)-**74** · 2H$_2$O occurred stereospecifically (100% de, 100% ee).

Compounds **53**, **54** · H$_2$O, **55**, **56** · H$_2$O, **56** · 2CH$_3$CN, **57–61**, **62** · CH$_3$CN, **63** · H$_2$O, **64–67**, **68** · H$_2$O, **69** · H$_2$O, **70**, **71** · 8H$_2$O, **72** · 3H$_2$O, **73** · H$_2$O, and **74** · 2H$_2$O were structurally characterized by single-crystal X-ray diffraction. Except for **70** [(Λ,*R*)-enantiomer], **71** · 8H$_2$O (*meso*-isomer), **72** · 3H$_2$O [(Δ,Δ,*R,R,R,R*)-enantiomer], and **74** · 2H$_2$O [(Λ,Λ,*S,S,S,S*)-enantiomer], the crystals are built up by pairs of enantiomers. Selected geometric parameters for the *Si*-coordination polyhedra of the compounds studied are listed in Table VII (**53–71**) and Table VIII (**72–74**). As can be seen from these data, quite similar structural features were observed. With only one exception (compound **65**), all *Si*-coordination polyhedra can be described as more or less distorted trigonal bipyramids, with the carboxylate oxygen atoms O1 (O5) and O3 (O7) in the axial sites. This is shown in Fig. 5 (example: **53**), Fig. 6 (example: **71** · 8H$_2$O), and Fig. 7 (examples: **72** · 3H$_2$O, **74** · 2H$_2$O). In contrast, the *Si*-coordination polyhedron of compound **65** (Fig. 5) is best described as a strongly distorted square pyramid, with the carbon atom C1 in the apical position. The Berry distortion (distortion of the ideal trigonal bipyramid toward the ideal square pyramid along the pathway of the Berry pseudorotation process) for this zwitterion amounts to 51.7%. Obviously, this strong distortion is the result of an intramolecular bifurcate N–H · · · O1/N–H · · · O4 hydrogen bond.[47] Generally, N–H · · · O hydrogen bonds play an important role in the crystal structures of most of the zwitterionic λ^5*Si*-silicates and λ^5*Si*,λ^5*Si'*-disilicates described in this section.

As can be seen from the data listed in Tables VII and VIII, the axial Si–O distances [Si–O1, Si–O3; Si1–O1, Si1–O3, Si2–O5, Si2–O7; 1.763(2)–1.841(3) Å] in the trigonal-bipyramidal structures are significantly longer than the equatorial ones [Si–O2, Si–O4; Si1–O2, Si1–O4, Si2–O6, Si2–O8; 1.6509(12)–1.685(3) Å]. Even in the strongly distorted square-pyramidal structure of **65**, the Si–O1 and Si–O3 distances [1.8085(12) Å, 1.7818(12) Å] are longer than the Si–O2 and Si–O4 distances [1.6731(11) Å, 1.6827(11) Å]. The Si–C distances amount to 1.868(4)–1.910(4) Å.

The experimentally established preference for the trigonal-bipyramidal structure, with the carboxylate oxygen atoms in the axial sites, is in good accordance with the results obtained by ab initio studies of the related anionic model species **75** (Fig. 8).[44] SCF/SVP geometry optimizations have demonstrated that the isomer **75a** (C_2 symmetry; local minimum) is significantly more stable than the isomers **75b** (C_1 symmetry; local minimum)

TABLE VII

Selected Bond Distances (Å) and Angles (°) for Compounds **53**, **54** · H$_2$O, **55**, **56** · H$_2$O, **56** · 2CH$_3$CN, **57–61**, **62** · CH$_3$CN, **63** · H$_2$O, **64–67**, **68** · H$_2$O, **69** · H$_2$O, **70**, and **71** · 8H$_2$O as Determined by Single-Crystal X-Ray Diffraction[a,b]

Compound	Si–O1	Si–O2	Si–O3	Si–O4	Si–C1	O1–Si–O3
53	1.822(2)	1.666(2)	1.795(2)	1.656(2)	1.893(2)	175.46(7)
54 · H$_2$O	1.7943(12)	1.6623(12)	1.7974(12)	1.6601(12)	1.883(2)	177.17(5)
55	1.7941(14)	1.656(2)	1.8132(14)	1.6733(14)	1.889(2)	176.98(5)
56 · H$_2$O	1.8003(8)	1.6636(8)	1.8029(8)	1.6764(8)	1.8915(11)	176.06(4)
56 · 2CH$_3$CN	1.8060(12)	1.6656(12)	1.7930(12)	1.6751(13)	1.896(2)	177.51(6)
57 (M I)	1.7907(12)	1.6771(13)	1.8334(12)	1.6669(12)	1.884(2)	168.46(6)
57 (M II)	1.8004(12)	1.6670(13)	1.8371(12)	1.6687(11)	1.884(2)	173.70(5)
58	1.804(2)	1.670(2)	1.832(2)	1.672(2)	1.871(4)	173.40(11)
59 (M I)	1.786(2)	1.685(3)	1.841(3)	1.663(3)	1.874(4)	170.1(2)
59 (M II)	1.810(4)	1.671(3)	1.836(4)	1.665(4)	1.868(4)	169.82(14)
60	1.7895(13)	1.6585(13)	1.7951(13)	1.6602(13)	1.902(2)	176.33(6)
61	1.795(2)	1.659(2)	1.826(2)	1.662(2)	1.875(2)	172.15(8)
62 · CH$_3$CN	1.7989(14)	1.6749(14)	1.7996(14)	1.6657(14)	1.882(2)	172.40(7)
63 · H$_2$O	1.8109(9)	1.6647(10)	1.8167(10)	1.6703(9)	1.8933(13)	178.79(5)
64	1.7875(14)	1.6664(14)	1.7916(14)	1.6614(14)	1.895(2)	176.40(7)
65	1.8085(12)	1.6731(11)	1.7818(12)	1.6827(11)	1.8741(14)	165.14(5)
66	1.8135(12)	1.6581(11)	1.7806(12)	1.6719(12)	1.889(2)	168.95(5)
67	1.7874(12)	1.6509(12)	1.7831(12)	1.6770(12)	1.882(2)	173.28(5)
68 · H$_2$O	1.794(2)	1.666(2)	1.802(2)	1.662(2)	1.888(4)	175.07(11)
69 · H$_2$O	1.824(2)	1.679(2)	1.763(2)	1.666(2)	1.888(3)	175.11(9)
70[c]	1.798(2)	1.655(2)	1.807(2)	1.653(2)	1.910(4)	176.35(7)
71 · 8H$_2$O[d]	1.794(4)	1.671(3)	1.777(4)	1.658(4)	1.884(6)	176.1(2)

[a] For the atomic numbering scheme, see Figs. 5 and 6. In the structures of **53–64** and **68–70**, the non-hydrogen atom bound to C1 (**53–56**, **60**, **63**, **64**, **68–70**: N atom; **57–59**, **61**, **62**: C atom) is pointing to O1. In the structures of **65–67**, the N-linked C atom of the phenyl group is pointing to O1.

[b] For compounds **57** and **59**, two crystallographically independent molecules (M I, M II) were observed.

[c] (Λ,R)-Configuration; the data for **70** given in Ref. 9 are not correct.

[d] *meso*-(Λ,Δ)-Configuration.

and **75c** (C_2 symmetry; local minimum). The energy differences amount to 22.3 kJ mol^{-1} (**75a/75b**) and 42.6 kJ mol^{-1} (**75a/75c**). The alternative square-pyramidal species **75d** (C_2 symmetry; transition state) and **75e** (C_s symmetry; transition state) are also significantly less stable than the energetically preferred isomer **75a**, the respective energy differences amounting to 29.6 kJ mol^{-1} (**75a/75d**) and 31.4 kJ mol^{-1} (**75a/75e**).

TABLE VIII

SELECTED BOND DISTANCES (Å) AND ANGLES (°) FOR COMPOUNDS **72**·3H$_2$O, **73**·H$_2$O, AND **74**·2H$_2$O AS DETERMINED BY SINGLE-CRYSTAL X-RAY DIFFRACTION[a]

Compound	Si1–O1 Si2–O5	Si1–O2 Si2–O6	Si1–O3 Si2–O7	Si1–O4 Si2–O8	Si1–C1 Si2–C2	O1–Si1–O3 O5–Si2–O7
72·3H$_2$O[b]	1.791(2)	1.675(2)	1.820(2)	1.665(2)	1.885(3)	175.56(9)
	1.797(2)	1.666(2)	1.825(2)	1.669(2)	1.884(3)	175.13(9)
73·H$_2$O[c]	1.7864(14)	1.6598(14)	1.8290(13)	1.6601(13)	1.887(2)	177.66(6)
	1.7943(14)	1.6766(13)	1.8043(14)	1.6577(13)	1.894(2)	175.97(6)
74·2H$_2$O[d]	1.824(2)	1.658(2)	1.780(2)	1.657(2)	1.902(2)	179.17(8)
	1.814(2)	1.664(2)	1.799(2)	1.656(2)	1.906(2)	177.46(8)

[a] For the atomic numbering scheme, see Fig. 7. In all structures, the nitrogen atoms bound to C1 and C2 are pointing to O1 and O5, respectively.
[b] (Δ,Δ,R,R,R,R)-Configuration.
[c] Racemate; (Δ,Δ,R,R,R,R/Λ,Λ,S,S,S,S)-configuration.
[d] (Λ,Λ,S,S,S,S)-Configuration.

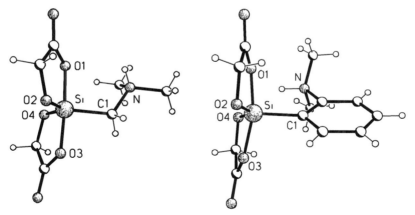

75

NMR studies (^{1}H, ^{13}C, ^{29}Si) demonstrated that the zwitterionic λ^5Si-silicates **53–60** and **63–70** and the biszwitterionic λ^5Si,λ^5Si'-disilicates **72–74**

FIG. 5.   Molecular structures of **53** (left) and **65** (right) in the crystal. For selected bond distances and angles, see Table VII.

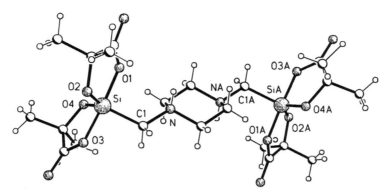

Fig. 6.   Molecular structure of **71** in the crystal of **71**·8H₂O. For selected bond distances and angle, see Table VII.

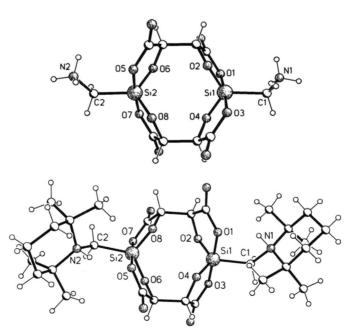

Fig. 7.   Molecular structures of **72** (solvate **72**·3H₂O, (Δ,Δ,R,R,R,R)-enantiomer; top) and **74** (solvate **74**·2H₂O, (Λ,Λ,S,S,S,S)-enantiomer; bottom) in the crystal. For selected bond distances and angles, see Table VIII.

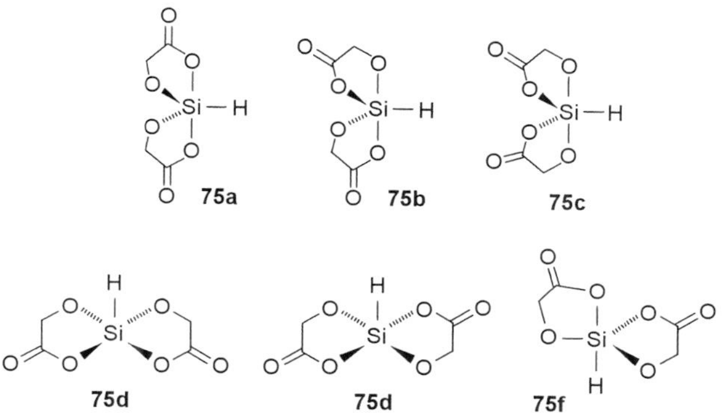

FIG. 8. Different geometries of the anionic model species **75** investigated by ab initio studies (local minima **75a–75c**; transition states **75d–75f**).

also exist in solution (**53–60, 63–70, 72, 73**: [D$_6$]DMSO; **74**: CD$_3$OD). The ^{29}Si chemical shifts are listed in Table IX (because of their poor solubility, no solution-state NMR data could be obtained for compounds **61**, **62** · CH$_3$CN, and **71** · 8H$_2$O). As can be seen from Table IX, the ^{29}Si chemical shifts determined in solution are very similar to the isotropic ^{29}Si chemical shifts obtained by solid-state ^{29}Si CP/MAS studies.

The solution-state NMR data of the zwitterions **53–60** and **63–69** are compatible with the presence of one particular species (such as the trigonal-bipyramidal isomer, with the carboxylate oxygen atoms in the axial positions) or with a rapid low-energy interconversion of different isomers (in this context, see ab initio studies of **75**). As the trigonal-bipyramidal structure, with the carboxylate oxygen atoms in the axial sites, is the energetically most favorable one, it is reasonable to assume that this particular species is also dominant in solution. Furthermore, there is experimental evidence that the respective (Λ)- and (Δ)-enantiomers of **53–60** and **63–69** are configurationally stable on the NMR time scale at room temperature. For example, diastereotopism was observed for the SiCH$_2$N protons (AB system) and the OCMe$_2$C protons (two separated singlets for the methyl groups) of compound **56**. As shown by VT ^{1}H NMR studies of **56** in [D$_6$]DMSO, CD$_3$CN, and CDCl$_3$, coalescence of the two methyl resonance signals occurs at higher temperatures. This behavior can be interpreted as a conversion of (Λ)-**56** and (Δ)-**56** into each other, the activation free enthalpies $\Delta G^{\ddagger}$ amounting to 76(1) kJ mol^{-1} ([D$_6$]DMSO), 75(1) kJ mol^{-1} (CD$_3$CN), and 82(2) kJ mol^{-1} (CDCl$_3$). As the coalescence temperature (76°C; [D$_6$]DMSO, c = 2.5–250 mmol L^{-1}) does not depend on the concentration, this (Λ)/(Δ)-isomerization is suggested to be an intramolecular process. This is in good accordance with the results obtained by ab initio

TABLE IX

^{29}Si Chemical Shifts for Compounds **53**, **54**·H$_2$O, **55**, **56**, **56**·H$_2$O, **57–60**, **63**·H$_2$O, **64–67**, **68**·H$_2$O, **69**·H$_2$O, **70**, **72**·3H$_2$O, **73**·H$_2$O, and **74**·2H$_2$O in Solution ([D$_6$]DMSO) and/or in the Crystal[a]

Compound	$\delta_{solution}$	$\delta_{crystal}$[b]	Compound	$\delta_{solution}$	$\delta_{crystal}$[b]
53	−90.3	−88.6	**64**	−99.3	−99.9
54·H$_2$O	−98.0	−96.4	**65**	−94.4	−91.8
55	−99.2	−98.1	**66**	−103.5	−103.2
56	−100.5	−100.1/−100.9	**67**	c	−102.6
56·H$_2$O	−100.5	−98.4	**68**·H$_2$O	−97.3	−95.3
57	−93.6	−89.5/−95.2	**69**·H$_2$O	−97.6	−100.7
58	−90.3	−90.5	**70**[d]	−94.6	e
59	−89.4	−88.4	**72**·3H$_2$O[f]	−91.7	e
60	−100.8	−99.3	**73**·H$_2$O[g]	−94.0	−92.2/−93.7
63·H$_2$O	−90.4	−87.6	**74**·2H$_2$O[h]	−92.9[i]	−93.0/−93.5

[a] Chemical shifts in ppm; spectra recorded at room temperature.

[b] Isotropic chemical shifts obtained by ^{29}Si CP/MAS NMR experiments.

[c] Experiment failed.

[d] (Λ,R)-Configuration.

[e] No experiment performed.

[f] (Δ,Δ,R,R,R)-Configuration.

[g] Racemate; (Δ,Δ,R,R,R/Λ,Λ,S,S,S)-configuration.

[h] (Λ,Λ,S,S,S)-Configuration.

[i] Solvent CD$_3$OD.

studies of the model species **75**. The experimentally established activation free enthalpies for **56** are comparable with the respective energy barrier calculated for **75** (70.8 kJ mol^{-1}): The *Si*-coordination polyhedron of the respective transition state **75f** (C_1 symmetry) is a distorted trigonal bipyramid, with one of the carboxylate oxygen atoms and the hydrogen atom in the axial sites ("twist mechanism").[44]

As shown for **65** and **66** in Scheme 16, compounds of this particular formula type can also undergo an intermolecular ligand exchange in solution ([D$_6$]DMSO, room temperature; detection of the thermodynamic equilibrium **65** + **66** $\rightleftarrows$ 2 **76**).[47] However, this ligand exchange is significantly slower than the intramolecular (Λ)/(Δ)-isomerization mentioned earlier.

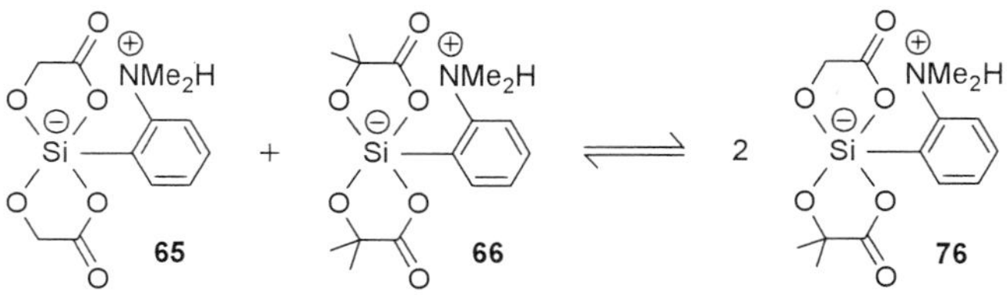

Scheme 16.

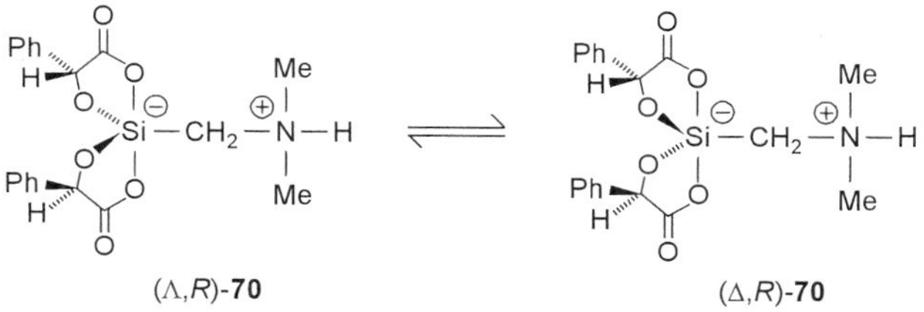

SCHEME 17.

After dissolution of **70** [(Λ,*R*)-enantiomer] in [D$_6$]DMSO at room temperature, two sets of resonance signals were observed in the ^{1}H and ^{13}C NMR spectra, indicating the presence of the two diastereomers (Λ,*R*)-**70** and (Δ,*R*)-**70** [thermodynamic equilibrium (Λ,*R*)-**70** ⇌ (Δ,*R*)-**70**; see Scheme 17]. In the ^{29}Si NMR spectrum, the expected two resonance signals are not resolved.

Interestingly, after dissolution of the optically active compounds **72**·3H$_2$O [(Δ,Δ,*R*,*R*,*R*,*R*)-enantiomer] and **74**·2H$_2$O [(Λ,Λ,*S*,*S*,*S*,*S*)-enantiomer] in [D$_6$]DMSO and CD$_3$OD, respectively, only one set of resonance signals was observed in the ^{1}H, ^{13}C, and ^{29}Si NMR spectra. This indicates the presence of only one particular species in solution. It is likely that the enantiomers (Δ,Δ,*R*,*R*,*R*,*R*)-**72** and (Λ,Λ,*S*,*S*,*S*,*S*)-**74** are configurationally stable. Obviously, the geometric requirements of the tetradentate tartrato(4−) ligands do not allow changes of the absolute configuration of the two chiral λ^5*Si*-silicate skeletons in these zwitterions.

E. Compounds Containing Two Bidentate Ligands of the Salicylato(2−) Type

The spirocyclic zwitterionic λ^5*Si*-silicates **77**,[52] **78**,[53] and **79**[53] contain two bidentate salicylato(2−) ligands. The chiral zwitterions are built up by two six-membered SiO$_2$C$_3$ ring systems that are connected by the silicon spirocenter, whereas all the other spirocyclic λ^5*Si*-silicates described in Section III contain two five-membered SiO$_2$C$_2$ (III,A to III,D) or SiO$_2$CN rings (III,F).

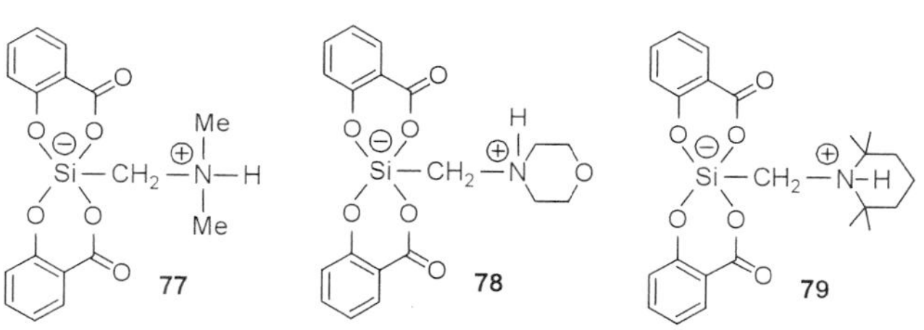

As illustrated for compounds **77** and **78** in Scheme 18, different methods were applied for the syntheses of **77**–**79** (**79** was obtained analogously to **78** according to method a). The racemic products **77a** · 0.7CH₃CN, **78** · CH₃CN, and **79** were isolated as crystalline solids. In addition, crystals of the racemic compound **77b** (an isomer of **77a**) were obtained. For the solvent-free compound **78** formation of enantiomorphic crystals was observed. The crystals studied by X-ray diffraction contained (just by accident) the (Δ)-enantiomer.

Compounds **77a** · 0.7CH₃CN, **77b**, **78** · CH₃CN, **78** [(Δ)-enantiomer], and **79** were structurally characterized by single-crystal X-ray diffraction. In all cases, the *Si*-coordination polyhedron was found to be a somewhat distorted trigonal bipyramid. This is shown for **77a** · 0.7CH₃CN and **77b** in Fig. 9. Except for **77b**, both axial positions of the compounds studied are occupied by the carboxylate oxygen atoms. Selected geometric parameters for **77a** · 0.7CH₃CN, **78** · CH₃CN, (Δ)-**78**, and **79** are listed in Table X. As can be seen from these data, the axial Si–O distances are significantly longer [1.758(2)–1.797(2) Å] than the equatorial ones [1.662(2)–1.6741(14) Å]. The Si–C distances amount to 1.887(3)–1.9058(14) Å. Similar Si–O and Si–C distances were observed for related compounds containing diolato(2−) ligands that derive from α-hydroxycarboxylic acids (in this context, see Tables VII and VIII in Section III,D).

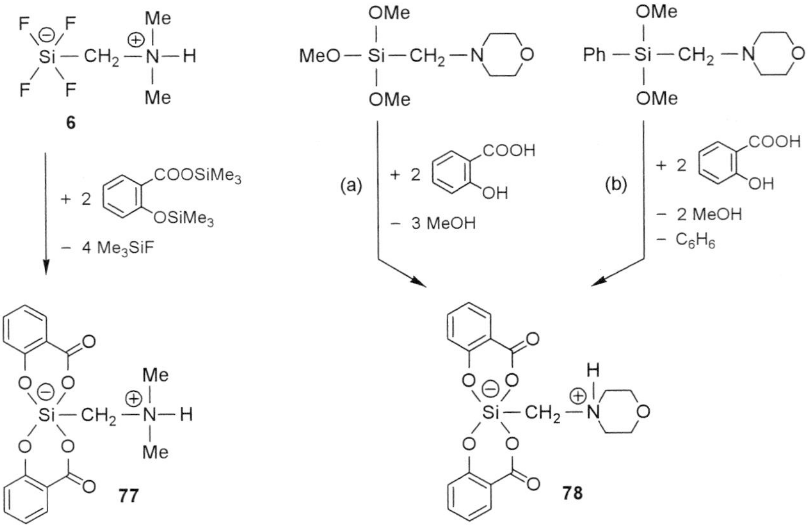

SCHEME 18.

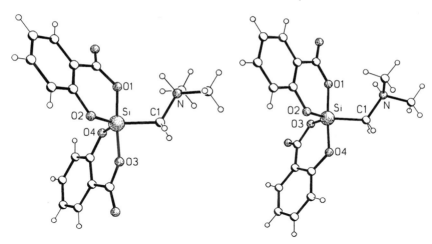

FIG. 9. Molecular structures of **77a** (solvate **77a** · 0.7CH$_3$CN, molecule I; left) and **77b** (right) in the crystal. For selected bond distances and angle for **77a** · 0.7CH$_3$CN, see Table X. Selected bond distances (Å) and angles (°) for **77b**: Si–O1, 1.782(2); Si–O2, 1.662(2); Si–O3, 1.697(2); Si–O4, 1.738(2); Si–C1, 1.880(2); O1–Si–O4, 175.41(9).

A quite different structure was observed for compound **77b**: The two carboxylate oxygen atoms occupy one axial and one equatorial position, leading to a significant differentiation between the two axial [1.738(2) Å, 1.782(2) Å] and between the two equatorial Si–O distances [1.662(2) Å, 1.697(2) Å] (in this context, see also legend for Fig. 9).

TABLE X

SELECTED BOND DISTANCES (Å) AND ANGLES (°) FOR COMPOUNDS **77a** · 0.7CH$_3$CN, **78** · CH$_3$CN, **78**, AND **79** AS DETERMINED BY SINGLE-CRYSTAL X-RAY DIFFRACTION[a,b]

Compound	Si–O1	Si–O2	Si–O3	Si–O4	Si–C1	O1–Si–O3
77a · 0.7CH$_3$CN (M I)	1.760(2)	1.670(2)	1.781(2)	1.668(2)	1.903(3)	175.21(11)
77a · 0.7CH$_3$CN (M II)	1.769(2)	1.666(2)	1.797(2)	1.662(2)	1.887(3)	174.63(9)
78 · CH$_3$CN	1.7830(9)	1.6732(10)	1.7709(9)	1.670(1)	1.9058(14)	174.46(5)
78[c]	1.762(4)	1.671(4)	1.791(4)	1.666(4)	1.903(5)	172.2(2)
79	1.786(2)	1.6741(14)	1.758(2)	1.664(2)	1.893(2)	174.25(8)

[a] For the atomic numbering scheme, see **77a** in Fig. 9. In all structures, the nitrogen atom bound to C1 is pointing to O1.

[b] For compound **77a** · 0.7CH$_3$CN, two crystallographically independent molecules (M I, M II) were observed.

[c] (Δ)-Configuration.

The existence of the two isomeric zwitterions **77a** and **77b** in the crystal-line state is in accordance with the results of ab initio studies with the anionic model species **80**.[52] The energy difference between the respective trigonal-bipyramidal alternatives (**a**: both carboxylate oxygen atoms in the axial sites; **b**: one carboxylate oxygen atom in an axial position, the other one in an equatorial site) amounts only to 4.0 kJ mol^{-1}.

80

Attempts to characterize compounds **77a** · 0.7CH$_3$CN, **78** · CH$_3$CN, **78**, and **79** by solution-state NMR experiments (^{1}H, ^{13}C, ^{29}Si) in [D$_6$]DMSO gave unsatisfactory results; the NMR data obtained were not in accordance with the existence of the respective zwitterions as the only species in solu-tion. However, NMR studies of **79** in CDCl$_3$ unequivocally demonstrated that this particular zwitterion exists in solution (because of their poor solubility, **77a** and **78** could not be studied in CDCl$_3$).

The isotropic ^{29}Si chemical shifts obtained for **77a** · 0.7CH$_3$CN, **78** · CH$_3$CN, **78**, and **79** by solid-state ^{29}Si CP/MAS NMR studies are as follows: **77a** (after removal of CH$_3$CN in vacuo), δ −120.0; **78** · CH$_3$CN, δ −123.1; **78**, δ −120.6; **79**, δ −120.4. The ^{29}Si chemical shift observed for **79** in the crystal is very similar to that observed in solution (CDCl$_3$, δ −121.7). Interestingly, the ^{29}Si NMR data of the zwitterions **77a**, **78**, and **79** deviate significantly from those determined for related zwitterionic λ^5Si-silicates containing diolato(2−) ligands that derive from α-hydroxycarboxylic acids (in this context, see Table IX in Section III,D).

F. *Compounds Containing Two Bidentate Ligands of the Acetohydroximato(2−) or Benzohydroximato(2−) Type*

The pentacoordinate silicon compounds **81**,[8,54] **82**,[54] **83**,[54] and **84**[55] are spirocyclic zwitterionic λ^5Si-silicates with an SiO_4C skeleton. The chiral zwitterions contain two diolato(2−) ligands that formally derive from aceto-hydroximic acid and benzohydroximic acid (tautomers of acetohydroxamic acid and benzohydroxamic acid).

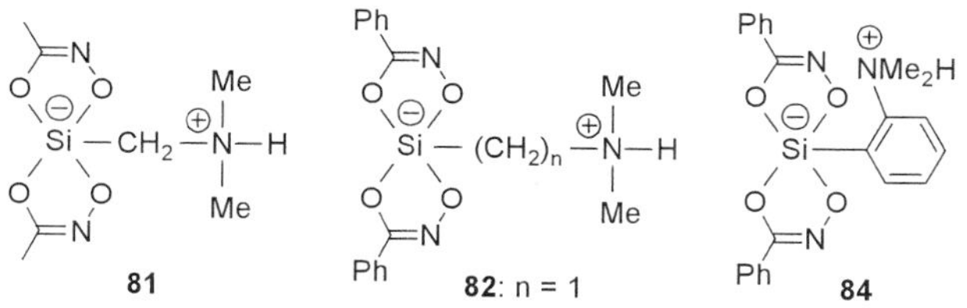

As illustrated for the synthesis of **81** in Scheme 19, compounds **81–84** were prepared by different methods (**82** was obtained according to methods a and b; **83** and **84** were synthesized by method a). All syntheses were carried out in acetonitrile at room temperature and the products isolated in good yields as crystalline solids. Method b involves a remarkable Si–C cleavage reaction.

The *Si*-coordination polyhedra of the chiral zwitterions **81–84** in the crystal were found to be distorted trigonal bipyramids, with the two carbon-linked oxygen atoms in the axial positions. This is illustrated for **81** in Fig. 10. The crystals of **81–84** are built up by pairs of (Λ)- and (Δ)-enantiomers. Selected geometric parameters for these compounds are listed in Table XI. As can be seen from these data, the axial Si–O distances [1.749(2)–1.801(4) Å] are significantly longer than the equatorial ones [1.683(2)–1.7182(14) Å]. The Si–C distances are 1.866(6)–1.897(2) Å.

NMR studies ([D$_6$]DMSO; ^{1}H, ^{13}C, ^{29}Si) have demonstrated that the

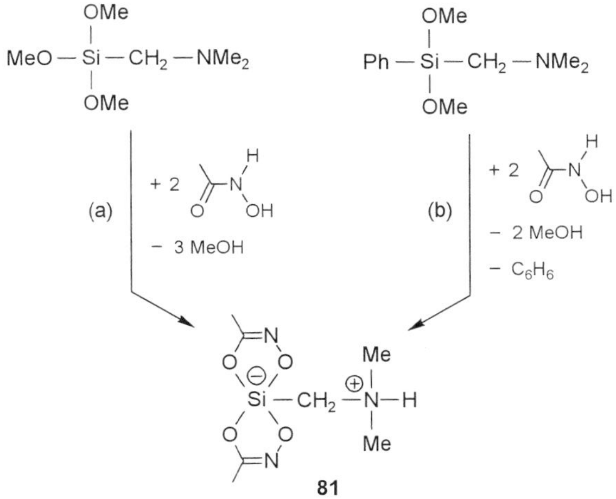

SCHEME 19.

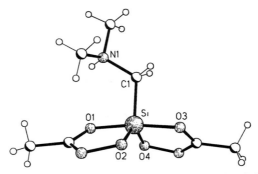

Fig. 10. Molecular structure of **81** in the crystal. For selected bond distances and angle, see Table XI.

zwitterions **81–84** also exist in solution. The presence of an AB spin system for the SiCH$_2$N moiety in the ^{1}H NMR spectrum of **81** at room temperature indicates that the (Λ)- and (Δ)-enantiomers are configurationally stable at the NMR time scale. As shown in Table XII, the ^{29}Si chemical shifts of **81–84** are very similar to the isotropic ^{29}Si chemical shifts determined for the crystalline phases by ^{29}Si CP/MAS NMR experiments.

IV

SPIROCYCLIC $\lambda^5 Si$-SILICATES WITH AN SiO_2N_2C SKELETON

The spirocyclic pentacoordinate silicon compounds **85**,[56] **86**,[39] and **87**[39] are zwitterionic $\lambda^5 Si$-silicates with an SiO_2N_2C skeleton. The chiral zwitterions

TABLE XI

SELECTED BOND DISTANCES (Å) AND ANGLES (°) FOR COMPOUNDS **81–84** AS DETERMINED BY SINGLE-CRYSTAL X-RAY DIFFRACTION[a,b]

Compound	Si–O1	Si–O2	Si–O3	Si–O4	Si–C1	O1–Si–O3
81[c]	1.770(2)	1.683(2)	1.766(2)	1.691(2)	1.883(2)	172.17(7)
82	1.7685(14)	1.6867(14)	1.7661(14)	1.7046(14)	1.897(2)	173.16(7)
83 (M I)	1.764(4)	1.698(4)	1.790(4)	1.694(4)	1.880(5)	171.0(2)
83 (M II)	1.760(4)	1.693(4)	1.801(4)	1.704(4)	1.866(6)	171.4(2)
84	1.7688(14)	1.7182(14)	1.749(2)	1.6859(14)	1.883(2)	161.68(7)

[a] For the atomic numbering scheme, see Fig. 10. In the structures of **81–83**, the non-hydrogen atom bound to C1 (**81, 82**: N atom; **83**: C atom) is pointing to O1. In the structure of **84**, the N-linked carbon atom of the phenyl group is pointing to O1.

[b] For compound **83**, two crystallographically independent molecules (M I, M II) were observed.

[c] The atomic numbering scheme for **81** given in Ref. 8 is not correct.

TABLE XII

^{29}Si Chemical Shifts for Compounds **81–84** in Solution ($[D_6]$DMSO) and/or in the Crystal[a]

Compound	$\delta_{solution}$	$\delta_{crystal}$[b]	Compound	$\delta_{solution}$	$\delta_{crystal}$[b]
81	−84.5	−83.9	**83**	−76.2	−74.1
82	−83.1	−83.5	**84**	−83.8	c

[a] Chemical shifts in ppm; spectra recorded at room temperature.
[b] Isotropic chemical shifts obtained by ^{29}Si CP/MAS NMR experiments.
[c] No experiment performed.

contain two 4-nitrobenzene-2-aminato-1-olato(2−) ligands, i.e., two five-membered SiONC$_2$ ring systems are connected by the silicon spirocenter. Compounds **85–87** are isoelectronic analogues of the zwitterionic $\lambda^5 Si$-silicates **41–43** (twofold O/NH replacement; see Section III,A).

As illustrated for compounds **85** and **86** in Scheme 20, different methods were used for the syntheses of **85–87** (**87** was prepared analogously to **86**). Compound **85** was synthesized at 40°C in acetonitrile under reduced pressure (removal of part of the solvent along with the dimethylamine formed), whereas **86** and **87** were prepared in acetonitrile at room temperature. All products were isolated in good yields as crystalline solids.

As demonstrated by single-crystal X-ray diffraction, the Si-coordination polyhedra of **85–87** are distorted trigonal bipyramids, with each of the axial positions occupied by the oxygen atoms. This is shown for compound **86** in Fig. 11. In all cases, the crystals are formed from pairs of (Λ)- and (Δ)-enantiomers. Selected geometric parameters for **85–87** are listed in Table XIII. As can be seen from the Si–O [1.8004(10)–1.829(6) Å], Si–N [1.741(7)–1.764(6) Å], and Si–C distances [1.867(8)–1.915(2) Å], the SiO_2N_2C frameworks of **85–87** are built up by five "normal" covalent bonds and do not involve a bonding system in the sense of the 4+1 coordination usually observed for pentacoordinate silicon species with Si–N bonds.

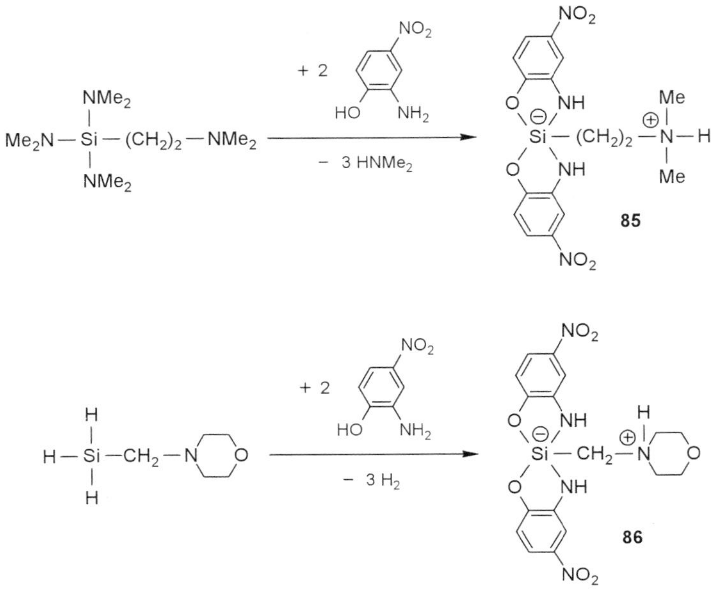

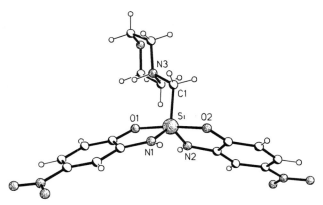

SCHEME 20.

NMR studies ([D₆]DMSO; ¹H, ¹³C, ²⁹Si) of **85** and **87** demonstrated that these zwitterions also exist in solution. Interestingly, only one set of resonance signals was found for these compounds at room temperature, whereas two sets of resonance signals were observed in the NMR spectra of the isoelectronic analogues **41** and **43**, indicating the existence of two

FIG. 11.   Molecular structure of **86** in the crystal. For selected bond distances and angle, see Table XIII.

TABLE XIII

SELECTED BOND DISTANCES (Å) AND ANGLES (°) FOR COMPOUNDS **85–87** AS DETERMINED
BY SINGLE-CRYSTAL X-RAY DIFFRACTION[a,b]

Compound	Si–O1	Si–O2	Si–N1	Si–N2	Si–C1	O1–Si–O2
85 (M I)	1.829(6)	1.806(6)	1.741(7)	1.751(7)	1.875(9)	171.8(3)
85 (M II)	1.821(6)	1.809(6)	1.753(7)	1.764(6)	1.867(8)	167.3(3)
86	1.8090(13)	1.8019(13)	1.7446(14)	1.754(2)	1.915(2)	174.39(6)
87	1.8004(10)	1.8004(10)	1.7436(13)	1.7476(11)	1.9086(14)	177.01(5)

[a] For the atomic numbering scheme, see Fig. 11. In all structures, the non-hydrogen atom bound to C1 (**85**: C atom; **86, 87**: N atom) is pointing to O1.

[b] For compound **85**, two cystallographically independent molecules (M I, M II) were observed.

isomeric species of the latter two compounds in solution (in this context, see Section III,A).

The ^{29}Si chemical shifts of **85–87** are as follows: **85**, δ −78.6 ([D$_6$]DMSO), δ −77.2 (solid state); **86**, δ −85.8 ([D$_6$]DMSO), δ −86.7 (solid state); **87**, δ −85.5 ([D$_6$]DMSO), δ −88.0 (solid state). As would be expected from the different electronegativities of nitrogen and oxygen, the ^{29}Si chemical shifts of **85** and **87** are somewhat different from those determined for their isoelectronic analogues: **41**, δ −75.04/75.13 ([D$_6$]DMSO), δ −75.1 (solid state); **43**, δ −81.9 ([D$_6$]DMSO), δ −80.6 (solid state).[39]

V

SPIROCYCLIC $\lambda^5 Si$-SILICATES WITH AN SiO_2S_2C SKELETON

The spirocyclic pentacoordinate silicon compounds **88**[57] and **89**[57] are zwitterionic $\lambda^5 Si$-silicates with an SiO_2S_2C skeleton. The chiral zwitterions contain two thioglycolato(2−) or two thiosalicylato(2−) ligands; i.e., two five-membered SiOSC$_2$ rings or two six-membered SiOSC$_3$ rings are connected by the silicon spirocenter. Compounds **88** and **89** represent thio analogs of the zwitterionic $\lambda^5 Si$-silicates **53** and **77a** (replacement of both alcoholate oxygen atoms by sulfur atoms; in this context, see Sections III,D and III,E).

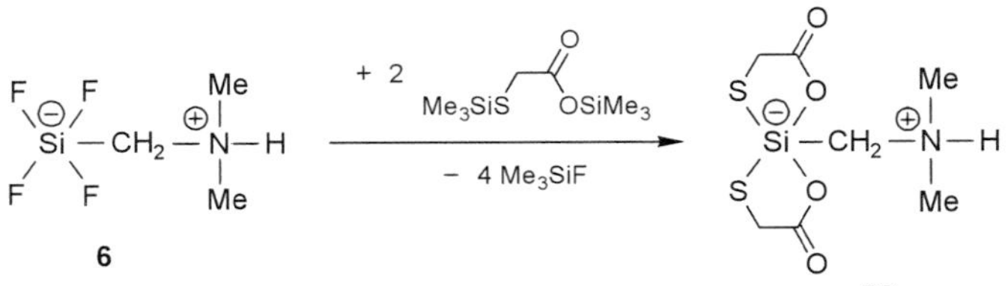

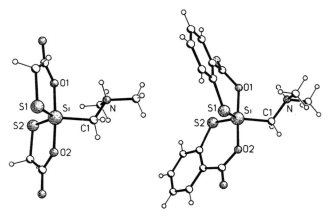

Scheme 21.

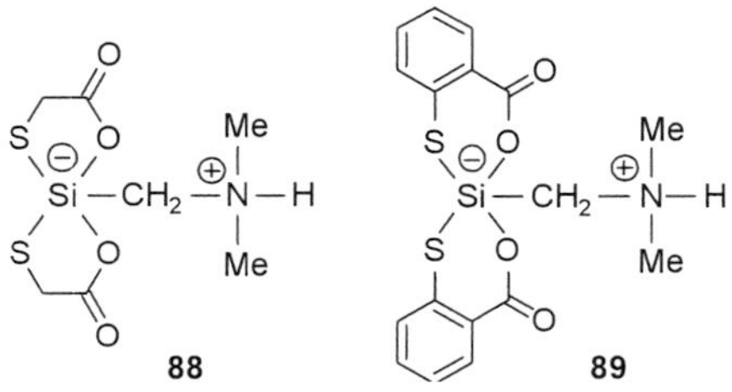

Compounds **88** and **89** were prepared by the method illustrated for the synthesis of **88** in Scheme 21. The reactions were carried out in acetonitrile at room temperature and the products isolated in good yields as crystalline solids.

Crystal structure analyses demonstrated that the *Si*-coordination polyhedra of **88** and **89** are distorted trigonal bipyramids, with each of the axial sites occupied by the carboxylate oxygen atoms. This is illustrated in Fig. 12.

Fig. 12.   Molecular structures of **88** (left) and **89** (right) in the crystal. For selected bond distances and angles, see Table XIV.

TABLE XIV

SELECTED BOND DISTANCES (Å) AND ANGLES (°) FOR COMPOUNDS **88** AND **89** AS DETERMINED BY SINGLE-CRYSTAL X-RAY DIFFRACTION[a,b]

Compound	Si–O1	Si–O2	Si–S1	Si–S2	Si–C1	O1–Si–O2
88 (M I)	1.809(4)	1.791(4)	2.157(2)	2.150(2)	1.902(5)	175.9(2)
88 (M II)	1.815(4)	1.762(4)	2.158(2)	2.144(2)	1.900(5)	176.6(2)
89	1.7657(14)	1.7600(14)	2.1590(8)	2.1638(9)	1.906(2)	177.15(8)

[a] For the atomic numbering scheme, see Fig. 12. In all structures, the nitrogen atom bound to C1 is pointing to O1.

[b] For compound **88**, two crystallographically independent molecules (M I, M II) were observed.

The crystals of both compounds contain pairs of (Λ)- and (Δ)-enantiomers. Selected geometric parameters for **88** and **89** are listed in Table XIV. Similar crystal structures were observed for the corresponding oxygen analogs **53** and **77**, respectively (see Sections III,D and III, E). As can be seen from the Si–O [1.7600(14)–1.815(4) Å], Si–S [2.144(2)–2.1638(9) Å], and Si–C distances [1.900(5)–1.906(2) Å], the SiO_2S_2C frameworks of **88** and **89** are best described as being built up by five "normal" covalent bonds rather than a bonding system in the sense of a 4+1 coordination. The Si–S distances are very similar to those observed for tetracoordinate silicon compounds with Si–S bonds.

The isotropic ^{29}Si chemical shifts of **88** and **89** in the solid state are as follows: **88**, δ −69.8; **89**, δ −98.8. As would be expected from the different electronegativities of sulfur and oxygen, these chemical shifts differ significantly from those determined for the corresponding oxygen analogs **53** (δ −88.6) and **77a** (δ −120.0).

VI

SPIROCYCLIC $\lambda^5 Si$-SILICATES WITH AN SiO_5 SKELETON

The spirocyclic pentacoordinate silicon compounds **90**,[45,58] **91**,[45,58] and **92**[59] are zwitterionic $\lambda^5 Si$-silicates with an SiO_5 skeleton. The chiral zwitterions contain two 2-methyllactato(2−) or two benzilato(2−) ligands. In these molecules, two five-membered SiO_2C_2 rings are connected by the silicon spirocenter. Compounds **90** and **91** represent isoelectronic analogs of the zwitterionic $\lambda^5 Si$-silicates **58** and **59** (O/CH_2 replacement; see Section III,D).

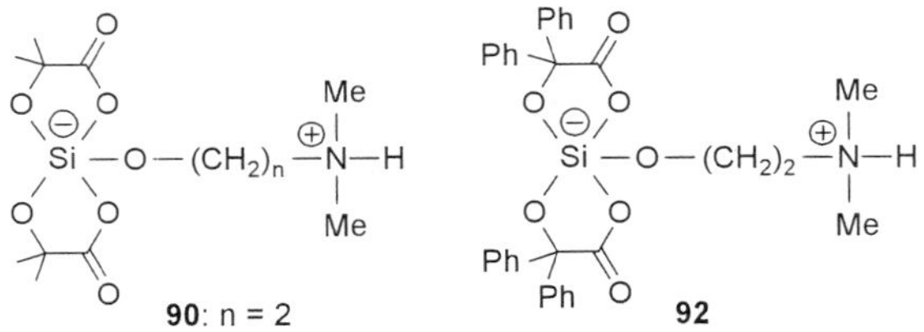

90: n = 2
91: n = 3

92

As illustrated for **90** in Scheme 22, different methods were used for the syntheses of compounds **90–92** (**91** was prepared according to methods a and b; **92** was obtained according to method a). The syntheses of **90–92** were carried out in boiling acetonitrile (in the case of **90** and **91**, part of the solvent along with the methanol was removed by distillation). Compound **92** was recrystallized from DMF to give the solvate **92 · DMF**. Compounds **90–92** and **92 · DMF** were isolated as crystalline solids.

Single-crystal X-ray diffraction studies have shown that the *Si*-coordination polyhedra of **90**, **91**, and **92 · DMF** are distorted trigonal bipyramids, with each of the axial positions occupied by the carboxylate oxygen atoms. This is shown for compound **90** in Fig. 13. All crystals contain pairs of (Λ)- and (Δ)-enantiomers. Selected geometric parameters for **90**, **91**, and **92 · DMF** are listed in Table XV. As can be seen from these data, the axial Si–O distances [1.773(2)–1.810(2) Å] are significantly longer than the equatorial ones [1.633(2)–1.673(2) Å]. Similar *Si*-coordination polyhedra

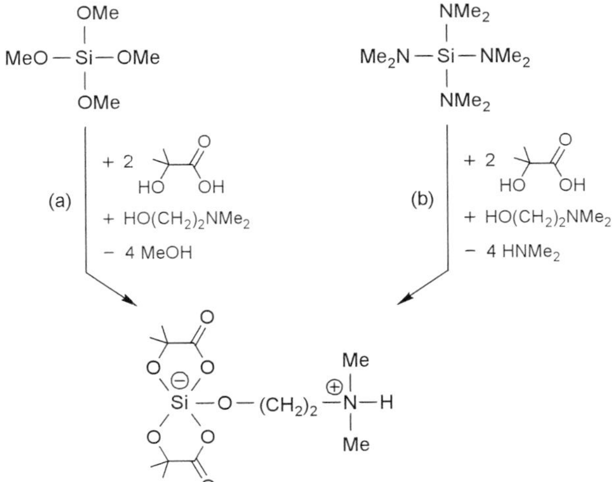

90

SCHEME 22.

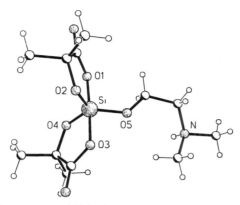

FIG. 13. Molecular structure of **90** in the crystal. For selected bond distances and angle, see Table XV.

were observed for compounds **58** and **59**, the isoelectronic derivatives of **90** and **91** (see Section III,D).

NMR studies ([D$_6$]DMSO; ^{1}H, ^{13}C, ^{29}Si) of **90–92** demonstrated that these zwitterions also exist in solution. The ^{29}Si chemical shifts are as follows: **90**, δ −111.6; **91**, δ −114.8; **92**, δ −113.9. These data are very similar to the isotropic ^{29}Si chemical shifts determined by solid-state ^{29}Si CP/MAS NMR experiments: **90**, δ −115.4; **91**, δ −115.2/116.2; **92**, δ −112.7; **92** · DMF, δ −114.2. As would be expected from the different electronegativities of oxygen and carbon, the ^{29}Si chemical shifts of **90** and **91** differ significantly from those of their isoelectronic methylene

TABLE XV

SELECTED BOND DISTANCES (Å) AND ANGLES (°) FOR COMPOUNDS **90**, **91**, AND **92** · DMF
AS DETERMINED BY SINGLE-CRYSTAL X-RAY DIFFRACTION[a,b]

Compound	Si–O1	Si–O2	Si–O3	Si–O4	Si–O5	O1–Si–O3
90	1.773(2)	1.653(2)	1.798(2)	1.659(2)	1.643(2)	176.83(8)
91 (M I)	1.785(2)	1.660(2)	1.804(2)	1.665(2)	1.633(2)	172.68(12)
91 (M II)	1.810(2)	1.668(2)	1.785(2)	1.662(2)	1.634(2)	172.02(12)
92	1.793(3)	1.673(2)	1.795(3)	1.663(2)	1.652(3)	174.96(8)

[a] For the atomic numbering scheme, see Fig. 13. In all structures, the C atom bound to O5 is pointing to O1.

[b] For compound **91**, two crystallographically independent molecules (M I, M II) were observed.

analogs **58** ([D$_6$]DMSO, δ -90.3; solid state, δ -90.5) and **59** ([D$_6$]DMSO, δ -89.4; solid state, δ -88.4).

The experimentally established geometries of the *Si*-coordination polyhe-dra of **90, 91,** and **92 · DMF** in the crystal are in good accordance with the results of ab initio studies of the related anionic model species **93** (Fig. 14).[45] SCF/SVP geometry optimizations have shown that the isomer **93a** (*C$_1$* symmetry; local minimum) is significantly more stable than the isomers **93b** (*C$_1$* symmetry; local minimum) and **93c** (*C$_1$* symmetry; local minimum). The energy differences amount to 20.4 kJ mol^{-1} (**93a/93b**) and 40.8 kJ mol^{-1} (**93a/93c**). The square-pyramidal species **93d** (*C$_1$* symmetry; transition state) and **93e** (*C$_s$* symmetry; transition state) are also significantly less stable than **93a**, the energy differences amounting to 35.1 kJ mol^{-1} (**93a/ 93d**) and 36.5 kJ mol^{-1} (**93a/93e**).

93

In the ^{1}H NMR spectra of **90** and **91** (solvent [D$_6$]DMSO), two separated singlets for the diastereotopic methyl groups of the 2-methyllac-tato(2$-$) ligands were observed at room temperature, indicating that the (Λ)- and (Δ)-enantiomers of these zwitterionic λ^5Si-silicates are configurationally stable on the NMR time scale under the experimental conditions used. Upon heating (VT ^{1}H NMR studies in the temperature range of 27°C to ca. 120°C), coalescence of the two singlets was observed, this temperature dependence being completely reversible on subsequent cooling. These results can be interpreted in terms of an intramolecular isomerization process that leads to a conversion of the respective (Λ)- and (Δ)-enantiomers into each other, the respective activation free enthalpies amounting to 79.7(4) kJ mol^{-1} (**90**) and 84.6(6) kJ mol^{-1} (**91**). The $\Delta G^{\ddagger}$ values for the isoelectronic analogs **58** [88.5(7) kJ mol^{-1}] and **59** [89(2) kJ mol^{-1}] were found to be somewhat higher. The experimentally established activation free enthalpies for **90** and **91** are comparable with the respective energy barrier calculated for the anionic model species **93** (85.3 kJ mol^{-1}): The *Si*-coordination polyhedron of the respective transition state **93f** (*C$_1$* symmetry) is a distorted trigonal bipyramid, with one of the carboxylate oxygen atoms and the OH group in the axial sites ("twist mechanism").[45]

VII

MONOCYCLIC λ^5Si-SILICATES WITH AN SiO_2FC_2 SKELETON

The pentacoordinate silicon compounds **94**,[23] **95**,[23] **96–98**,[60] **99**,[60,61] **100**,[61,62] **101–103**,[60] **104**,[61,62] **105**,[62] **106**,[62] and **107**[61,63] are monocyclic zwitterionic λ^5Si-silicates with an SiO_2FC_2 skeleton. The chiral zwitterions each contain one bidentate diolato(2−) ligand that formally derives from 1,2-dihydroxybenzene, salicylic acid, glycolic acid, oxalic acid, benzohydroximic acid (tautomer of benzohydroxamic acid), 2-methyllactic acid, or (S)-mandelic acid.

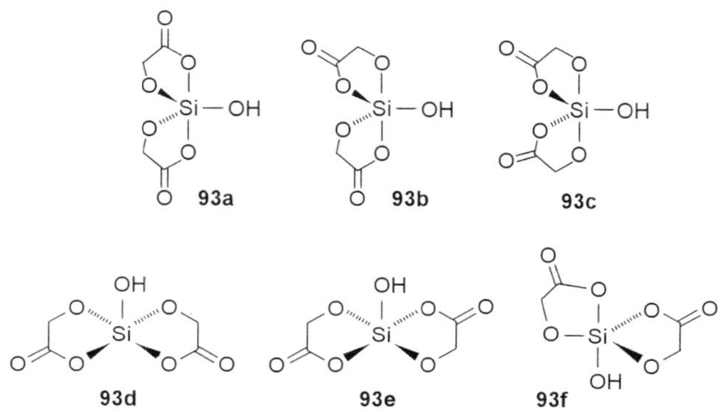

Fig. 14. Different geometries of the anionic model species **93** investigated by ab initio studies (local minima **93a–93c**; transition states **93d–93f**).

Compounds **94–107** were prepared by the method illustrated for the synthesis of **96** in Scheme 23. As shown by crystal structure analyses, the *Si*-coordination polyhedra of **94**, **95** · CH$_3$CN, **96–105**, **106** · $\frac{1}{2}$CH$_3$CN, and **107** are distorted trigonal bipyramids, with the axial sites occupied by a fluorine atom and an oxygen atom (carboxylate oxygen atom in the case of **95**, **96**, **99**, **100**, and **103–107**; carbon-linked oxygen atom in the case of **98** and **102**). This is shown for compounds **94**, **95** · CH$_3$CN, and **96–98** in Fig. 15. The crystals of **94**, **95** · CH$_3$CN, and **96–104** are built up by pairs of (*A*)- and (*C*)-enantiomers, whereas the crystals of the diastereomerically and enantiomerically pure compounds **105**, **106** · $\frac{1}{2}$CH$_3$CN, and **107** contain a single stereoisomer [**105**, (*A,S*)-configuration; **106** · $\frac{1}{2}$CH$_3$CN, (*C,S*)-configuration; **107**, (*A,S,S*)-configuration]. Selected geometric parameters for compounds **94**, **95** · CH$_3$CN, **96–105**, **106** · $\frac{1}{2}$CH$_3$CN, and **107** are listed in Table XVI.

Solution-state NMR studies (^{1}H, ^{13}C, ^{19}F, ^{29}Si) at room temperature have shown that the zwitterions **94–104** are also present in solution. The ^{29}Si chemical shifts were found to be very similar to the isotropic ^{29}Si chemical

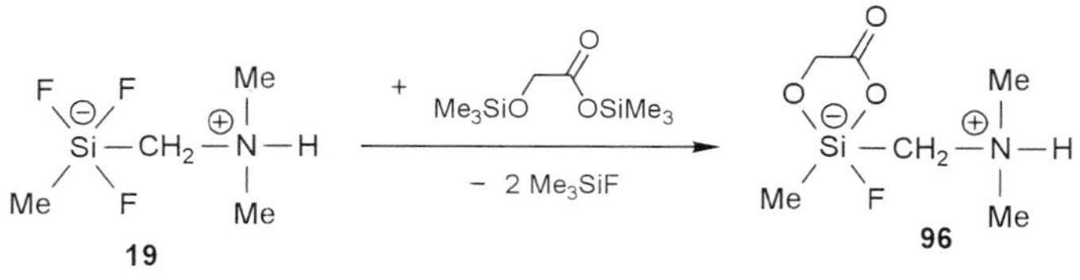

Scheme 23.

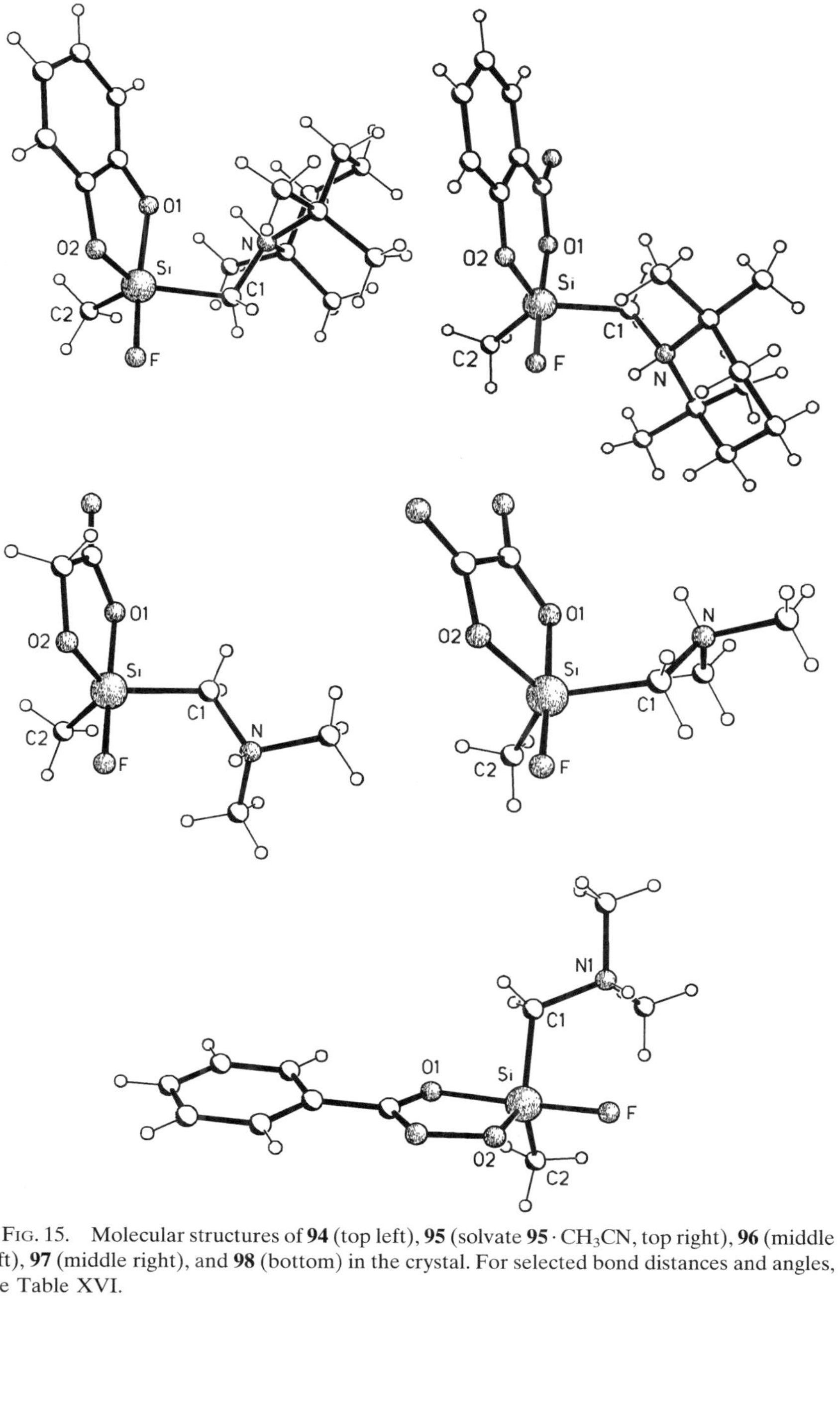

FIG. 15. Molecular structures of **94** (top left), **95** (solvate **95** · CH_3CN, top right), **96** (middle left), **97** (middle right), and **98** (bottom) in the crystal. For selected bond distances and angles, see Table XVI.

TABLE XVI

Selected Bond Distances (Å) and Angles (°) for Compounds **94**, **95** · CH_3CN, **96–105**, **106** · $\frac{1}{2}CH_3CN$, and **107** as Determined by Single-Crystal X-Ray Diffraction[a,b]

Compound	Si–F	Si–O1	Si–O2	Si–C1	Si–C2	F–Si–O1
94	1.673(2)	1.874(2)	1.710(2)	1.914(2)	1.863(3)	170.35(8)
95 · CH_3CN	1.7094(8)	1.8234(10)	1.6859(10)	1.8992(13)	1.863(2)	176.86(5)
96	1.7053(11)	1.8679(13)	1.6827(13)	1.904(2)	1.864(2)	173.68(6)
97	1.6661(14)	1.918(2)	1.7392(14)	1.906(2)	1.854(2)	172.43(5)
98	1.6844(12)	1.8478(14)	1.703(2)	1.898(2)	1.862(2)	170.29(7)
99	1.6885(11)	1.8734(13)	1.6730(14)	1.904(2)	1.873(2)	175.81(6)
100	1.7023(7)	1.8605(8)	1.6728(8)	1.9221(10)	1.9188(10)	172.51(3)
101	1.6824(8)	1.8710(9)	1.728(1)	1.8915(11)	1.8611(13)	172.62(4)
102	1.6712(12)	1.8783(14)	1.706(2)	1.913(2)	1.8658(2)	170.94(7)
103	1.7016(9)	1.8647(11)	1.6680(11)	1.911(2)	1.844(2)	173.59(4)
104	1.7045(7)	1.8595(9)	1.6861(9)	1.9433(12)	1.9295(13)	171.41(4)
105 (M I)[c]	1.681(3)	1.868(4)	1.666(4)	1.899(6)	1.841(7)	174.5(2)
105 (M II)[c]	1.677(4)	1.867(4)	1.666(4)	1.903(6)	1.844(7)	174.8(2)
105 (M III)[c]	1.685(3)	1.872(4)	1.661(4)	1.914(6)	1.872(6)	174.9(2)
106 · $\frac{1}{2}CH_3CN$ (M I)[d]	1.6673(10)	1.8876(12)	1.6832(12)	1.924(2)	1.931(2)	171.17(5)
106 · $\frac{1}{2}CH_3CN$ (M II)[d]	1.7047(11)	1.8725(11)	1.6695(12)	1.924(2)	1.921(2)	170.56(5)
107 (M I)[e]	1.702(3)	1.855(3)	1.676(4)	1.931(5)	1.866(6)	174.4(2)
107 (M II)[e]	1.715(3)	1.849(4)	1.668(4)	1.920(5)	1.870(6)	174.4(2)

[a] For the atomic numbering scheme, see Fig. 15.

[b] For compound **105**, three crystallographically independent molecules (M I, M II, M III) were observed; for compounds **106** · $\frac{1}{2}CH_3CN$ and **107**, two crystallographically independent molecules (M I, M II) were observed.

[c] (*A,S*)-Configuration.

[d] (*C,S*)-Configuration.

[e] (*A,S,S*)-Configuration.

shifts determined by solid-state ^{29}Si CP/MAS experiments (Table XVII). In all cases, the $SiCH_2N$ protons were found to be diastereotopic, indicating that the respective (*A*)- and (*C*)-enantiomers of **94–104** are configurationally stable on the NMR time scale. This assumption is supported by $^{19}F\{^1H\}$ NMR studies of **103** (CDCl₃). While the $^{19}F\{^1H\}$ NMR spectrum of **103** under achiral conditions shows only one resonance signal, a set of two signals was observed upon addition of 0.75 mol equiv of the solvation agent (*R*)-2,2,2-trifluoro-1-(9-anthryl)ethanol [(*R*)-TFAE].[61] This phenomenon results from the existence of diastereomeric solvates in solution [(*A*)-**103** · (*R*)-TFAE and (*C*)-**103** · (*R*)-TFAE].

^1H NMR studies of the diastereomerically and enantiomerically pure compound (*A,S,S*)-**107** (CD₂Cl₂, −20°C) demonstrated that this zwitterion

TABLE XVII

^{29}Si Chemical Shifts and $^{1}J_{SiF}$ Coupling Constants for Compounds **94**, **95** · CH$_3$CN, **96–105**, **106** · ½CH$_3$CN, and **107** in Solution and in the Crystal[a]

Compound	$\delta_{solution}$ ($^{1}J_{SiF}$ [Hz])	$\delta_{crystal}$ ($^{1}J_{SiF}$ [Hz])[b]
94	−74.8 (245.7)[c]	−78.0 (253.8)
95 · CH$_3$CN	−88.0 (247.6)[c]	−91[d]
96	−74.0 (248.2)[e]	−73.2 (220.9)
97	−78.6 (263.1)[e]	−74.8 (261.6)
98	−75.5 (255.4)[e]	−73.8 (281.0)
99	−74.2 (244.5)[c]	−74.8 (268.2)
100	−79.4 (267.6)[f]	−80.1 (268.4)
101	−79.1 (266.1)[e]	−78.6 (278.3)
102	−75.0 (254.0)[c]	−74.1 (263.1)
103	−81.7 (243.5)[c]	−81.4 (247.2)
104	−86.3 (270.3)[g]	−85.9 (282.9)
105[h]	−78.2 (245.7)[i]	−80.6 (247.1)
106 · ½CH$_3$CN[j]	−83.2 (271.0)[k]	[n]
107[l]	−79.7 (253.0)[m]	−81.2 (246.0)

[a] Chemical shifts in ppm; spectra recorded at room temperature.
[b] Isotropic chemical shifts obtained by ^{29}Si CP/MAS NMR experiments.
[c] Solvent CDCl$_3$.
[d] Broad resonance signal.
[e] Solvent [D$_6$]DMSO.
[f] Solvent CD$_3$CN.
[g] Solvent CD$_2$Cl$_2$.
[h] (A,S)-Configuration.
[i] Solvent CD$_3$CN, spectrum recorded at −20°C.
[j] (C,S)-Configuration.
[k] Solvent CD$_2$Cl$_2$, spectrum recorded at −50°C.
[l] (A,S,S)-Configuration.
[m] Solvent CD$_2$Cl$_2$, spectrum recorded at −40°C.
[n] No experiment performed.

undergoes isomerization in solution yielding the corresponding diastereomer (C,S,S)-**107** [detection of the thermodynamic equilibrium (A,S,S)-**107** ⇌ (C,S,S)-**107**; ratio of diastereomers ca. 1 : 1] (Scheme 24).[61] This particular isomerization process (epimerization) involves an inversion of absolute configuration of the chiral trigonal-bipyramidal $\lambda^5 Si$-skeleton of the two diastereomers. According to kinetic studies, an intramolecular process is assumed for this epimerization. An analogous isomerization process (enantiomerization) would lead to a conversion of the respective (A)- and (C)-enantiomers of **94–104** into each other [e.g., (A)-**96** ⇌ (C)-**96**; Scheme 24].

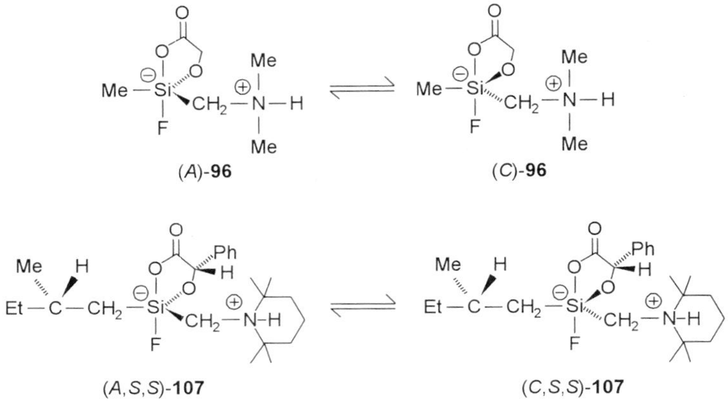

SCHEME 24.

VIII

CONCLUSIONS

The chemistry of zwitterionic pentacoordinate silicon compounds (zwitterionic λ^5Si-silicates) has been developed systematically over the past 10 years. These compounds contain a silicate moiety with five covalent silicon–element bonds and an onium group. In the zwitterions studied so far, this onium moiety is an ammonium group. The pentacoordinate (formally negatively charged) silicon atom and the tetracoordinate (formally positively charged) nitrogen atom in these neutral molecules are separated by a spacer, such as a $(CH_2)_n$ or $O(CH_2)_n$ moiety. Compared to ionic λ^5Si-silicates with five covalent silicon–element bonds and compared to neutral pentacoordinate silicon compounds with a 4+1 coordination to silicon, zwitterionic λ^5Si-silicates are characterized by some very special features that arise from the zwitterionic nature of these compounds.

As a general rule, these zwitterionic λ^5Si-silicates are high-melting crystalline solids that are almost insoluble in nonpolar organic solvents and mostly even exhibit poor solubility in polar organic solvents. The crystals of these compounds consist of highly polar molecular building blocks that undergo intermolecular dipolar interactions.

Different methods have been used for their syntheses, some of these methods involving remarkable Si–C cleavage reactions. Up to now, zwitterionic λ^5Si-silicates with SiF_4C, SiF_3C_2, SiO_4C, SiO_2N_2C, SiO_2S_2C, SiO_5, and SiO_2FC_2 skeletons have been synthesized.

Generally, the products isolated directly from the reaction mixtures (or obtained by subsequent recrystallization) were well-crystallized solids, suitable for characterization by single-crystal X-ray diffraction and solid-state NMR spectroscopy. However, zwitterionic $\lambda^5 Si$-silicates do not only provide a lot of structural information about pentacoordination in the solid state. As the zwitterions are neutral molecules, they can also be studied by NMR spectroscopy in solution to get information about their structure and their dynamic behavior (dynamic processes involving the pentacoordinate silicon atom). In contrast to ionic $\lambda^5 Si$-silicates, experimental problems resulting from cation/anion interactions do not exist in the case of zwitterionic species.

Most of the zwitterionic compounds studied so far are chiral, with a chiral $\lambda^5 Si$-silicate skeleton. Most of them have been isolated as racemic mixtures and in some cases as enantiomerically pure compounds, some of the optically active compounds being configurationally stable in solution. With these experimental investigations, in combination with computational studies, a new research area concerning the stereochemistry of molecular pentacoordinate silicon compounds has been developed.

Up to now, the studies on zwitterionic $\lambda^5 Si$-silicates have mainly focused on their synthesis and structural characterization. In future investigations, the chemical and physical properties of these compounds should be investigated in more detail. In this context, the biological properties of zwitterionic $\lambda^5 Si$-silicates are also of great interest. In general, zwitterionic compounds are characterized by unique properties that arise from the charge separation in these molecules. It will be a challenge to develop the chemistry of zwitterionic $\lambda^5 Si$-silicates further and to explore their potential for practical applications systematically.

ACKNOWLEDGMENTS

R.T. expresses his sincere thanks to his co-workers and colleagues, without whose contributions this article could not have been written; their names are cited in the references. In addition, financial support of our work by the *Deutsche Forschungsgemeinschaft* and the *Fonds der Chemischen Industrie* and support with chemicals by the *Bayer AG* (Leverkusen and Wuppertal-Elberfeld) and the *Merck KGaA* (Darmstadt) are gratefully acknowledged.

REFERENCES

(1) Tandura, S. N.; Voronkov, M. G.; Alekseev, N. V. *Top. Curr. Chem.* **1986**, *131*, 99–189.
(2) Sheldrick, W. S. In *The Chemistry of Organic Silicon Compounds;* Patai, S.; Rappoport, Z., Eds.; Wiley: Chichester, 1989; Part 1, pp. 227–303.
(3) Corriu, R. J. P.; Young, J. C. In *The Chemistry of Organic Silicon Compounds;* Patai, S.; Rappoport, Z., Eds.; Wiley: Chichester, 1989; Part 2, pp. 1241–1288.
(4) Holmes, R. R. *Chem. Rev.* **1990**, *90*, 17–31.
(5) Chuit, C.; Corriu, R. J. P.; Reye, C.; Young, J. C. *Chem. Rev.* **1993**, *93*, 1371–1448.

272 REINHOLD TACKE *et al.*

(6) Tacke, R.; Becht, J.; Lopez-Mras, A.; Sperlich, J. *J. Organomet. Chem.* **1993**, *446*, 1–8.
(7) Wong, C. Y.; Woollins, J. D. *Coord. Chem. Rev.* **1994**, *130*, 175–241.
(8) Tacke, R.; Dannappel, O.; Mühleisen, M. In *Organosilicon Chemistry II—From Molecules to Materials;* Auner, N.; Weis, J., Eds.; VCH: Weinheim, 1996; pp. 427–446.
(9) Tacke, R.; Dannappel, O. In *Tailor-made Silicon-Oxygen Compounds—From Molecules to Materials;* Corriu, R.; Jutzi, P., Eds.; Vieweg: Braunschweig/Wiesbaden, 1996; pp. 75–86.
(10) Lukevics, E.; Pudova, O. A. *Chem. Heterocycl. Compd. (Engl. Transl.)* **1996**, *32*, 1381–1418.
(11) Holmes, R. R. *Chem. Rev.* **1996**, *96*, 927–950.
(12) Kost, D.; Kalikhman, I. In *The Chemistry of Organic Silicon Compounds;* Rappoport, Z.; Apeloig, Y., Eds.; Wiley: Chichester, 1998; Vol. 2, Part 2, pp. 1339–1445.
(13) Dathe, C. Doctoral Thesis, Technische Universität Dresden, 1966.
(14) Müller, R. *Organomet. Chem. Rev.* **1966**, *1*, 359–377.
(15) Müller, R. *Z. Chem.* **1984**, *24*, 41–51.
(16) (a) Schomburg, D.; Krebs, R. Unpublished results. (b) Krebs, R. Doctoral Thesis, Technische Universität Braunschweig, 1987.
(17) Strohmann, C.; Tacke, R.; Mattern, G.; Kuhs, W. F. *J. Organomet. Chem.* **1991**, *403*, 63–71.
(18) Tacke, R.; Becht, J.; Mattern, G.; Kuhs, W. F. *Chem. Ber.* **1992**, *125*, 2015–2018.
(19) Tacke, R.; Lopez-Mras, A.; Becht, J.; Sheldrick, W. S. *Z. Anorg. Allg. Chem.* **1993**, *619*, 1012–1016.
(20) Tacke, R.; Becht, J.; Dannappel, O.; Ahlrichs, R.; Schneider, U.; Sheldrick, W. S.; Hahn, J.; Kiesgen, F. *Organometallics* **1996**, *15*, 2060–2077.
(21) Tacke, R.; Pfrommer, B.; Lunkenheimer, K.; Hirte, R. *Organometallics* **1998**, *17*, 3670–3676.
(22) Tacke, R.; Becht, J.; Lopez-Mras, A.; Sheldrick, W. S.; Sebald, A. *Inorg. Chem.* **1993**, *32*, 2761–2766.
(23) Tacke, R.; Neugebauer, R. E.; Pülm, M. Unpublished results.
(24) (a) Tacke, R.; Becht, J.; Pülm, M. Unpublished results. (b) Becht, J. Doctoral Thesis, Universität Karlsruhe, 1994.
(25) Tacke, R.; Lopez-Mras, A.; Sperlich, J.; Strohmann, C.; Kuhs, W. F.; Mattern, G.; Sebald, A. *Chem. Ber.* **1993**, *126*, 851–861.
(26) (a) Tacke, R.; Heermann, J.; Willeke, R. Unpublished results. (b) Heermann, J. Doctoral Thesis, Universität Würzburg, 1998.
(27) Sperlich, J.; Becht, J.; Mühleisen, M.; Wagner, S. A.; Mattern, G.; Tacke, R. *Z. Naturforsch.* **1993**, *48b*, 1693–1706.
(28) (a) Tacke, R.; Ulmer, B.; Pfrommer, B. Unpublished results. (b) Pfrommer, B. Doctoral Thesis, Universität Würzburg, 1998.
(29) (a) Tacke, R.; Richter, I.; Pfrommer, B. Unpublished results. (b) Pfrommer, B. Doctoral Thesis, Universität Würzburg, 1998.
(30) Tacke, R.; Heermann, J.; Pülm, M.; Gottfried, E. *Monatsh. Chem.* **1999**, *130*, 99–107.
(31) Tacke, R.; Sperlich, J.; Strohmann, C.; Frank, B.; Mattern, G. *Z. Kristallogr.* **1992**, *199*, 91–98.
(32) Tacke, R.; Mühleisen, M.; Lopez-Mras, A.; Sheldrick, W. S. *Z. Anorg. Allg. Chem.* **1995**, *621*, 779–788.
(33) Tacke, R.; Wiesenberger, F.; Lopez-Mras, A.; Sperlich, J.; Mattern, G. *Z. Naturforsch.* **1992**, *47b*, 1370–1376.
(34) Tacke, R.; Lopez-Mras, A.; Sheldrick, W. S.; Sebald, A. *Z. Anorg. Allg. Chem.* **1993**, *619*, 347–358.
(35) Tacke, R.; Sperlich, J.; Strohmann, C.; Mattern, G. *Chem. Ber.* **1991**, *124*, 1491–1496.

(36) Dannappel, O.; Tacke, R. In *Organosilicon Chemistry II—From Molecules to Materials;* Auner, N.; Weis, J., Eds.; VCH: Weinheim, 1996, pp. 453–458.

(37) Mühleisen, M.; Tacke, R. In *Organosilicon Chemistry II—From Molecules to Materials;* Auner, N.; Weis, J., Eds.; VCH: Weinheim, 1996, pp. 447–451.

(38) (a) Tacke, R.; Mühleisen, M. Unpublished results. (b) Mühleisen, M. Doctoral Thesis, Universität Karlsruhe, 1994.

(39) (a) Tacke, R.; Pfrommer, B. Unpublished results. (b) Pfrommer, B. Doctoral Thesis, Universität Würzburg, 1998.

(40) Pülm, M.; Tacke, R. *Organometallics* **1997**, *16*, 5664–5668.

(41) Tacke, R.; Pülm, M. Unpublished results.

(42) (a) Tacke, R.; Heermann, J.; Pfrommer, B. Unpublished results. (b) Heermann, J. Doctoral Thesis, Universität Würzburg, 1998. (c) Pfrommer, B. Doctoral Thesis, Universität Würzburg, 1998.

(43) Tacke, R.; Richter, I.; Pülm, M. Unpublished results.

(44) Tacke, R.; Bertermann, R.; Biller, A.; Dannappel, O.; Pülm, M.; Willeke, R. *Eur. J. Inorg. Chem.* **1999**, 795–805.

(45) Tacke, R.; Pfrommer, B.; Pülm, M.; Bertermann, R. *Eur. J. Inorg. Chem.* **1999**, 807–816.

(46) (a) Tacke, R.; Heermann, J.; Pülm, M. Unpublished results. (b) Heermann, J. Doctoral Thesis, Universität Würzburg, 1998.

(47) Tacke, R.; Lopez-Mras, A.; Jones, P. G. *Organometallics* **1994**, *13*, 1617–1623.

(48) Mühleisen, M.; Tacke, R. *Chem. Ber.* **1994**, *127*, 1615–1617.

(49) Tacke, R.; Heermann, J.; Pülm, M. *Organometallics* **1997**, *16*, 5648–5652.

(50) Mühleisen, M.; Tacke, R. *Organometallics* **1994**, *13*, 3740–3742.

(51) Tacke, R.; Mühleisen, M.; Jones, P. G. *Angew. Chem.,* **1994**, *106*, 1250–1252; *Angew. Chem., Int. Ed. Engl.* **1994**, *33*, 1186–1188.

(52) Pülm, M.; Willeke, R.; Tacke, R. In *Organosilicon Chemistry IV—From Molecules to Materials;* Auner, N.; Weis, J., Eds.; Wiley-VCH: Weinheim, in press.

(53) Tacke, R.; Heermann, J.; Pülm, M.; Richter, I. *Organometallics* **1998**, *17*, 1663–1668.

(54) Biller, A.; Pfrommer, B.; Pülm, M.; Tacke, R. In *Organosilicon Chemistry IV—From Molecules to Materials;* Auner, N.; Weis, J., Eds.; Wiley-VCH: Weinheim, in press.

(55) (a) Tacke, R.; Heermann, J.; Willeke, R. Unpublished results. (b) Heermann, J. Doctoral Thesis, Universität Würzburg, 1998.

(56) Pfrommer, B.; Tacke, R. *Eur. J. Inorg. Chem.* **1998**, 415–418.

(57) Tacke, R.; Willeke, R. Unpublished results.

(58) Pfrommer, B.; Tacke, R. In *Organosilicon Chemistry IV—From Molecules to Materials;* Auner, N.; Weis, J., Eds.; Wiley-VCH: Weinheim, in press.

(59) Tacke, R.; Mühleisen, M. *Inorg. Chem.* **1994**, *33*, 4191–4193.

(60) (a) Willeke, R.; Neugebauer, R. E.; Pülm, M.; Dannappel, O.; Tacke, R. In *Organosilicon Chemistry IV—From Molecules to Materials;* Auner, N.; Weis, J., Eds.; Wiley-VCH: Weinheim, in press. (b) Tacke, R.; Dannappel, O.; Neugebauer, R. E.; Pülm, M.; Willeke, R. Unpublished results.

(61) Neugebauer, R. E.; Bertermann, R.; Tacke, R. In *Organosilicon Chemistry IV—From Molecules to Materials;* Auner, N.; Weis, J., Eds.; Wiley-VCH: Weinheim, in press.

(62) Tacke, R.; Paschold, T.; Neugebauer, R. E.; Pülm, M.; Burschka, C. Unpublished results.

(63) Tacke, R.; Neugebauer, R. E.; Willeke, R. Unpublished results.

ADVANCES IN ORGANOMETALLIC CHEMISTRY, VOL. 44

Transition Metal Complexes of Vinylketenes

SUSAN E. GIBSON*

Centre for Chemical Synthesis
Department of Chemistry
Imperial College of Science, Technology, and Medicine
South Kensington, London, United Kingdom SW7 2AY
and
Department of Chemistry
Kings College
The Strand, London, United Kingdom WC2R 2LS

MARK A. PEPLOW

Centre for Chemical Synthesis
Department of Chemistry
Imperial College of Science, Technology, and Medicine
South Kensington, London, United Kingdom SW7 2AY

* Née Thomas.

275

I

INTRODUCTION

The study of the transition metal complexes of η^4-vinylketene ligands began with the first isolation of such a compound by King in 1963.[1] During the following thirty five years, they have risen from the status of "mere" curios to become useful precursors in the synthesis of organic compounds. They have also been considered as important intermediates in certain reaction mechanisms, particularly the Dötz annulation.[2] The rate of development of the chemistry of these compounds has risen dramatically in the last decade, particularly following the instigation of a comprehensive survey of their reactivity.[3] Previously, the synthesis and characterization of the complexes had been the main focus of literature discussion.

Before embarking on a discussion of the chemistry in this area, it is useful to consider some structural aspects of η^4-vinylketene complexes. Over the course of the research into these compounds, it has become clear that there are three major canonical forms that contribute to the bonding. Authors generally choose to represent the complexes with one particular canonical form, using both crystallographic evidence and reactivity studies to select one structure as the best representation. The η^4-vinylketene form may be represented in several ways (**1,1′,2**), but fundamentally implies that the metal moiety is engaged in bonding to varying degrees with all four carbon atoms of the butadienylidene backbone. Structures **1** or **1′** are often used when subsequent reactions of the vinylketene ligand are under consideration. The other main alternative is the $\eta^3 : \eta^1$-allylacyl form (**3**), which has been invoked on the basis of crystallographic studies showing the similarity in bond lengths of the terminal C_3-fragment with analogous π-allyl systems. Certain reactions result in a formal insertion into the metal–acyl bond, and for this reason the $\eta^3 : \eta^1$-allylacyl form is the most illustrative representation on these occasions. Finally, the 1,4-σ:2,3-π (or $\eta^1 : \eta^2 : \eta^1$) structure (**4**) has been most frequently used to show that the decomplexation of the central C_2 (π-bonded) fragment is a mechanistically viable process, a transformation that has been suggested as the first step of several reactions of vinylketene complexes. The most explicit representation of the vinylketene complex is **2**, and although it is a representation that falsely suggests a purely σ-bonded structure, it may be useful

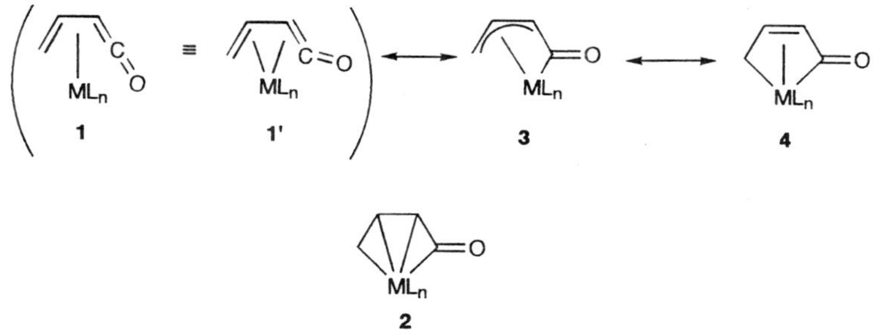

SCHEME 1.

when considering the movement of electrons around the system. Wherever possible, we have represented the vinylketene complexes in the manner preferred by the original authors. Similarly, the authors' definition of the bonding mode of the complexes has also been used, despite a degree of ambiguity in the application of the "η" notation. The $\eta^3 : \eta^1$ formalism is widely used throughout the literature despite the fact that it describes an allylacyl representation, which would usually suggest η^4 as the correct definition (the complex still has a hapticity of four, whatever its canonical representation). In keeping with this common usage, structure 4 has been described as $\eta^1 : \eta^2 : \eta^1$, although confusingly it is often referred to in the literature as merely η^2. Consequently, the reader should be aware of the potential for misdirection in this notation system.

These structural considerations have provided a convenient way to delineate the scope of this review. We shall only be considering vinylketene complexes that fall within the bonding parameters laid out above. Complexes such as 5,[4] 6,[5] and 7[6] are all η^2-complexes and as such will not be covered here, unless they have direct relevance to related η^4-complexes. Similarly, free vinylketenes[7] will not be discussed, except where their chemistry has been related to their complexed analogues. The review includes, to the best of our

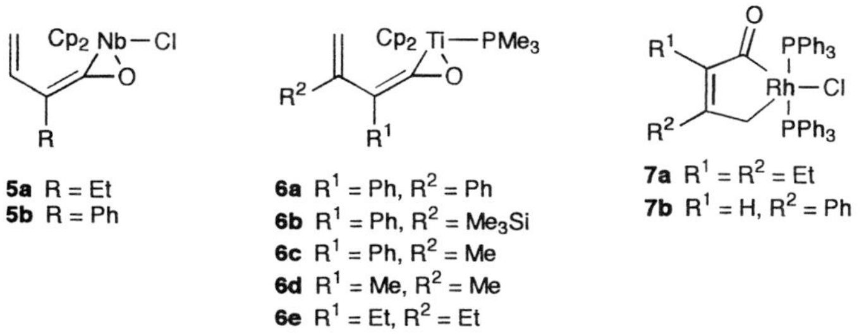

5a R = Et
5b R = Ph

6a R^1 = Ph, R^2 = Ph
6b R^1 = Ph, R^2 = Me$_3$Si
6c R^1 = Ph, R^2 = Me
6d R^1 = Me, R^2 = Me
6e R^1 = Et, R^2 = Et

7a R^1 = R^2 = Et
7b R^1 = H, R^2 = Ph

SCHEME 2.

knowledge, *every* reported η^4-vinylketene complex ever isolated. It also deals selectively with reports of reactions in which a η^4-vinylketene complex has been proposed as a key intermediate, and the experiments that have been used to trap these intermediate vinylketene complexes from reaction mixtures.

II

η^4-(VINYLKETENE) CHROMIUM(0) COMPLEXES

A. Introduction

The most common citation of η^4-vinylketene complexes in the literature is to be found in mechanistic discussions of the Dötz annulation.[8] In the years since the reaction was discovered,[2] there has been an enormous amount of research conducted to determine a definitive mechanism. Nevertheless, certain transformations in the mechanism are still open to speculation, the most recent reassessment having been postulated by Solà.[9] It is not within the scope of this review to comprehensively discuss a subject which has been extensively covered elsewhere.[8,10,11] However, the reactions of η^4-vinylketene complexes of chromium, and indeed of all the other metals in this review, are so intrinsically linked with the Dötz annulation that we must acquaint ourselves with its intricacies. We shall see the same structures occurring in mechanisms time after time, and the same classes of organic molecules being isolated as final products.

B. Annulation of Alkynes with Fischer Carbenes

The Dötz reaction was first reported[2] for an aryl carbene of the type **8**, and the resulting benzannulation may be summarized conveniently as shown in Scheme 3. The major product of the reaction is usually the naphthol **9**, the result of a formal [3 + 2 + 1] cycloaddition.[10] In cases where CO insertion does not take place, five-membered rings (**10**) may also be formed.

Note that the benzannulation[8,10,12,13] represents a special case of a more general reaction: The aryl group may be replaced by hydrogen,[14] alkyl,[15–19] alkenyl,[19–21] $(CH_2)_n$-tethered alkynyl,[22–24] polyalkynyl,[25] heteroaromatic,[26,27] cyclopropyl,[28] or iminyl[29] groups. Mixtures of products are almost always seen, and as such the range of organic products accessible from this reaction is multitudinous. Studies on the effects of varying the metal,[12] the heteroatom,[11,30–32] the aryl substituents,[13] and the alkyne substrate[11] have all been conducted. Important investigations into the effects of changes in the stereochemistry of the reaction intermediates have also been carried out.[27]

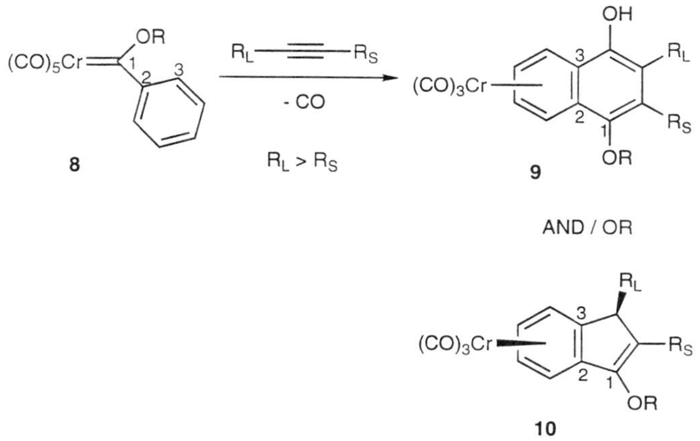

SCHEME 3.

There seems little doubt that the first, rate-determining step of the reaction is the reversible dissociation of a carbonyl ligand from the parent carbene complex (**8**), which is then followed by coordination of the alkyne into the vacant site to give complex **11**.[33] For many years, it was assumed[8,34] that a [2 + 2] cyclization then occured to give the chromacyclobutene **12**: Complexes of this type are known for other metals (see Section V,A). However, Hofmann *et al.* have calculated[35,36] that this is an energetically unfavorable intermediate with respect to the η^3-vinylcarbene (allylidene) complex **13**. It has been suggested by Wulff[12] that the chromacyclobutene may be present as a solvent-stabilized complex that allows the interconversion of *E*- and *Z*-isomers of the vinylcarbene. Only the *E*-isomer is configured to allow naphthol formation, and for the sake of brevity we shall follow its progress alone.

A particularly interesting piece of evidence concerning the nature of this sequence has been presented by Barluenga *et al.*[21] When the vinylcarbene complex **14** was heated, decarbonylation afforded the alkene-stabilized complex **15**. Upon treatment with dimethyl acetylenedicarboxylate, the alkyne-insertion product **16** was isolated. This complex proved to be unstable in solution at room temperature and decomposed readily to **17**, the expected product of a Dötz reaction with an aminovinylcarbene.

It is at this point that opinion diverges on the progress of the reaction. The original Dötz mechanism[8] postulates the insertion of a carbonyl ligand into the vinylcarbene *E*-**13** to give the η^4-vinylketene complex *E*-**18**. This may then cyclize to give the cyclohexadienone complex **19**, which yields the naphthol complex **20** after simple tautomerism. It should be noted that a density functional study[37] has suggested that the η^4-vinylketene complex

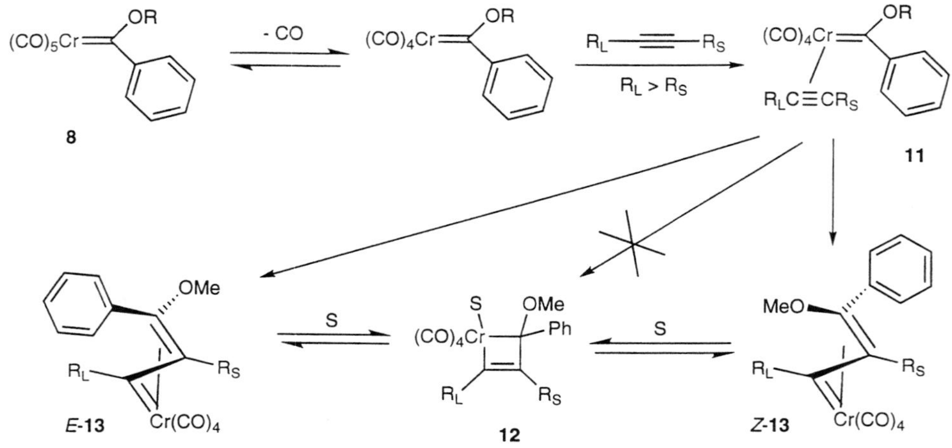

SCHEME 4.

E-**18** is only a stable intermediate in the reactions of *phenyl*carbenes (to give naphthols). It was calculated that vinylcarbenes are capable of passing directly from an η^3-vinylcarbene complex to the appropriate cyclohexadienone complex, before ultimately yielding a phenol.

The most significant alternative to this mechanism was proposed by Casey,[34] who suggests that cyclization may take place *before* carbonyl inser-

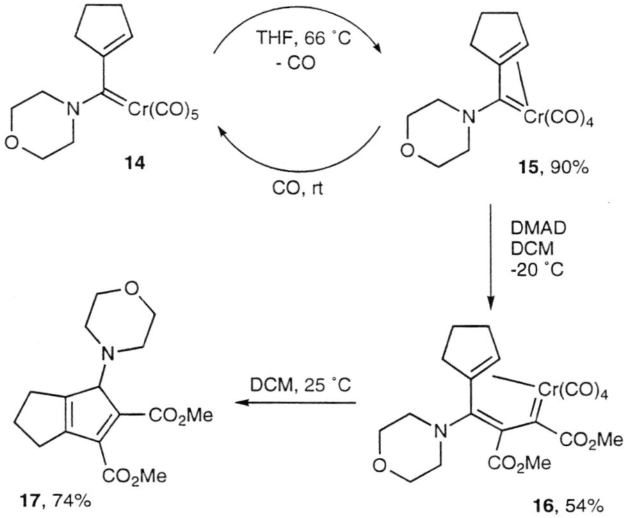

SCHEME 5.

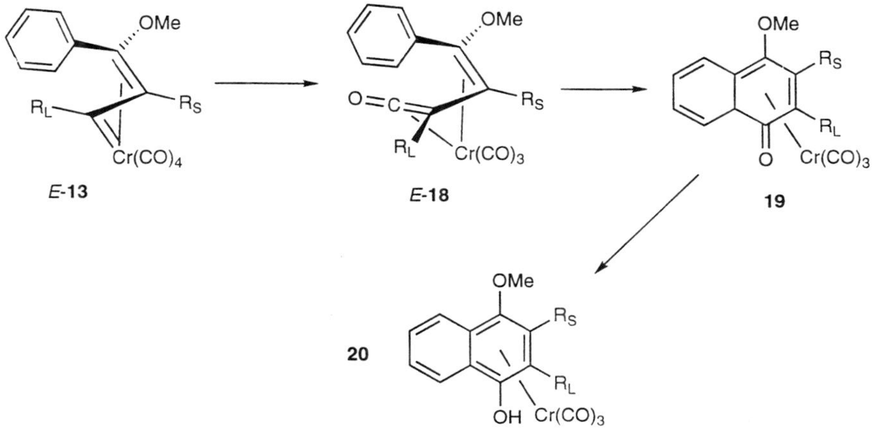

SCHEME 6.

tion, affording the chromacycle **21**. This may then undergo carbonylation to give complex **22**, which subsequently deinserts the chromium moiety to once again give the cyclohexadienone complex **19**. Cyclohexadieneone complexes are generally accepted as the final intermediate in these reactions, and Wulff et al. have successfully isolated[12] (albeit in very low yield) a cyclohexadienone–molybdenum complex from the reaction between a molybdenum carbene and 3-hexyne. Gibson (née Thomas) has also isolated[38,39] a cyclohexadienone complex from the reaction between a vinylketene–alkyne adduct and aluminum(III) chloride under a carbon monoxide atmosphere.

The vast majority of the circumstantial evidence concerning this reaction points toward the presence of η^4-vinylketene complexes such as E-**18** at some point along the reaction coordinate. Moreover, the chromacyclohexadiene pathway cannot fully explain all of the data concerning the benzannulation reaction.[13] The accumulation of evidence for the intermediacy of η^4-vinylketenes has been recorded elsewhere,[8,10,11,13,27] but will be considered here briefly to give an impression of the general methods that have been used to elucidate the reactivity of vinylketene complexes.

Analysis of product distributions has often been used to evince the presence of η^4-vinylketenes in the Dötz reaction. Wulff has studied the reactions of a wide variety of substituted aryl chromiumcarbenes and drawn conclusions from the data that point directly toward η^4-vinylketene complexes as key intermediates.[13] He has also isolated[40] bicyclic lactone products (e.g., **23**) characteristic of a van Halban–White cyclization[41–44] from the reaction of vinylcarbenes and ketoalkynes. These bicyclic lactones are not generally

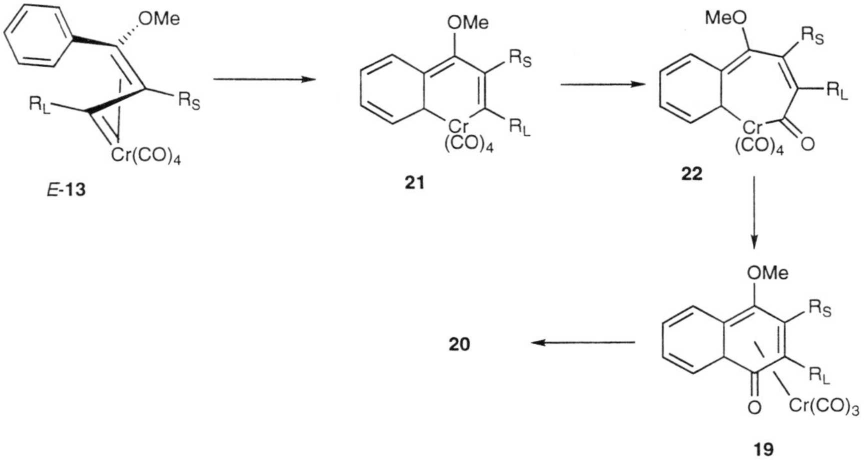

SCHEME 7.

seen as products of the Dötz reaction,[40] possibly because they require the presence of an unusual cross-conjugated vinylketene (**24**). The chromium cannot be bound in an η^4-configuration at this stage, as it would inhibit the internal rotation necessary for cyclization.

Mori has reported that in the reaction of chromium carbene **25** with an alkyne containing a tethered 4-amidobutyne unit (**26**), a postulated vinylketene complex (**27**) is intercepted by nucleophilic amide attack, yielding a mixture of lactams (**28** and **29**).[15] The expected naphthol **30** was also isolated in low yield.

Many other reactions designed to trap intermediate vinylketene complexes are known. Dötz has used alkynes with a pendant alcohol to produce the butyrolactones *E*-**31** and *Z*-**31** in a 70:30 ratio and 34% yield.[16] These are formed by the nucleophilic attack of the pendant alcohol on the ketene

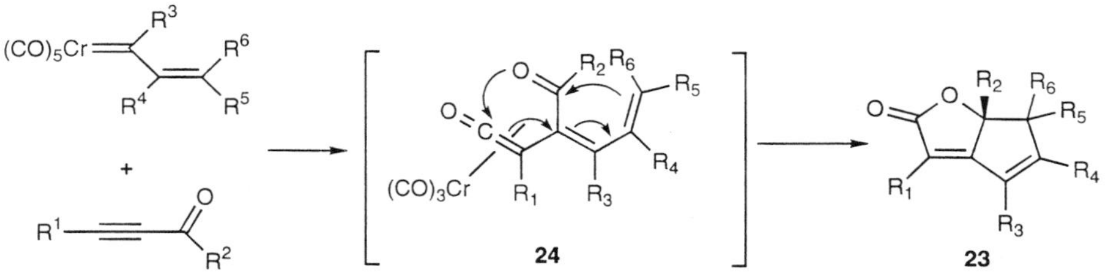

SCHEME 8.

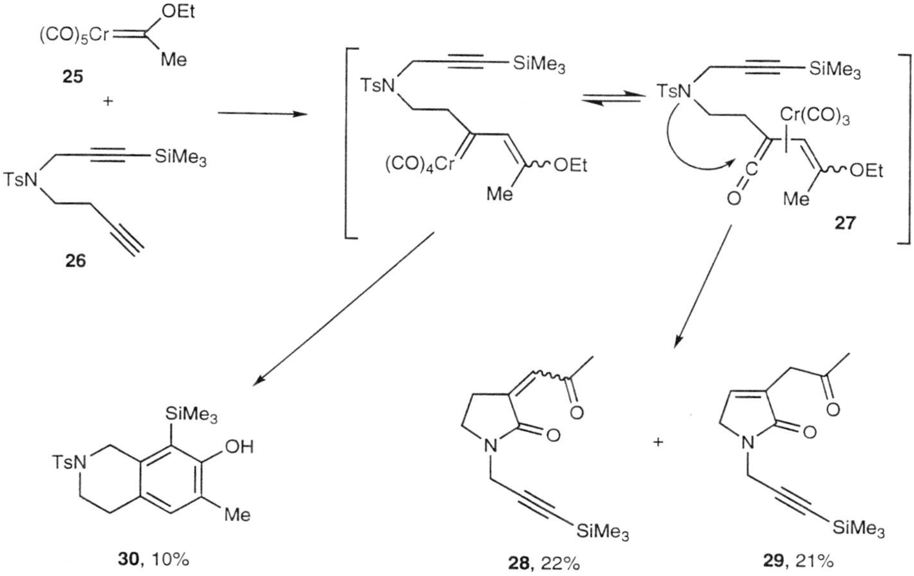

SCHEME 9.

carbonyl carbon (C-1). Notice that the cyclobutenone **32** is the product of a formal 4π electrocyclic reaction of the corresponding free vinylketene.

Similarly, Yamashita has utilised free alcohols to trap the vinylketene intermediates formed in the reactions of aryl and heteroaromatic carbenes, resulting in β, γ-unsaturated esters.[26,45] If a secondary amine is used in the

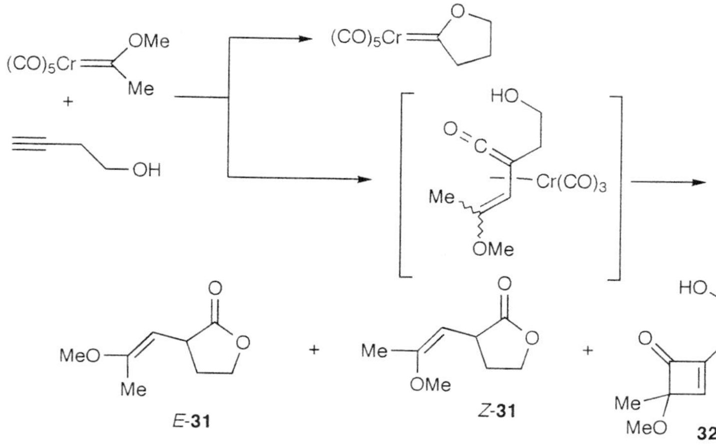

SCHEME 10.

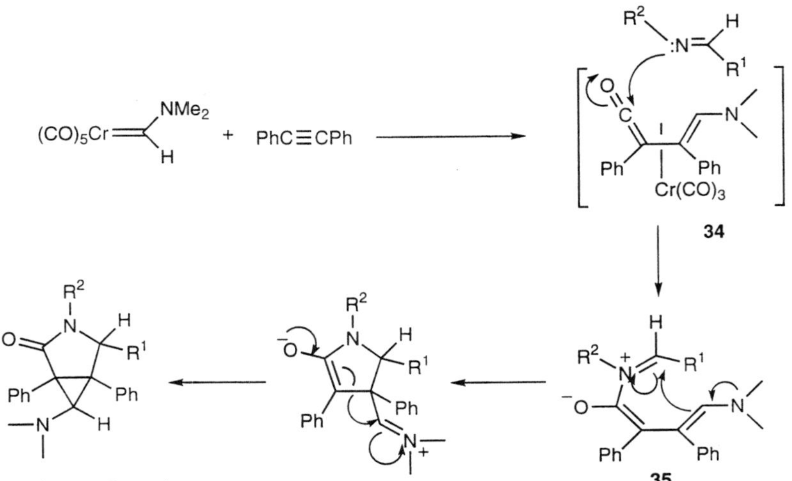

33.a R^1 = H, R^2 = Bu^t 75%
33.b R^1 = Ph, R^2 = Me 77%

SCHEME 11.

same way, the expected β, γ-unsaturated amide is isolated.[26] Hegedus has shown that bicyclic lactams (**33**) may be constructed by the reaction of an imine with the η^4-vinylketene intermediate (**34**) formed from the reaction of a dimethylaminocarbene with diphenylacetylene.[14] Interestingly, Hegedus has also carried out the reaction between an imine and an η^4-(vinylketene)iron(0) complex, isolating compounds analogous to the postulated intermediate **35** (see Section VI,J).[46]

Alkenes may be used to trap intermediate vinylketene complexes (e.g., **36**) by undergoing a [2 + 2] cycloaddition to form 4,5-bicyclic systems (e.g., **37**).[18,24] Prior to this report by Wulff, there had only been one previous example of a [2 + 2] cycloaddition to a complexed ketene system.[47]

Carbonyl insertions into metallocarbenes have previously been observed for several different metals, including iron[48] (see Section VI,C) and manganese.[49] Indeed, carbonyl insertions into chrominum and tungsten diphenyl-carbenes have been shown to be viable processes.[50] Most importantly, Wulff has isolated[51] an η^4-vinylketenecobalt (I) complex from the reaction between a cobalt carbene and an acetylene, a transformation that necessitates such a carbonyl insertion (see Section V,B).

Perhaps the most compelling evidence for the intermediacy of η^4-vinylketene complexes in the Dötz annulation has been provided by the isolation of both free and complexed vinylketenes from the reaction mixture. Dötz himself has reported[52,53] the isolation of both arene-complexed (**38**) and

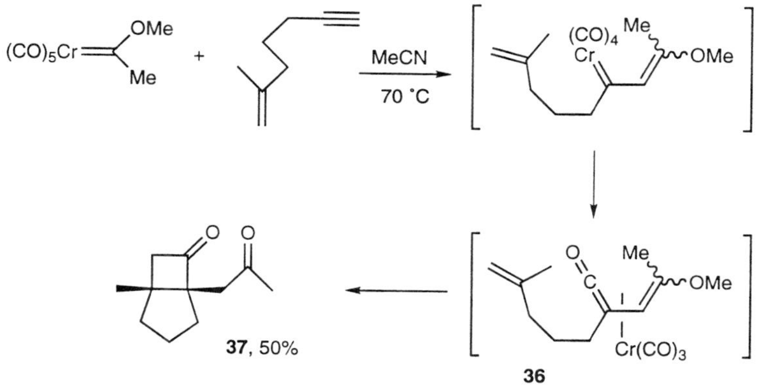

SCHEME 12.

free vinylketenes (**39**) from the reaction of a pentacarbonyl [methoxy-(aryl)carbene]chromium(0) species (**40**) with bis(trimethylsilyl)acetylene.

In an analogous reaction, Wulff isolated[54] the free vinylketene **41** in good yield from the reaction of pentacarbonyl[methoxy(2,6-dimethylphenyl) carbene]chromium(O) with trimethylsilylacetylene. The E-stereochemistry of the product was found to be unchanged even after heating to 200°C for 24 h.[54]

It has long been known that silyl substitution has a great stabilizing effect on free vinylketenes,[55,56] although the reasons for this are not fully understood.[54] The generation of free vinylketenes from cyclobutenones[55-60] has proved to be very effective and has also allowed many theoretical studies on the kinetics[61] and molecular orbital characteristics[62-64] of this interconversion to be carried out. Liebeskind has shown[58,59] that free vinyl-ketenes generated from the ring opening of cyclobutenones may react with

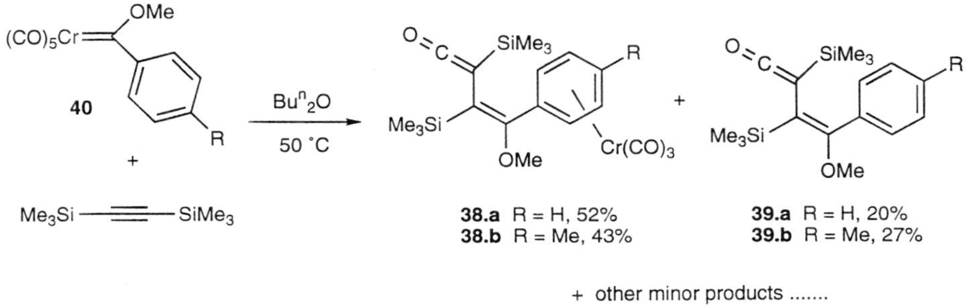

38.a R = H, 52%
38.b R = Me, 43%

39.a R = H, 20%
39.b R = Me, 27%

+ other minor products ....

SCHEME 13.

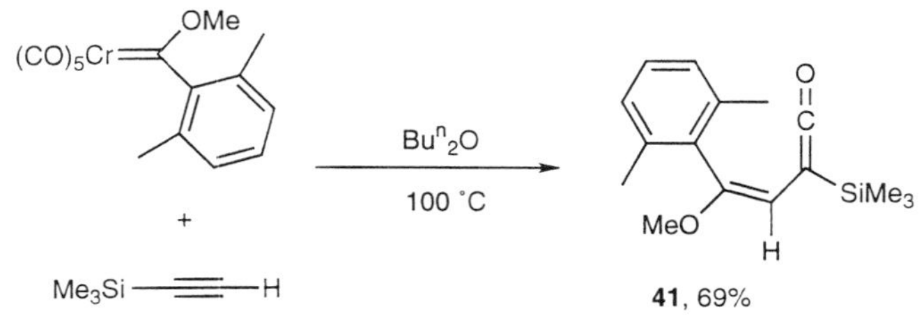

SCHEME 14.

alkynes to yield annulation products that match those from Dötz reactions that would involve the same *complexed* vinylketene intermediates.

The single best piece of evidence for the intermediacy of vinylketene complexes is however the isolation and characterization by Wulff[22] of amine-stabilized η^4-vinylketenechromium(0) complexes (**42**) from the thermolysis of the chromium carbene complexes **43**, containing a tethered alkyne functionality. This was the first time that a d^6-η^4-vinylketene complex of any group 6 metal had been isolated.

Crucially, it was found[22] that upon reaction of the vinylketene complex **42.a** with 1-pentyne, the same product distribution was seen as for the direct thermolysis of 1-pentyne with the precursor carbene **43.a**. The analysis was simplified by reduction of the crude reaction mixture with McMurry's reagent to produce a mixture of the isomeric indanols **44–46**.

The indanols **44** and **45** can only be the products of a formal [4 + 2] cycloaddition[23] of the vinylketene complex **42.a** with 1-pentyne. Note that upon reaction of **42.b** with diethylpropynylamine a formal [2 + 2] cycloaddition[65] is seen to take place, yielding the cyclobutenone **47** along with a tricarbonylchromium complex, tentatively identified as **48**.[66,67] As one would expect, the vinylketene complex **42.b** underwent 1,2-additions with pyrrolidine and sodium methoxide in methanol, yielding **49** and **50**, respectively. The CO-insertion step leading to vinylketene formation is reversible in some systems,[51,68,69] but there is no evidence of this for complex **42.a**. Heating a benzene solution of complex **42.a** at 80°C under an atmosphere

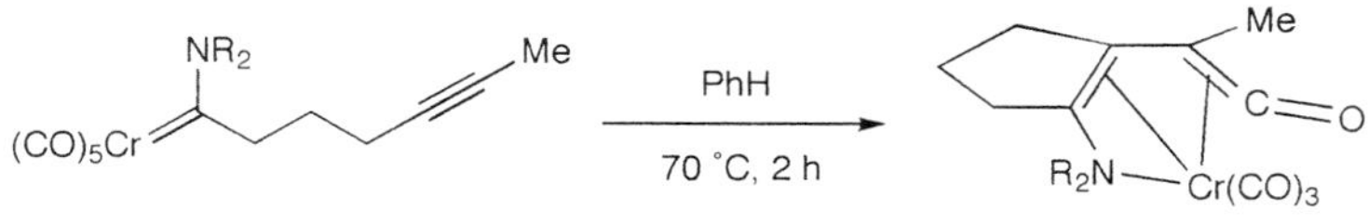

43.a $R_2 = (CH_2)_4$
43.b $R_2 = Me_2$

42.a $R_2 = (CH_2)_4$, 52%
42.b $R_2 = Me_2$, 43%

SCHEME 15.

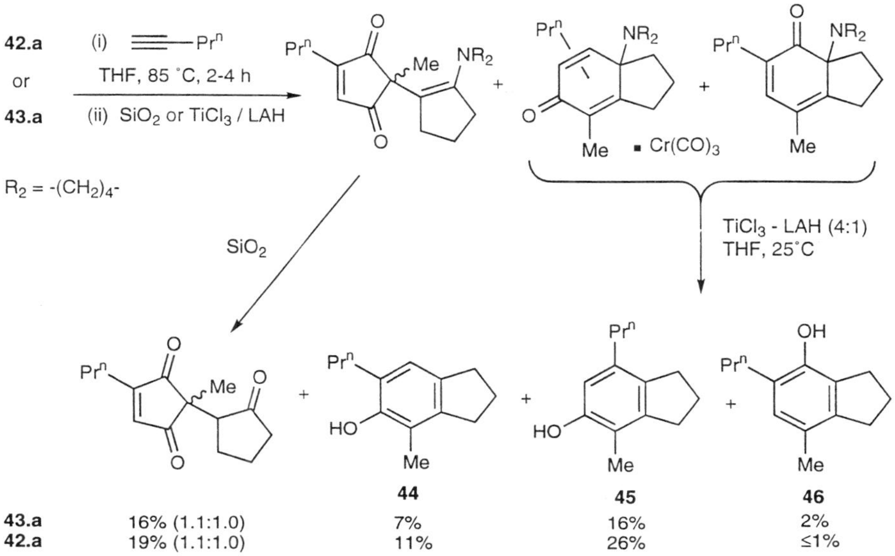

42.a (i) ≡—Pr^n

or   THF, 85 °C, 2-4 h

43.a (ii) SiO$_2$ or TiCl$_3$ / LAH

R_2 = -(CH$_2$)$_4$-

SiO$_2$

TiCl$_3$ - LAH (4:1)
THF, 25°C

		44	45	46
43.a	16% (1.1:1.0)	7%	16%	2%
42.a	19% (1.1:1.0)	11%	26%	≤1%

SCHEME 16.

of $^{13}CO_{(g)}$ led to a 14% recovery of **42.a**, in which there had been no isotope incorporation into the ketene carbonyl or the carbonyl ligands.

III

η^4-(VINYLKETENE)MOLYBDENUM AND TUNGSTEN COMPLEXES

A. Introduction

Sharing the same periodic group as chromium has meant that many of the Dötz annulation reactions seen in Section II,B have been carried out using analogous molybdenum[12,19,40,70] and tungsten[12,19,25,71] carbenes. Geoffroy has unsuccessfully attempted[71] to utilize the improved stability of certain tungsten complexes to isolate Dötz reaction intermediates, in order to distinguish between the mechanisms proposed by Dötz and Casey (see Section II,B). Wulff has published[12] a detailed summary of the effect of the metal on the product distribution in the Dötz reaction, and so no further discussion is required here. Instead, we shall focus on the small but varied selection of syntheses of η^4-(vinylketene)molybdenum and tungsten complexes.

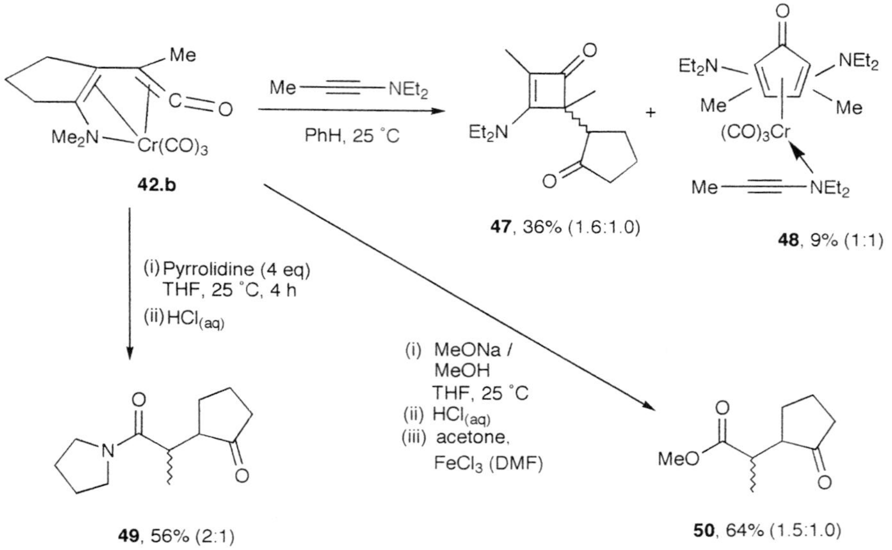

SCHEME 17.

B. *Synthesis by Carbonyl Insertion into an η^3-Carbene Complex*

Asaro and Mayr have reported[72] that the η^3-vinylcarbene (allylidene) tungsten complex[73] **51** reacts with 2 equiv of sodium diethyldithiocarbamate to yield the η^4-(vinylketene)tungsten(0) complex **52.a** in about 1 h. The same procedure was used to generate **52.b** and **52.c** from the appropriate dithiocarbamates. NMR studies showed that the products existed as mixtures of isomers, arising from the two possible arrangements of the dithiocarbamate ligands (**52.I** and **52.II**). X-ray crystallography[72,74] showed a complex the bonding of which is much closer to η^4-vinylketene than $\eta^3 : \eta^1$-allylacyl in character. When complex **52** was heated in d_8-THF under an atmosphere of $^{13}CO_{(g)}$, incorporation at both the ketene carbonyl and the carbonyl ligand was seen, proving that the vinylcarbene carbonylation process is reversible in this system.

The mechanism proposed[72] involves initial nucleophilic attack at the carbene carbon by the dithiocarbamate anion, effectively resulting in addition across the metal–carbon bond. Rearrangements of the dithiocarbamate ligands then form an $\eta^3 : \eta^1$-allyldithiocarbamate species: Complex **53** was isolated from the reaction mixture of **51** with the diethyldithiocarbamate and identified by X-ray crystallography.

Cleavage of the carbon–sulfur bond in the allyldithiocarbamate unit and

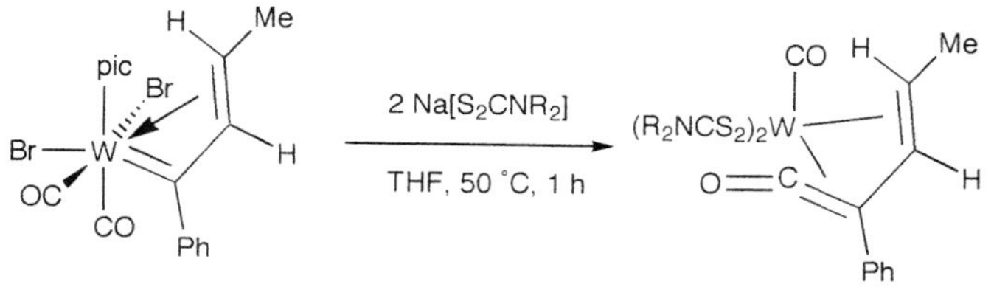

SCHEME 18.

pic = 4-picoline

52.a R = Et, 90.4%
52.b R = Me, 96.5%
52.c R = Ph, 39.5%

chelation of the dithiocarbamate fragment then induces carbonylation to yield the vinylketene complex **52.a**.

C. Synthesis from Cyclopropenes

The insertion of a metal carbonyl fragment into a cyclopropene ring was used in the earliest known synthesis of an η^4-vinylketene complex[1] (see Section VI,B), and this method has been applied to a wide variety of transition metal systems (see Sections IV,B, V,C, and VI,B). In 1978, Binger reported that both hexacarbonylmolybdenum and hexacarbonyltungsten failed to react with 3, 3-dimethylcyclopropene, whereas other metal carbonyls such as diironnonacarbonyl (see Section VI,B) and dicarbonylcyclopentadienyl(tetrahydrofuran)manganese (see Section IV,B) had undergone the expected insertion to yield η^4-vinylketene complexes.[75] However, in 1984 Templeton showed that various molybdenum and tungsten dithiolate complexes (**54**) inserted into cyclopropene and 1-methylcyclopropene to afford a range of vinylketene products (**55**).[76] No yields were reported, although the complexes **55.f–55.i** were prepared in greater than 50% yield. The

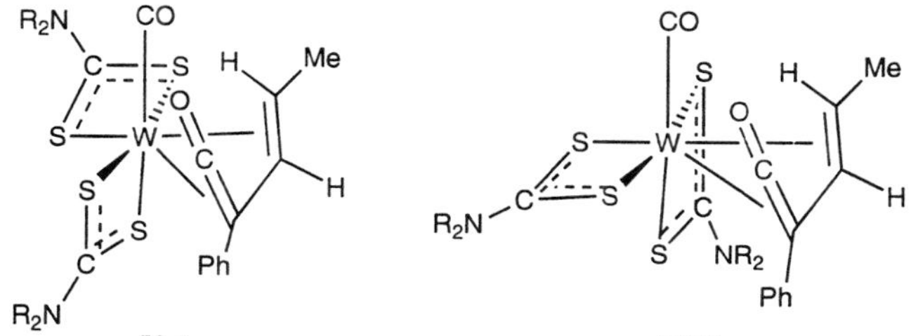

SCHEME 19.

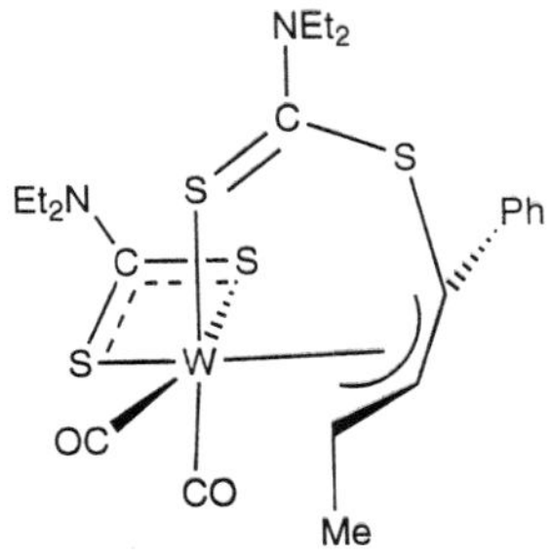
NEt2
C
S
S
Et2N
Ph
C
S
S
W
OC
CO
Me

53

SCHEME 20.

molybdenum complexes were extremely unstable, and satisfactory analytical data were not provided for complexes **55.a** and **55.b**.

D. *Synthesis from (1,2,3-η^3)-trans-Butadienyl Complexes*

In 1992, Green reported the unexpected synthesis of an η^4-vinylketenemolybdenum (II) complex while investigating the reactivity of η^4-butadienylmolybdenum (II) species.[77] He had previously found that upon deprotonation of the cationic η^4-butadienemolybdenum(II) complex **56**, the neutral η^3-pentadienyl complex **57** was formed.[78] He also established that this was in fact a fully reversible process, and that upon treatment with

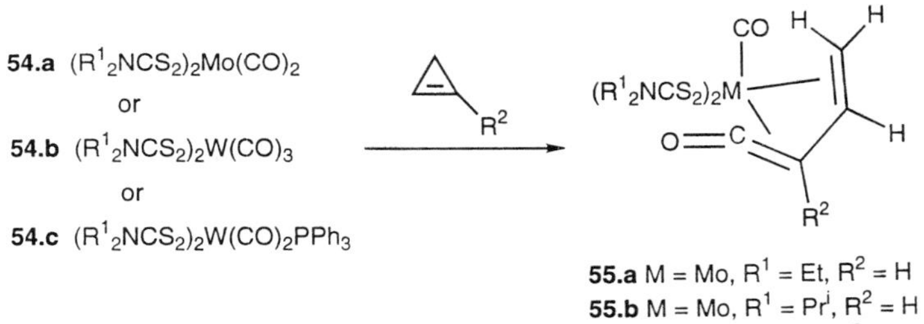
54.a (R1 2NCS2)2Mo(CO)2
or
54.b (R1 2NCS2)2W(CO)3
or
54.c (R1 2NCS2)2W(CO)2PPh3
R2
CO
H
H
(R1 2NCS2)2M
O
C
H
R2

55.a M = Mo, R^1 = Et, R^2 = H
55.b M = Mo, R^1 = Pr^i, R^2 = H
55.c M = W, R = Me, R^2 = H
55.d M = W, R = Et, R^2 = H
55.e M = W, R = Ph, R^2 = H
55.f M = W, R = Me, R^2 = Me
55.g M = W, R = Et, R^2 = Me
55.h M = W, R = Pr^i, R^2 = Me
55.i M = W, R = C_4H_4, R^2 = Me

SCHEME 21.

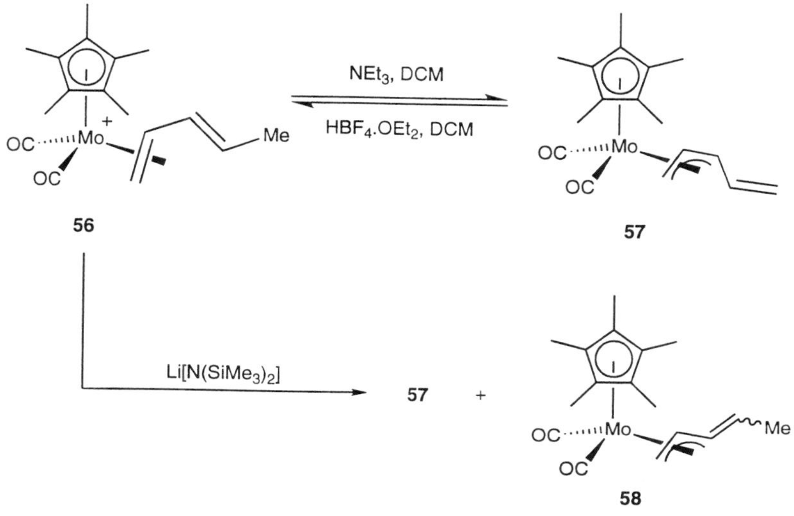

SCHEME 22.

tetrafluoroboric acid, **57** reverts to the η^4-butadiene precursor **56**. When triethylamine was substituted by the stronger amide base Li[N(SiMe$_3$)$_2$] in the reaction with **56**, the expected product **57** was isolated along with another neutral species tentatively identified as the (1,2,3-η)-*trans*-butadienyl complex (**58**). The first example of a (1,2,3-η)-*trans*-butadienyl complex had been reported by Nesmayanov[79] in 1976, and these complexes have subsequently received attention from several different groups.[80–84]

In an attempt to further elucidate the chemistry of these η^3-butadienyl complexes, similar reactions were attempted with the analogous complex of an unsubstituted butadiene ligand. It was found[77] that upon treatment of the cationic complex **59** with Li[N(SiMe$_3$)$_2$] the neutral (1,2,3-η^3)-butadienyl species **60** could be isolated in 27% yield. Complex **60** may also be obtained by the desilylation of cation **61** with tetrabutylammonium fluoride. However, on treatment of complex **60** with acid, the reverse reaction did not occur. Instead, the η^4-vinylketene species **62** was isolated, the trifluoromethanesulfonate ligand being subsequently replaced by iodide (complex **63**) for ease of analysis and crystallographic identification. This provided evidence that the vinylketene ligand had an extremely delocalized bonding structure, with significant contributions from both the corresponding η^3 : η^1-allylacyl and η^1 : η^2 : η^1-vinylketene structures (see Section I).

The unsubstituted nature of the butadiene ligand in **60** is undoubtedly the cause of this marked change in reactivity. Whereas deprotonation of

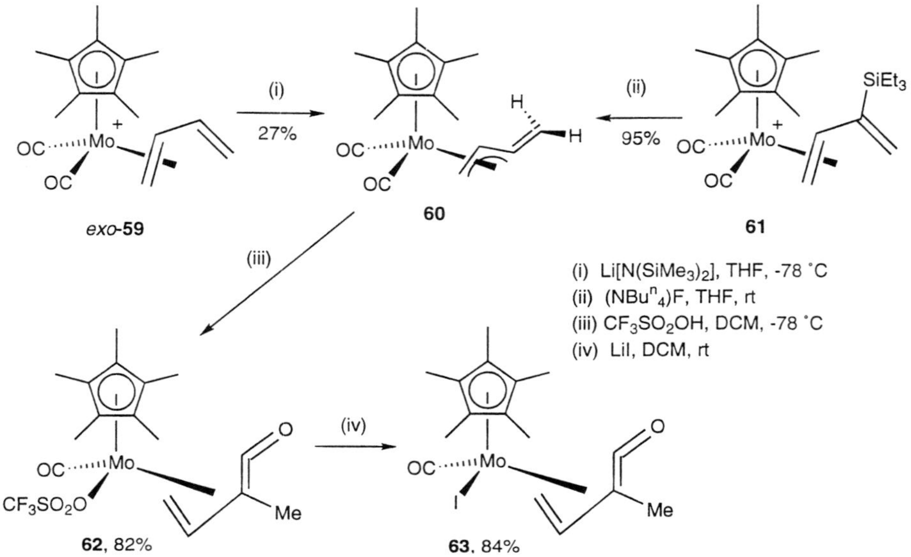
exo-59
(i)
27%
60
(ii)
95%
61
SiEt3
(iii)
(i) Li[N(SiMe3)2], THF, -78 °C
(ii) (NBu^n4)F, THF, rt
(iii) CF3SO2OH, DCM, -78 °C
(iv) LiI, DCM, rt
(iv)
62, 82%
63, 84%

SCHEME 23.

56 occurs at the terminal methyl group to yield the neutral η^3-pentadienyl complex **57**, deprotonation of *exo*-**59** occurs first on one of the methyl groups of the pentamethylcyclopentadienyl ligand, which then undergoes an *intramolecular* deprotonation on the C-2 position of the butadiene ligand, thus forming the (1,2,3-η^3)-*trans*-butadienyl complex (**60**). It should be noted that this deprotonation will only occur if the butadiene ligand is in the *exo* orientation, which is in a mobile equilibrium[78,85] with the *endo* isomer.

Complex *exo*-**60** is then protonated to give the η^3-vinylcarbene complex *exo*-**64**, which subsequently inserts carbon monoxide in the well-established manner (see Sections II,B, V,B, VI,B, VI,C, VI,E, VI,J, and VII), affording the 16-electron species *endo*-**65**. Anion trapping of the unsaturated species finally yields the vinylketene complex *endo*-**62**.

It is of note that *exo*-**60** provides *endo*-**62** exclusively, implying that at some point an isomerism takes place, although no evidence is provided to support this postulate. Thus, *endo*-**64** and *exo*-**65** are suggested as possible intermediates.

Preliminary studies[77] into the reactivity of complex *endo*-[62] have shown that treatment with triethylamine regenerates the *exo*-(1,2,3-η^3)-*trans*-butadienyl complex *exo*-**60** in almost quantitative yield, suggesting that the transformation *endo*-**65** to *endo*-**62** is reversible. The reactions of complex **62** with nucleophiles also suggest that this step is reversible, as this reactivity

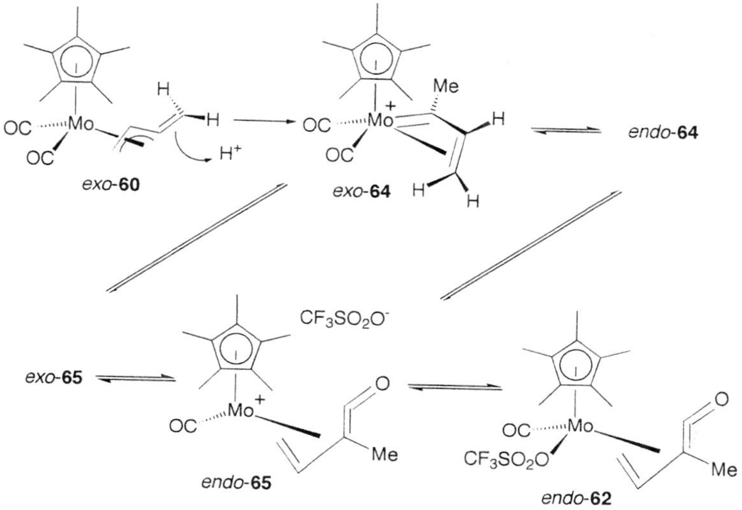

Scheme 24.

more closely resembles that generally associated with η^3-vinylcarbenes. Treatment of **62** with triphenylphosphine resulted not in a ligand substitution, but in a nucleophilic attack by the phosphine resulting in the formation of the phosphonium-substituted η^3-allyl complex **66.b**. This is in direct contrast to η^4-vinylketene complexes, which undergo nucleophilic attack at the ketene carbonyl carbon (C-1).[86–90] When the iodo-complex **63** was treated with triphenylphosphine in the presence of silver tetrafluoroborate the same product (**66.a**) was isolated. These observations suggest that the η^4-vinylketene may equilibrate with its η^3-vinylcarbene analogue in solution. Nucleophilic attack then occurs at the carbenoid carbon of this species, a process that has some precedent.[71]

In 1996, Liu reported the preparation of three η^4-vinylketenetungsten(0)

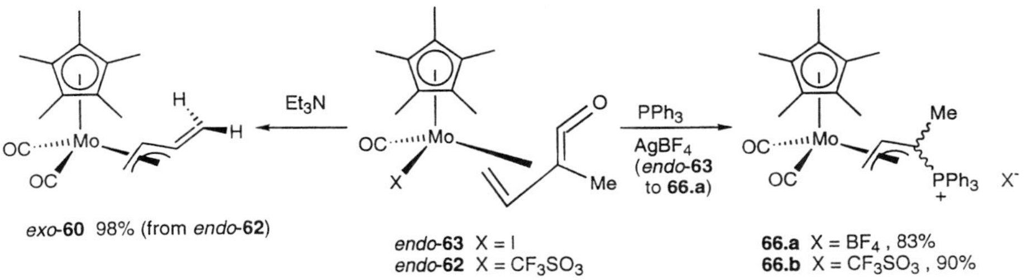

Scheme 25.

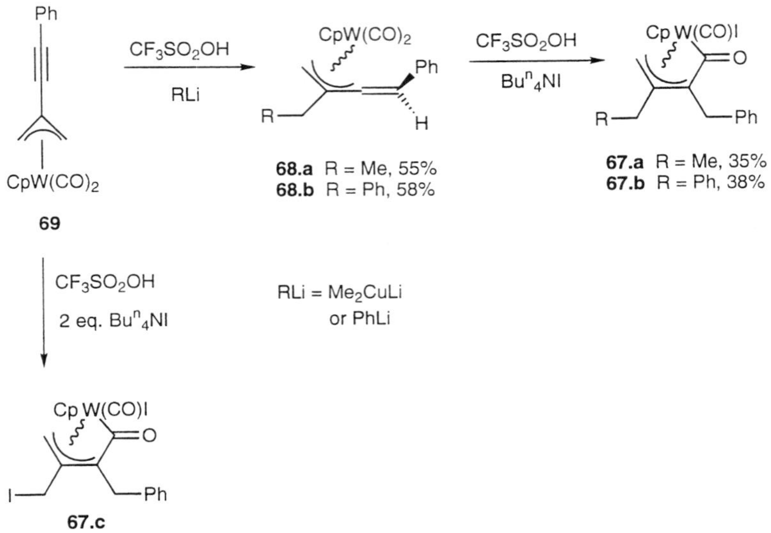

SCHEME 26.

complexes (**67.a**, **67.b**, and **67.c**)[91] using the same method as Green.[77] It was shown that η^3-*trans*-butadienyltungsten complexes (**68**) could be prepared by performing a protonation/nucleophilic attack sequence on dicarbonylcyclopentadienyl[η^3-2-(phenylethynyl)allyl]tungsten(II) (**69**). Once isolated, these complexes were then protonated in the presence of iodide to yield the η^4-vinylketene complexes **67.a** and **67.b**. For crystallographic

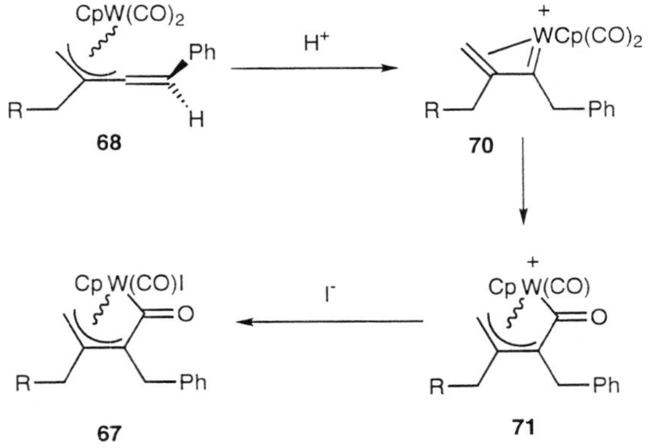

SCHEME 27.

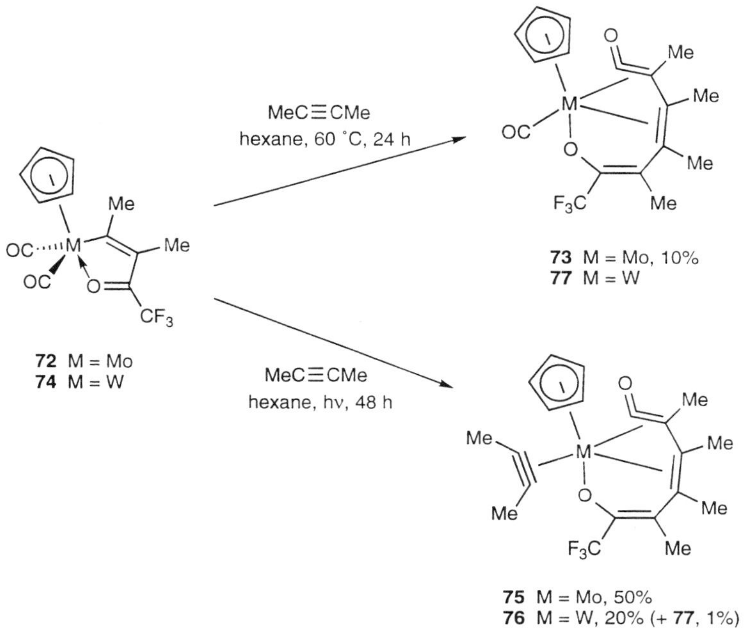

73 M = Mo, 10%
77 M = W

72 M = Mo
74 M = W

MeC≡CMe
hexane, 60 °C, 24 h

MeC≡CMe
hexane, hv, 48 h

75 M = Mo, 50%
76 M = W, 20% (+ **77**, 1%)

SCHEME 28.

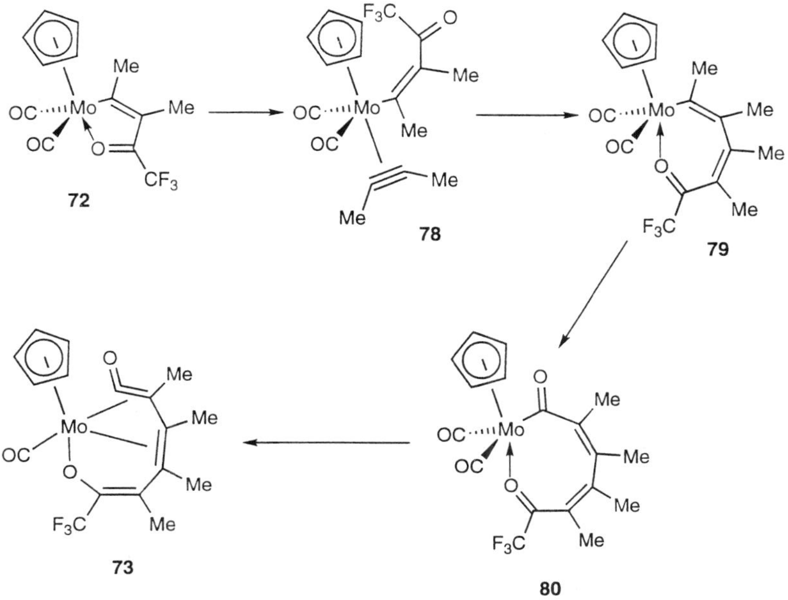

72

78

79

73

80

SCHEME 29.

purposes, the complex **67.c** was synthesized directly from the η^3-allyl complex **69** by reaction with trifluoromethane sulfonic acid in the presence of 2 equiv of iodide. X-ray analysis showed that complex **67.c** had a structure most consistent with the $\eta^3 : \eta^1$-allylacyl formalism (see Section I).

Following the precedent set by Green,[77] the reaction is assumed to proceed through the η^3-vinylcarbene complex **70**, which is formed by the protonation of the *CHPh* carbon of **68**. This then undergoes carbonyl insertion to afford the 16-electron complex **71**, whose coordination sphere is subsequently saturated by iodide, affording the η^4-vinylketene product (**67**).

E. *Synthesis from (1,4)-η^2-Vinylketone Complexes*

As part of a study of the reactions of metallacyclic γ-ketovinyl complexes of molybdenum and tungsten with acetylenes, directed toward the synthesis of complexed γ-lactones, Stone has reported[92] the isolation of several vinylketene complexes. When complex **72** was heated with 2-butyne, one molecule of the alkyne was incorporated into the complex with concomitant carbonylation. X-ray analysis of the product (**73**) has shown unequivocally that the C-1 to C-4 vinylketene fragment is bonded in a planar, η^4-configuration. In contrast to the thermal reaction, ultraviolet irradiation of **72** or **74** in the presence of 2-butyne affords the complexes **75** and **76**, respectively, where the lone carbonyl remaining after alkyne insertion had been replaced by a third molecule of the alkyne.

Stone proposed a mechanism involving substitution at the coordination site occupied by the oxygen of the vinylketone ligand (in **72**) by 2-butyne to give complex **78**. The vinylketone chain then migrates onto the coordinated alkyne (affording **79**) before a second migration onto a carbonyl ligand yields a transient 16-electron complex that may be stabilized by scavenging a carbonyl ligand from other reactants (**80**). The final step is thought to proceed via an unprecedented electrocyclic rearrangement, forming two new carbon–carbon double bonds available for metal complexation, thus yielding the vinylketene product **73**.

IV

η^4-(VINYLKETENE)MANGANESE(I) AND η^4- (VINYLKETENE)VANADIUM(I) COMPLEXES

A. *Introduction*

Among the metals investigated, manganese and vanadium have been the least explored in the synthesis of η^4-(vinylketene) complexes, with only three different compounds of this type having been isolated. They are grouped together here because they have all been prepared by insertion of a metal moiety into a cyclopropene ring, arguably the most common method of vinylketene synthesis for the greatest variety of transition metals (vide infra).

B. Synthesis by Cyclopropene Insertion

The reactions between cyclopropenes and carbon monoxide in the presence of transition metals have been of some use in synthesis,[93] and in 1978 Binger initiated a study of the reactions between metal carbonyls and cyclopropenes in order to elucidate the generality of these reactions.[75] It was found that dicarbonyl η^5-cyclopentadienyl(tetrahydrofuran)manganese(I) reacted with 3,3-dimethylcyclopropene at 0°C to produce η^4-(vinylketene) complex **81** in fair yield. The only other transition metal in Binger's study that was found to react with 3,3-dimethylcyclopropene in this manner was iron (see Section VI,B).

A similar investigation by Weiss[94] showed that 1,2,3-triphenylcyclopropene undergoes photolysis-induced insertion with tricarbonyl(cyclopentadienyl)manganese(I) in tetrahydrofuran to give complex **82**. Presumably dicarbonyl η^5-cyclopentadienyl(tetrahydrofuran)manganese(I) is the active solvent species initially formed in this reaction, which is therefore expected to proceed as shown in Scheme 30. The analogous reaction with tetracarbonyl(cyclopentadienyl)vanadium(I) was also performed, forming the vinylketene complex **83** in good yield. It is surprising that the chemistry of this complex has not been developed further, as there is so little known about the η^4-(vinylketene) complexes of early transition metals.

Weiss has also carried out the analogous insertion reaction with dicarbonyl(cyclopentadienyl)cobalt(I), which yielded a mixture of products (see Section V.C)[94].

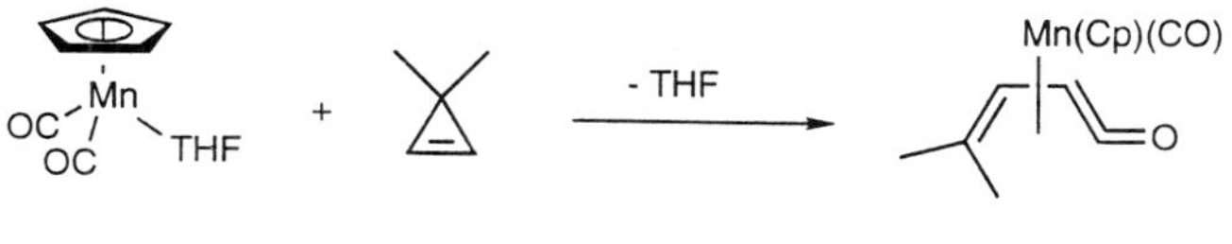

SCHEME 30.

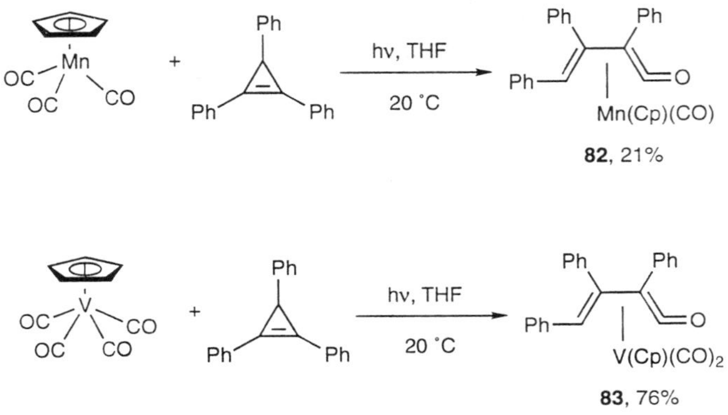

82, 21%

83, 76%

Scheme 31.

V
η^4-(VINYLKETENE)COBALT(I) COMPLEXES

A. Introduction

Although the preparative chemistry of (vinylketene)cobalt(I) complexes is relatively limited in the literature, the methods used include all the major procedures that have been more widely exploited in the analogous chromium and iron systems. There are many similarities between the intermediates involved in the synthesis of vinylketene complexes of iron, chromium, and cobalt, but as the metal is varied the complexes containing analogous ligands often exhibit significant differences in stability and reactivity (see Sections II and VI). Comparison of such species has often been an important aim of the research in this area. The (vinylketene)cobalt(I) complexes have also been shown to be synthetically useful precursors to a variety of naphthols, 2-furanones, α-pyrones, phenols,[6,22,95] β, γ-unsaturated esters,[51] and furans.[51,96a]

B. Synthesis from Carbenes and Subsequent Reactivity

In 1986, Wulff reported the first synthesis of an η^4-vinylketene complex from the reaction between a carbene and an alkyne[51] the former had of course been postulated as key intermediates in analogous reactions of chromium[8] and iron[97] carbenes (see Sections II.B and VI.C, respectively). On treatment of the tricarbonylcobalt carbene (**84**) with 3-hexyne under mild conditions, the η^4-vinylketene complex (**85**) and the γ-keto unsaturated

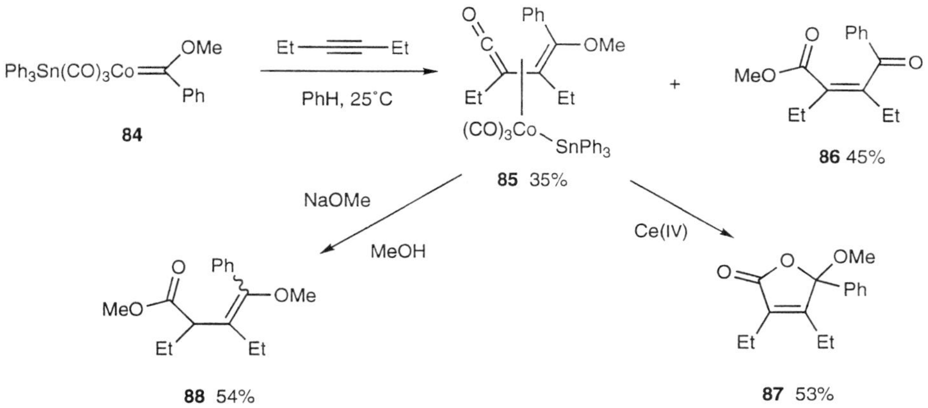

SCHEME 32.

ester (**86**) were isolated. The reaction was found to be entirely independent of the concentration of the alkyne and inhibited by carbon monoxide.

The η^4-vinylketene complex (**85**) could be oxidatively decomplexed with Ce(IV) to afford the lactone (**87**). Although no reaction was observed with methanol (unlike a postulated chromium analogue[16,18,26]), treatment with sodium methoxide produced the expected β, γ-unsaturated ester (**88**). Thermolysis of complex **85** afforded no trace of the naphthol that one would expect[33] from a proposed chromium vinylketene complex with the same *syn* relationship between the phenyl group and the ketene moiety. Instead, only the furan (**89.a**) was seen. Indeed, upon exhaustive reaction of tricarbonylcobalt carbenes (**84** and **90**) with different alkynes, the furans (**89.a–d**) were isolated as the exclusive products in moderate to excellent yields.

The mechanism of these transformations is proposed to proceed via the initial [2 + 2] addition of the alkyne to the carbene to produce a cobaltacyclobutene species (**91**), which may then insert carbon monoxide

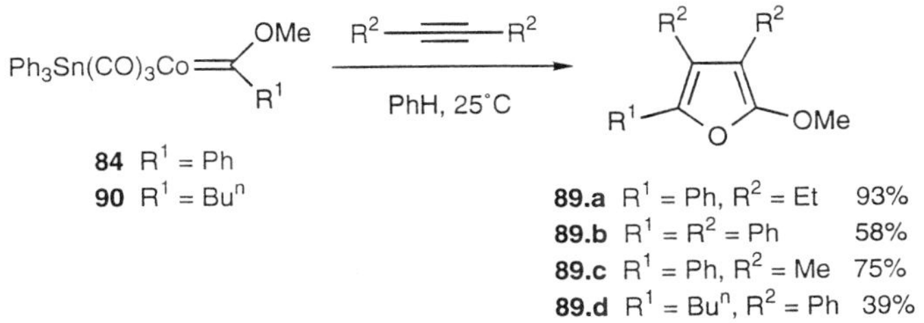

SCHEME 33.

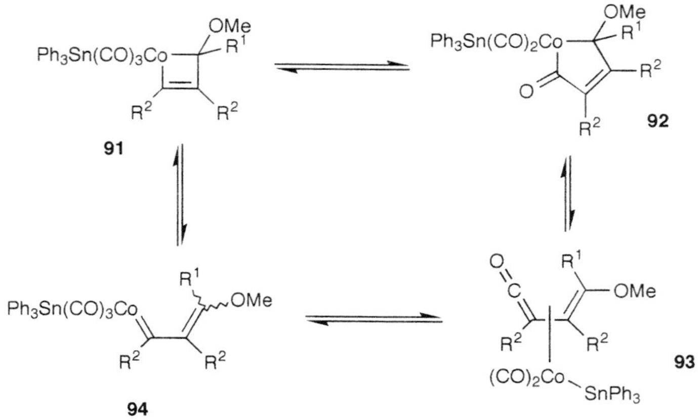

SCHEME 34.

to give the 16-electron species (92) before completing its coordination sphere to give the η^4-vinylketene complex (93). Alternatively, the vinylcarbene complex (94) may form prior to carbon monoxide insertion. Although such formal [2 + 2] additions have subsequently been discredited for the analogous chromium carbene reactions,[24,35,36] cobaltacyclobutenes of this kind have actually been isolated.[96] Subsequent conversion of complex 92 to the furan is suggested to proceed via methoxy migration (92 to 95), methoxy/acyl reductive elimination (95 to 96) and addition of the ester carbonyl to the carbene carbon (96 to 97).[51]

The utility of this reaction was demonstrated in the synthesis of bovolide (98).[51]

C. Synthesis from Cyclopropenes

Although the original synthesis of an η^4-vinylketene complex[1] (see Section VI,B) had used the method of metal insertion into a cyclopropene ring, it was more than 20 years before it was applied to metals such as

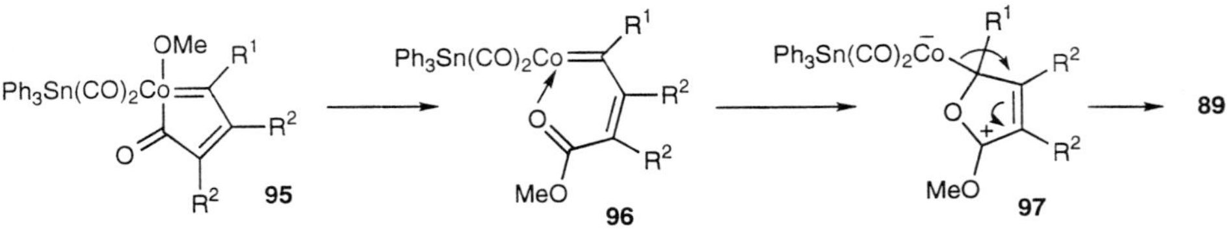

SCHEME 35.

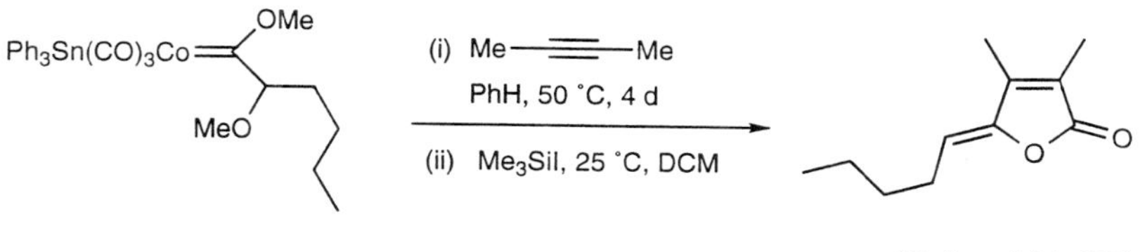

SCHEME 36.

cobalt, managanese, and vanadium (see Section IV,B). Weiss has reported[86] that the reaction of dicarbonylcyclopentadienyl cobalt with 1,2,3-triphenyl-cyclopropene afforded the isomeric vinylketene complexes **99** and **100**, along with the indanone **101**.

Reaction of the same cyclopropene with octacarbonyldicobalt resulted in a 14.6% yield of an inseparable mixture of the dimeric species (**102**) in which the two vinylketene moieties may adopt a *cis* (**102.a** "auf Deckung") or *trans* (**102.b** "auf Lücke") configuration.

In a later paper by Weiss,[68] the methodology was extended to a more complex cyclopropene, and an intermediate cobaltacyclobutene (**103**) was proposed. In an analogous insertion reaction with nonacarbonyldiiron, a vinylcarbene complex was isolated along with the expected vinylketene complex (see Section VI,B). However, no such vinylcarbene cobalt complex was isolated, even when cyclopentadienyl bis(ethene) cobalt was used in place of dicarbonylcyclopentadienyl cobalt, and the only product isolated was the vinylketene complex **104**, represented here in the $\eta^3 : \eta^1$-allylacyl structure.

D. Synthesis from Cobaltacyclobutenes and Subsequent Reactivity

As mentioned earlier (Sections V,B and V,C), cobaltacyclobutenes have been proposed as intermediates in the synthesis of vinylketene cobalt complexes. The isolation of a suitable cobaltacyclobutene by O'Connor[96] (by

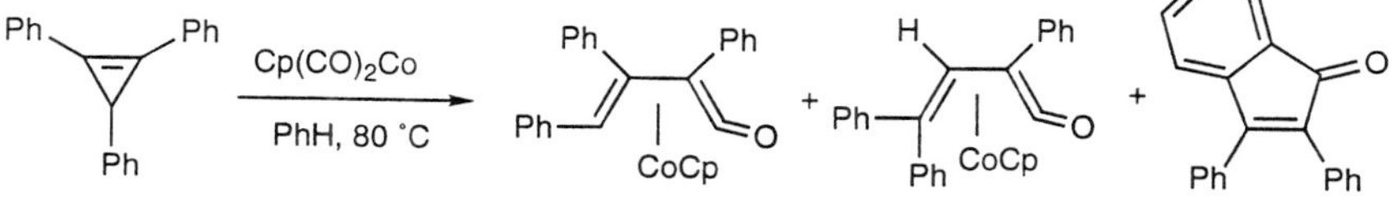

SCHEME 37.

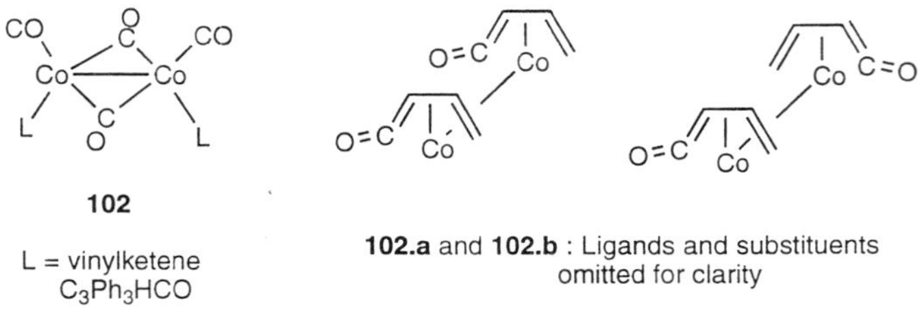

102

L = vinylketene
C$_3$Ph$_3$HCO

102.a and **102.b** : Ligands and substituents
omitted for clarity

SCHEME 38.

reaction of the appropriate cobalt–alkyne complex with ethyl diazoacetate) allowed the further investigation of these complexes. It was found[96a] that upon thermolysis of complex **105** under an atmosphere of carbon monoxide, the η^4-vinylketene complex **106**, the η^4-pyrone complex **107**, and furan **108** were formed, although the furan was not isolated from the crude reaction mixture. Further thermolysis of either **106** or **107** under an atmosphere of carbon monoxide led to an equilibrium mixture, which slowly converted to furan **108** and dicarbonylcyclopentadienyl cobalt.

O'Connor proposed a mechanism involving deinsertion of carbon monoxide from the vinylketene complex **106** to form the new cobaltacyclobutene **109**. The cobalt may then undergo a 1,3-shift to the carbonyl of the ester group to create the oxycobaltacycle **110**, before deinsertion of the cobalt moiety forms the furan **108**. Alternatively, **109** may rearrange to the vinyl-carbene **111**, which then undergoes ester-carbonyl attack on the carbene carbon to form the zwitterionic species **112**, which finally aromatizes to yield the furan **108**. Notice that this latter postulate is identical to the final steps of the mechanism formulated by Wulff (see Section V,B) for the reaction between a cobalt carbene and an alkyne, in which a cobaltacyclobutene is a key intermediate.[51]

E. Synthesis from Cyclobutenones

It has already been seen that cobalt may insert into cyclopropenes to yield, after further insertion of a carbonyl ligand, η^4-vinylketene complexes

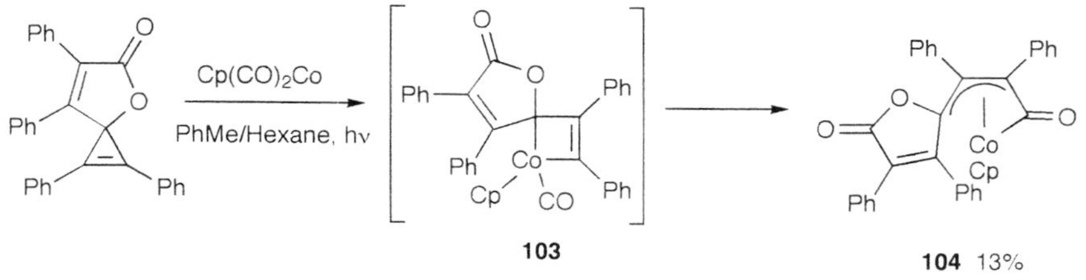

103

104 13%

SCHEME 39.

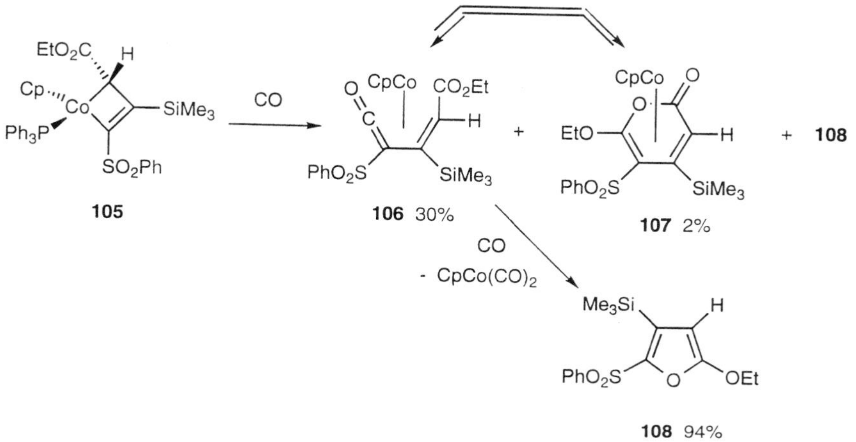

SCHEME 40.

(see Section V,B). In a similar procedure, Liebeskind has shown that insertion into cyclobutenones (**113**) also affords vinylketene complexes, in which the ketene carbonyl originates not from a carbonyl ligand, but instead from the organic substrate[6,95]

This reaction was only seen to proceed using the electron-rich complex $(\eta^5\text{-}C_9H_7)Co(PPh_3)_2$, all attempts with $ClCo(PPh_3)_3$ and $CpCo(CO)_2$ having failed, or availing such low yields of products that they could only be tentatively assigned. Similarly, when R^1 (Schemes 42, 43) was substituted by alkyl groups, low-yielding mixtures of products were seen. Although in

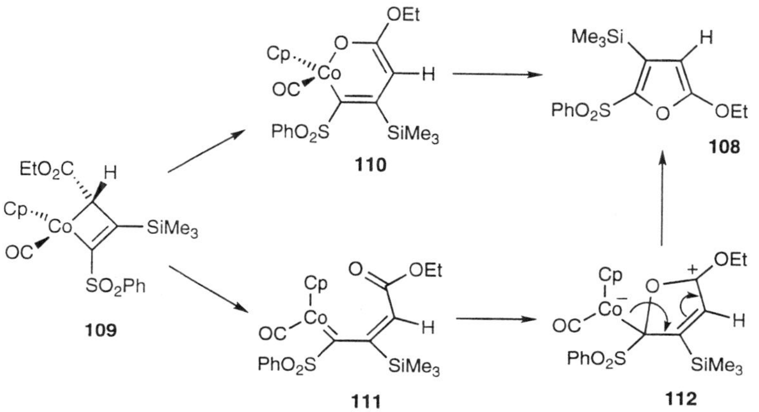

SCHEME 41.

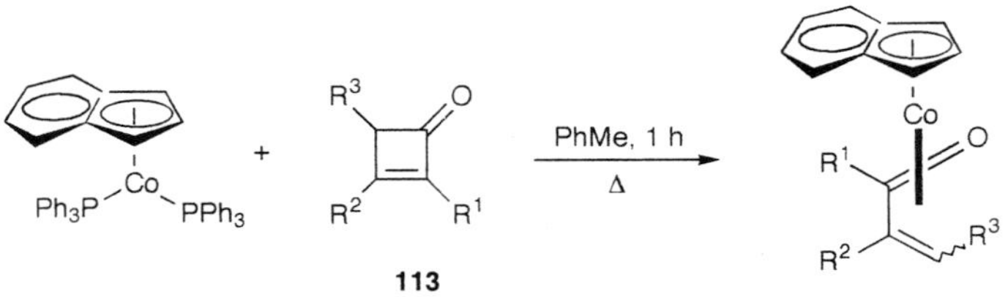

	R¹	R²	R³	anti/syn	Yield
114.a	H	Ph	H		32%
114.b	H	Ph	Me	95:5	80%
114.c	H	SiMe₃	Me	78:22	81%
114.d	H	Buᵗ	Me	>98:2	42%
114.e	H	Buᵗ	Ph	>98.2	17%

SCHEME 42.

all cases the *anti* isomers predominate in the product mixture, it was only when $R^2 = Bu^t$ that it was the sole product. Thermolysis of the *anti* isomer of **114.b** resulted in an enrichment of the *syn* isomer, as confirmed by n.O.e. measurements.

It should be noted that upon reaction with an electron-rich cyclobutenone $(R^1 = R^3 = H, R^2 = OEt)$, the major product formed was a cobaltacyclopentenone, which may also be considered to be an η^2-vinylketene complex. A similar η^2-structure was isolated after heating **114.a** with a large excess of triphenylphosphine, which replaces the ligand site vacated by the central C_2 unit. Interestingly, such η^2-vinylketene complexes are the expected products from the analogous insertion of rhodium into cyclobutenones (e.g., **7**).

Liebeskind favors a mechanism for the formation of **114** that cites the η^2-alkene complex **115** as a crucial intermediate. The cobalt moiety coordinates on the opposite face to the R^1 substituent, which means that upon disrotatory opening[98] of the carbocycle the R^1 group rotates inwards, accounting for the favored *anti* stereochemistry. The possibility of the vinylke-

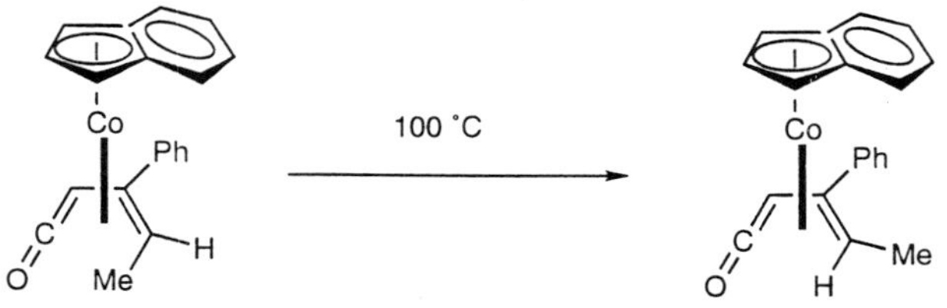

anti **114.b**   *syn* **114.b**

SCHEME 43.

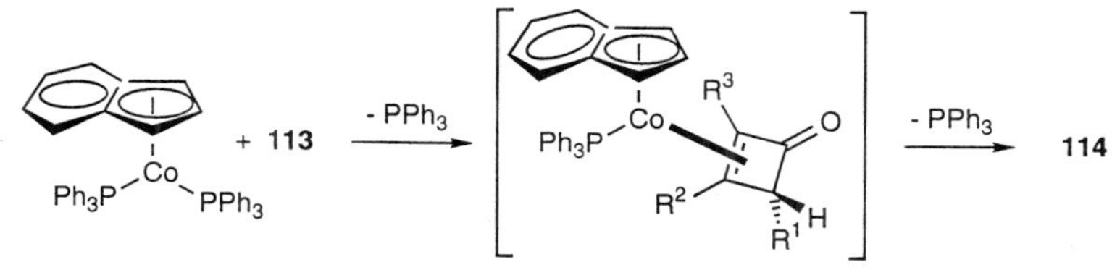

SCHEME 44.

tene fragment being formed by electrocyclic ring opening prior to cobalt coordination may be rejected on the grounds that the reaction also proceeds at low temperature, and in the presence of 2-propanol, which would trap any free ketene. The possibility of a cobaltacyclobutenone being a key intermediate was also rejected, as addition of such a compound to the reaction mixture resulted in no elevation in the yield of **114** and recovery of the unchanged cobaltacyclobutenone.

This chemistry was developed with the specific intention of application to the synthesis of phenols. There had only been one previous example of the addition of alkynes to η^4-vinylketene complexes yielding phenols,[22] despite a cornucopia of other organic fragments having been isolated from such reactions (see Sections II, V, and VI). The best results were obtained on reaction of the 3-phenylvinylketene complex **114.a** with several alkynes.

Upon substitution of the 4-position (**114.b**, R^3 = Me), the cyclohexadienone complex **118** was also isolated from the product mixture obtained with dimethyl acetylenedicarboxylate. Only the cyclohexadienone complex

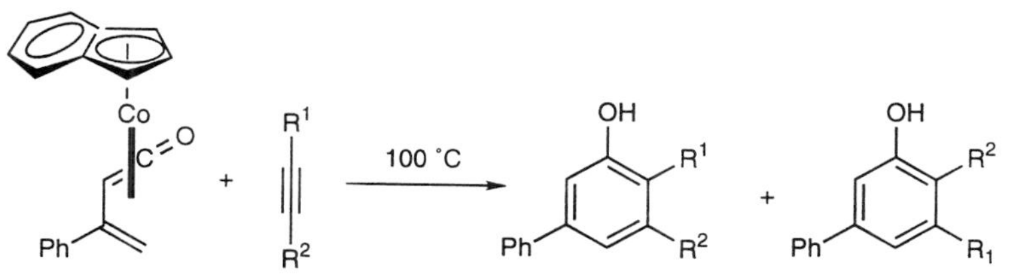

		R^1	R^2	116:117	Yield
116.a		Et	Et		65%
116.b, 117.a		H	Bu^n	71:29	63%
116.c		CO$_2$Me	CO$_2$Me		45%
116.d, 117.d		CO$_2$Et	Me	46:54	92%

SCHEME 45.

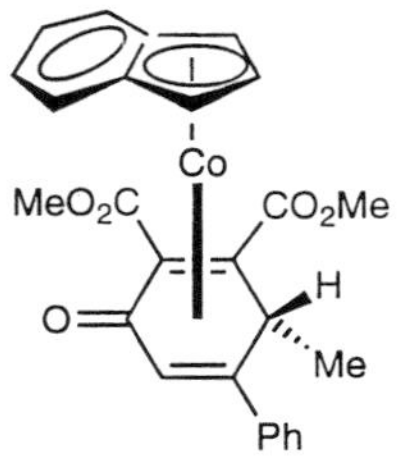

118

SCHEME 46.

with *exo* methyl stereochemistry was isolated, presumably as the *endo* isomer converts to the phenol more rapidly due to steric strain.

These results led to the proposal of the following mechanism. Decomplexation of the central C_2 fragment allows coordination of the alkyne (intermediate **119**), which then inserts to form the metallacycle **120**. Deinsertion (reductive eliminate of the cobalt moiety allows ring closure to give the cyclohexadienone complex **121**, which upon decomplexation yields the desired phenol. The regiochemistry of the alkyne insertion determines the ratio of **116**:**117** (for simplicity, only the sequence leading to **116** has been shown).

In a final experiment, it was shown that the 4-phenylvinylketene complex

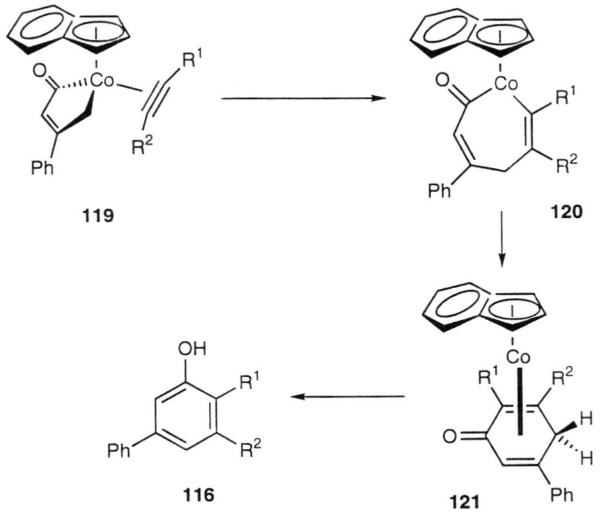

SCHEME 47.

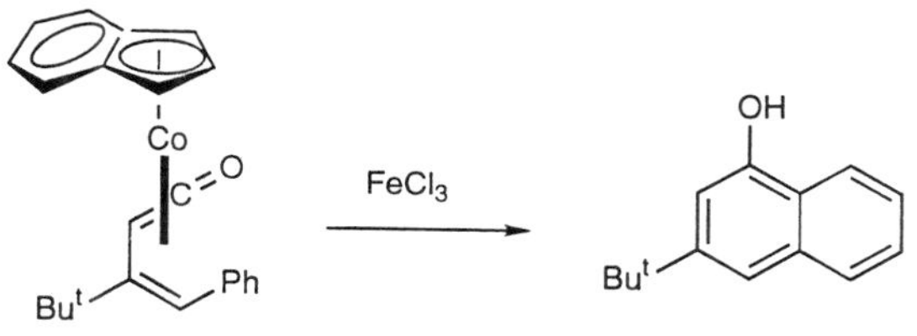

114.e → **122**, 84%

SCHEME 48.

114.e undergoes a benzannulation reaction analogous to the Dötz reaction, yielding naphthol **122**. Although oxidation, and not heating alone, was required for ring formation, this was the first example of an isolated η^4-vinylketene complex being converted to a naphthol.

Free vinylketenes are known to undergo benzannulation reactions.[99] It is worth recalling (see Section II,B) that Liebeskind has demonstrated[58,59] that free vinylketenes generated from the ring opening of cyclobutenones may react with alkynes to yield annulation products that match those obtained from Dötz reactions presumed to involve the same *complexed* vinylketene intermediates. Not only does this observation provide great support for a vinylketene intermediate along the course of such annulation reactions, it also prompts investigations into the exact nature of the stabilizing effects of metal complexation on vinylketene species.

F. Synthesis from a Free Vinylketene

Vollhardt's investigations[100] into electrocyclic transformations on a CpCo template produced the following unusual result. After photolysis of $CpCo(CO)_2$ in the presence of the tosyl hydrazone of *trans*-4-phenyl-3-buten-2-one, the only isolated product was the vinylketene complex **123**. Note that the tricarbonyliron analogue of this complex has also been isolated.[3,87] The mechanism of formation was not discussed, but it seems likely that the ketene carbonyl originated as a carbonyl ligand that replaced the hydrazone moiety, perhaps via a vinylcarbene intermediate.

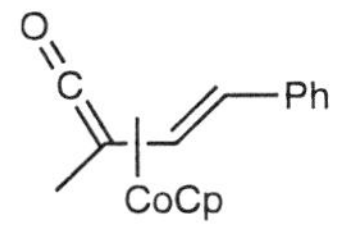

123, 40%

SCHEME 49.

VI

η^4-(VINYLKETENE)IRON(0) COMPLEXES

A. *Introduction*

The richest area of synthetic η^4-vinylketene chemistry is found in their complexes with iron(0) moieties. Their preparation has been achieved by methods discussed earlier (see all previous sections), including insertion into a cyclopropene substrate[1,68,75,101–105] and carbonylation of η^3-vinylcarbene complexes.[48,106,107] Many other methods rely on the *in situ* generation of vinylcarbene complexes (see Sections II,B, III,B, III.D, and V.B). The diversity of preparative methods has allowed the isolation of a great number of η^4-(vinylketene)iron(0) complexes, most of which have proved to be extremely stable crystalline solids, and not particularly air-sensitive.[87] Consequently, while much of the work reviewed so far has been concerned primarily with the isolation and characterization of vinylketene complexes, in this section we shall see a shift in attention to the subsequent reactions of these complexes, providing more unusual organometallic complexes[89,108] or useful organic fragments.[109] The relative abundance and stability of η^4-(vinylketene)iron(0) complexes has allowed their use in mechanistic studies,[106] providing many analogues of intermediates proposed for the Dötz annulation (see Section II,B). Indeed, much of the mechanistic evidence for this reaction has been provided by inference with the related iron species.

B. *Synthesis by Insertion into Cyclopropenes*

It is appropriate that the first isolated η^4-vinylketene complex[1] was prepared using the most common method in this area, and with the vinylketene moiety coordinated to the metal that has provided the widest variety of such complexes. In 1963, King reported that upon reaction of 1,1,3-trimethylcyclopropene with triiron dodecacarbonyl, bright yellow crystals of a compound of composition $C_{10}H_{10}FeO_4$ were isolated in 5% yield.[1] He proposed that the compound had a vinylketene structure (**124** or **125**), spectroscopic evidence being insufficient to distinguish between them. King pointed out the similar stability of this compound to other tricarbonyl(diene)iron(0) complexes, and that in his proposed structure the ketene carbonyl must have resulted from a carbonyl insertion step. Both of these observations are key to the chemistry of such compounds.

It was not until 1979 that he improved this synthesis and provided an unequivocal X-ray structure of the product. The cyclopropene is extremely

124   **125**

SCHEME 50.

volatile, and so use of diethyl ether (rather than benzene) at reflux, along with a $-78°C$ condenser, allowed the isolation of **124** in approximately 60% yield. The use of diiron nonacarbonyl, a more labile source of $[Fe(CO)_4]$, is also significant. The structure of **124** was consistent with the insertion of $[Fe(CO)_4]$ into the most sterically hindered (but also the most electron-rich) carbon–carbon bond, possibly to form the intermediate ferracyclobutene **126**. Although a cursory study of the reactivity of **124** was carried out, the most important findings were that the ketene carbonyl was unreactive toward hydrazine, sodium borohydride, and lithium aluminum hydride.

It is quite possible that this renewed effort by King had been spurred on by the publication of papers by Weiss[103] and Binger[75] during the previous year. As part of a study on the reactions of cyclopropenes with several different metal carbonyls, Binger found that when diiron nonacarbonyl was treated with an excess of 3,3-dimethylcyclopropene at 30–35°C in acetone, the η^4-vinylketene complex **127** was formed in good yield (70%).[75] Similarly, **127** could be synthesized by the room-temperature photolysis of iron pentacarbonyl with the cyclopropene, albeit in lower yield (17%). It was found that yields dropped dramatically in other solvents such as benzene or diethyl ether. It was also observed that **127** was unreactive toward ethereal HCl and aqueous H_2SO_4, and that oxidation with Ce(IV) provided only unidentifiable tars. However, reaction with triphenylphosphine highlighted the potential for alternative reaction pathways. In benzene, ligand replacement to form the dicarbonyltriphenylphosphine complex **128** was seen to occur,

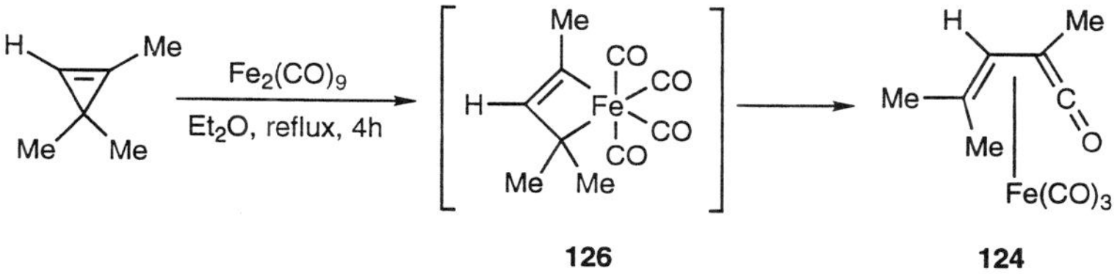

126   **124**

SCHEME 51.

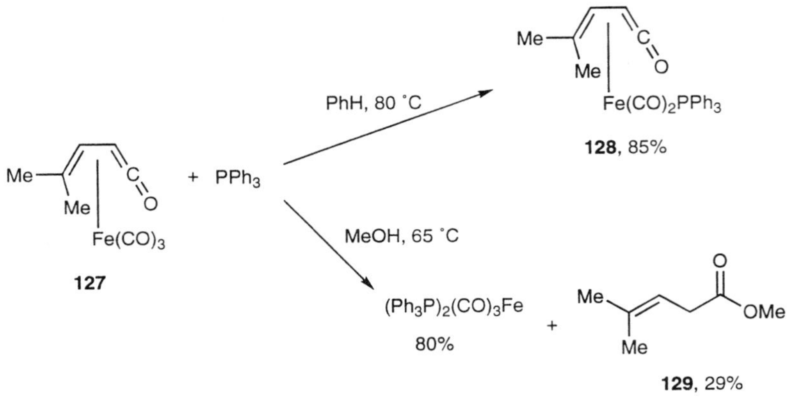

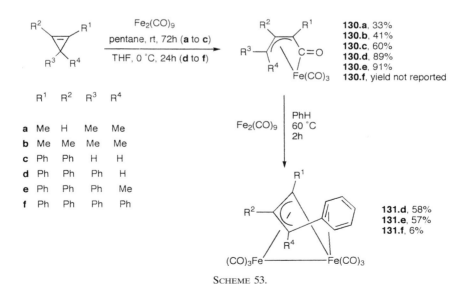

SCHEME 52.

while reaction in methanol afforded tricarbonyl-bis(triphenylphosphine) iron(0) and the β,γ-unsaturated ester **129**.

In a complementary study, Weiss reacted equimolar amounts of diiron nonacarbonyl and various cyclopropenes, yielding the products **130.a–130.f** in fair to good yield.[103] Notice that (on crystallographic evidence), he prefers to represent the products in the $\eta^3:\eta^1$-allylacyl structure (see Section **1**). Once isolated, the phenyl-substituted vinylketene complexes were reacted

SCHEME 53.

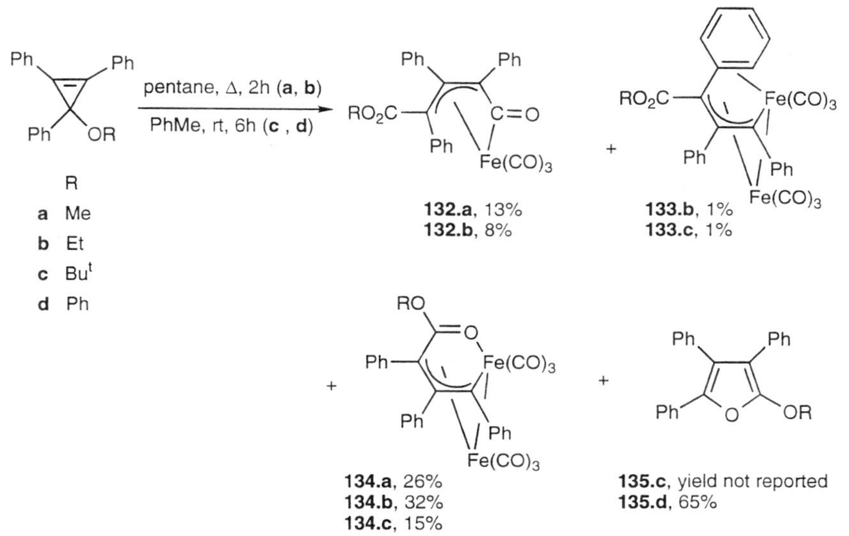

SCHEME 54.

with a further equivalent of diiron nonacarbonyl to afford **131.d–131.f**, containing an Fe–Fe bond.

Weiss later extended this work[102] to incorporate a system where R^4 is an alkoxy functionality, and found that along with the expected product **132**, the diiron species **133** and **134** could also be isolated, along with the substituted furans **135.c** and **135.d**. Complexes **133.b** and **134.a** were characterized by X-ray crystallography, and it is suggested by Weiss that the furan may be formed by the deinsertion of the ketone-complexed $Fe(CO)_3$ moiety. This could be regarded as akin to attack by a free carbene on the oxygen of the carbonyl portion of the ligand (**136**).

In a further extension of this methodology, Weiss reported that spirocyclic

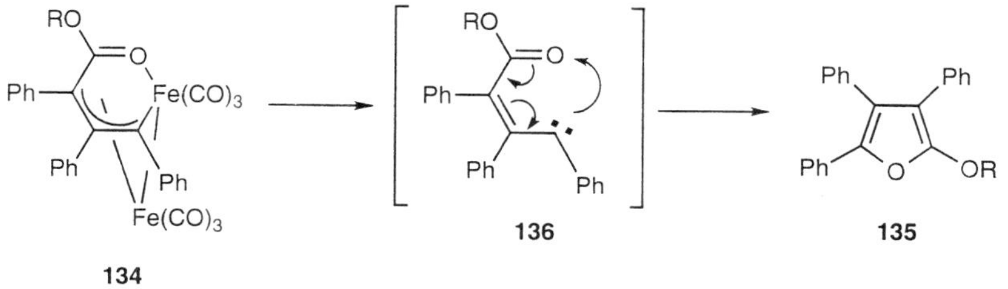

SCHEME 55.

lactones containing a cyclopropene ring (137) would also undergo a similar insertion reaction with diiron nonacarbonyl.[68] In addition to the expected vinylketene product (138), he also isolated a vinylcarbene complex (139) that was unusually stable for a carbene not containing a heteroatom bonded to the carbene carbon. Vinylcarbene 139 was air-stable and did not decompose in boiling benzene. Furthermore, upon reaction with carbon monoxide an insertion occurs to form the vinylketene complex 138, a reaction that is reversible at 60°C. It should be noted that although the vinylketene complex was originally represented (as reproduced here) as an η^4 structure, later crystallographic evidence[68a] prompted Weiss to suggest an $\eta^3 : \eta^1$-allylacyl structure as a more appropriate representation. This reaction had been seen previously in a manganese coordinated system[49] and has been implicated as an intermediate step of the Fisher–Tropsch synthesis.[110] Similar reactivity studies were carried out with an analogous cobalt system, although slightly different results were obtained (see Section V,C).

Weiss studied[68a] the reactivity of both new complexes, and found that a variety of phosphines and phosphites would also convert the vinylcarbene complex 139 into the corresponding vinylketene complex (140), capturing one of the carbonyl ligands from the coordination sphere of the metal to become the ketene carbonyl. Only in the case of triphenylphosphine was the dicarbonyl(phosphine)vinylcarbene complex (141) isolated, which then required addition of carbon monoxide to convert it to the dicarbonyl(triphenylphosphine)vinylketene complex 140.a. This interconversion was reversible and proceeded quantitatively.

When a second equivalent of trimethylphosphine was reacted with 140.e, the ketene functionality of the vinylketene ligand was displaced leaving only the C=C bond coordinated in an η^2 fashion (142). This product could also be isolated by the direct treatment of vinylcarbene 139 with 2 equiv

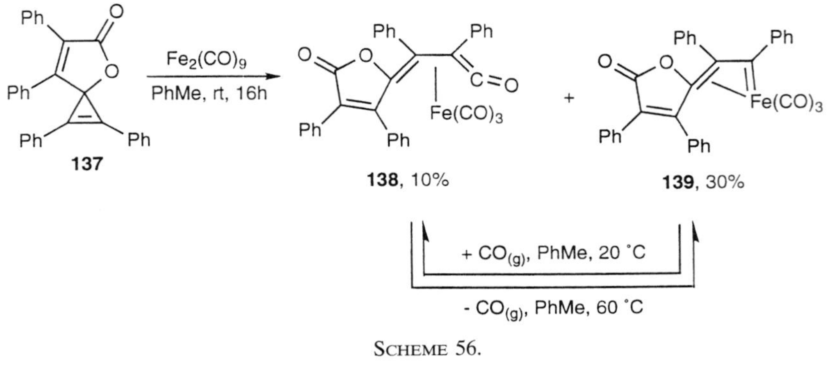

SCHEME 56.

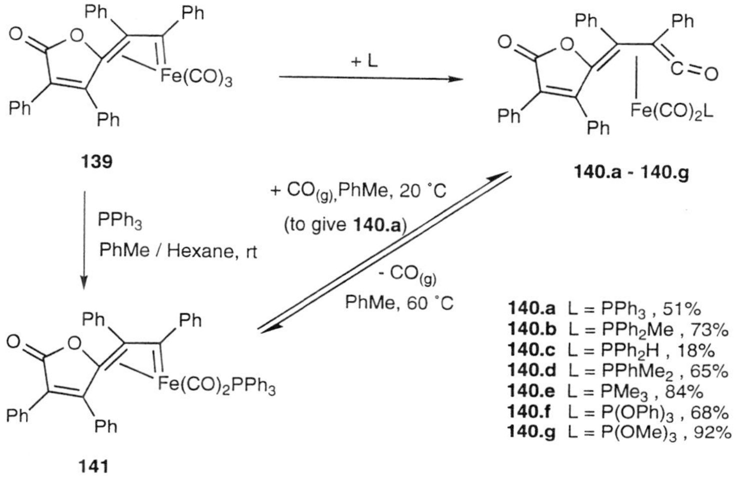

Scheme 57.

of trimethylphosphine, although the yield for this particular reaction was not reported. The two phosphine ligands were shown to have a *trans* orientation.

The vinylketene complex **140.g** was shown to react toward electrophiles at the ketene oxygen. Although iodomethane had no effect, reaction with trimethyloxonium tetrafluoroborate caused methylation to produce the cationic complex **143** in fair yield.

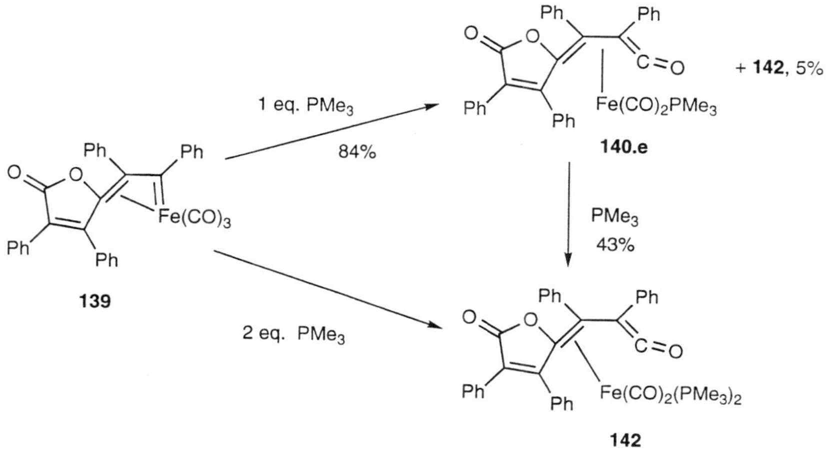

Scheme 58.

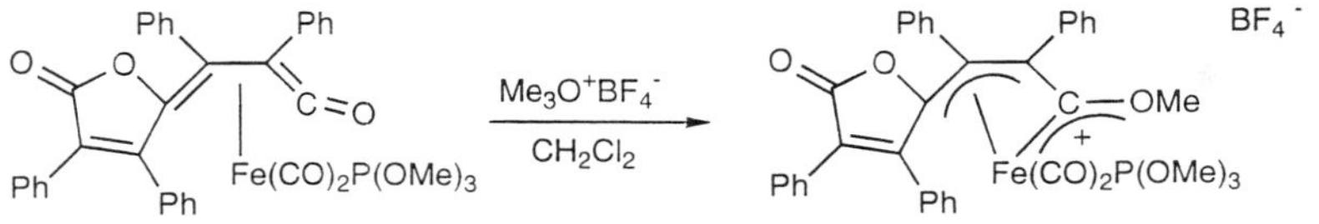

SCHEME 59.

Similar synthetic studies to those performed on the alkoxy-substituted cyclopropenes by Weiss[102] were carried out by Franck-Neumann[101,105] in 1981. In this case, the cyclopropenes surveyed had a carbomethoxy functionality attached to the double bond (**144**), which often resulted in the formation of a mixture of products (**145.a–145.g** and **146.a–146.g**).

Analogous experiments were performed on cyclopropene systems containing an acetyl substituent (**147**). Note that in this case only one regioisomer is isolated (**148**), although in the case where the R substituent is *iso*propyl it was not possible to unequivocally define the product as **148.a** or **148.a′**. Interestingly, the same regiochemistry is observed through the series of alkyl substituents, but is reversed when a vinyl group is present, yielding **148.d**.

Both sets of observations confirm that the ring-opening process is subject to both steric and electronic influences, the combination of which produces some subtle effects in product distribution. Ultimately, the regiochemistry of the product must be decided *after* a cyclopropene–$Fe(CO)_4$ complex is formed, but probably *during* the cleavage of one of the C–C bonds in the cyclopropene ring.

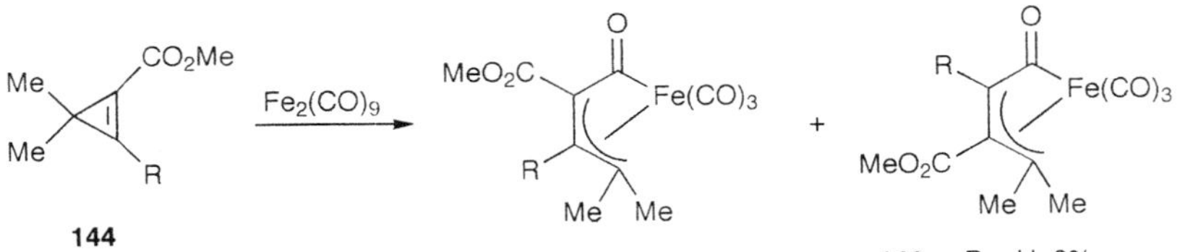

145.a R = H, 63%
145.b R = CO_2Me, 62%
145.c R = Me, 40%
145.d R = Pr^n, 35%
145.e R = Pr^i, 0%
145.f R = Ph, 49%

145.g R = [structure], 60%

146.a R = H, 0%
146.b R = CO_2Me, 0%
146.c R = Me, 42%
146.d R = Pr^n, 46%
146.e R = Pr^i, 81%
146.f R = Ph, 24%

146.g R = [structure], 5%

SCHEME 60.

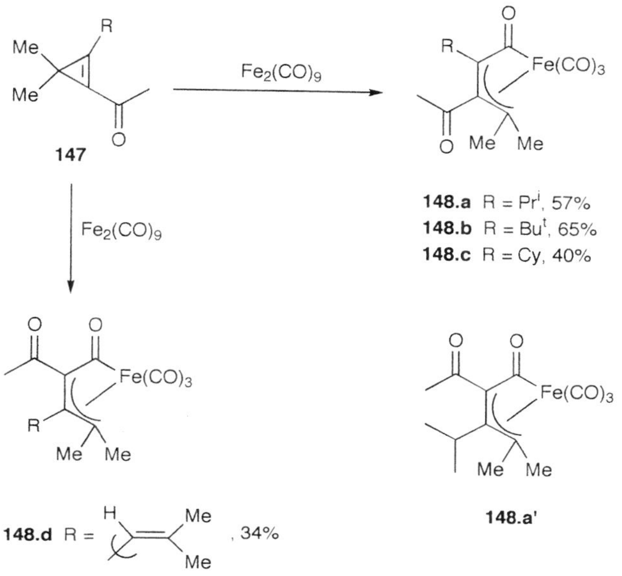

SCHEME 61.

The further reactivity of these complexes was then examined, and it was found that treatment of **148.a** with sodium borohydride resulted in a quantitative and entirely selective reduction of the uncomplexed organic carbonyl functionality to give complex **149**.

The thermolysis of certain members of the carbomethoxy series (**146.c** and **145.c**) results in a decarbonylation reaction that affords the exclusively *E*-diene complexes **150** or the *E/Z*-diene complex mixtures **151**, respectively. These products may be formed by direct thermolysis of the precursor cyclopropene (**144**) in the presence of diiron nonacarbonyl, isolated in 80% overall yield and in a approximate ratio of $3:2:1$ (**150**: *Z*-**151**: *E*-**151**), a product distribution matching that obtained from the isolated vinylketene

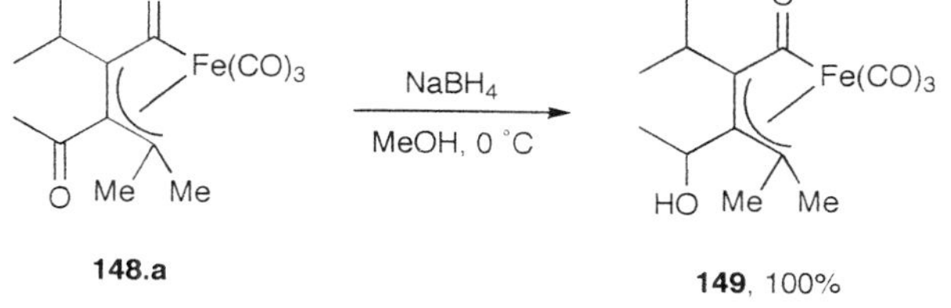

SCHEME 62.

complexes. Indeed, when the complexes **146.d–146.f**, **148.a**, and **148.b** were heated in benzene, the same classes of diene products were seen.

The greater lability of complex **146.c** (compared to **145.c**), as evinced by the much shorter reaction time, is typical of those that bear a carbomethoxy or acetyl substituent at the central carbon of an η^3-allylic ligand. The temperature required for complete decarbonylation of complexes of type **146** and **148** increases with the size of the R-substituent, which suggests a mechanism involving hydride transfer.[111] This would also explain the observed activating effect of the centrally located carbomethoxy group in **146.c**, which would clearly labilize the methyl proton shown explicitly in **146**.

In complexes where the carbomethoxy or acyl group is adjacent to the metal-complexed acyl group (e.g., **145.a**), photolysis affords a 1:1 mixture of isoprenic diene complexes (*E*-**152** and *Z*-**152**) directly. The formation of an intermediate allylketene complex (**153**) may be demonstrated by methanol trapping, followed by aerial oxidation to quantitatively yield the diester **154**.

In further reactivity studies, Franck-Neumann has shown[112] that a range of products such as **155** may be isolated by the trimethylamine *N*-oxide–promoted decomplexation of vinylketene complexes.

As before, when the carbomethoxy group is bonded to the central carbon of the allyl portion of the vinylketene complex, the increased lability of a methyl proton allows the formation of a second product, in this case the diene complex **150**. In the cases where R = Pr^i(**146.e** and **148.a**), decomplexation of the tricarbonyliron moiety allows the thermodynamically favored

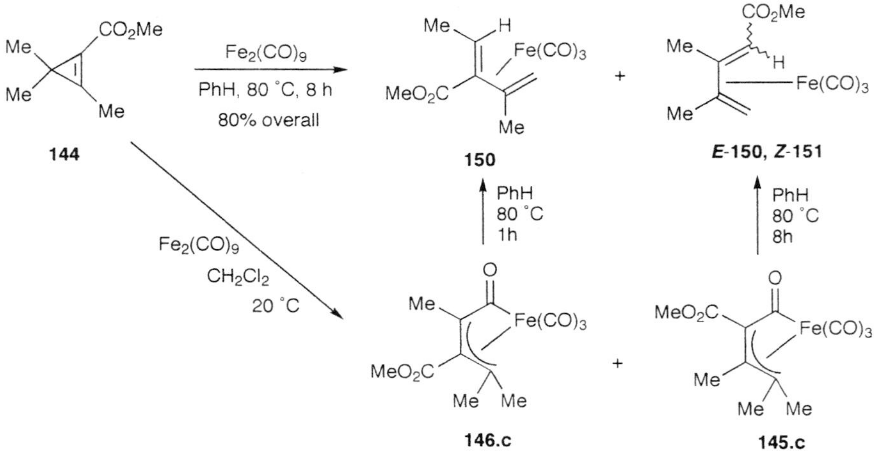

SCHEME 63.

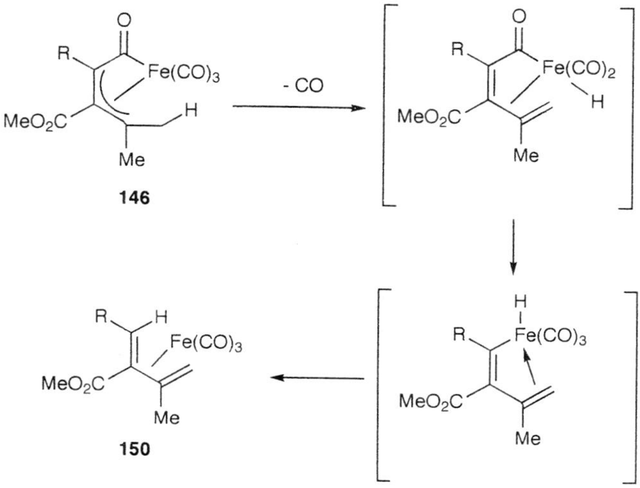

SCHEME 64.

conjugation of double bonds to occur via a H-transfer process, affording **157** and **159**, respectively.

Decomplexation of the carbomethoxylated compounds **145.g**, **145.c**, and **145.b** could also be carried out using an excess of methanolic iron(III) chloride, which at lower temperatures yielded the expected diesters (**160**)

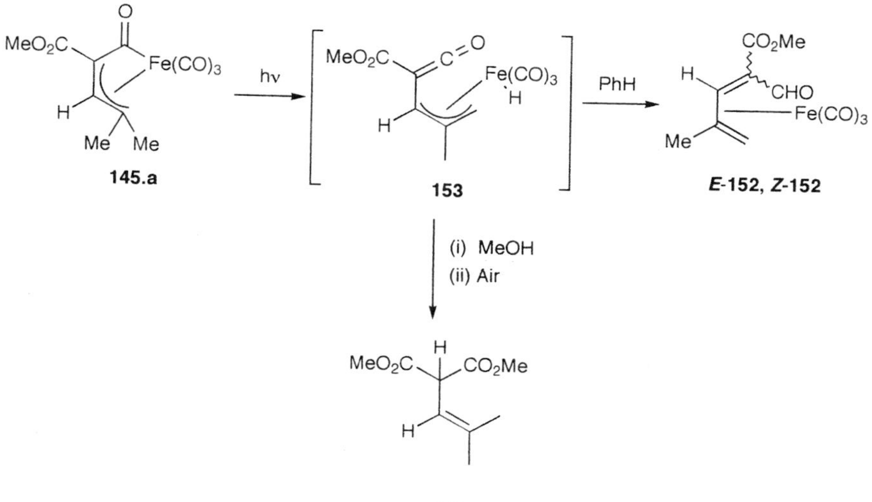

SCHEME 65.

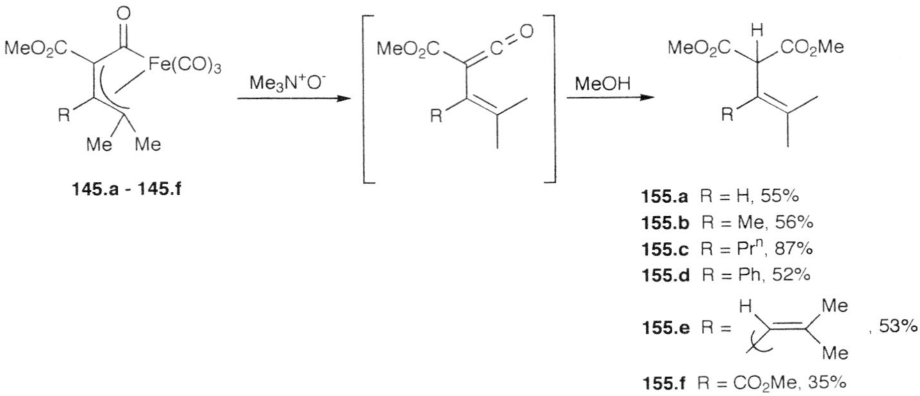

155.a R = H, 55%
155.b R = Me, 56%
155.c R = Pr^n, 87%
155.d R = Ph, 52%

155.e R = [structure], 53%

155.f R = CO$_2$Me, 35%

SCHEME 66.

as the sole products. However, upon continued reaction at elevated temperatures, the cyclic lactones **161** were isolated. The chloride **162** was isolated from the reaction of **160.b** with methanolic iron(III) chloride and could be subsequently converted to lactone **161.b** by heating.

In 1987, Nitta reported the formation of an unexpected vinylketene complex from the reaction of an azido-substituted cyclopropene with diiron nonacarbonyl.[104] They had previously investigated the chemical behavior of the complexed nitrene intermediates that result from the reaction of organic azides and iron carbonyls[113] and were interested in replicating the thermal isomerization of 3-azido-1,2,3-triphenylcyclopropene (**163**) into 4,5,6-triphenyl-1,2,3-triazine   using   a   metal   carbonyl-promoted   re-

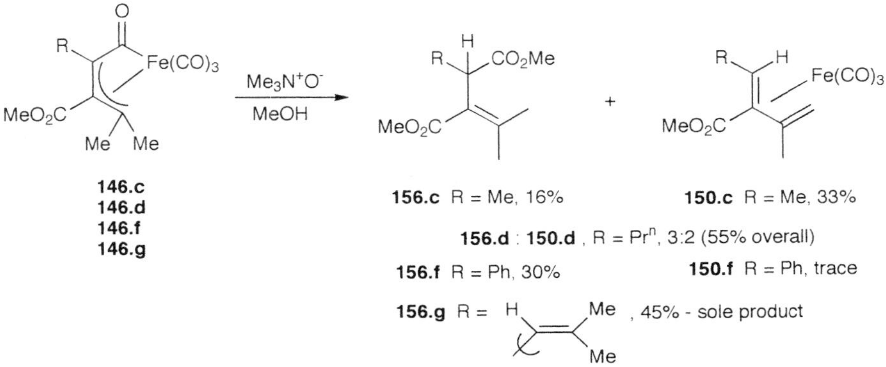

146.c
146.d
146.f
146.g

156.c R = Me, 16%   150.c R = Me, 33%

156.d : 150.d , R = Pr^n, 3:2 (55% overall)

156.f R = Ph, 30%   150.f R = Ph, trace

156.g R = [structure] , 45% - sole product

SCHEME 67.

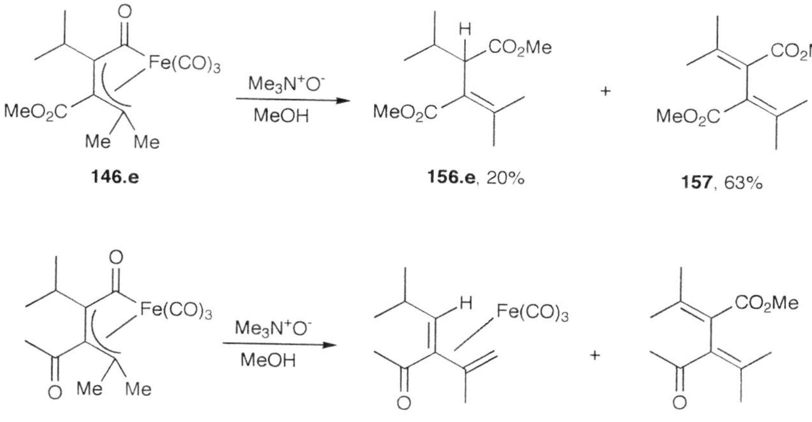

SCHEME 68.

arrangement. They found that only upon treatment of **163** with diiron nonacarbonyl *in methanol* (rather than benzene or acetonitrile) did the mixture of products formed include the vinylketene complex **164**. Nitta proposes than this observation may be accounted for if the azide is initially displaced by methanol to give the methoxycyclopropene **165**, before ring opening by diiron nonacarbonyl in the usual way.

C. Synthesis from η^3-Vinylcarbene Complexes

The synthesis of vinylketene complexes from vinylcarbene species occurs by the carbonylation of the carbene carbon, a process that has been postulated to occur in various organometallic transformations[110,114,115] (see Sections II,B and VI,J). However, it was not until 1978 that Mitsudo and Watanabe showed that this process could take place between isolated species, rather than postulated intermediates.[48] They had previously isolated the first η^3-(vinylcarbene)iron species **166**[116] and found[48,106] that upon treatment with a coordinating ligand the vinylketene complexes (**167**) were formed, the X-ray structure of which suggested[48] an $\eta^3 : \eta^1$-allylacyl bonding arrangement. It is ironic that the mechanism of such a well-researched reaction should still not be entirely certain.[106,117] Mitsudo and Watanabe were in fact the first to present an unequivocal, X-ray evinced structure for an η^4-(vinylketene)iron(0) complex, although both King[1] and Hoffmann[118] had previously reported such compounds (see Sections VI,B and VI,F, respectively). It is worth noting here that Weiss has conducted stud-

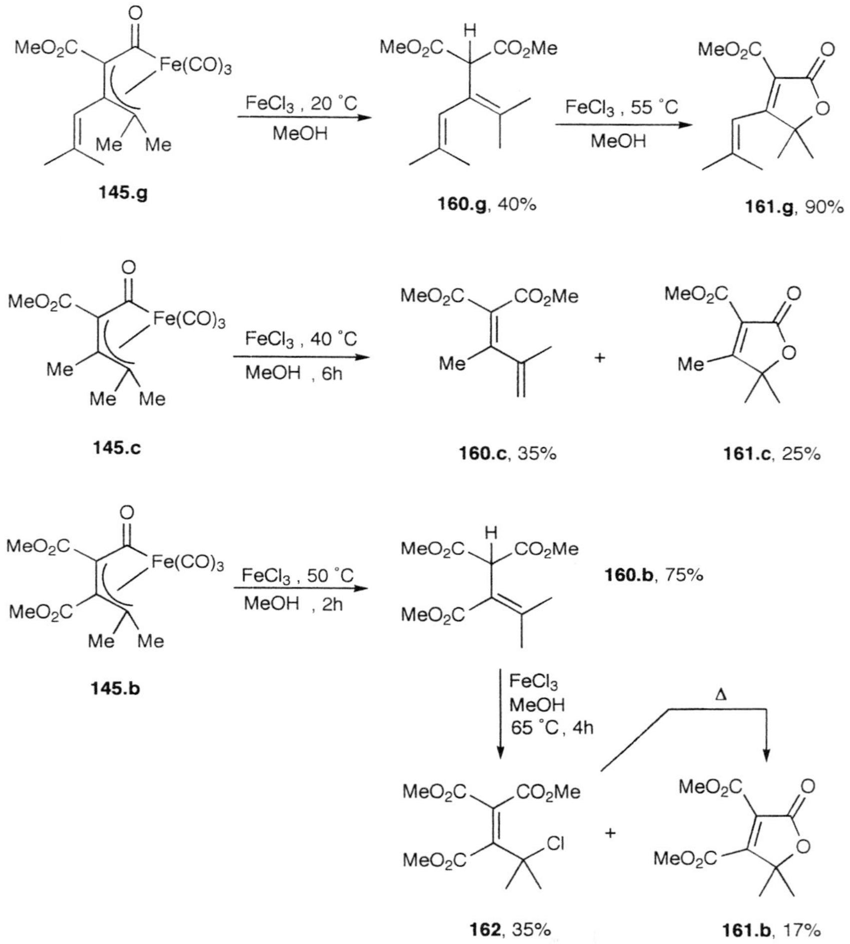

SCHEME 69.

ies[68] on the interconversion of η^3-(vinylcarbene)- and η^4-(vinylketene)iron complexes, which have already been discussed in Section VI,B.

Treatment of complexes 167.c and 167.d with a further equivalent of a coordinating ligand was found to yield substituted ferracyclopentenones (168), in which the L and L' groups adopt a *cis* geometry. The product 168.a could also be synthesized directly from the precursor vinylcarbene complex by treatment with 2 equiv of the appropriate ligand. Note that when 167.a or 167.b was treated with PPh$_3$ or CO$_{(g)}$ (80 atm), only starting

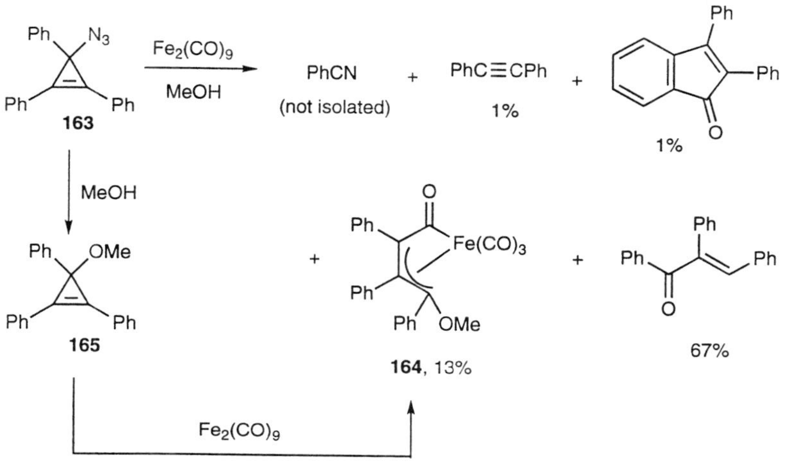

SCHEME 70.

materials were recovered. When **167** and **168** are compared, it appears that a decarbonylation of the acyl group and a carbonylation of the terminal methylene group must both occur during this transformation, making this the first reported example of such a decarbonylation/carbonylation re-arrangement of a vinylketene ligand.

When the initial vinylcarbene complex is substituted with a second methoxycarbonyl group (complex **169**), a different reactivity pattern is observed. Addition of methyldiphenylphosphine or dimethylphenylphosphine to **169** results in formation of the expected vinylketene complex **170**. However, the analogous reaction with triphenylphosphine yielded a complex mixture at room temperature, and upon heating the simple ligand-substituted product **171** is formed. When **169** is reacted with carbon monoxide, the pyrone complex **172** is formed. Finally, reaction of the vinylketene

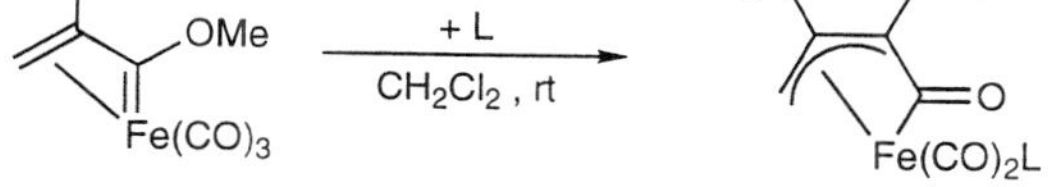

166

167.a L = CO , (1 atm : 38%, 80 atm : 49%)
167.b L = PPh$_3$, 94%
167.c L = PPh$_2$Me, 68%
167.d L = PPhMe$_2$, 71%

SCHEME 71.

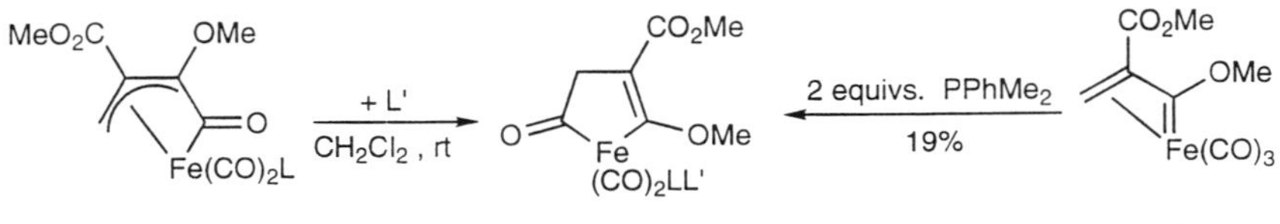

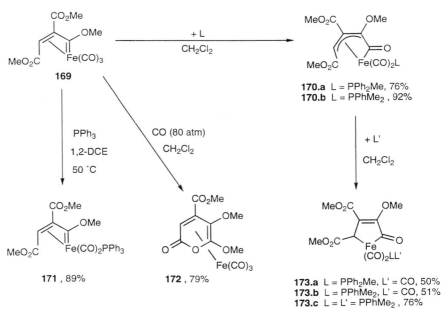

167.c L = PPh₂Me
167.d L = PPhMe₂

168.a L = L' = PPh₂Me, 13%
168.b L = PPhMe₂, L' = CO, 65%
168.c L = PPhMe₂, L' = PPhMe₂, 45%

166

SCHEME 72.

complexes with a second equivalent of a coordinating ligand affords the ferracyclopentenones, **173**.

Note that the regiochemistry of **173** is different from that of the cyclopentenones seen previously (**168**), arguably due to the fact that a decarbonylation/carbonylation does *not* occur in this instance. Once decomplexation of the central C₂ fragment of the vinylketene ligand has taken place, coordination to the free site by the available ligand may occur directly if the Fe–CHR bond is sufficiently stable (when R is an electron-withdrawing carbomethoxy group, **170**), or the decarbonylation/carbonylation pro-

169

170.a L = PPh₂Me, 76%
170.b L = PPhMe₂ , 92%

171 , 89%

172 , 79%

173.a L = PPh₂Me, L' = CO, 50%
173.b L = PPhMe₂, L' = CO, 51%
173.c L = L' = PPhMe₂ , 76%

SCHEME 73.

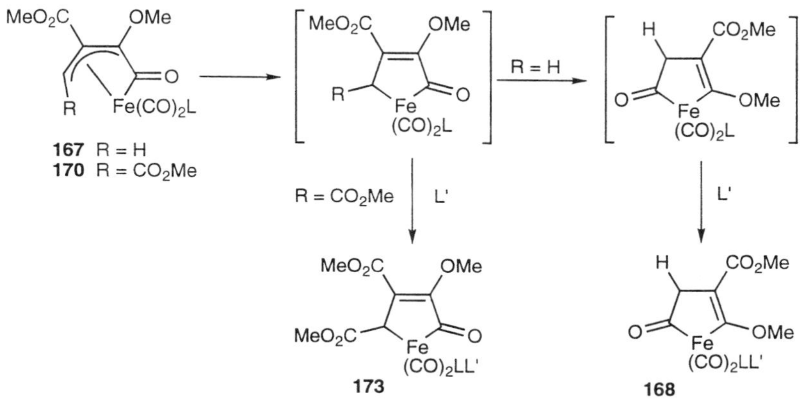

SCHEME 74.

cess may occur before ligand "quenching" (when R is a hydrogen atom with no stabilizing properties, **167**).

As part of ongoing research into the behavior of (vinylcarbene)iron complexes,[119,120] Mitsudo and Watanabe found that the trifluoromethyl-substituted vinylcarbene **174** exhibited a reactivity different from that of both **166** and **169**.[107] Upon treatment of the complex **174** with triphenylphosphine the vinylketene complex **175** is formed, a reaction identical to that seen in the series of vinylcarbene complexes **166** (R = H). However, when the vinylcarbene **174** is exposed to a high pressure of carbon monoxide, it is converted cleanly to the ferracyclopentenone **176**. Remember that when the vinylcarbene complex **166** (R = H) was treated in the same manner, conversion stopped at the vinylketene complex **167**: Even when exposed to a pressure of 80 atmospheres of $CO_{(g)}$, no further reaction was seen to occur. An electron donating ligand (L = PR_3) is required for conversion to cyclopentenone structure **168**. Conversely, when the more electron-rich vinylcarbene **169** is exposed to carbon monoxide in the same manner, the pyrone complex **172** is formed.

The unusual pyrone complex **172** is thought[106] to be formed through

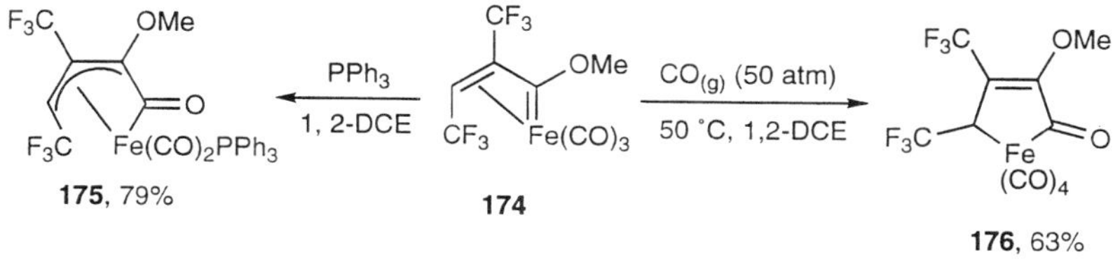

SCHEME 75.

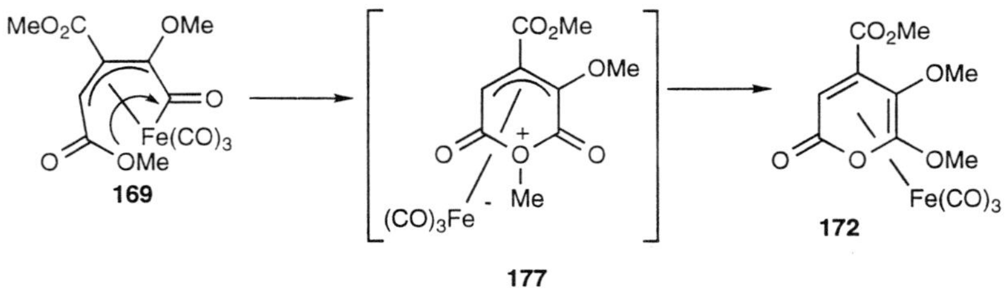

SCHEME 76.

nucleophilic attack by the methoxycarbonyl group on the ketene carbonyl in complex **169**. Subsequent methyl transfer within the resultant oxonium π-allyl complex (**177**) affords the observed pyrone complex.

Similar pyrone complexes were isolated by Semmelhack[97a] as the products of the reaction between tetracarbonyl[ethoxy(alkyl)carbene]iron(0) complexes and various acetylenes. Vinylketene complexes are proposed as key intermediates in the mechanism of this conversion, which closely matches analogous reactions with cobalt carbenes[51] (see Section V,B), while showing crucial differences with the analogous reaction of a chromium carbene (see Section II,B).

D. Synthesis from Acylferrates (via η³-Vinylcarbene Complexes)

While attempting to prepare an η^1-(vinylcarbene)iron complex[121] by the alkylation or acylation of an α,β-unsaturated acylferrate, Mitsudo and Watanabe found[122] that the major isolated product was in fact an η^4-vinylketene complex (**178**), formed presumably by the carbonylation of an intermediate η^1-vinylcarbene, which may then undergo olefin coordination to the vacant metal site. All attempts to isolate such intermediates, or to observe them by ^{13}C NMR spectroscopy, failed. Only in the reaction between potassium tetracarbonyl(η^1-cinnamoyl)ferrate (**179.a**) and pivaloyl chloride (**180.b**) was a side product (**181**) isolated in appreciable yield. In other reactions, only a trace (<1%) of such a compound was detected by spectroscopy. The bis(triphenylphosphine)iminium(1+) (PPN) salts of **179.a** and **179.b** also reacted with 2 equiv of ethyl fluorosulfonate to give **178.g** and **178.h** in 21 and 37% yield, respectively. All products were somewhat unstable to silica gel, hence the low isolated yields in some cases.

The side product is formed by the addition of a [Fe(CO)₃] unit to the postulated η^3-vinylcarbene (**182**), generated *in situ* from the decarbonylation of a vinylketene complex. Indeed, such a diiron species may be synthesized by direct addition of diiron nonacarbonyl to the vinylketene precursor **178.a**, or by direct thermolysis of **178.a** in toluene.

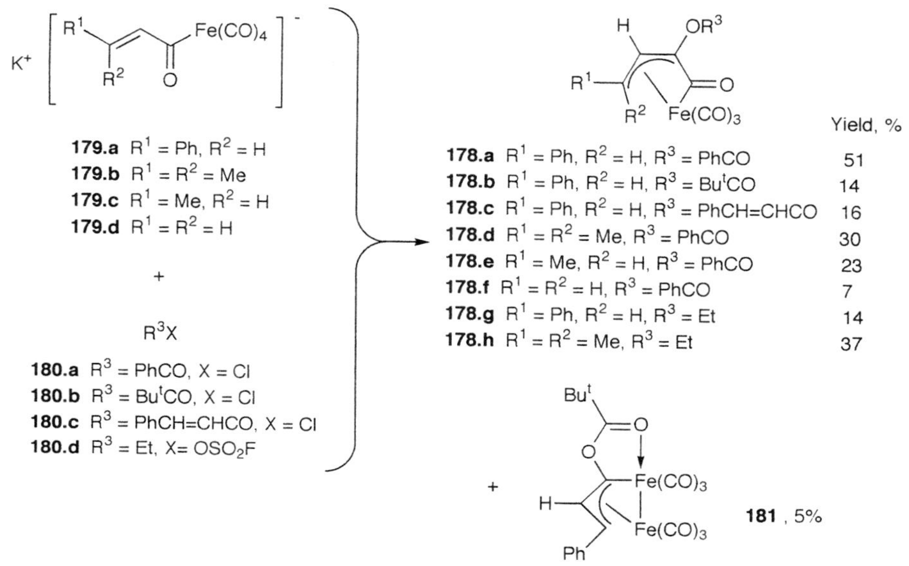

$$K^+ \left[\begin{array}{c} R^1 \diagdown \diagup Fe(CO)_4 \\ R^2 \quad O \end{array} \right]^-$$

179.a $R^1 = Ph$, $R^2 = H$
179.b $R^1 = R^2 = Me$
179.c $R^1 = Me$, $R^2 = H$
179.d $R^1 = R^2 = H$

+

R^3X

180.a $R^3 = PhCO$, $X = Cl$
180.b $R^3 = Bu^tCO$, $X = Cl$
180.c $R^3 = PhCH=CHCO$, $X = Cl$
180.d $R^3 = Et$, $X = OSO_2F$

		Yield, %
178.a $R^1 = Ph$, $R^2 = H$, $R^3 = PhCO$		51
178.b $R^1 = Ph$, $R^2 = H$, $R^3 = Bu^tCO$		14
178.c $R^1 = Ph$, $R^2 = H$, $R^3 = PhCH=CHCO$		16
178.d $R^1 = R^2 = Me$, $R^3 = PhCO$		30
178.e $R^1 = Me$, $R^2 = H$, $R^3 = PhCO$		23
178.f $R^1 = R^2 = H$, $R^3 = PhCO$		7
178.g $R^1 = Ph$, $R^2 = H$, $R^3 = Et$		14
178.h $R^1 = R^2 = Me$, $R^3 = Et$		37

+

181, 5%

SCHEME 77.

However, upon thermolysis of **178.d** or **178.h**, isomeric mixtures of the butadiene complexes **183.a** and **183.b** were formed. Since intramolecular hydrogen transfer within $(3,3$-dimethyl-$\eta^3 : \eta^1$-allylacyl)iron complexes is well precedented[101] (see Section VI,B), it seems likely that this process is responsible for diene complex formation. Note that only the Z-diene complex was isolated from the reaction mixture, a surprisingly stereoselective result.

E. Synthesis from η^1-Alkynylcarbenes (via η^3-Vinylcarbene Complexes)

In 1991, Park reported[123] the first synthesis of iron alkynylcarbene complexes (**184**), involving the nucleophilic attack of a lithium acetylide on pentacarbonyl iron, followed by electrophilic quench. With such compounds in hand, he proceeded to investigate their reactivity[123,124] and found that upon addition of cyclopentadiene to the alkynylcarbene complexes **184**, the products formed were η^4-vinylketene complexes (**185**). During column chromatography, some of these products (**185.a** and **185.b**) were transformed into the tricarbonyl(norbornadiene)iron derivatives **186**. Others (**185.c** and **185.d**, not shown) were hydrolyzed as part of the workup procedure, to afford pure samples of the norbornadiene complexes **186.c** and

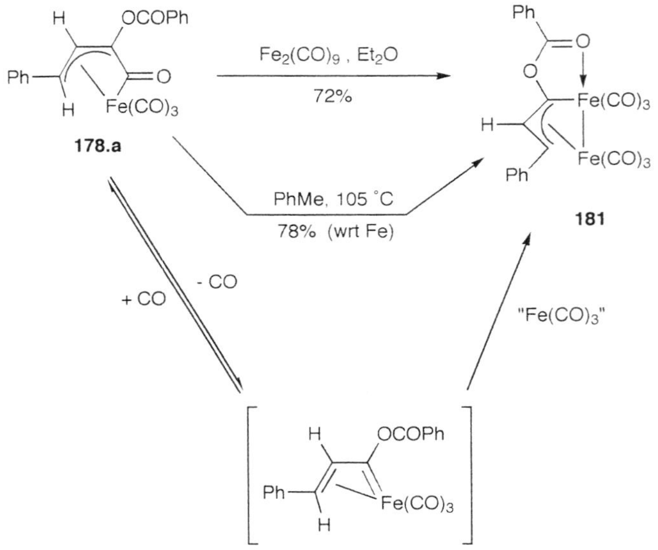

SCHEME 78.

186.d. An isolated sample of **185.e** could be converted to **186.e** in quantitative yield by stirring with silica gel in hexane. When complex **184.a** was reacted with cyclopentadiene at 25°C, the carbene complex **187** was isolated in 92% yield. At this temperature, the analogous η^1-carbenes formed from compounds **184.b–184.e** were insufficiently stable to be isolated. At 25°C, **187** slowly converts to **185.a** via a CO-dissociation process, leading to the postulated vinylcarbene complex **188**. Carbene **187** may be converted directly to **185.a** by raising the reaction temperature to 40°C.

It has been proposed[123] that the vinylketene complex **185** is in reversible equilibrium with the vinylcarbene species **188**, which may then protonate in the presence of silica gel to afford the cationic complex **189**. Reductive

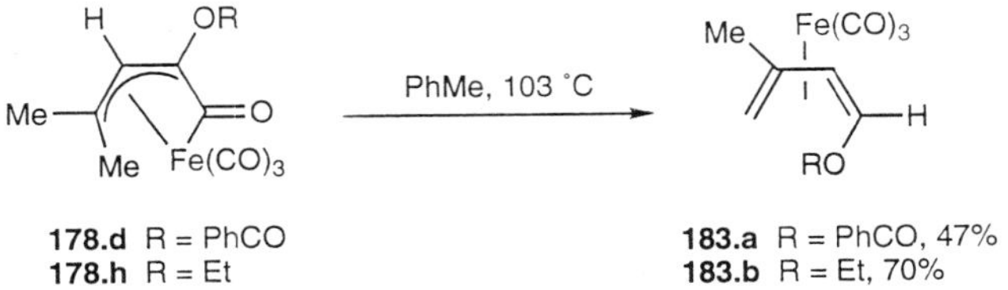

178.d R = PhCO
178.h R = Et

183.a R = PhCO, 47%
183.b R = Et, 70%

SCHEME 79.

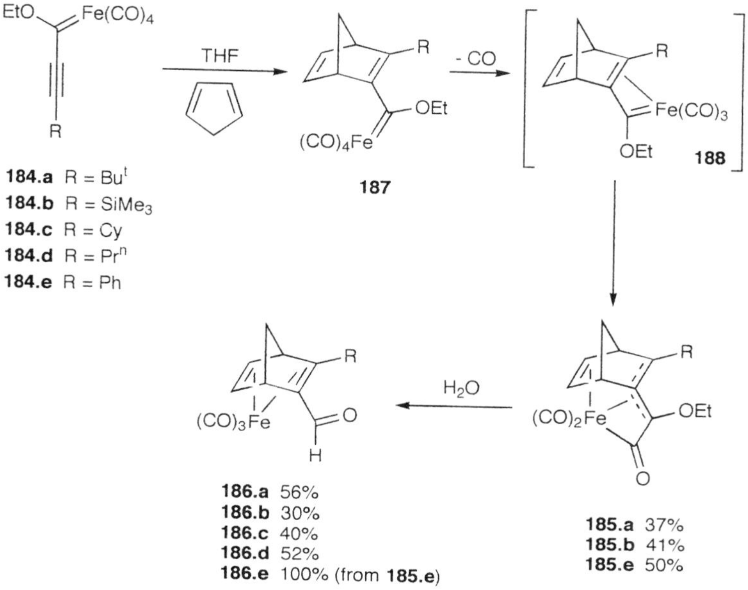
EtO
Fe(CO)4
R
THF
187
- CO
OEt
(CO)4Fe
Fe(CO)3
188
184.a R = Bu^t
184.b R = SiMe3
184.c R = Cy
184.d R = Pr^n
184.e R = Ph
H2O
(CO)3Fe
(CO)2Fe
O
H
186.a 56%
186.b 30%
186.c 40%
186.d 52%
186.e 100% (from 185.e)
185.a 37%
185.b 41%
185.e 50%

SCHEME 80.

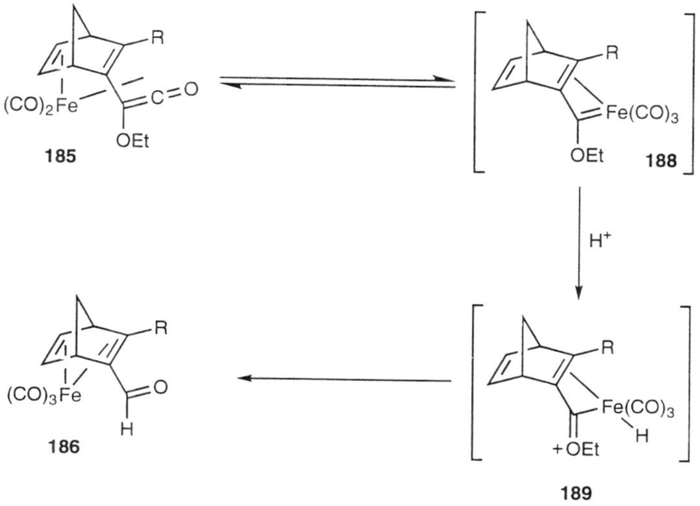
R
(CO)2Fe
C=O
OEt
185
Fe(CO)3
188
H+
(CO)3Fe
186
+OEt
H
189

SCHEME 81.

185.b $\xrightarrow[\text{acetone}]{\text{Me}_3\text{N}^+\text{O}^-}$

190, 48%

SCHEME 82.

elimination is then accompanied by loss of the ethyl group before recoordination of the alkene to give **186**.

Decomplexation of the vinylketene complex **185.b** with trimethylamine *N*-oxide affords the ethyl ester **190** in 48% yield.

In a similar study, Park has shown[125] that cyclic dienes other than cyclopentadiene will also undergo a [4 + 2] cycloaddition reaction with the (trimethylsilyl)alkynylcarbene complex (**184.b**). When **184.b** was reacted with 1,3-cyclohexadiene in THF, the diene complex **191** was isolated along with the vinylketene complex **192**, which was prone to hydrolysis during column chromatography as expected.[123] When **192** was stirred with silica gel and water in hexane, an almost quantitative conversion to the aldehyde complex **193** was observed.

However, a crucial difference was seen in analogous reactions with the (*tert*-butyl)alkynylcarbene complex **184.a**, carried out under a pressure of carbon monoxide. A [4 + 2] cycloaddition takes place, but in this case the ene fragment is not the alkyne functionality, but the carbene–alkyne bond. The mechanism presented by Park implies that this is due to the preliminary

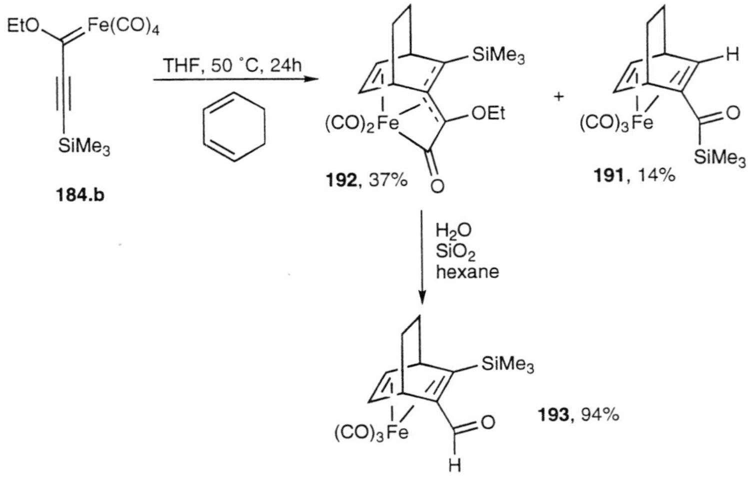

SCHEME 83.

[1,3] migration of the ethoxy group to the alkyne carbon, presumable forming an allenylidene species. The central C_2 double bond of this intermediate then participates in the cycloaddition reaction. This reaction ultimately yields the tricarbonyl(η^4-α-pyrone)iron complexes **194**. Such pyrone complexes have been isolated from high-pressure carbonylations[106] of η^3-vinylcarbene complexes (see Section VI,C), and it seems likely that a similar mechanism is involved in their formation in this case.

The isolation of **195** from the reaction of **184.a** with 1,3-cyclohexadiene at relatively elevated temperatures (135°C, as opposed to the usual reaction temperature of 75°C) has led Park to propose the following mechanism for the formation of **194.a** and **195**. A [4 + 2] cycloaddition gives the vinylcarbene **196**, before carbonylation leads to the vinylketene complex **197** in the precedented way (see Sections II,B, VI,B, VI,J). Migration of the ethoxy group, followed by another carbonylation of the carbene **198**, affords vinylketene **199**, which undergoes reversible ethoxy migration (**199** and **200**). At lower temperatures, nucleophilic attack by the ethoxy group on the ketene carbonyl ultimately yields the pyrone **194.a**, in the manner discussed in Section VI,C. At higher temperatures, loss of a carbonyl ligand from the iron moiety and alkene coordination to the vacant site yields the vinylketene complex **195**.

It is interesting to note that such pyrone complexes have also been isolated from the carbonylation of products obtained from the reaction of **184.a** with a secondary amine.[126] Park invokes two vinylketene complexes (**201** and **202**) as key intermediates in the transformation, although none

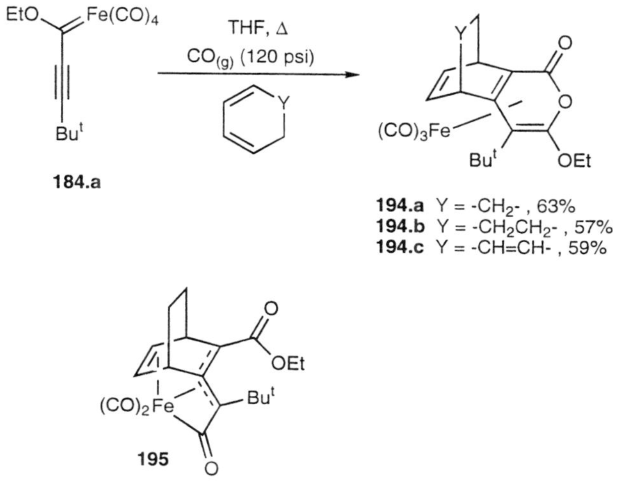
EtO
Fe(CO)4
Bu^t
184.a
THF, Δ
CO(g) (120 psi)
Y
O
O
(CO)3Fe
Bu^t
OEt
194.a Y = -CH2- , 63%
194.b Y = -CH2CH2- , 57%
194.c Y = -CH=CH- , 59%
O
OEt
(CO)2Fe
Bu^t
195
O

SCHEME 84.

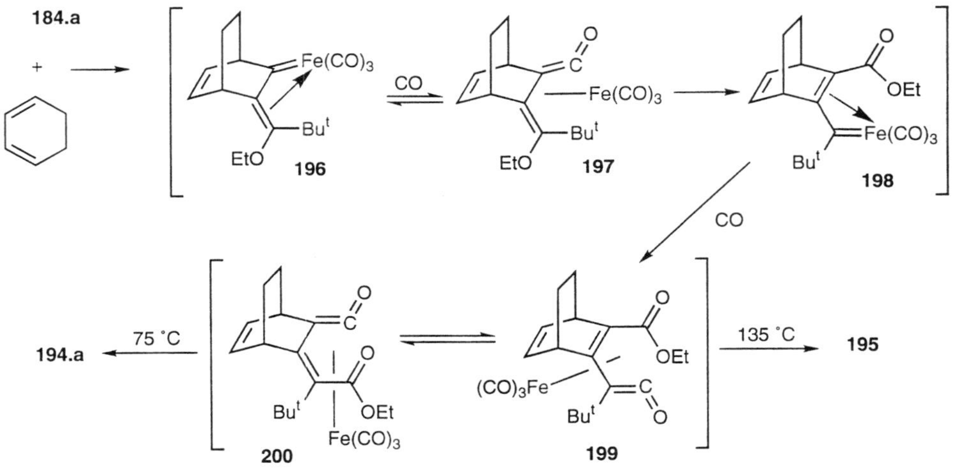

Scheme 85.

were isolated from the product mixtures. Other than the dimethylamine migration, inducing formation of the furan complex **203**, the transformations of the other intermediates have been sufficiently discussed already (see this section, and Section II,B). Note that the carbenes **204** were isolable, in very high yield, from the reaction mixture and could be subsequently transformed to the pyrones (**205**). This certainly gives credence to the mechanisms already proposed for the formation of pyrones from vinylketene precursors.

F. Synthesis from 2-Methoxyallyl Halides

The earliest alternative to cyclopropene insertion as a viable vinylketene synthesis was published by Hoffmann[118] in 1972. Upon reaction of **206.a** with diiron nonacarbonyl, the vinylketene complex **207** was isolated in low yield. The analogous bromide and iodide substrates formed the π-allyl complexes **208**, although a trace of **207** was isolated from the reaction of the bromide **206.b**.

G. Synthesis from a Free Ketene

Herrmann has reported[127,128] that irradiation of an ethereal mixture of iron pentacarbonyl and diphenylketene leads to the synthesis of the red vinylketene complex **209**, containing the unusual feature of partial involve-

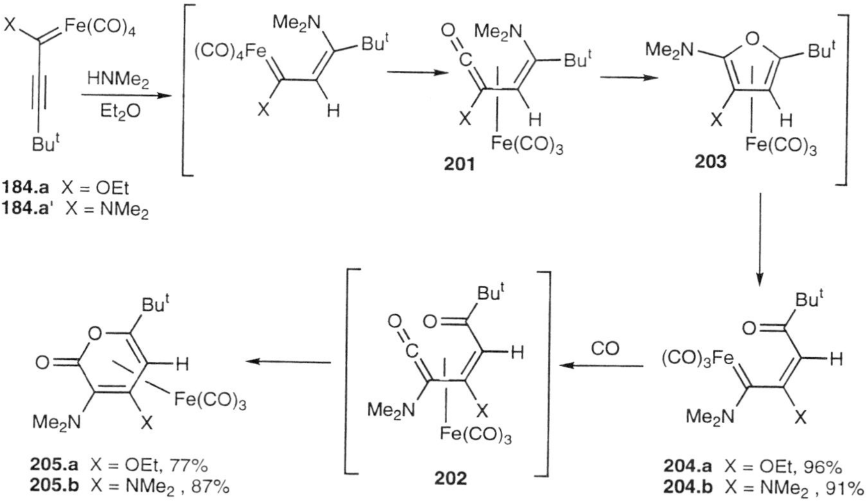

SCHEME 86.

ment of a phenyl ring in π-allyl bonding. In an exchange reaction with $^{13}CO_{(g)}$, incorporation into all carbonyl ligand sites and the ketene carbonyl was observed. Upon reaction of **209** with diiron nonacarbonyl, the black diiron species **210** was formed.[128] Heating the vinylketene complex **209** under an atmosphere of ethene resulted in addition of the C_2 fragment with concomitant loss of $CO_{(g)}$ to form the diene complex **211**.[128]

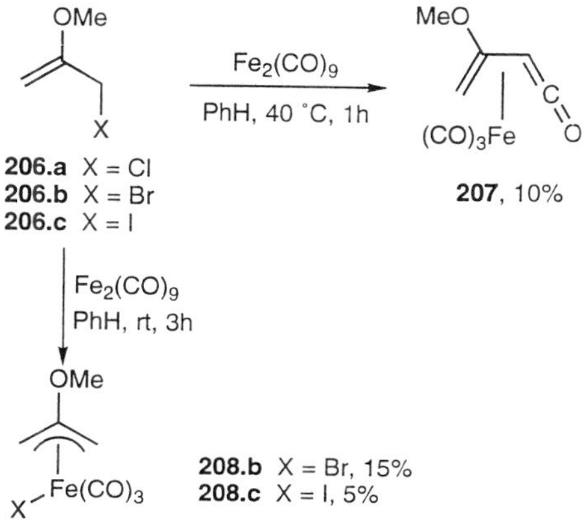

SCHEME 87.

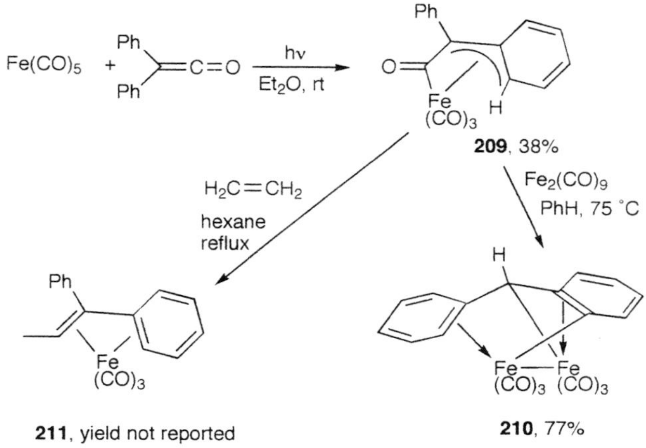

SCHEME 88.

H. Synthesis from a Propargyl–Iron Complex

During a study of the reactivity of transition metal–propargyls, Wojcicki found[129] that upon reaction of **212** with diiron nonacarbonyl, the $\eta^3:\eta^1$-allylacyl complex **213** was formed. The mechanism[130] is thought to proceed by formation of the iron–acetylene complex **214**, which then rearranges to a zwitterionic η^2-allene complex (**215**) bearing a Fe(CO)$_3$ group as a negative terminus. Nucleophilic attack by this Fe(CO)$_3$ group on the allene (to give **216**) and subsequent alkene coordination gives the final product **213**.

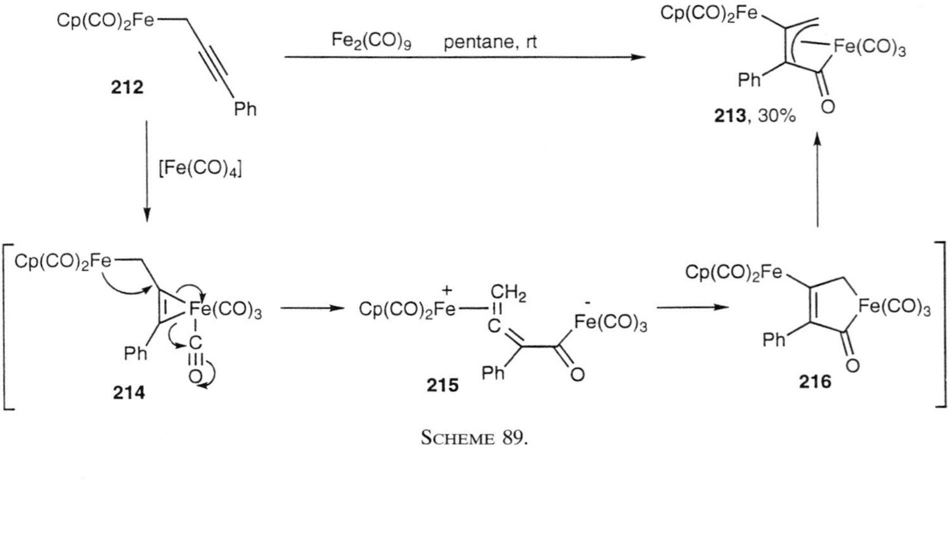

SCHEME 89.

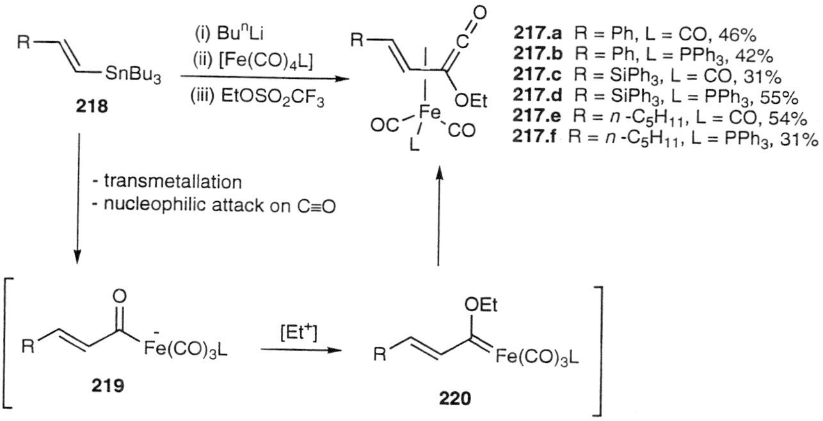

SCHEME 90.

I. Synthesis from Vinylstannanes (via η^1-Vinylcarbene Complexes)

As part of a large study into the synthesis and reactivity of the η^4-vinylketene complexes of iron (see Section VI,J), Gibson (née Thomas) has reported[131] the synthesis of the ethoxy-substituted vinylketenes 217. Transmetallation of the vinylstannane 218 with Bu^{n}Li affords a vinyllithium species, which may then attack a carbonyl ligand on the iron substrate in a nucleophilic fashion. The acylferrate complex 219 is then alkylated to form the η^1-vinylcarbene complex (220), in a manner analogous to the method of Mitsudo and Watanabe (see Section VI,D).[122] This carbene may then undergo carbonyl insertion, and complexation of the pendant alkene, to yield the product vinylketene complex 217. The nucleophilic addition of a vinyllithium species to hexacarbonylchromium has been shown[132] to proceed by the same route, but then terminates at the η^1-vinylcarbene stage. This suggests that the insertion of carbon monoxide into a metal–carbon double bond has a higher activation energy for chromium than for iron.[131]

J. Synthesis from η^4-Vinylketone Complexes (via η^3-Vinylcarbene Complexes)

In 1989, Thomas reported[3] the novel synthesis of tricarbonyl(η^4-vinylketene)iron(0) complexes (221) from the corresponding η^4-vinylketones (222). Nucleophilic attack by methyllithium on a carbonyl ligand is thought to produce the anionic complex 223, which then carbonylates to give the η^2-

alkene complex **224**. Nucleophilic attack by the oxy-anion on the free
ketone and subsequent loss of the resultant acetate fragment affords the
η^3-vinylcarbene complex **225**, which then carbonylates to produce the vinylketene products.[3,39,87,89,133,134] There is much evidence to support this mechanism.[87] Other alkyllithium reagents (including Bu^{t}Li) may be used, although
the best results are usually seen with MeLi. In all cases, stoichiometric
amounts are required for the reaction to go to completion. If the reaction
is carried out under an inert atmosphere, rather than carbon monoxide, it
appears that after initial attack by the alkyllithium, the iron-coordinated
acyl fragment may transfer to C-4 of the 1-oxadiene unit, which after
protonation yields a free 1,4-diketone.[135] In the absence of an alkyllithium
reagent, no reaction is observed.[87] Also, lithium alkanoates have been
extracted from the crude reaction mixtures in accordance with the transformation of **224** to **225**. Of course, there have been many reports of the direct
conversion of η^3-vinylcarbenes to η^4-vinylketenes by reaction with carbon
monoxide.[48,68,106,107]

In order to elucidate the nature of this mechanism further, Gibson (née
Thomas) studied[134,136] the conversion of an optically pure sample of vinylketone **222.e** to the corresponding vinylketene (**221.e**). The vinylketone was
resolved by carbonyl ligand substitution with (+)-neomenthyldiphenyl-
phosphine, followed by subsequent separation of the resultant diastereomers to yield an optically enriched product.[136,137] When **222.e**, of known
e.e., was treated with methyllithium under an atmosphere of carbon monoxide, the expected vinylketene complex **221.e** was isolated and was found

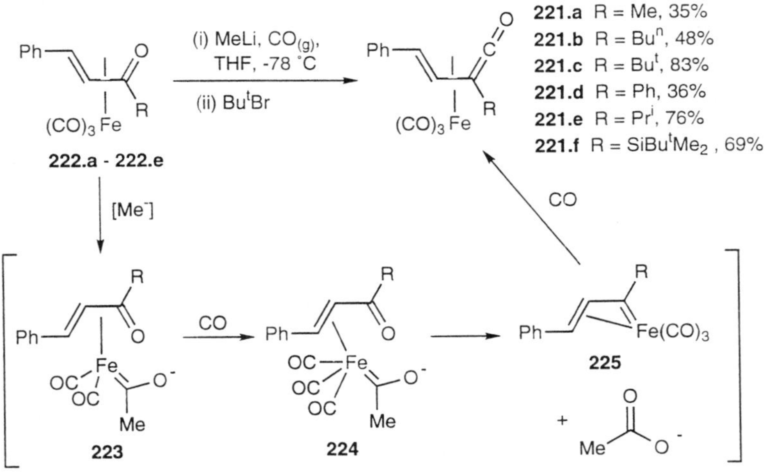

SCHEME 91.

to have the same e.e., thus illustrating that having lost no stereochemical information during the course of the reaction, the presence of a symmetrical intermediate along the reaction coordinated was forbidden. This excludes the η^1-vinylcarbene **226** as a possible intermediate, adding weight to the hypothesis that **225** is the most likely key intermediate. Once isolated, **221.e** could be further enriched by fractional recrystallization to give an overall yield of 57% (96% e.e.). Although this was a useful method for the generation of enantiomerically enriched samples of vinylketene complexes, a much more efficient kinetic resolution procedure was developed soon afterwards (see later discussion).[138]

Further evidence for the intermediacy of an η^3-vinylcarbene has been provided by García-Mellado and Alvarez-Toledano.[139] They have shown that lithium dimethylcuprate may be used in place of an alkyllithium reagent *in the absence of carbon monoxide,* necessary CO being scavenged from other metal–carbonyl species; hence the low yields observed from these reactions.

They have also shown that the dimeric bis-π-allyl complex **227** is formed, under different conditions, from **221.g**, **228**, and **229**, suggesting that all three reactions proceed through a common intermediate, the most likely candidate being the η^3-vinylcarbene complex **230**.

Thomas has also synthesized vinylketene complexes containing both sulfur[140] (**231.a**) and silicon[141] (**231.b, 231.c**) bonded to the C-4 position, in place of the phenyl substituent seen in complexes **221**.

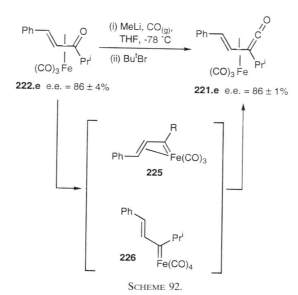

SCHEME 92.

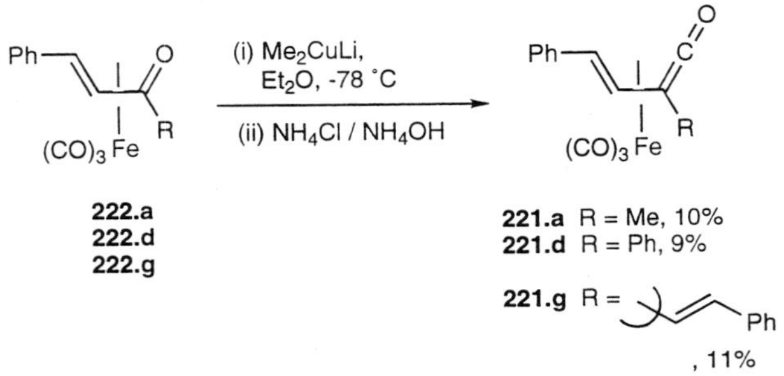

SCHEME 93.

Having established a very effective method for the synthesis of tricarbonyl(η^4-vinylketene)iron(0) complexes, Thomas has subsequently undertaken the most comprehensive study on the reactivity of these complexes to date. The reactions of **221** with phosphoramidate anions,[90,134] coordinating ligands such as phosphines[3] and isonitriles,[69,87,89,135,142,143] a variety of nucleophiles,[86,89,135,142] phosphonoacetate anions,[88,89] alkynes,[108,109,144,145] and alkenes[146,147] have all been investigated. Crucially, Thomas has also developed a method[138] for the kinetic resolution of the vinylketene complexes (**221**) that ultimately yields enantiomerically pure samples of the complex. This

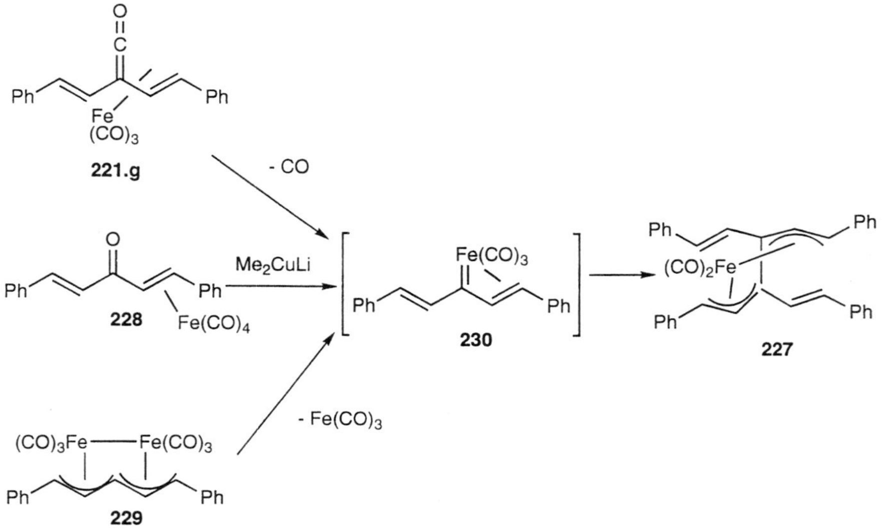

SCHEME 94.

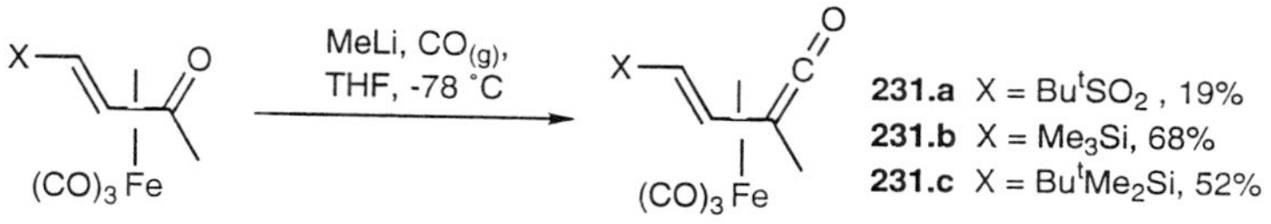
X
O
(CO)3 Fe
MeLi, CO(g),
THF, -78 °C
X
O
C
(CO)3 Fe
231.a X = ButSO2 , 19%
231.b X = Me3Si, 68%
231.c X = ButMe2Si, 52%

SCHEME 95.

has allowed studies to be carried out that probe the mechanisms of several of these reactions.[38,39,134,148]

Let us first consider the reaction between vinylketene complexes and phosphoramidate anions.[90,134] Deprotonation of the appropriate phosphoramidate with Bu^{n}Li affords the corresponding phosphoramidate anion (**232.a–232.c**), which undergoes a Wadsworth–Emmons type reaction on the ketene carbonyl to afford the ketenimine complexes **233**. Ketenimine complexes may also be prepared by the thermolysis of the vinylketene complexes (**221**) with isonitriles (vide infra).[69]

Having explored the scope of this reaction, it was then possible to use an optically active phosphoramidate (**232.d**), derived from (*S*)-α-methylbenzylamine, to perform a kinetic resolution procedure.[134,138] Upon reaction of **221** with 0.5 equiv of the phosphoramidate anion **232.d**, selective reaction with the *R*-isomer of **221** resulted in an enhancement in the proportion of *S*-**221** along with the expected ketenimine products (**233**). The absolute configuration of *R*-**233.d** was determined by X-ray crystallography, and thus all other absolute configurations assigned by comparison of CD and HPLC data.[39,138]

As seen in the conversion of an optically enriched vinylketone complex to the corresponding vinylketene complex (vide supra), the availability of enantiomerically pure samples of the vinylketene complexes **221** allowed the subsequent reactions of these complexes to be "screened" for the presence of symmetrical intermediates along the reaction coordinate. Their

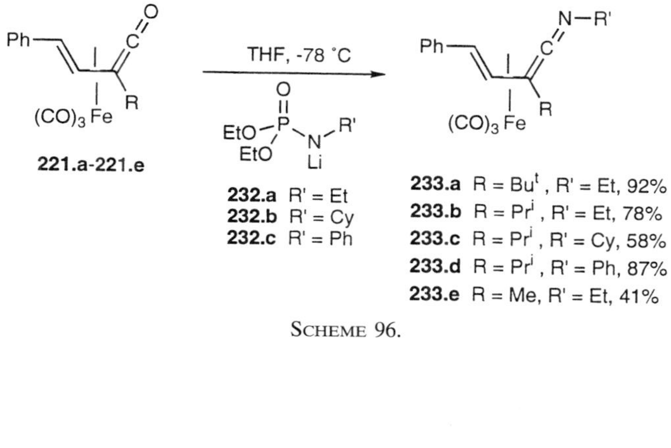
Ph
O
C
R
(CO)3 Fe
221.a-221.e
THF, -78 °C
O
P
EtO
EtO
N
R'
Li
232.a R' = Et
232.b R' = Cy
232.c R' = Ph
Ph
N—R'
C
R
(CO)3 Fe
233.a R = But , R' = Et, 92%
233.b R = Pri , R' = Et, 78%
233.c R = Pri , R' = Cy, 58%
233.d R = Pri , R' = Ph, 87%
233.e R = Me, R' = Et, 41%

SCHEME 96.

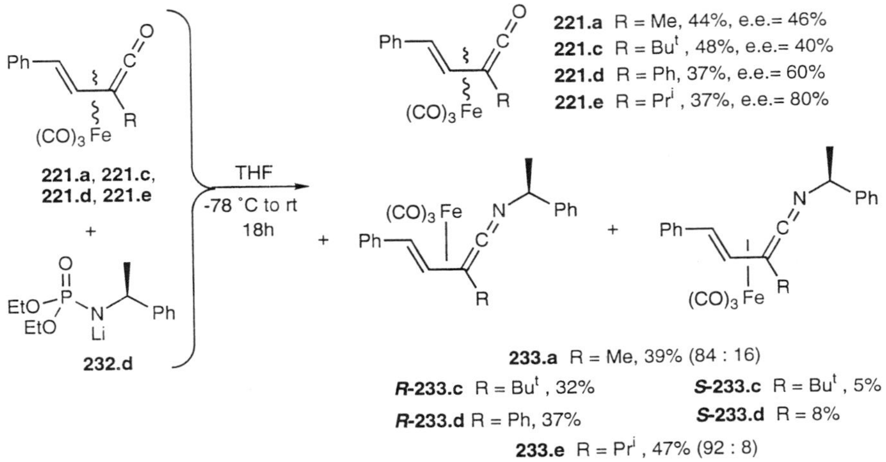

Ph
(CO)3Fe
R
221.a, 221.c,
221.d, 221.e
+
EtO
EtO
P
N
Li
Ph
232.d
THF
-78 °C to rt
18h
221.a R = Me, 44%, e.e.= 46%
221.c R = Bu^t, 48%, e.e.= 40%
221.d R = Ph, 37%, e.e.= 60%
221.e R = Pr^i, 37%, e.e.= 80%
233.a R = Me, 39% (84 : 16)
R-233.c R = Bu^t, 32%
S-233.c R = Bu^t, 5%
R-233.d R = Ph, 37%
S-233.d R = 8%
233.e R = Pr^i, 47% (92 : 8)

SCHEME 97.

presence is indicated by the loss of stereochemical information during the conversion of **221** to the various reaction products. All the relevant studies, and their mechanistic consequences, will be discussed in the latter part of this section.

The vinylketene complex **221.a** was shown[3] to undergo ligand replacement when treated with triphenylphosphine to yield the dicarbonyl(triphenylphosphine)vinylketene complex **234**. This complex was also isolated from the reaction of vinylketone **222.a** with methyllithium in the presence of triphenylphosphine, indicating that a phosphine, as well as carbon monoxide, could induce the migration of a coordinated carbonyl ligand to form a ketene carbonyl. Mitsudo, Watanabe, and Weiss have all shown that this process occurs in the reaction between phosphines and isolated η^3-vinylcarbene complexes (see Sections VI,B and VI,C).[48,68,106]

The reactions of the vinylketene complexes (**221**) with various isonitriles has been studied extensively.[69,87,89,135,142,143] Initially, a ligand replacement

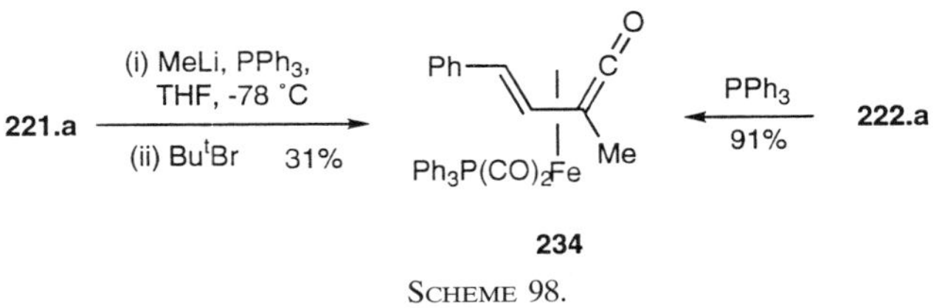

221.a
(i) MeLi, PPh3,
THF, -78 °C
(ii) Bu^{t}Br 31%
Ph
Me
Ph3P(CO)2Fe
PPh3
91%
222.a
234

SCHEME 98.

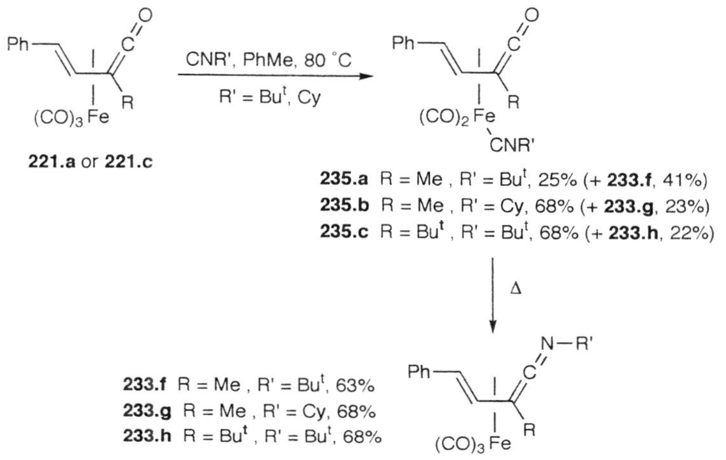
Ph
(CO)3Fe
R
221.a or 221.c
CNR', PhMe, 80 °C
R' = Bu^t, Cy
(CO)2Fe
CNR'
235.a R = Me , R' = Bu^t, 25% (+ 233.f, 41%)
235.b R = Me , R' = Cy, 68% (+ 233.g, 23%)
235.c R = Bu^t , R' = Bu^t, 68% (+ 233.h, 22%)
Δ
N—R'
233.f R = Me , R' = Bu^t, 63%
233.g R = Me , R' = Cy, 68%
233.h R = Bu^t , R' = Bu^t, 68%

SCHEME 99.

reaction takes place to form the dicarbonyl(isonitrile) complexes **235**. Further heating causes the rearrangement to the corresponding ketenimine **233**.

It has been shown[69,87,89] that this rearrangement is an intramolecular process, by first isolating two different dicarbonyl(isonitrile) complexes (**235.b** and **235.c**) and heating them together. No "scrambling" of the CNR′ groups was observed, indicating that isonitrile decomplexation does not occur under these reaction conditions. The yield of **233.h** was lower than expected, because of recovery of some **235.c** from the reaction mixture. The yield was later reported[135] as 58%, based on **235.c** recovery.

Since the conversion of **235** to **233** requires only heating, it is possible to synthesise the vinylketenimine complex directly from the vinylketene merely by extending the reaction time sufficiently.[69,87,89] Note that **233.j** and **233.k** were obtained as 1:1 mixtures of diastereoisomers. The subsequent reactivity of the vinylketenimine complexes,[87,89,135,143] particularly their utility in the asymmetric synthesis of quaternary carbon centers,[143] has also been investigated, but is beyond the scope of this review.

235.b + **235.c** $\xrightarrow{\text{PhMe, 80 °C, 16h}}$ **233.g** (61%) + **233.h** (47%)

SCHEME 100.

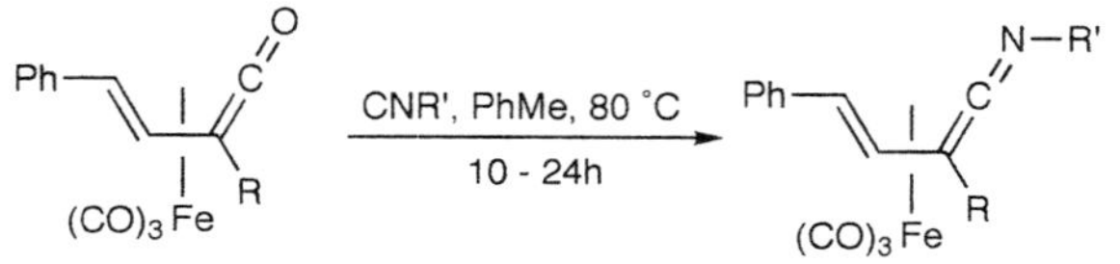

SCHEME 101.

In general, vinylketene complexes **221** undergo reaction with nucleophiles at the ketene carbonyl carbon (C-1), yielding β,γ-unsaturated ketones (**236**).[86] It is interesting to note that the analogous vinylketenimine complexes undergo nucleophilic attack at C-2.[86,87,89,135,142]

Alvarez-Toledano has investigated[149] the reactivity of tertiary amines toward vinylketene complexes of type **221** and found that a number of unusual transformations take place. The formation of **237** and **238** is not directly linked to the presence of the amine, but instead relies on a thermolytic coupling of two carbene complexes formed from the vinylketene ligand, with additional loss of carbon monoxide. It is thought that **239** is formed by the deprotonation of the labile C-4 proton, followed by subsequent coordination of a capping $Fe(CO)_3$ group.

Reaction under the same conditions with dimethylallylamine produced a mixture of seven identifiable products.[149] Complexes **237**, **238**, and the dimethylallylamine analogue of **239** (**239'**) were all present, along with the organic products **240–243**. Amides **240** and **241** are formed through initial attack by the amine on the ketene carbonyl carbon (C-1), followed by

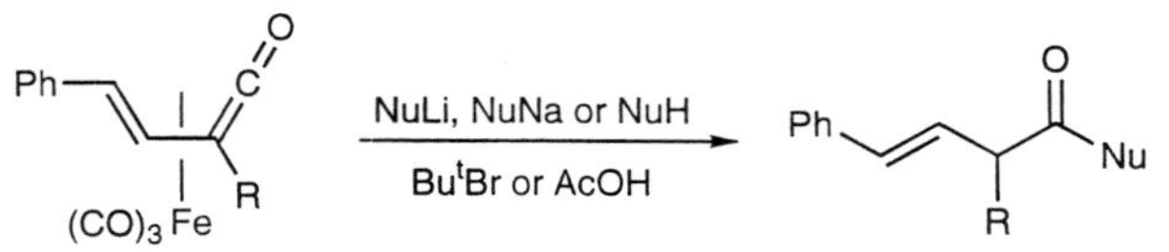

SCHEME 102.

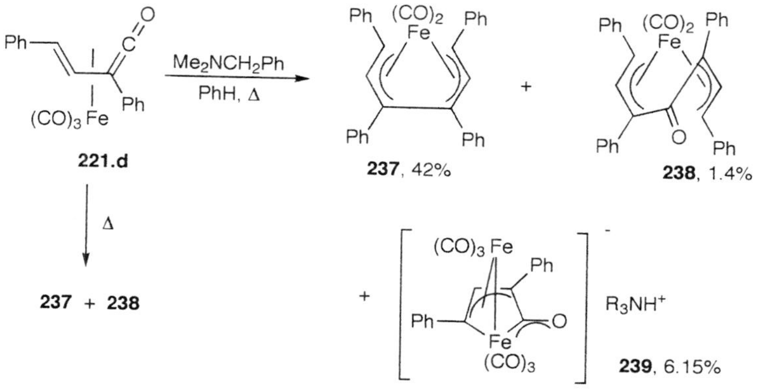

SCHEME 103.

coordination of the allyl group to the iron moiety. Subsequent re-arrangement and loss of the iron fragment yields **240** and **241**. If decarbony-lation of the vinylketene complex occurs first (to give a vinylcarbene), then attack by dimethylallylamine takes place by initial coordination of the alkene to the unsaturated coordination sphere of the iron moiety. Subsequent rearrangement, and deinsertion of $Fe(CO)_3$, yields **242** and **243**.

Alvarez-Toledano has also investigated[150] the reaction of vinylketene complexes **221.d** and **221.g** with Davy's reagent, a dimeric species that generates a phosphorus-sulfur yield *in situ*. This then attacks the ketene carbonyl in a nucleophilic fashion, generating the intermediate **244**, which collapses to yield the product (**245**).

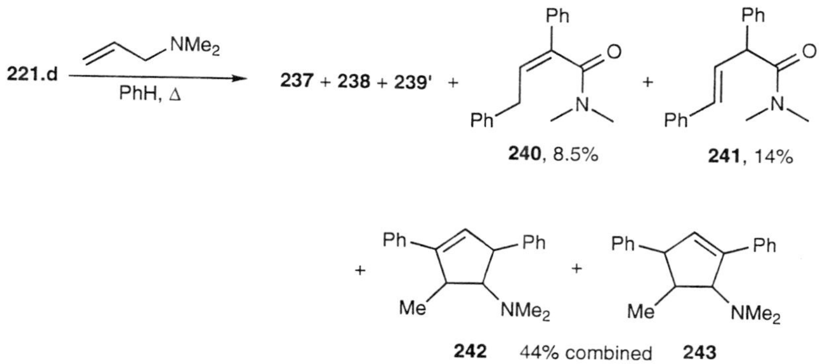

SCHEME 104.

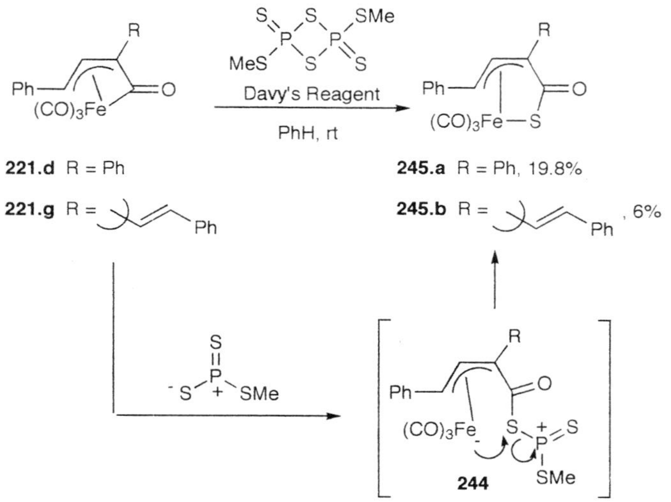

SCHEME 105.

In 1991, Thomas reported[88a] that the reaction between vinylketene complexes (**221**) and several phosphonoacetate anions generated vinylallene complexes (**246**), in some cases with extremely high stereoselectivity.[88,89] This Wadsworth–Emmons type reaction occurs via attack by the phosphonoacetate carbanion at the ketene carbonyl carbon, and product ratios clearly depend on the steric bulk of the R and R′ substituents. The relative stereochemistry of the major isomers of **246** were determined by X-ray analysis. Upon oxidation of the vinylallene complexes with iron(III) chloride, a range of substituted furanones were isolated.[88b,89]

The Thomas investigation then turned to the reactivity of vinylketene complexes (**221**) with alkynes.[108,109,144] Initially, electron-poor alkynes were used, and in the cases **247.d–247.g** complete regiocontrol is exhibited over the insertion, with the more sterically demanding substituent being located on the carbon nearest to the iron moiety.

It has already been seen (see Section II,B and V,E) that vinylketene complexes of cobalt[6,95] and chromium[22] react with alkynes to produce cyclopentadienones, indanones, and substituted phenols. It has been shown[108,109,144,145] that similar products may be derived from the alkyne adducts **247**. Indeed, when alkyl or aryl substituted alkynes were reacted with vinylketene complex **221.e**, a mixture of organic and organometallic products was isolated. In the cases where the alkyne is attached to an aromatic substituent, the expected alkyne adduct (**247.h** and **247.i**) is isolated in low yield. However, when the vinylketene complex was treated

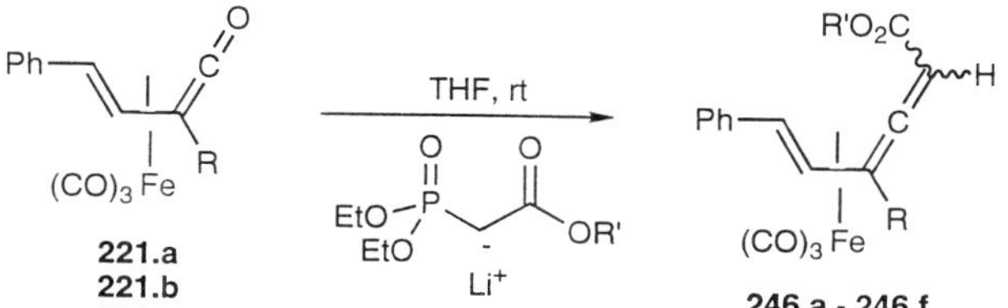

221.a
221.b
221.c

246.a - 246.f

	Stereoisomeric ratio major : minor	Yield of major isomer
246.a R = Me , R' = Me	70 : 30	(25%)
246.b R = Bu^n , R' = Me	50 : 50	(38%)
246.c R = Bu^t , R' = Me	70 : 30	(30%)
246.d R = Me , R' = Bu^t	98 : 2	(71%)
246.e R = Bu^n, R' = Bu^t	80 : 20	(43%)
246.f R = Bu^t, R' = Bu^t	85 : 15	(39%)

SCHEME 106.

with 1-octyne, a mixture of cyclopentadienone **248.a** and phenol **249.a** was formed. The analogous cyclopentadienones **248.b** and **248.c** could be synthesized by prolonged thermolysis of either **221.e** (with the appropriate alkyne) or **247.h/247.i**, although the corresponding phenol was never observed in these reaction mixtures.

The final part of this particular alkyne study[108,109,144] involved the electron-rich alkynes ethyl ethynyl ether and diethylpropynylamine. Upon reaction of **221.a** or **221.e** with ethyl ethynyl ether, the expected adducts **247.j** and **247.k** were formed. Note that the electron-donating substituent is located

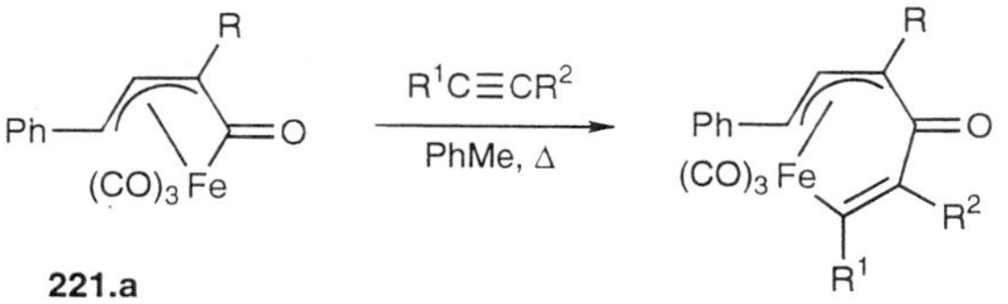

221.a
221.c
221.e

247.a R = Me, R^1 = CO_2Me, R^2 = CO_2Me, 69%
247.b R = Pr^i, R^1 = CO_2Me, R^2 = CO_2Me, 93%
247.c R = Bu^t, R^1 = CO_2Me, R^2 = CO_2Me, 97%
247.d R = Pr^i, R^1 = CO_2Me, R^2 = H, 61%
247.e R = Bu^t, R^1 = CO_2Me, R^2 = H, 68%
247.f R = Pr^i, R^1 = COMe, R^2 = H, 52%
247.g R = Bu^t, R^1 = COMe, R^2 = H, 39%

SCHEME 107.

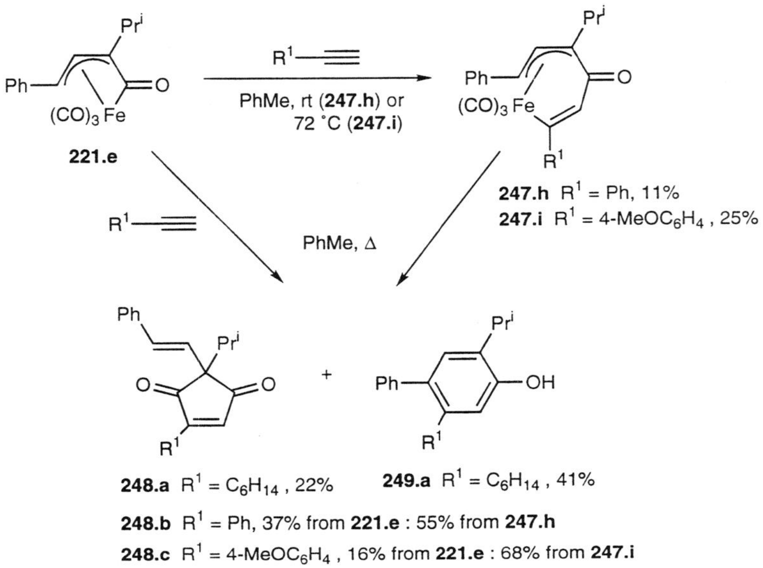

SCHEME 108.

on the carbon adjacent to the iron moiety, as was the case for the electron-withdrawing substituents seen earlier. This seems to suggest that the regio-selectivity is primarily a steric, rather than an electronic, effect. The adduct formed from reaction of **221.c** or **221.e** with diethylpropynylamine is unusual in that the alkene residue is sufficiently electron-rich to displace a carbonyl ligand from the coordination sphere of the iron, ultimately yielding **250.a** and **250.b**. Upon prolonged thermolysis either of the appropriate vinylke-tene complex in the presence of the alkyne, or of the alkyne adduct alone, the phenol **249** is formed as the sole product, with no trace of a correspond-ing cyclopentadienone. Further studies on the reactivity of alkyne adducts **247** have also been conducted.[145]

Shortly after this reaction was reported, Hegedus found[46] that imines undergo a similar insertion process. The products (**251**) are structurally very similar to the alkyne adducts (**247**), and decomplexation with iodine, followed by treatment with trimethylamine N-oxide, afforded a variety of substituted pyridones in good yield.

In 1994, Thomas reported[146,147] that alkenes also underwent an addition reaction with vinylketene complexes that differed crucially in the loss of the ketene carbonyl fragment. Complexes **252.a–252.d** were isolated as yellow crystalline solids. Clearly this suggests that the process occurs by a mechanism different from the alkyne insertion, and this will be discussed

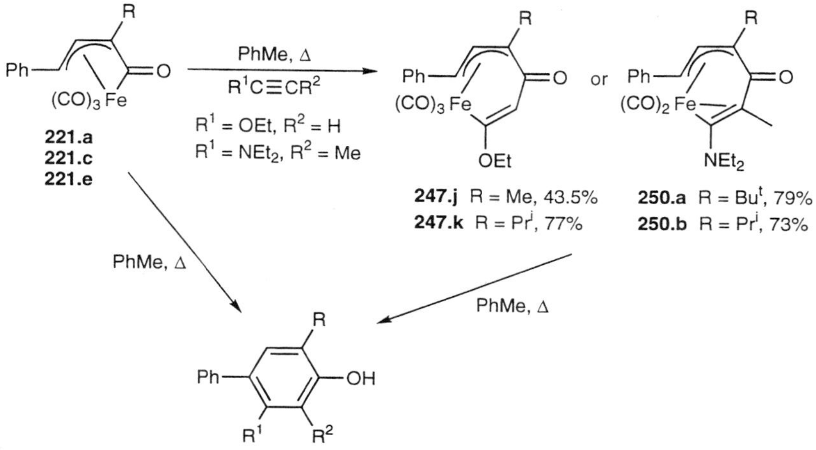

249.b R = Me, R^1 = OEt, R^2 = H, 58% from **221.a** : 64% from **247.j**
249.c R = Pr^i, R^1 = OEt, R^2 = H, 71% from **221.e** : 57% from **247.k**
249.d R = Pr^i, R^1 = NEt_2, R^2 = Me, 59% from **221.e** : 73% from **250.b**
249.e R = Bu^t, R^1 = NEt_2, R^2 = Me, 72% from **221.c** : 76% from **250.a**

SCHEME 109.

later. Decomplexation of the adducts with Ce(IV) led primarily to substituted cyclopropanes,[147] the product of a formal extrusion of the iron moiety with carbon–carbon bond formation between C-1 and C-3.

As mentioned earlier, the availability of enantiomerically pure samples of vinylketenes **221** has allowed Gibson (née Thomas) to investigate the mechanisms of many of the transformations discussed herein. By monitoring the preservation (or otherwise) of the stereochemical information between reactant and products, vital information about the nature of the reaction

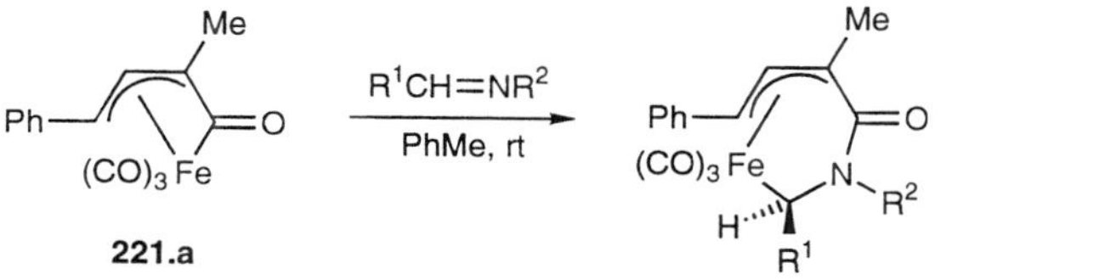

251.a R^1-R^2 = $O(CH_2)_2CH_2$, 87%
251.b R^1-R^2 = $CH_2(CH_2)_2CH_2$, 63%
251.c R^1-R^2 = $(CH_2)_3$, 46%
251.d R^1 = OMe, R^2 = CH_2Ph, 58%

SCHEME 110.

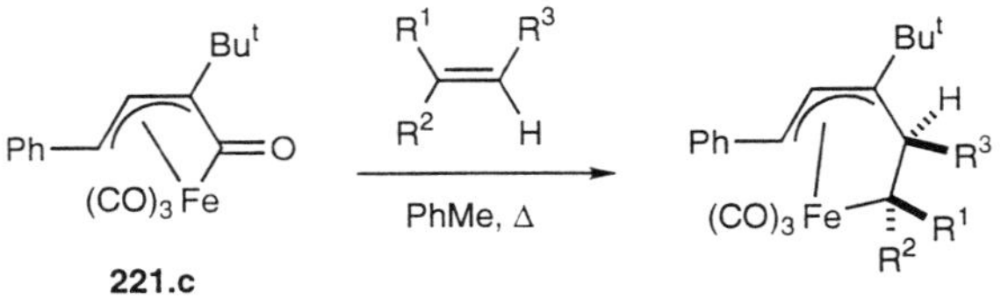

252.a $R^1 = CO_2Me$, $R^2 = H$, $R^3 = CO_2Me$, 53%
252.b $R^1 = H$, $R^2 = CO_2Me$, $R^3 = CO_2Me$, 54%
252.c $R^1 = H$, $R^2 = COMe$, $R^3 = CO_2Me$, 57%
252.d $R^1 = H$, $R^2 = CF_3$, $R^3 = CO_2Me$, 64%

SCHEME 111.

intermediates could be gleaned. In the reports published before 1996, η^1-vinylcarbene complexes (e.g., **226**, and more generally **253**) were often quoted[69,87,89,147] by Thomas as key intermediates in reactions with compounds that had some coordinating ability, including isonitriles and alkenes. These were postulated to form via the decarbonylation of the vinylketene complex **221** to give the η^3-vinylcarbene complex **225**, a well-precedented reaction. The styryl portion of the ligand could then decomplex, to be replaced by the incoming reactant L (forming **253**). Subsequent insertion of L into the iron–carbon carbene bond, followed by recoordination of the styryl fragment, would yield the final product (ketenimine **233** or alkene

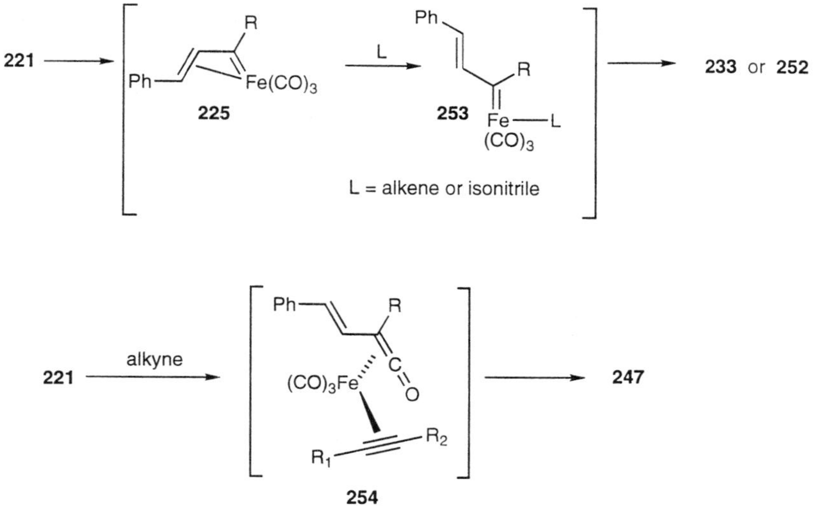

SCHEME 112.

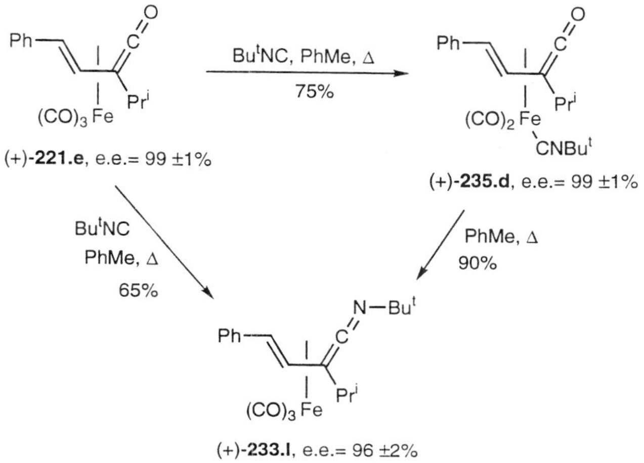

Scheme 113.

adduct **252**). In the case of the reaction of **221** with alkynes, it was suggested[144] that the styryl portion would decomplex first, to be replaced by the coordinating alkyne, producing an η^2-vinylketene complex (**254**). It is important to notice that **253** is a symmetrical intermediate assuming Berry pseudorotation or iron–carbene rotation occurs. Although **254** is *not* symmetrical, it resembles **253** in an important way: the styryl unit has again become decomplexed as a fundamental part of the mechanism.

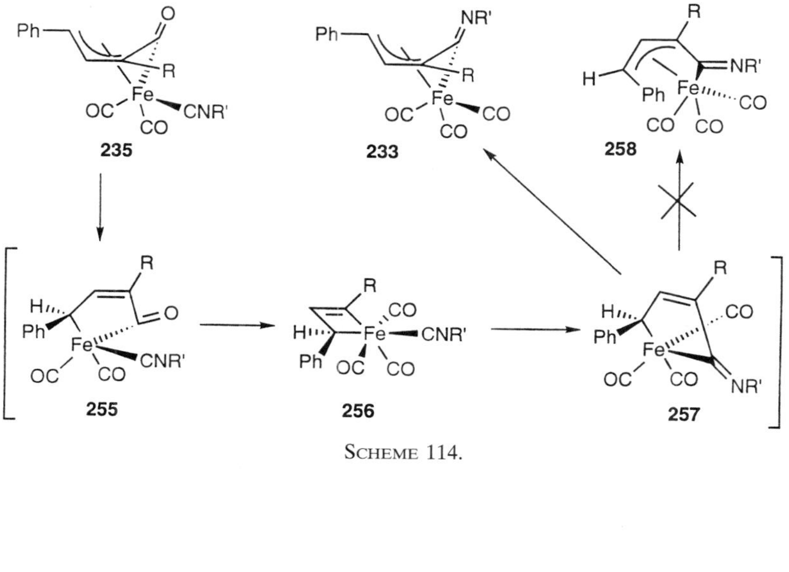

Scheme 114.

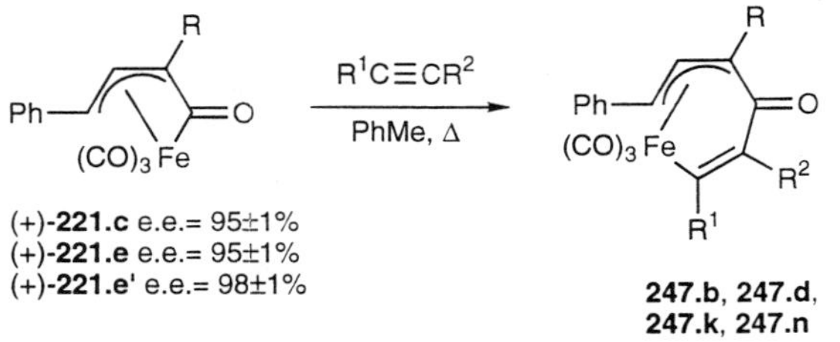

(+)-**221.c** e.e.= 95±1%
(+)-**221.e** e.e.= 95±1%
(+)-**221.e'** e.e.= 98±1%

**247.b, 247.d,
247.k, 247.n**

247.b R = Pr^i, R^1 = CO$_2$Me, R^2 = CO$_2$Me, 52%   e.e.= 99±1% (from **221.e'**)
247.d R = Pr^i, R^1 = CO$_2$Me, R^2 = H, 70%   e.e.= 97±1% (from **221.e**)
247.k R = Pr^i, R^1 = OEt, R^2 = H, 57%   e.e.= 95±2% (from **221.e**)
247.n R = Bu^t, R^1 = OEt, R^2 = H, 57%   e.e.= 93±2% (from **221.c**)

SCHEME 115.

In 1996, Gibson (née Thomas) reported[148] that the reactions of enantio-merically enriched vinylketene complex (+)-**221.e** with Bu^{t}NC proceeded with complete retention of stereochemistry. The complexes **235.d** and **233.l** were found to have the same absolute stereochemistry as (+)-**221.e** by comparison of CD spectra. The fact that no loss of stereochemical information was observed led to the discounting of an η^1-vinylcarbene complex (**253**) as a potential intermediate, and so it was proposed that at no point during the reaction did the styryl fragment decomplex from the iron center, thus preserving the stereochemical integrity of the system.

An alternative mechanism for the conversion of dicarbonyl(isonitrile) complex **235** to vinylketenimine complex **233** has therefore been formulated, the key point of which is that the iron decomplexes from the central C$_2$ fragment of the vinylketene ligand, and *not* the terminal styryl section. Thus, complex **255** is formed, allowing the incorporation of the ketene carbonyl into the coordination sphere as a carbonyl ligand (**256**), before insertion of the isonitrile into the newly formed Fe–C bond (**257**). Finally, recoordination of the C$_2$ fragment occurs stereoselectively to yield the ketenimine **233**. Recoordination of the central alkene portion on the opposite face of the complex (to give **258**) is disfavored, because of the high-energy steric interaction between the phenyl ring and the iron moiety that results from this process.

An analogous study was carried out[38] to investigate the alkyne insertion mechanism. Reacting optically enriched samples of vinylketene complexes **221.c** and **221.e** with three different alkynes, the expected adducts (**247.b**, **247.d**, **247.k**, and **247.n**) were isolated, and their e.e. measured by HPLC or ^{1}H NMR methods. It was shown that there had been no loss of stereochemical information, and comparison of CD spectra (along with X-ray

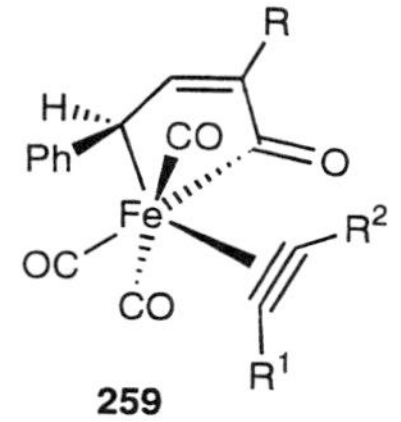

259

SCHEME 116.

evidence) confirmed that the adducts possessed the same absolute stereo-chemistry as the precursor vinylketenes. As part of this study,[38] it was found that the complexes **247.k** and **247.n** could be converted to the corresponding cyclohexadieneone complexes by treatment with aluminum(III) chloride under an atmosphere of carbon monoxide. Such products have previously been postulated as intermediates in the Dötz reaction (see Section II,B) and have been isolated from the reactions of cobalt and chromium vinylke-tenes with alkynes (see Sections II,B and V,E).

This information confirmed that a symmetrical intermediate could not lie along the reaction coordinate, although the alkyne complex **254** was still theoretically a viable intermediate. However, the stereochemical study

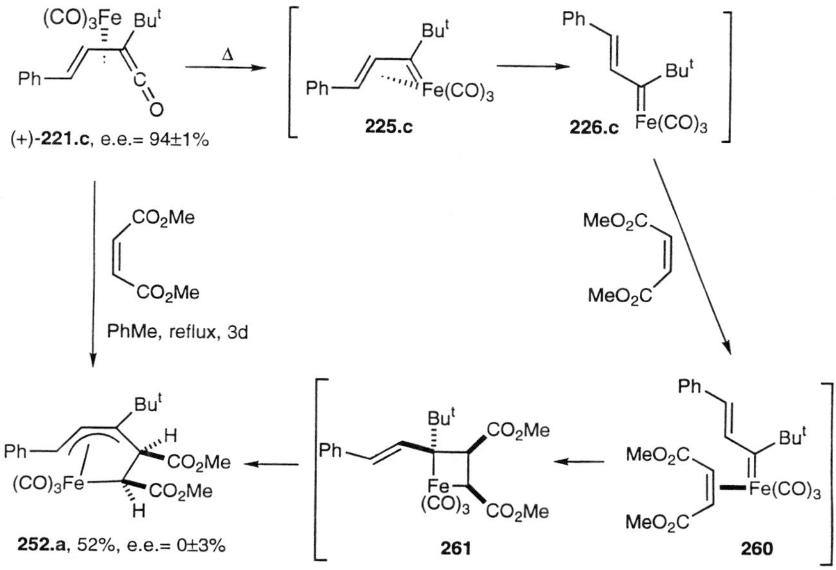

SCHEME 117.

into the mechanism of ketenimine formation[148] had suggested that dissociation of the central C_2 fragment of the vinylketene backbone was the most likely transformation in ligand exchange reactions of vinylketene complexes such as **221**. Consequently, the complex **259** was proposed as the key intermediate instead. Note that it has the same coordination pattern as the isonitrile complex **255**, preserving the stereochemical information inherent to the starting material at C-5.

Interestingly, the reaction between optically enriched vinylketene complex **221.c** and dimethyl maleate has been shown[39] to lead to racemization in the adduct **252.a**. This implies that the original mechanism proposed[146,147] for this transformation, which proceeds via the symmetrical η^1-vinylcarbene **226.c**, is the most likely route. Thermolytic decarbonylation of the vinylketene complex leads to the η^3-vinylcarbene **225.c**, which then coordinates the alkene (via **226.c**) to give the alkene complex **260**. Carbon–carbon bond formation leads to the ferracyclobutane **261**, which then saturates the coordination sphere of the iron by recomplexation of the styryl fragment to yield the product **252.a**.

VII

CONCLUSION

It is clear that one of the most valuable aspects of the chemistry of vinylketene complexes is that the ligands in question are more often than not extremely reactive in their uncomplexed state. Use of the complexed vinylketene allows easy access to a class of synthon that would be at best difficult to handle, and at worst unobtainable. As we have seen, a wide variety of organic fragments are available from the reactions of such complexed species. The relative stability of many of the vinylketene complexes has also allowed their isolation from reaction mixtures, leading to insights into the mechanisms of various organic and organometallic transformations.

The stabilizing effect of metal complexation has great import when considering the use of catalytic amounts of a transition metal reagent to accelerate organic transformations that pass through unstable vinylketene species. It has been shown that annulation reactions take place with certain classes of vinylketenes in both their free and their complexed states. The extension of this methodology to incorporate a catalytic transition metal, or indeed the application of this concept to alternative organic transformations, seems likely to yield some important results.

The fact that such a variety of transition metals have been complexed to vinylketene ligands has some intrinsic value, in that some insight into aspects of periodicity may be gleaned. At the present moment, little is known about the reactivity of the vinylketene complexes of vanadium and manganese, and so it would be foolhardy to make sweeping statements about variations in reactivity right across the transition metal series. However, we have seen that while (vinylketene)chromium complexes are extremely reactive, and consequently difficult to isolate, the molybdenum and (particularly) tungsten analogues of such compounds are much more stable. Similarly, as we move across the periodic table, we see that the iron and cobalt vinylketene complexes have an increased stability. However, many more reactivity studies are needed to build a fuller picture of periodic trends. It seems pertinent to suggest that the vanadium complex **83**, formed in high yield from a relatively simple reaction, would be a prime candidate for future research.

Many of the syntheses we have seen within this review depend on the carbonylation of a vinylcarbene complex for the generation of the vinylketene species. The ease of this carbonylation process is controlled, to some degree, by the identity of the metal. The electronic characteristics of the metal will clearly have a great effect on the strength of the metal–carbon double bond, and as such this could be a regulating factor in the carbene–ketene transformation. It is interesting to note the comparative reactivity of a (vinylcarbene)chromium species with its iron analogue: The former is a fairly stable species, whereas the latter has been shown to carbonylate readily to form the appropriate (vinylketene)iron complex.

Methods for the generation and use of enantiomerically pure vinylketene complexes have been well developed in recent years. This has allowed stereocontrol to be exerted over subsequent reactions to yield optically active organic or organometallic products. It has also allowed investigations to be carried out into the reaction mechanisms of several fundamental processes. This area is currently ripe for further exploration and would be of particular interest in its application to other metal systems. We trust that the work presented here will serve as a platform for further new and exciting discoveries.

REFERENCES

(1) King, R. B. *Inorg. Chem.* **1963**, *2*, 642.
(2) Dötz, K. H. *Angew. Chem. Int. Ed. Engl.* **1975**, *14*, 644.
(3) Alcock, N. W.; Danks, T. N.; Richards, C. J.; Thomas, S. E. *J. Chem. Soc., Chem. Commun.* **1989**, 21.
(4) Kerr, M. E.; Bruno, J. W.; *J. Am. Chem. Soc.* **1997**, *119*, 3183.
(5) Meinhart, J. D.; Santarsiero, B. D.; Grubbs, R. H. *J. Am. Chem. Soc.* **1986**, *108*, 3318.
(6) Huffman, M. A.; Liebeskind, L. S.; Pennington, W. T. *Organometallics* **1992**, *11*, 255.
(7) (a) Bibas, H; Koch, R.; Wentrup, C. *J. Org. Chem.* **1998**, *63*, 2619; (b) Collomb, D.;

Doutheau, A. *Tetrahedron Lett.* **1997**, *38*, 1397; (c) Bibas, H; Wong, M. W.; Wentrup, C. J. *Am. Chem. Soc.* **1995**, *117*, 9582.

(8) Dötz, K. H. *Angew. Chem. Int. Ed. Engl.* **1984**, *23*, 587.

(9) Torrent, M.; Duran, M.; Solà, M. *Chem. Commun.* **1998**, 999.

(10) Wulff, W. D. *Comprehensive Organic Synthesis,* Trost, B. A., Ed.; Pergamon: Oxford, 1991; Vol. 5, p. 1065.

(11) Wulff, W. D. *Advances in Metal-Organic Chemistry,* Vol. 1; Liebeskind, L. S., Ed.; JAI Press: Greenwich, CT; 1989.

(12) Wulff, W. D.; Bax, B. M.; Brandvold, T. A.; Chan, K. S.; Gilbert, A. M.; Hsung, R. P. *Organometallics,* **1994**, *13*, 102.

(13) Bos, M. E.; Wulff, W. D.; Miller, R. A.; Chamberlin, S.; Brandvold, T. A. *J. Am. Chem. Soc.* **1991**, *113*, 9293.

(14) Hegedus, L. S.; Miller Jr., D. B. *J. Org. Chem.* **1989**, *54*, 1241.

(15) Ochifuji, N.; Mori, M. *Tetrahedron Lett.* **1995**, *36*, 9501.

(16) Dötz, K. H.; Sturm, W. *J. Organomet. Chem.* **1985**, *285*, 205.

(17) Dötz, K. H.; Dietz, R. *J. Organomet. Chem.* **1978**, *157*, C55.

(18) Wulff, W. D.; Kaesler, R. W. *Organometallics* **1985**, *4*, 1461.

(19) Challener, C. A.; Wulff, W. D.; Anderson, B. A.; Chamberlin, S.; Faron, K. L.; Kim, O. K.; Murray, C. K.; Xu, Y-C.; Yang, D. C.; Darling, S. D. *J. Am. Chem. Soc.* **1993**, *115*, 1359.

(20) Dötz, K. H.; Glänzer, J. *J. Chem. Soc., Chem. Commun.* **1993**, 1036.

(21) Barluenga, J.; Aznar, F.; Martin, A.; García-Granda, S.; Pérez-Carreño, E. *J. Am. Chem. Soc.* **1994**, *116*, 11191.

(22) Anderson, B. A.; Wulff, W. D. *J. Am. Chem. Soc.* **1990**, *112*, 8615.

(23) Xu, Y-C.; Challener, C. A.; Dragisich, V.; Brandvold, T. A.; Peterson, G. A.; Wulff, W. D.; Williard, P. G. *J. Am. Chem. Soc.* **1989**, *111*, 7269.

(24) Kim, O. K.; Wulff, W. D.; Jiang, W. *J. Org. Chem.* **1993**, *58*, 5571.

(25) Bao, J.; Dragisich, V.; Wenglowsky, S.; Wulff, W. D. *J. Am. Chem. Soc.* **1991**, *113*, 9873.

(26) Yamashita, A.; Scahill, T. A. *Tetrahedron Lett.* **1982**, *23*, 3765.

(27) McCallum, J. S.; Kunng, F-A.; Gilbertson, S. R.; Wulff, W. D. *Organometallics,* **1988**, *7*, 2346.

(28) Tumer, S. U.; Herndon, J. W.; McMullen, L. A. *J. Am. Chem. Soc.* **1992**, *114*, 8394.

(29) Dragisich, V.; Murray, C. K.; Warner, B. P.; Wulff, W. D.; Yang, D. C. *J. Am. Chem. Soc.* **1990**, *112*, 1251.

(30) Chan, K. S.; Peterson, G. A.; Brandvold, T. A.; Faron, K. L.; Challener, C. A.; Hyldahl, C.; Wulff, W. D. *J. Organomet. Chem.* **1987**, *334*, 9.

(31) Yamashita, A. *Tetrahedron Lett.* **1986**, *27*, 5915.

(32) Yamashita, A.; Toy, A.; Watt, W.; Muchmore, C. R. *Tetrahedron Lett.* **1988**, *29*, 3403.

(33) Fischer, H.; Mühlemeier, J.; Märkl, R.; Dötz, K. H. *Chem. Ber.* **1982**, *115*, 1355.

(34) Casey, C. P. *Reactive Intermediates*; Jones, M., Jr.; Moss, R. A., Eds.; Wiley: New York, 1981; Vol. 2, p. 135.

(35) Hofmann, P.; Hämmerle, M. *Angew. Chem. Int. Ed. Engl.* **1989**, *28*, 908.

(36) Hofmann, P.; Hämmerle, M.; Unfried, G. *New J. Chem.* **1991**, *15*, 769.

(37) Gleichmann, M. M.; Dötz, K. H.; Hess, B. A. *J. Am. Chem. Soc.* **1996**, *118*, 10551.

(38) Benyunes, S. A.; Gibson (née Thomas), S. E.; Peplow, M. A. *Tetrahedron: Asymmetry* **1997**, *8*, 1535.

(39) S. A. Benyunes, S. E. Gibson (née Thomas) and M. A. Peplow, manuscript in preparation.

(40) Brandvold, T. A.; Wulff, W. D.; Rheingold, A. L. *J. Am. Chem. Soc.* **1990**, *112*, 1645.

(41) van Halban, H.; Geigel, H. *Z. Phys. Chem.,* **1920**, *96*, 233.

(42) van Halban, H.; Schmid, H.; Hochweber, M. *Helv. Chim. Acta* **1947**, *30*, 1135.

(43) van Halban, H.; Schmid, H.; Hochweber, M. *Helv. Chim. Acta* **1948**, *31*, 1899.
(44) Cannon, J. R.; Patrick, V. A.; Raston, C. L.; White, A. H. *Aust. J. Chem.* **1978**, *31*, 1265.
(45) Yamashita, A. *J. Am. Chem. Soc.* **1985**, *107*, 5823.
(46) Reduto dos Reis, A. C.; Hegedus, L. S. *Organometallics 1995, 14*, 1586.
(47) Herrmann, W. A. *Angew. Chem. Int. Ed. Engl.* **1974**, *13*, 335.
(48) Mitsudo, T.; Sasaki, T.; Takegami, Y.; Watanabe, Y. *J. Chem. Soc., Chem. Commun.* **1978**, 252.
(49) Herrmann, W. A.; Planck, J. *Angew. Chem. Int. Ed. Engl.* **1978**, *17*, 525.
(50) Fischer, H. *Angew. Chem. Int. Ed. Engl.* **1983**, *22*, 874.
(51) Wulff, W. D.; Gilbertson, S. R.; Springer, J. P. *J. Am. Chem. Soc.* **1986**, *108*, 520.
(52) Dötz, K. H. *Angew. Chem. Int. Ed. Engl.* **1979**, *18*, 954.
(53) Dötz, K. H.; Fügen-Köster, B. *Chem. Ber.* **1980**, *113*, 1449.
(54) Tang, P.-C.; Wulff, W. D. *J. Am. Chem. Soc.* **1984**, *106*, 1132.
(55) Huang, W.; Fang, D.; Temple, K.; Tidwell, T. T. *J. Am. Chem. Soc.* **1997**, *119*, 2832.
(56) Colomvakos, J. D.; Egle, I.; Ma, J.; Pole, D. L.; Tidwell, T. T.; Warkentin, J. *J. Org. Chem.* **1996**, *61*, 9522.
(57) Adams, N.; Dillon, E. A.; Dillon, J. L.; Gao, Q. *Tetrahedron Lett.* **1997**, *38*, 2231.
(58) Sun, L.; Liebeskind, L. S. *J. Org. Chem.* **1995**, *60*, 8194.
(59) Birchler, A. G.; Liu, F.; Liebeskind, L. S. *J. Org. Chem.* **1994**, *59*, 7737.
(60) Tidwell, T. T.; Zhao, D-C. *J. Am. Chem. Soc.* **1992**, *114*, 10980.
(61) Mayr, H.; Huisgen, R. *J. Chem. Soc., Chem. Commun.* **1976**, 57.
(62) Bibas, H; Wong, M. W.; Wentrup, C. *Chem. Eur. J.* **1997**, *3*, 237.
(63) Koch, R.; Wong, M. W.; Wentrup, C. *J. Org. Chem.* **1996**, *61*, 6809.
(64) Kikuchi, O. *Bull. Chem. Soc. Jpn.* **1982**, *55*, 1669.
(65) Barbaro, G.; Battaglia, A.; Giorgianni, P. *J. Org. Chem.* **1987**, *52*, 3289.
(66) King, R. B.; Harmon, C. A. *Inorg. Chem.* **1976**, *15*, 879.
(67) LePage, T.; Nakasuji, K.; Breslow, R. *Tetrahedron Lett.* **1985**, *26*, 5919.
(68) (a) Valéri, T.; Meier, F.; Weiss, E. *Chem. Ber.* **1988**, *121*, 1093: (b) Klimes, J.; Weiss, E. *Angew. Chem. Int. Ed. Engl.* **1982**, *21*, 205.
(69) Richards, C. J.; Thomas, S. E. *J. Chem. Soc., Chem. Commun.* **1990**, 307.
(70) Semmelhack, M. F.; Ho, S.; Steigerwald, M.; Lee, M. C. *J. Am. Chem. Soc.* **1987**, *109*, 4397.
(71) Garrett, K. E.; Sheridan, J. B.; Pourreau, D. B.; Feng, W. C.; Geoffroy, G. L.; Staley, D. L.; Rheingold, A. L. *J. Am. Chem. Soc.* **1989**, *111*, 8383.
(72) Asaro, M. F.; Mayr, A.; Kahr, B.; van Engen, D. *Inorg. Chim. Acta* **1994**, *220*, 335.
(73) Mayr, A.; Asaro, M. F.; Glines, T. J.; van Engen, D.; Tripp, G. M. *J. Am. Chem. Soc.* **1993**, *115*, 8187.
(74) Mayr, A.; Asaro, M. F.; Glines, T. J. *J. Am. Chem. Soc.* **1987**, *109*, 2215.
(75) Binger, P.; Cetinkaya, B.; Krüger, C. *J. Organomet. Chem.* **1978**, *159*, 63.
(76) Templeton, J. L.; Herrick, R. S.; Rusik, C. A.; McKenna, C. E.; McDonald, J. W.; Newton, W. E. *Inorg. Chem.* **1985**, *24*, 1383.
(77) Benyunes, S. A.; Deeth, R. J.; Fries, A.; Green, M.; McPartlin, M.; Nation, C. B. M. *J. Chem. Soc., Dalton Trans.* **1992**, 3453.
(78) Benyunes, S. A.; Binelli, A.; Green, M.; Grimshire, M, J. *J. Chem. Soc., Dalton Trans.* **1991**, 895.
(79) Nesmeyanov, A. N.; Kolobova, N. E.; Zlotina, I. B.; Lokshin, B. V.; Leshcheva, I. F.; Znobina, G. K.; Anisimov, K. N. *J. Organomet. Chem.* **1976**, *110*, 339.
(80) Guilieri, F.; Benaim, J. *J. Organomet. Chem.* **1984**, *276*, 367.
(81) Bruce, M. I.; Liddell, M. J.; Snow, M. R.; Swincer, A. G. *Organometallics* **1990**, *9*, 96.
(82) Hughes, R. P.; Lambert, J. M. J.; Rheingold, A. L. *Organometallics* **1985**, *4*, 2055.
(83) Brisdon, B. J.; Deeth, R. J.; Hodson, A. G. W.; Kemp, C. M.; Mahon, M. F.; Molloy, K. C. *Organometallics* **1991**, *10*, 1107.

(84) Fischer, R. A.; Fischer, R. W.; Herrmann, W. A.; Herdtwerk, E. *Chem. Ber.* **1989**, *122*, 2035.

(85) Faller, J. W.; Rosan, A. M. *J. Am. Chem. Soc.* **1977**, *99*, 4858.

(86) Hill, L.; Richards, C. J.; Thomas, S. E. *J. Chem. Soc., Chem. Commun.* **1990**, 1085.

(87) Alcock, N. W.; Richards, C. J.; Thomas, S. E. *Organometallics* **1991**, *10*, 231.

(88) (a) Hill, L.; Saberi, S. P.; Slawin, A. M. Z.; Thomas, S. E.; Williams, D. J. *J. Chem. Soc., Chem. Commun.* **1991**, 1290; (b) Saberi, S. P.; Thomas, S. E. *J. Chem. Soc., Perkin Trans. 1* **1992**, 259.

(89) Hill, L.; Richards, C. J.; Thomas, S. E. *Pure Appl. Chem.* **1992**, *64*, 371.

(90) Benyunes, S. A.; Gibson (née Thomas), S. E.; Stern, J. A. *J. Chem. Soc., Perkin Trans. 1* **1995**, 1333.

(91) Cheng, C-Y.; Hsieh, C-H.; Lee, G-H.; Peng, S-M.; Liu, R-S. *Organometallics* **1996**, *15*, 1565.

(92) Green, M.; Nyathi, J. Z.; Scott, C.; Stone, F. G. A.; Welch, A. J.; Woodward, P. *J. Chem. Soc. Dalton* **1978**, 1067.

(93) Binger, P.; Schuchardt, U. *Angew. Chem.* **1975**, *87*, 715.

(94) Jens, K-J.; Weiss, E. *Chem. Ber.* **1984**, *117*, 2469.

(95) Huffman, M. A.; Liebeskind, L. S. *J. Am. Chem. Soc.* **1990**, *112*, 8617.

(96) (a) O'Connor, J. M.; Ji, H-L. *J. Am. Chem. Soc.* **1993**, *115*, 9846; (b) O'Connor, J. M.; Ji, H.; Iranpour, M. *J. Am. Chem. Soc.* **1993**, *115*, 1586.

(97) (a) Semmelhack, M. F.; Tamura, R.; Schnatter, W.; Springer, J. *J. Am. Chem. Soc.* **1984**, *106*, 5363; (b) Semmelhack, M. F.; Park, J. *Organometallics,* **1986**, *5*, 2350.

(98) Pettit, R.; McKennis, J. S.; Case, R.; Slegeir, W. *J. Am. Chem. Soc.* **1974**, *96*, 287.

(99) Danheiser, R. L.; Brisbois, R. G.; Kowalczyk, J. J.; Miller, R.F. *J. Am. Chem. Soc.* **1990**, *112*, 3093.

(100) King Jr., J. A.; Vollhardt, P. C. *J. Organomet. Chem.* **1994**, *470*, 207.

(101) Franck-Neumann, M.; Dietrich-Buchecker, C.; Khemiss, A. *Tetrahedron Lett.* **1981**, *22*, 2307.

(102) Klimes, J.; Weiss, E. *Chem. Ber.* **1982**, *115*, 2606.

(103) Dettlaf, G.; Behrens, U.; Weiss, E. *Chem. Ber.* **1978**, *111*, 3019.

(104) Kobayashi, T.; Murata, N.; Nitta, M. *Bull. Chem. Soc. Jpn.* **1987**, *60*, 3062.

(105) Franck-Neumann, M.; Dietrich-Buchecker, C.; Khemiss, A. K. *J. Organomet. Chem.* **1981**, *220*, 187.

(106) Mitsudo, T.; Watanabe, H.; Sasaki, T.; Takegami, Y.; Watanabe, Y. *Organometallics* **1989**, *8*, 368.

(107) Fujita, K.; Masuda, H.; Mitsudo, T.; Nagano, S.; Suzuki, T.; Watanabe, Y. *Organometallics,* **1995**, *14*, 4228.

(108) Morris, K. G.; Saberi, S. P.; Slawin, A. M. Z.; Thomas, S. E.; Williams, D. J. *J. Chem. Soc., Chem. Commun.* **1992**, 1778.

(109) Morris, K. G.; Saberi, S. P.; Thomas, S. E. *J. Chem. Soc., Chem. Commun.* **1993**, 209.

(110) Henrici-Olivé, G.; Olivé, S. *Angew. Chem. Int. Ed. Engl.* **1976**, *15*, 136.

(111) Hoffman, N. E.; Puthenpurackal, T. *J. Org. Chem.* **1965**, *30*, 420.

(112) Franck-Neumann, M.; Dietrich-Buchecker, C.; Khemiss, A. K. *J. Organomet. Chem.* **1982**, *224*, 113.

(113) Nitta, M.; Kobayashi, T. *Bull. Chem. Soc. Jpn.* **1984**, *57*, 1035.

(114) Stevens, A. E.; Beauchamp, J. L. *J. Am. Chem. Soc.* **1978**, *100*, 2584.

(115) Muetterties, E. L.; Stein, J. *Chem. Rev.* **1979**, *79*, 479.

(116) Mitsudo, T.; Nakanishi, H.; Inubushi, T.; Morishima, I.; Watanabe, Y.; Takegami, Y. *J. Chem. Soc., Chem. Commun.* **1976**, 416.

(117) Bodnar, T. W.; Cutler, A. R. *J. Am. Chem. Soc.* **1983**, *105*, 5926.

(118) Hill, A. E.; Hoffmann, H. M. R. *J. Chem. Soc., Chem. Commun.* **1972**, 574.

(119) Nakatsu, K.; Mitsudo, T.; Nakanishi, H.; Watanabe, Y.; Takegami, Y. *Chem. Lett.* **1977**, 1447.

(120) (a) Mitsudo, T; Watanabe, Y.; Nakanishi, H.; Morishima, I.; Inubushi, T.; Takegami, Y. *J. Chem. Soc., Dalton* **1978,** 1298; (b) Mitsudo, T.; Nakanishi, H.; Inubushi, T.; Morishima, I.; Watanabe, Y.; Takegami, Y. *J. Chem. Soc., Chem. Commun.* **1976,** 416.

(121) Mitsudo, T.; Ogino, Y.; Komiya, Y.; Watanabe, H.; Watanabe, Y. *Organometallics* **1983**, 2, 1202.

(122) Mitsudo, T.; Ishihara, A.; Kadokura, M.; Watanabe, Y. *Organometallics* **1986**, 5, 238.

(123) Park, J.; Kang, S.; Whang, D.; Kim, K. *Organometallics* **1991**, 10, 3413.

(124) Park, J.; Kang, S.; Whang, D.; Kim, K. *Organometallics* **1992**, 11, 1738.

(125) Park, J.; Kang, S.; Won, C.; Whang, D.; Kim, K. *Organometallics* **1993**, 12, 4704.

(126) Park, J.; Kim, J. *Organometallics* **1995**, 14, 4431.

(127) Herrmann, W. A.; Gimeno, J.; Weichmann, J. *J. Organomet. Chem.* **1981,** 213, C26.

(128) Bkouche-Waksman, I.; Ricci Jr., J. S.; Koetzle, T. F.; Weichmann, J.; Herrmann, W. A. *Inorg. Chem.* **1985**, 24, 1492.

(129) Young, G. H.; Willis, R. R.; Wojcicki, A. *Organometallics* **1992**, 11, 154.

(130) Wojcicki, A.; Shuchart, C. E. *Coord. Chem. Rev.* **1990**, 105, 35.

(131) Gibson (née Thomas), S. E.; Kipps, M.; Stanley, P. D.; Ward, M. F.; Worthington, P. A. *Chem. Commun.* **1996**, 263.

(132) Wulff, W. D.; Bauta, W. E.; Kaesler, R. W.; Lankford, P. J.; Miller, R. A.; Murray, C. K.; Yang, D. C. *J. Am. Chem. Soc.* **1990**, 112, 3642.

(133) Ibbotson, A.; Thomas, S. E.; Tustin, G. J. *Tetrahedron* **1992**, 48, 7629.

(134) Benyunes, S. A.; Gibson (née Thomas), S. E.; Jefferson, G. R.; Potter, P. C. V.; Peplow, M. A.; Rahimian, E.; Smith, M. H.; Ward, M. F. *Organic Synthesis via Organometallics*; Helmchen, G., Ed; Vieweg: Wiesbaden, 1997; p. 75.

(135) Hill, L.; Richards, C. J.; Thomas, S. E. *Pure Appl. Chem.* **1990**, 62, 2057.

(136) Benyunes, S. A.; Gibson (née Thomas), S. E. *Chem. Commun.* **1996**, 43.

(137) Marcuzzi, A.; Linden, A.; Rentsch, D.; von Phillipsborn, W. *J. Organomet. Chem.* **1992,** 429, 87.

(138) Benyunes, S. A.; Gibson (née Thomas), S. E.; Ward, M. F. *Tetrahedron: Asymmetry* **1995**, 6, 2517.

(139) García-Mellado, O.; Gutiérrez-Pérez, R.; Alvarez-Toledano, C.; Toscano, R. A.; Cabrera, A. *Polyhedron* **1997**, 16, 2979.

(140) Ibbotson, A.; Reduto dos Reis, A. C.; Saberi, S. P.; Slawin, A. M. Z.; Thomas, S. E.; Tustin, G. J.; Williams, D. J. *J. Chem. Soc., Perkin Trans. 1* **1992**, 1251.

(141) Gibson (née Thomas), S. E.; Tustin, G. J. *J. Chem. Soc., Perkin Trans. 1* **1995**, 2427.

(142) Alcock, N. W.; Pike, G. A.; Richards, C. J.; Thomas, S. E. *Tetrahedron: Asymmetry* **1990**, 1, 531.

(143) Richards, C. J.; Thomas, S. E. *Tetrahedron: Asymmetry* **1992**, 3, 143.

(144) Morris, K. G.; Saberi, S. P.; Salter, M. M.; Thomas, S. E.; Ward, M. F.; Slawin, A. M. Z.; Williams, D. J. *Tetrahedron* **1993**, 49, 5617.

(145) Saberi, S. P.; Salter, M. M.; Thomas, S. E.; Slawin, A. M. Z.; Williams, D. J. *J. Chem. Soc., Perkin Trans. 1* **1994**, 167.

(146) Saberi, S. P.; Thomas, S. E.; Ward, M. F.; Worthington, P. A.; Slawin, A. M. Z.; Williams, D. J. *J. Chem. Soc., Chem. Commun.* **1994**, 2169.

(147) Gibson (née Thomas), S. E.; Saberi, S. P.; Slawin, A. M. Z.; Stanley, P.; Ward, M. F.; Williams, D. J.; Worthington, P. *J. Chem. Soc., Perkin Trans. 1* **1995**, 2147.

(148) Benyunes, S. A.; Gibson (née Thomas), S. E.; Peplow, M. A. *Chem. Commun.* **1996**, 1757.

(149) Alvarez-Toledano, C.; Cano, A. C.; Toscano, R. A.; Parlier, A.; Rudler, H. *Bull. Soc. Chim. Fr.* **1993**, 130, 660.

(150) Cano, A. C.; Toscano, R. A.; Bernés, S.; García-Mellado, O.; Alvarez-Toledano, C.; Rudler, H. *J. Organomet. Chem.* **1995**, 496, 153.

Subject Index

A

Ab initio studies
  acyclic λ^5Si-silicates, 227
  spirocyclic zwitterionic λ^5Si-silicates, with
    bidentate ligands
    benzene-1,2-diolato(2–), 232–233
    ethene-1,2-diolato(2–), 237–238
    naphthalene-2,3-diolato(2–), 232–233
Abstraction, hydrogen
  intramolecular, by alkyl radical, 81
  reactivity trends, 79
Acetohydroximato(2–) bidentate ligand, in
  spirocyclic λ^5Si-silicates
  NMR studies, 255–256
  synthesis, 255
Acidic reactions, acyclic λ^5Si-silicates,
  227–228
Acrylonitrile, reaction with digermenes,
  138
Acyclic (ammonioorganyl)tetrafluorosili-
  cates, synthesis, 223
Acyclic (ammonioorganyl)trifluoro(organyl)
  silicates, synthesis, 223
Acyclic λ^5Si-silicates
  ab initio studies, 227
  acidic reaction, 227–228
  geometry optimizations, 226
  NMR studies, 226–227
  physical properties, 227
  synthesis, 223
Acylferrates, for vinylketene synthesis,
  324–325
Acyl radical, reaction with Sn hydrides,
  96–97
Alcohols, reaction with germenes, 125
Alkenes reactions
  with digermenes, 136
  with germenes, 126–127
Alkoxyl radical, reaction with Sn hydrides,
  101
α-Alkoxy radical, reaction with Sn hydrides, 95–96

Alkyl radical
  intramolecular hydrogen abstraction, 81
  reactions with Sn hydrides, 90–91, 94–99
Alkynes, annulation with Fischer carbenes,
  278–287
η^1-Alkynylcarbenes, in vinylketene synthe-
  sis, 325–326, 328–330
o-(Allyloxy)phenyl radical, Arrhenius func-
  tion, 77
Aluminum, terphenyl derivatives, 23–25
Annulation, alkynes with Fischer carbenes,
  278–287
Antimony, terphenyl derivatives, 47
Arrhenius function, *o*-(allyloxy)phenyl radi-
  cal, 77
Arsenic, terphenyl derivatives, 45–47
Arylboron dihalide, reduction, 20
Aryl ligands, bulkiness, 2–3
Azides, reaction with germylene, 146

B

BDE, *see* Bond dissociation energy
Benzannulation, alkynes, 278
Benzene-1,2-diolato(2–) bidentate ligand,
  in spirocyclic λ^5Si-silicates
  ab initio studies, 232–233
  chemical properties, 235
  geometry optimization, 232–233
  NMR studies, 233–234
  synthesis, 230
  X-ray diffraction analysis, 230–233
Benzilic acid, in spirocyclic λ^5Si-silicates,
  241
Benzohydroximato(2–) bidentate ligand, in
  spirocyclic λ^5Si-silicates
  NMR studies, 255–256
  synthesis, 255
Benzoyloxyl radical, reaction with Sn hydrides, 99, 101
Benzyl radical, reaction with Sn hydrides, 94
Beryllium compounds, terphenyl derivatives, 10–12

Cumulative List of Contributors
for Volumes 1–36

Abel, E. W., **5,** 1; **8,** 117

Aguiló, A., **5,** 321

Akkerman, O. S., **32,** 147

Albano, V. G., **14,** 285

Alper, H., **19,** 183

Anderson, G. K., **20,** 39; **35,** 1

Angelici, R. J., **27,** 51

Aradi, A. A., **30,** 189

Armitage, D. A., **5,** 1

Armor, J. N., **19,** 1

Ash, C. E., **27,** 1

Ashe, A. J., III, **30,** 77

Atwell, W. H., **4,** 1

Baines, K. M., **25,** 1

Barone, R., **26,** 165

Bassner, S. L., **28,** 1

Behrens, H., **18,** 1

Bennett, M. A., **4,** 353

Bickelhaupt, F., **32,** 147

Birmingham, J., **2,** 365

Blinka, T. A., **23,** 193

Bockman, T. M., **33,** 51

Bogdanović, B., **17,** 105

Bottomley, F., **28,** 339

Bradley, J. S., **22,** 1

Brew, S. A., **35,** 135

Brinckman, F. E., **20,** 313

Brook, A. G., **7,** 95; **25,** 1

Bowser, J. R., **36,** 57

Brown, H. C., **11,** 1

Brown, T. L., **3,** 365

Bruce, M. I., **6,** 273, **10,** 273; **11,** 447;
  12, 379; **22,** 59

Brunner, H., **18,** 151

Buhro, W. E., **27,** 311

Byers, P. K., **34,** 1

Cais, M., **8,** 211

Calderon, N., **17,** 449

Callahan, K. P., **14,** 145

Canty, A. J., **34,** 1

Cartledge, F. K., **4,** 1

Chalk, A. J., **6,** 119

Chanon, M., **26,** 165

Chatt, J., **12,** 1

Chini, P., **14,** 285

Chisholm, M. H., **26,** 97; **27,** 311

Chiusoli, G. P., **17,** 195

Chojinowski, J., **30,** 243

Churchill, M. R., **5,** 93

Coates, G. E., **9,** 195

Collman, J. P., **7,** 53

Compton, N. A., **31,** 91

Connelly, N. G., **23,** 1; **24,** 87

Connolly, J. W., **19,** 123

Corey, J. Y., **13,** 139

Corriu, R. J. P., **20,** 265

Courtney, A., **16,** 241

Coutts, R. S. P., **9,** 135

Coville, N. J., **36,** 95

Coyle, T. D., **10,** 237

Crabtree, R. H., **28,** 299

Craig, P. J., **11,** 331

Csuk, R., **28,** 85

Cullen, W. R., **4,** 145

Cundy, C. S., **11,** 253

Curtis, M. D., **19,** 213

Darensbourg, D. J., **21,** 113; **22,** 129

Darensbourg, M. Y., **27,** 1

Davies, S. G., **30,** 1

Deacon, G. B., **25,** 237

de Boer, E., **2,** 115

Deeming, A. J., **26,** 1

Dessy, R. E., **4,** 267

Dickson, R. S., **12,** 323

Dixneuf, P. H., **29,** 163

Eisch, J. J., **16,** 67

Ellis, J. E., **31,** 1

Emerson, G. F., **1,** 1

Epstein, P. S., **19,** 213

Erker, G., **24,** 1

Ernst, C. R., **10,** 79

Errington, R. J., **31,** 91

Cumulative Index
for Volumes 37–44

ISBN 0-12-031144-5